Newnes
Communications Technology Handbook

Dedicated to my grandchildren, Amy, Jonathan, Phillip and Simon, with the hope that their lives may be enriched and enhanced by the rapidly developing spread of communications technology.

Newnes
Communications Technology
Handbook

Geoff Lewis
BA, MSc MRTS, MIEIE

BH NEWNES

Newnes
An imprint of Butterworth-Heinemann Ltd
Linacre House, Jordan Hill, Oxford OX2 8DP

ℛ A member of the Reed Elsevier plc group

OXFORD LONDON BOSTON
MUNICH NEW DELHI SINGAPORE SYDNEY
TOKYO TORONTO WELLINGTON

First published 1994

British Library Cataloguing in Publication Data
Lewis, Geoffrey E.
   Newnes Communications Technology Handbook
   I. Title
   621.382

ISBN 0 7506 1729 2

Library of Cataloguing in Publication Data
Lewis, Geoffrey E.
   Newnes communications technology handbook/Geoffrey E. Lewis
   p.   cm
   Includes bibliographical references and index.
   ISBN 0 7506 1729 2
   1. Telecommunication–Handbooks, manuals, etc. I. Title.
   TK5101.L483   93-45442
   621.382–dc20   CIP

Typeset by TecSet Ltd, Wallington, Surrey
Printed and bound in Great Britain by Bath Press, Avon

Contents

Preface

The traditional boundaries between communications, information processing, television and computing systems have disappeared. Today, the term telecommunications virtually represents an all-embracing technology. The reasons for this convergence are related to the rapid development and expansion of electronic and semiconductor technologies that have occurred over the recent past.

This book owes its origins to the difficulties experienced in finding suitable reference information needed to understand the terminology and jargon of this expanding field.

The book is basically presented as a set of monographs organised in an alphabetical manner. Each section provides an introduction to the topic, followed by more detailed descriptions and explanations. Many of the miscellaneous terms associated with the subject are separately listed and each section ends with a listing of references suggested for further reading.

Two distinct routes to the location of information have been provided. The extensive contents listing gives a fast access to general information about a particular topic, while the conventional index provides more specific guidance. In addition, extensive cross-referencing has been used to provide a link between the converging topic areas.

Because a mathematical expression can often explain a system behaviour in a very concise manner, these concepts have been included where it is felt to be necessary. However, in many cases, these expressions can be glossed over at first reading without losing too much of the context.

Finally, the book is presented as an aid to all the students, technicians, engineers, systems managers and those who work on the fringes of telecommunications, that have a need to understand new areas of this rapidly expanding business.

Geoff Lewis

1 Analogue systems and concepts

Analogue systems are characterised by electrical signals that vary continuously between two extreme amplitude levels. There is thus an infinity of actual levels that such signals can assume. Compared with other forms, analogue signals usually occupy minimum bandwidth.

1.1 Amplifiers

Amplifiers can be sub-divided into many different types, each designed for some specific purpose. But depending upon the time period for which the output current flows with respect to a changing input signal, they can be assigned to one of the classifications shown in Fig 1.1.

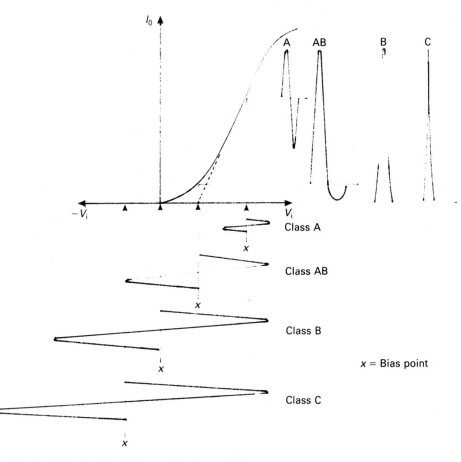

Figure 1.1 Amplifier classification by transfer characteristic.

Class A

The output current flows for the whole or 360° of the input sinusoidal signal. The amplifier is biased to the mid-point of the straight part of its characteristic as shown, to ensure the minimum of distortion. As a large signal or power amplifier, the efficiency as measured by the ratio:

$$(\text{ac signal power output}/\text{dc power input}) \times 100\%$$

is theoretically 50%. In a practical case however, an efficiency in excess of 35% is rarely achieved. Because of the high degree of linearity, these amplifiers are commonly used for audio or low frequency applications.

Class B

The output current flows for only 180°, or half of the sinusoidal input signal. The amplifier being biased to cut-off causes such distortion that it effectively acts as a half-wave rectifier. The theoretical efficiency is 78.5% ($\pi/4 \times 100\%$), and in practice this rarely exceeds 65%. In low frequency and audio applications, Class B is used in push–pull, using two stages, one for the amplification of each half cycle. Even so, it is often necessary to take care to minimise the effect of cross-over distortion. For radio frequency applications, this class can be used with a tuned circuit load, where the natural flywheel action replaces the missing half cycle.

Class AB

The output current flows for more than 180° of the input cycle as the amplifier is biased to the projected cut-off point. This is defined as the bias point at which the amplifier current would fall to zero if the characteristic had been linear. The theoretical efficiency is somewhat less than that for Class B, but when used in a push–pull circuit, the cross-over distortion is reduced. The class is very commonly used for audio power amplification or as a radio frequency amplifier with a tuned circuit load.

Class C

These amplifiers are biased beyond the cut-off point so that output current flows for much less than 180° of the input cycle. The typical angle of flow, which is defined by the ratio:

$$\frac{\text{Time for which output current flows}}{\text{Period of the input signal}} \times 360°$$

is less than 120°. Theoretically the efficiency approaches 100% as the angle of flow approaches zero. In practice, this value can be as high as 90%. Because of the severe distortion that Class C operation generates, this class can only be used with a tuned circuit load, i.e., as a radio frequency amplifier, or as a harmonic generator, with

the tuned circuit extracting the required harmonic frequency from the distortion components.

Class D

For audio and low frequency applications, the requirements of low distortion and high power output produce a conflict. The low efficiency of Class A is due to the excessive power dissipation in the output stages. Operation has to remain within the safe working area to protect the amplifier from thermal damage (see Fig. 1.2).

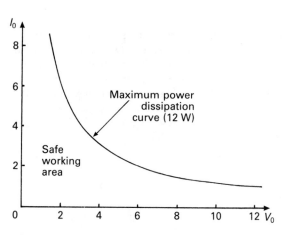

Figure 1.2   Maximum power dissipation curve.

If the amplifier can be operated in either the saturated or cut-off state, then the power dissipation will be low, because either the voltage across the active device or the current through it will be very low. Further, if the amplifier can be switched rapidly between these states, the power dissipation during the transition through the active region will also be very low. The output power is controlled by varying the time spent in these two extreme states, this depending upon the instantaneous amplitude of the signal to be amplified. Thus Class D operation is a form of pulse width or duration modulation, which can achieve a practical efficiency as high as 97%.

Typically the switching frequency exceeds 50 kHz, but significantly higher values may be used. This has the advantages of smaller filters to remove the switching frequency from the output, lower radiation of unwanted interference, reduced intermodulation distortion and a wider bandwidth.

Class E

When a switching mode is used for RF power amplifiers and the transition periods between the On and Off states become an appreciable part of the period of the signal being processed, a significant power loss arises. Class E overcomes this problem by including *LC* filters between the switcher and the load to delay the current transition relative to the voltage swing. This ensures that either the voltage or the current is simultaneously zero during the switching period.

Theoretically the efficiency for Class E can be 100% with practical values in excess of 90% at VHF frequencies being achieved.

Class E is equally applicable to thermionic or semiconductor RF power devices and has a low sensitivity to component tolerances

Current dumping amplifiers

This system was devised to provide audio frequency power amplification with very low distortion. Operation is based on the premise that a system that provides a distortionless no load voltage and an output impedance that is independent of the load, must be distortionless. The basic circuit is shown in Fig. 1.3, where A_1 is a small Class A amplifier that is capable of providing the total output voltage swing, but with a very limited current capability. The emitter follower transistors Tr_1, and Tr_2 are designed to *dump* current into the load under the control of A_1. In this way, the two sections combine to provide the total output power. The very low level of distortion is achieved by the use of the feedback networks. Capacitor C_1 ensures that the integrator circuit which has a response that extends to well beyond the maximum operating frequency, provides a gain that falls with rising frequency. R_2 provides a feedback voltage that depends upon the impedance of L_1 and thus has an opposite effect to that produced by C_1.

A further feedback path provided by R_3 operates over the dumper stages.

1.2 Distortions (see also under Measurements, p. 152)

The information carried by an analogue signal is contained within the signal's instantaneous amplitude, frequency and phase. Any phenomenon that disturbs this relationship, distorts the waveform and reduces the intelligibility of the signal. Since this is an unwanted feature of the signal processing, distortion has a noise-like effect. This effect can be quantified either by measurement or analysis and can be expressed as a kind of signal to noise ratio, usually in the form:

$$\frac{\text{Magnitude of distortion component}}{\text{Magnitude of undistorted signal}} \times 100\%$$

Amplitude or non-linear distortion

In a practical amplifier, the dynamic transfer characteristic is always to some extent non-linear. That is, equal incremental inputs do not produce equal incremental outputs. This is particularly noticeable in large signal power amplifiers, where a saturation condition arises. A useful measure of a system's performance in this respect is shown by the 1 dB compression point as indicated in Fig. 1.4. This point represents the output power level where the signal is 1 dB below the value that would obtain if the system were linear. In the case shown, 1 dB compression occurs at an input level of 7 dBm.

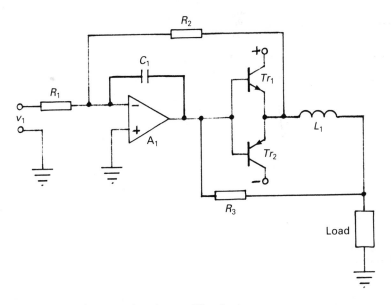

Figure 1.3 Basic current dumping amplifier circuit.

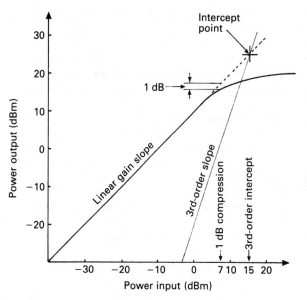

Figure 1.4 Measurement of intermodulation and non-linearity.

Attenuation distortion

In any transmission system, some frequencies will be attenuated more than others. If the transmitted signal contains frequency components that are so attenuated, then the wave shape of the original signal will become distorted.

Cross-over distortion

When push–pull amplifiers are biased to operate in Class B, the non-linearity near to the origin causes distortion during the zero crossing period as shown in Fig. 1.5, the distortion component being predominantly the third harmonic.

Group delay distortion

Any non-linearity in the phase versus frequency response of a circuit or system, results in *group delay*, which is defined by the slope of this characteristic and given by: group delay $T_D = \delta\beta/\delta\omega$, where β is the phase angle in radians and ω is the angular velocity $= 2\pi f$. The same relationship for f in Hz and phase in degrees is given by $T_D = \Delta\phi/\Delta f \times 2.778 \times 10^{-3}$.

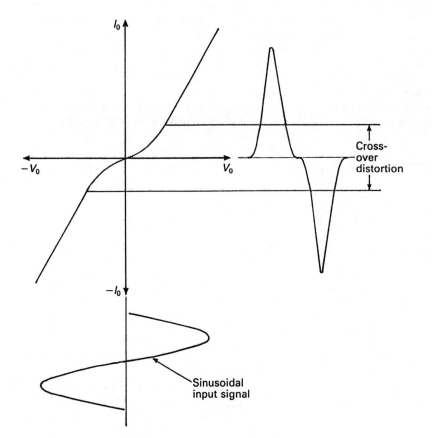

Figure 1.5 Class B push–pull operation and cross-over distortion.

The effect of group delay on a modulated wave is to generate *envelope distortion*. The spectrum of a modulated wave has components, f_c and $f_c \pm n f_m$ (see p.186), where n is an integer representing the harmonics of f_m. Any non-linear phase shift will thus distort the modulation envelope. In the case of audio signals, this is not of great importance, because the ear is fairly tolerant of this form of distortion. However, for video signals, the phase shift represents a time shift of the component frequencies so that the result becomes obvious to the eye. Differential group delay is defined as the group delay variation over the total bandwidth being considered.

Harmonic distortion

Due to the non-linearity of an amplifier's transfer characteristic, frequency components that were not present at the input will appear at the output. The magnitude of each unwanted component normally diminishes as the order of the harmonic, in most cases the second- and third-order components being most troublesome.

Intermodulation distortion

When a signal that contains two or more components passes through a non-linear system, a kind of modulation effect results in the production of sum and difference frequencies plus various beat notes that are not harmonically related to the original signal. In the case of audio systems, these produce discordant sounds. The spectrum of second- (2IM) and third-order (3IM) intermodulation products are shown in Fig. 1.6. Except in the case of very wide band systems, the 2nd-order terms $(f_2 - f_1, f_2 + f_1)$ will fall well outside the pass band and only the 3rd-order terms $(2f_1 - f_2$ and $2f_2 - f_1)$ will be troublesome.

Figure 1.4 uses the third-order intercept point to equate 3IM with the linear response of a power amplifier that has a nominal gain of 10 dB. Over the linear part of the curve, the gain slope is 1 dB/dB, but

with increasing input, a point is reached at which saturation occurs due to over driving. The output signal now contains fundamental and harmonic components. If the signal level is now slowly reduced, the harmonic content falls faster than the fundamental. The 3rd harmonic falls at 3 dB/dB as opposed to the 1 dB/dB of the fundamental. The two slopes intersect at the intercept point. If this point is known for a given fundamental output power, the level of the third-order term can be evaluated. In the example shown, the intercept occurs at an input of 15 dBm.

Phase distortion

When the signal propagation time through a system is frequency dependent, the different components of a complex wave will arrive at a load at relatively different times. This will result in change of wave shape that in audio systems is not very important as the ear is not offended by such phase distortion. However, in the case of video signals, this can lead to obvious picture impairments. For zero phase distortion, the phase and group velocities of the complex wave should be constant over the range of frequencies of interest.

Total harmonic distortion (THD)

This defines the distortion factor for a system. If D_2, D_3, D_4, etc., represent the distortion components at 2nd, 3rd, 4th harmonic, etc., then the total harmonic distortion D, is given by:

$$D = \sqrt{(D_2^2 + D_3^2 + D_4^2 + ...)}$$

1.3 Feedback amplifiers

Signal feedback from output to input is a technique used to modify the performance of an amplifier. For example, for any given amplifier, there is a gain × bandwidth product that is a constant and by using feedback, gain can be traded for bandwidth and vice versa. This effect is shown in Fig. 1.8. The magnitude of the sample or fraction obtained from the output signal is given by

$$b = v_{fb}/v_o$$

where v_{fb} and v_o are feedback and output voltages, respectively (see Fig. 1.7).

v_{fb} can be used in one of two ways. As (regenerative) positive feedback (PFB), or (degenerative) negative feedback (NFB), where it either aids (in phase) or opposes (anti-phase) the original input signal v_i. The former leads to increased gain with reduced bandwidth

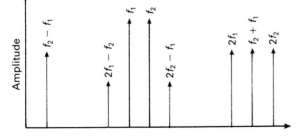

Figure 1.6 Frequency spectrum of 2nd- and 3rd-order intermodulation products.

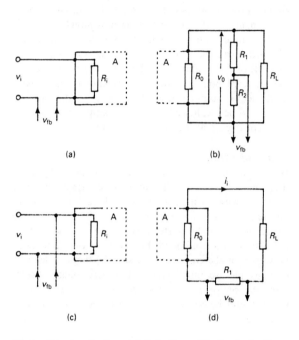

(a)   (b)

(c)   (d)

Figure 1.7 Application and derivation of feedback signal v_{fb}.

and ultimately to oscillations, while the latter has the opposite effect.

If the amplifier open-loop gain (gain before feedback) is A, then the closed-loop gain is given by:

$$A' = A/(1 \pm \beta A)$$

plus for NFB and minus for PFB. In a practical situation, both A and β may be complex quantities. As the gain bandwidth product is a constant, the bandwidth will be modified in the same proportion as the change of gain.

Application of feedback signal

Figure 1.7 shows how the feedback signal can be applied and derived. (At (a) and (c) the signal is applied either in series or in shunt (parallel) with v_i the input signal. No matter whether the feedback is positive or negative, the resulting effect on the input resistance R_i will be the same.

Generation of feedback signal

At (b) and (d), v_{fb} is either proportional to the load voltage or the load current. Either $v_{fb} = v_o \times R_2/(R_1 + R_2)$ or $v_{fb} = i_L \times R_1$. In either case, the values of R_1 and R_2 are chosen to be high enough or low enough so as not to unduly load the amplifier output. Again the effect on the

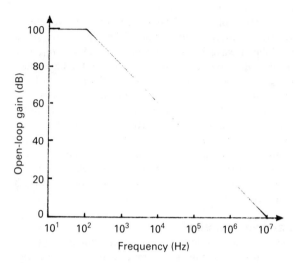

Figure 1.8 The gain $\times$ bandwidth product and frequency response of an operational amplifier.

amplifier output resistance R_o is the same for both PFB and NFB.

Effects of feedback

Neglecting any loading effect transmitted through the feedback loop, to a first approximation, the effect on both R_o and R_i is as follows:

for voltage derived feedback, R_o becomes R_o/D
for current derived feedback R_o becomes $R_o.D$
for series applied feedback R_i becomes $R_i.D$
for shunt applied feedback R_i becomes R_i/D.

where $D = 1 + \beta A$, and is sometimes known as the densensitising factor.

Negative feedback improves the amplifier performance by reducing the sensitivity to drift in component tolerances due to heat and time, the level of distortion generated, the response to noise, the degree of phase shift that it generates, and increases its bandwidth.

Feedforward amplifiers

This is a technique that can be used to minimise the distortion produced in a power amplifier. The output signal is sampled and this is then compared, after making due allowance for any phase shift, with the input signal. Any difference between the two must be representative of the distortion that is occurring and this is then used to generate a correcting signal that is fed forward to be added to the output signal to correct the distortion.

Gyrators

This is a circuit arrangement that can be used to synthesise an inductor, without using a physical coil. An ideal gyrator is a non-reciprocal two-port device whose input and output impedances are related by $Z_i = R^2/Z_o$ where R is known as the resistance of gyration.

Circuits can be designed around operational amplifiers which allow inductances of up to 1 H to be obtained from a capacitor of 1 nF and is effective up to a frequency of more than 1 MHz.

1.4 Operational amplifiers (Op Amp)

These circuits were originally designed to carry out mathematical operations, but with the introduction of integrated circuit (IC) versions, their characteristics and versatility allow them to be used in very many different applications. The basic IC is fabricated around a four-block structure. The input stage utilises a long-tailed pair differential amplifier and this is followed by a gain block, buffer, and output driver stages. The basic circuit of the long-tailed pair is shown in Fig. 1.9, in which the base bias networks have been omitted in the interests of simplicity. As shown, the circuit provides for both balanced inputs and outputs. When unbalanced inputs need to be catered for, one of the bases can be grounded via a suitable value of capacitor. The common resistor in the emitter lead which acts as a constant current device, may be replaced with a suitably biased transistor. As the emitter current is thus constant, any increase in the collector current of one side, is balanced by a corresponding reduction in the other. The output signal, which depends upon the current difference at the two collectors, is stabilised against temperature drift problems because such increases or decreases in collector current due to a rise in temperature become self-cancelling. The gain block usually consists of compound stages again designed around long-tailed pairs and this is followed by emitter followers to provide buffering. The output driver stage often uses complementary push–pull emitter followers to provide a very low output impedance. As dc coupling is used throughout, the device has a gain that extends down to zero frequency (dc). Due to the symmetry of the circuit and the extensive use of long-tailed pairs, the temperature drift in characteristics is very small and the common mode rejection ratio is very high. (See also Characteristics and parameters, p. 8) The ideal Op Amp would have input and output resistances (R_i and R_o) that were infinity and zero, respectively, together with infinite gain and bandwidth. A practical device may have open loop values of R_i as high as 10^{12} ohms, while R_o can be less than 100 ohms, together with a gain of 100 dB at 10 Hz and a gain bandwidth product of 10 MHz as indicated by Fig. 1.8.

Basic configurations

Inverting amplifier. This circuit, shown in Fig. 1.10, is sometimes known also as a virtual earth amplifier. This arises for the following reasons. Because the gain is so very high, the signal at the inverting input is correspondingly small and close to zero. The non-inverting input is at earth potential and so the differential input is also practically zero. Therefore the inverting input must also be practically at earth potential. Point X is therefore described as a virtual earth point. Because of this, any input signal v_i, must produce a current that flows through R_2 and R_1, to provide the output signal v_o. Because of the signal inversion, the stage gain with NFB is given by:

$$A = -R_1/R_2.$$

The circuit input resistance can be shown to be equal to R_2. The output resistance R_o will be reduced approximately by the same factor as the gain reduction. The new bandwidth can be calculated from the gain $\times$ bandwidth

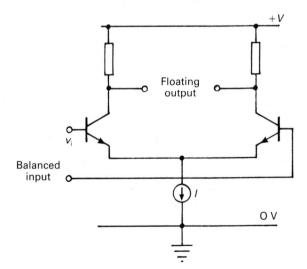

Figure 1.9 Basic circuit of long-tailed pair amplifier.

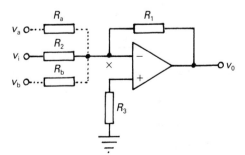

Figure 1.10 Inverting operational amplifier.

product. In order to minimise the temperature drift in the input offset voltage (see also Characteristics and parameters), the resistor R_3 is chosen to be the nearest preferred value to the parallel combination of R_1 and R_2. Thus the amplifier parameters are now chiefly dependent upon the external components and not the IC.

Since point X is virtually an earth point, it can be used to sum several signals, as is also shown in Fig. 1.10. The overall mathematical operation then becomes:

$$v_o = -\{v_a(R_1/R_a) + v_1(R_1/R_2) + v_b(R_1/R_b)\}$$

Non-inverting amplifier By rearranging the external components, the amplifier can be made non-inverting as shown in Fig. 1.11. The stage gain is now given by $A = (1 + R_1/R_2)$; R_3 again being equal to the parallel combination of R_1 and R_2.
;By replacing R_1 with a short circuit and R_2 with an

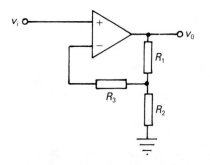

Figure 1.11   Non-inverting Op Amp.

open circuit, the amplifier becomes a unity gain voltage follower, with very high input and very low output resistances.

If the negative sign produced by the circuit in Fig. 1.10 creates a problem, then the configuration shown in Fig. 1.12 can be used.

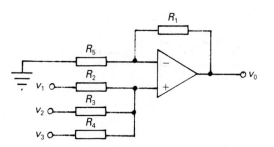

Figure 1.12   Non-inverting adder.

Characteristics and parameters

Common and differential mode gains. The closed-loop relationship between the differential input potential at the two inputs and the output voltage is described as the differential mode gain A_d. If the two input terminals are connected in parallel and supplied with the same input signal relative to earth, then theoretically, the output signal should be zero. But in practice, due to imperfections within the IC, there will be a small output. The relationship between this and the input signal is described as the common mode gain A_c.

Common mode rejection ratio (CMRR). This is a useful figure of merit for comparing the characteristics of differential amplifiers and is the ratio of A_d to A_c, or the difference if these values are expressed in dB. In practice, CMRR, which should in theory be infinity, can easily reach 100 dB.

Input offset voltage and current. Ideally when the differential input signal is zero, then the output should also be zero. However, due to imperfections within the IC, the output will have a non-zero value. The magnitude of the input dc voltage that is required to achieve this condition is described as the input offset voltage. The input offset current is then simply the difference between the two resulting currents at the input terminals.

Offset temperature drift. Due to the temperature coefficient of the semiconductor material, both the input offset voltage and current will be temperature sensitive. This is described in terms of change in value per °C. At a temperature of 25°C, these values can be as low as $1\,\mu\text{V}/°\text{C}$ and $1\,\text{nA}/°\text{C}$. R_3 in Figs 1.10 and 1.11 is included to minimise this effect.

Output offset voltage. This is the value of the output voltage relative to earth when the two input terminals are grounded.

Overload recovery and settling times. This is a measure of the device's response to large drive signals. The overload recovery time, which is typically less than 1 ms, is the time taken for the circuit to recover to a rated output voltage level after being in a saturated condition. The settling time is usually quoted as the time taken for the circuit output to reach 0.01% of its final value in response to a step input signal. Settling times as low as $1\,\mu\text{s}$ are common.

Power supply rejection ratio (PSRR). This expresses the dependence of the input offset voltage on variation in supply voltage level. PSRR is usually less significant than the temperature dependence and a value of around $10 \mu V/V$ can be expected.

Slew rate and full power response/bandwidth. An operational amplifier responds to high level signals somewhat slower than small signals. This is due to the circuit capacitances that require time to change their state of charge. For this reason the full power bandwidth is not likely to exceed about 4 MHz. The slew rate defines the maximum rate of change of output voltage for a large step input signal and is typically greater than $50 V/\mu s$. If this rate is exceeded by too large a driving signal, then distortion results. The relationship between full power bandwidth and slew rate is given by:

$$f_m = \text{slew rate}/(2\pi V_{\text{out.max}})$$

where f_m is the upper 3 dB frequency.

Charge amplifiers

Capacitive transducers function on the principle of converting a measurement into an equivalent electrical charge. Such a device may thus be represented by an equivalent circuit of a fixed voltage V, in series with a capacitor. Movement then produces a change of capacitance that results in a change of charge. Amplifiers designed to work in such conditions require a very high input impedance so as to draw negligible current from the transducer. The NFB path is provided by a small capacitor C_f, usually shunted with a resistor to ensure good low frequency response. For the inverting amplifier configuration, the gain sensitivity is given by:

$$v_o/C = -V/C_f$$

The alternative non-inverting amplifier circuit provides a gain that is marginally higher.

Current conveyors

This is a class of amplifier with two inputs X and Y and one output Z, such that the output current depends only upon the input current at X or its voltage equivalent at Y. The circuits have frequency dependent negative resistance (FDNR) that make them suitable for such applications as voltage controlled oscillators, filters, differentiators, integrators, and analogue computing devices.

Current mirrors or repeaters

These are compound circuits, usually in IC form, designed so that an output current closely follows or mirrors an input current. This is achieved with a circuit that has a very low input impedance and very high output impedance. The technique is commonly used in

an attempt to avoid the constraints imposed by the gain bandwidth product of the amplifier.

Differential amplifier

Due to the nature of the input transducer, many instrumentation systems require an amplifier with a differential input and single-ended output. The Op Amp arrangement shown in Fig. 1.13 meets this requirement.

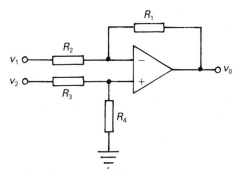

Figure 1.13 Differential Op Amp.

If the resistors have the following relationship:

$$R_2/R_1 = R_3/R_4$$

then the output voltage v_o is given by

$$v_o = R_1/R_2[v_1 - v_2]$$

Exponential or logarithmic amplifier. If the NFB component is replaced by the exponential resistance of a diode, the amplifier has a logarithmic gain characteristic. The basic circuit configuration is shown in Fig. 1.14, the gain being given by

$$v_o = k_1 \ln(k_2 v_i)$$

where k_1 and k_2 are constants.

If the diode and resistor are interchanged, the amplifier acquires an anti-logarithmic response.

Fast half-wave rectifier

The circuit configuration shown in Fig. 1.15 is that of a high speed (high frequency) half-wave rectifier. While v_i is positive, D_1 is cut off and D_2 is conductive so that the input is effectively grounded and the output v_o is zero. When v_i is negative, D_1 is On and D_2 cut off. The circuit then operates as an inverting amplifier with $v_o = -(R_2/R_1)v_i$. The upper frequency of operation

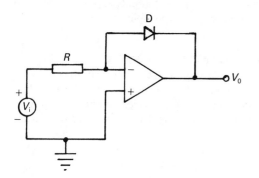

Figure 1.14   Exponential or logarithmic amplifier.

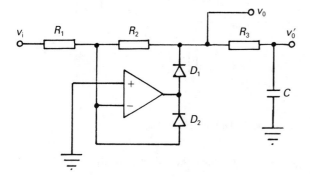

Figure 1.15   Fast half-wave rectifier.

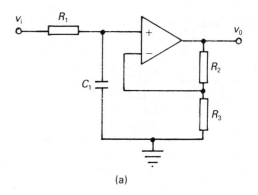

(a)

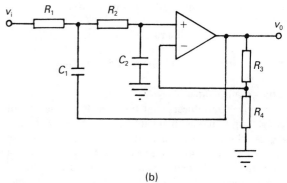

(b)

Figure 1.16   Butterworth filters: (a) 1st order; (b) 2nd order.

depends upon the slew rate, particularly close to the zero crossing, because at this point the diode junctions need to switch rapidly between $+600$ mV and -600 mV. The circuit can be converted to non-inverting simply by grounding the point v_i and transferring the input to the non-inverting terminal.

Filters (active) (see also p. 123)

Active filters can be constructed by connecting low-pass RC filter components to a unity gain, voltage follower circuit. If the R and C elements are interchanged, the filter takes on high-pass characteristics. Figure 1.16 shows how 1st- and 2nd-order Butterworth filters can be constructed. Filters with Chebyshev and other characteristics can also be synthesised.

Third- and higher order filters are constructed by cascading 1st- or 2nd-order filter sections as needed. Band-pass filters can be constructed either by using LC components in the circuit or by combining the outputs of low- and high-pass RC sections with a summing circuit. Because of the gain that is available with Op Amps, high Q band-pass active filters can be constructed in this way.

Instrumentation amplifiers

These circuits often have to operate with very small input signals and in electrically noisy surroundings. The control applications require that such amplifiers should have a high degree of linearity combined with a very small temperature drift. Circuits are thus often assembled from unity gain buffer amplifiers at the inputs to ensure a good CMRR with a very high input impedance. Buffered output stages are also common to ensure a very low output impedance. To minimise the response to noise, such amplifiers will often need to be shielded to avoid the effects of stray magnetic and electrostatic fields.

Isolation amplifiers

These are a special type of instrument amplifier, often used for medical/hospital applications such as patient monitoring or treatment. In these cases, very high common mode voltages due to transients and long signal leads can arise. The circuits operate in regions of high radio frequency interference (RFI) or electromagnetic interference (EMI). The chief requirements are very high input impedances with an input to output isolation that is in the order of 10^{12} ohms. Such a value can be achieved either by using transformer or opto-coupling between the separate input and output ICs.

Multiplying amplifiers

Certain types of Op Amp are available as analogue multipliers. One input signal is applied conventionally while the second is applied as a controlling bias to the long tail constant current transistor of the input stage. When one input is dc and the other bi-polar, the operation is described as two quadrant multiplication. Some of these devices are available as four quadrant multipliers, where both inputs can be bi-polar.

Peak and average detector circuits

The circuit of Fig. 1.17 is that of a positive peak detector, based on a unity gain voltage follower. Whilst v_i is greater than the voltage v_o and that at the inverting input, the diode D is conductive and causes C to charge to the peak level of the input signal. When D is cut off, C is isolated from v_i. The capacitor can be discharged or reset by the use of a pulsed shunt FET switch or by being discharged into a load. Reversing the diode polarity will change the circuit into a negative peak detector. The circuit can be simply converted into an average detector by adding a cascaded RC low-pass filter section. Then if the input signal v_i is an AM wave, the filter will remove the carrier component and v_o is suitable for driving such circuits as an AGC system.

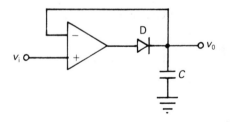

Figure 1.17 Positive peak detector.

Precision clamp circuit

In the circuit shown in Fig 1.18, the voltage V_R is used as a reference level against which the input signal v_i is to be clamped. Whilst v_i is less than V_R, the amplifier is acting as a unity gain voltage follower with the output positive and diode D forward biased. The output voltage v_o is thus equal to V_R, the clamping level. If v_i is less than v_R, D is cut off and if R_L is much greater than R, then v_o is still approximately equal to V_R.

Precision rectifier circuit

A conventional diode circuit only conducts as long as the forward bias voltage exceeds a threshold level. This precludes the use of silicon devices for rectifying signals of less than about 600 mV amplitude. By using the

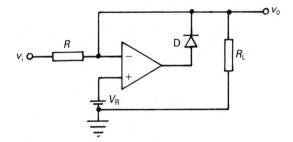

Figure 1.18 Precision clamp.

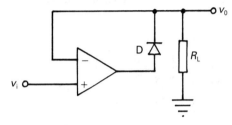

Figure 1.19 Precision rectifier.

circuit shown in Fig. 1.19, this threshold value is in effect divided by the gain of the Op Amp, so that rectification can take place down to levels as low as $50\,\mu\text{V}$. When v_i is negative the diode D is cut off and so the only current that flows through the load is the very small leakage current of the diode.

Sample and hold circuit

In analogue servo control systems it is often necessary to provide a varying dc voltage that is proportional to some error condition. Such a circuit is shown in Fig. 1.20 where, for example, the input signal may be a sawtooth waveform of varying time duration. If the FET is gated by accurately timed positive pulses, the voltage that is charging the capacitor C, will be proportional to the instantaneous value of the sawtooth and will be representative of its instability. Whilst the FET is cut off, the charge on C will leak away through the circuit load, in readiness for the next charging pulse.

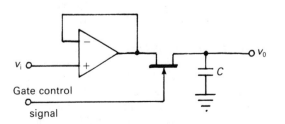

Figure 1.20 Sample and hold circuit.

Switches or transmission gates

Although these are not operational amplifier devices, they are often used in combination as shown in Fig. 1.21. The gates are typically used to multiplex analogue signals in switching applications and are commonly associated with voltage follower circuits. Using CMOS technology, these devices are fast, with switching speeds typically less than 75 ns, with an OFF isolation better than 70 dB and a transmission loss of less than 3 dB, up to frequencies above 100 MHz. The switching action is controlled by the digital inputs C, which may be either complementary as shown or by a single logic pulse.

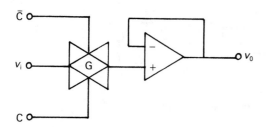

Figure 1.21 Analogue or CMOS gate.

Transconductance amplifiers

These are essentially voltage to current convertors and are particularly suited to such applications as the driving of cathode ray tube scanning assemblies. Two configurations are shown in Fig. 1.22. For cases where the load is not grounded at one end, the arrangement in (a) can be used. With v_i as the time varying input signal, the current in the load Z_L is v_i/R. This current also flows through the amplifier source so this circuit must be capable of supporting the necessary drive. When one side of the load has to be at earth potential, the arrangement shown in (b) can be used. If $R_1/R_2 = R_3/R_4$, then the load current is $-v_i/R_2$.

Transresistance amplifiers

Generally, this circuit, shown in Fig. 1.22(c), is used as a current to voltage convertor and is particularly useful with transducers that are current generators. Because of the virtual earth at the inverting input of the Op Amp, the current through R_2 is practically zero. The output voltage is therefore given by $v_o = -I_i/R_1$.

Very low frequency amplifiers

In certain instrumentation or control situations, it is necessary to amplify very low frequency signals whose amplitudes may be varying over a period of minutes or even hours. The use of conventional ac amplifiers in such cases is not practical due to the value and the size of suitable coupling capacitors. One alternative is to use

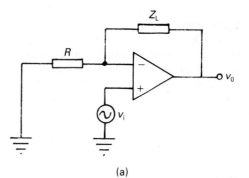

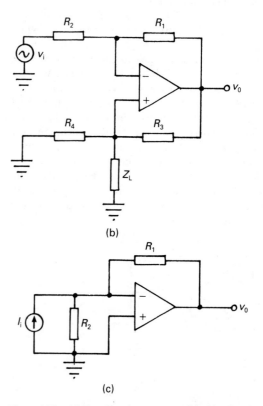

Figure 1.22 (a) Transconductance amplifier for floating load. (b) Transconductance amplifier for grounded load. (c) Transresistance amplifier.

dc coupled amplifiers, but these can suffer from the effects of parameter drift with temperature variation and from noise. The problem can be resolved by using *chopper stabilised amplifiers*. The slowly varying signal is chopped or sampled at a suitable rate, perhaps 10 kHz, so that the signal becomes modulated on to the switching frequency. This new signal can now easily be amplified using ac techniques. The complex signal is then demodulated to recover the amplified dc component. The stability that is achieved with this method, is due to the synchronous nature of the modulation–demodulation process.

Virtual earth amplifier

A concept devised to simplify the analysis of operational amplifiers.

Zero crossing detector circuit

If the reference voltage of a comparator circuit is set to zero, then an input signal will cause the output to change state rapidly as it passes through zero. This change will only be limited by the slew rate. Thus if the input signal is a sinusoid, then the output will be a square wave. If the time constant of CR in Fig. 1.23 is short compared with the period of the input signal, then v_o, the output, will become differentiated to produce positive and negative pulses as the input signal passes through zero. By adding a diode D, the negative pulses will be removed and the voltage pulses across the load R_L will represent the start of each input cycle. Such a circuit can therefore be used to provide marker or synchronising pulses.

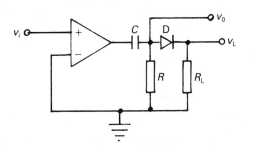

Figure 1.23 Zero crossing detector.

1.5 Oscillators

Oscillators are amplifiers that provide their own input through the use of positive feedback. The stage gain A' under this condition is given by $A' = A/(1 - \beta A)$, where A is the gain without feedback and β is the fraction of the output signal used for feedback. Thus as βA increases, the stage gain rises and the bandwidth falls. When $\beta A = 1$, the gain reaches infinity and the bandwidth falls to a single frequency. The dc power supply is the source of the output signal. Because the oscillator converts dc power into an ac form, it acts as an energy convertor which is not 100% efficient. Thus like the amplifier, it should work into a matched load and be capable of dissipating heat. The feedback loop either contains, or is coupled to, a *frequency-determining network*. Provided that this has filter-like characteristics, the amplifier can be made to oscillate.

Criteria for oscillations

The criteria for maintaining oscillations are that the *loop phase shift* should be zero or multiples of 360° (2π rad) and that the *loop gain* should not be less than unity. This latter point is sometimes described as the *Barkhausen criterion*. In a practical oscillator, the loop gain needs to be marginally greater than unity, not only to provide for the output signal, but also to allow for any drift of component tolerances that might increase the loop attenuation. The oscillation-maintaining criteria can be summarised as:

$$A\angle\theta \times \beta\angle\phi \geq 1/0° \text{ or } |A|\underline{/\theta + \phi} \geq 1/0°$$

Provided that these criteria are met, then the circuit will oscillate near to the natural frequency of the frequency-determining network, only marginally modified by the input and output impedances of the amplifier.

Excessive feedback

In special cases, the loop gain is made much greater than unity. This use of excessive feedback results in an almost violent change of state in the amplifier and generates non-sinusoidal wave shapes.

1.5.1 Frequency controlling elements

Crystals

When certain crystalline substances such as quartz, Rochelle salts or tourmaline are mechanically stressed, they produce electrical charges. Conversely, the application of an electric field produces mechanical strain. If this field is an alternating one, then the molecular structure will vibrate. If the frequency is adjusted it will be found that at some particular value, the vibration will reach a maximum amplitude. This phenomenon is known as the piezo (or pressure) electric effect which gives rise to a resonant characteristic. Frequency controlling elements are cut in slab form to accurate dimensions from the parent crystal and then highly polished.

Crystal cuts. The crystals which exhibit these properties are of long hexagonal form with three major axes. The longitudinal or optical or Z axis is not important in this respect. The X or electrical axis lies parallel to the parallel sides, while the Y or mechanical axis is normal to this. When an electrical potential is applied along the X axis, mechanical vibration occurs along the Y axis. The various cuts are defined by their angular relationship to these axes. X and Y cut crystals have their large surfaces normal to the X and Y axes, respectively. Other sub-cuts are defined by rotating the slab around one edge along the X axis at a chosen angle to the Z axis. These are usually defined as AT,

BT, CT, DT, or FT cuts. Each particular cut has a unique temperature coefficient of frequency stability, with the AT cut being the most stable.

Crystal equivalent circuits and characteristics. The resonant effect can be equated with the response of a circuit containing inductance, capacitance and resistance as shown in Fig. 1.24(a), where:

> L represents the inertia of the crystal structure:
> C is inversely proportional to the stiffness of the slab; and
> R represents the frictional losses in the crystal.

The parallel component C_o in Fig. 1.24(a) is due to the self-capacitance of the electrodes that are plated on to the crystal to provide electrical connections. Since the losses are very small, each crystal can have a very high Q factor, an in circuit or loaded value of 5000 being typical. Reference to Fig. 1.24(b), shows that the device has an impedance that is resistive at two frequencies: a series resonance at f_s, given by $f_s = 1/(2\pi\sqrt{LC})$ (resonant) when C_o is neglected, and a second parallel one at f_p (anti-resonant) when C_o is included. The small separation between f_s and f_p is given by:

$$f_p/f_s = \sqrt{1 + C/C_o}$$

Crystal limitations. Because the resonant frequency depends upon the crystal dimensions, the thickness in particular, the upper frequency is restricted due to fragility. Experimental devices have been reported that will oscillate up to almost 900 MHz in the fundamental mode. However, crystals are more commonly restricted to about 70 MHz in fundamental mode and perhaps 500 MHz in the seventh overtone or harmonic mode. By adding a small series or shunt capacitor, it is possible to slightly trim the actual operating frequency.

Dielectric resonators

These are small disc-shaped devices that are formed from barium or zirconium titanate compounds by a sintering process. Like quartz crystals these have a resonant characteristic that is dimensionally controlled, but capable of operating at frequencies up to 18 GHz in a fundamental mode. The device can be tuned over a frequency range of about 1% relative to the resonant frequency, by using a small metal disc to provide a variable air gap to earth.

Inductors and capacitors

These components in parallel form the basis of very many electronic circuits and the effects of resonance are well known. The temperature stability of frequency depends upon the stability of the inductor and capacitor. The figure of merit, Q factor, has an important bearing on the circuit behaviour. A high Q is desirable because its frequency-selective properties help to improve harmonic suppression. The resonant frequency is given by

$$f_o = (1/2\pi)\sqrt{(1/LC - R^2/L^2)}\ \text{Hz}$$

where L is the value of the inductance in henries,
  C is the value of the capacitor in farads, and
  R is the series loss resistance of the inductor in ohms.
For $Q > 50$, this amplifies to

$$f_o = 1/(2\pi\sqrt{LC})\ \text{Hz}$$

Resonance at a given frequency could be produced by an almost infinite range of values for L and C. However, it is important that circuits work into loads of optimum impedance for maximum transfer of

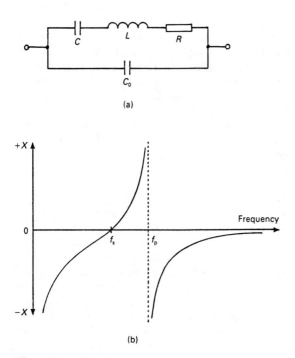

(a)

(b)

Figure 1.24   (a) Equivalent circuit of crystal; (b) reactance/frequency characteristic.

energy. The dynamic impedance of the parallel tuned circuit is given by

$$Z_D = L/(CR) \text{ ohms}$$

As L is reduced, the Q factor tends to rise, but only up to a point. When C is made larger, the dielectric losses increase. Therefore there is often an optimum L to C ratio for best results.

If the impedances of any circuits coupled to such a network become significant, this will have a loading effect that will produce a shift in frequency, i.e. generate frequency pulling. Therefore to maintain an adequate working Q factor, the coupling between circuits should be light.

For microwave operation the inductive component is commonly replaced by resonant lines that can be tuned to specific wavelengths.

Resistance capacity components
These components are used to maintain positive feedback by generating a phase shift, the number of component pairs being dependent upon the degree of phase shift required. These are usually restricted to low frequency applications (less than about 250 kHz). The frequency of operation depends mainly upon the time constant (CR seconds) of the components.

Surface acoustic wave components (see also p. 125)
Any filter component that has the correct amount of phase shift can be used in a positive feedback loop to generate oscillations. These devices that are a solid-state version of many filter configurations are ideal. Although less commonly used than crystals, it is anticipated that frequencies from about 20 MHz up to 2 GHz could be generated using these devices.

Noise and distortion in oscillators
Any unwanted signal component in the oscillator output can be considered as noise. Non-linearity in the amplifier will produce waveform distortion and introduce harmonics. Unexpected stray capacitance can produce parasitic resonances that are not harmonically related to the wanted signal frequency. Instability in the power supply can create amplitude modulation. Temperature drift can cause component value variations that lead to instability of frequency. Random noise from a variety of sources can produce phase modulation or phase noise. This in particular, broadens the frequency spectrum of the oscillator. Instead of producing just a single frequency, side band components are introduced (see also p. 185).

Class A bias is used for those oscillators required to provide the minimum of distortion, while Class C is used for harmonic generators, where the required overtone can be extracted with a suitable filter.

1.5.2 Oscillator circuits

Crystal oscillators

Butler oscillator This two-stage oscillator, shown in Fig. 1.25, is in effect a grounded or common base amplifier followed by an emitter follower. Positive feedback is achieved by the crystal that couples the two in-phase emitters. The circuit oscillates quite accurately close to the series resonant frequency of the crystal. A small degree of frequency variability can be achieved by adding a small tuning capacitor in series with the crystal. The circuit as shown provides a low impedance output but this can easily be modified by including either an RF choke or a suitably tuned circuit as a collector load for Tr_2.

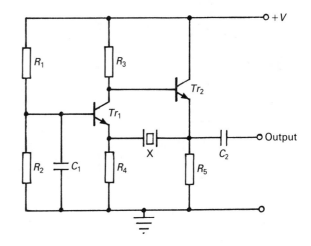

Figure 1.25 Butler crystal oscillator.

Miller oscillator. This is the equivalent of the tuned drain/tuned gate oscillator but with the crystal acting as the gate tuned circuit. Positive feedback is provided via the Miller capacitance, $C = C_{dg}(1 + A)$, where C_{dg} is the interelectrode capacitance between drain and gate. The circuit, shown in Fig 1.26, oscillates very close to the parallel resonant mode of the crystal, and the very high shunt impedance presented in this mode is better suited to use with FET devices than with bipolar transistors.

Pierce oscillator. This circuit, shown in Fig. 1.27, is unique among oscillators in that it can function without any tuned circuit. Reference to Fig. 1.24(b) shows that above series resonance, the crystal impedance is inductive. This combines with the self-feedback capacitance to produce oscillations close to f_s when the

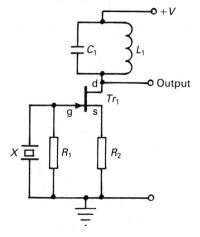

Figure 1.26   Miller crystal oscillator.

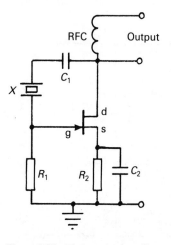

Figure 1.27   Pierce crystal oscillator.

feedback is a maximum. A small capacitance between the gate and electrode and ground can be used to introduce a small degree of tuning.

Dielectric resonator oscillators

The circuit depicted in Fig. 1.28 might be described as the microwave equivalent of the Pierce oscillator. The dielectric resonator disc is placed as shown, in the electromagnetic field produced by the transmission line section connected to the drain electrode. This field is coupled by the disc into the gate line. The line lengths are such that they have an inductive impedance to produce positive feedback at the resonant frequency of the disc. The operating frequency can be tuned over a range of about 100 MHz in 10 GHz, by perturbing the resonator's magnetic field by varying an air gap between the disc and the metallic enclosure of the circuit.

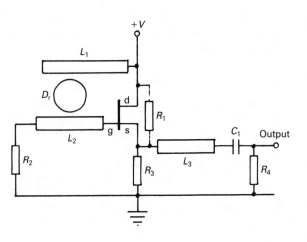

Figure 1.28   Dielectric resonator oscillator.

Inductance–capacitance oscillators

Clapp oscillator. The circuit shown in Fig. 1.29 uses a series tuned circuit to control the frequency of oscillation and an amplifier that operates in the common drain mode due to the inclusion of the bypass capacitor C_4. Capacitors C_2 and C_3 form a potential divider network to couple the tuned circuit into the three electrodes of the amplifier. The circuit oscillates at slightly above the series resonant frequency of L_1C_1, making the impedance inductive and providing positive feedback. Because of the very low impedance of the resonant circuit, this oscillator is relatively immune to drift in the parameters of the amplifier and thus forms a very stable variable frequency oscillator.

Colpitts oscillator. The circuit of Fig. 1.30 depicts a popular RF oscillator that has good stability. The

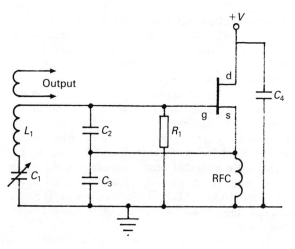

Figure 1.29   Clapp oscillator.

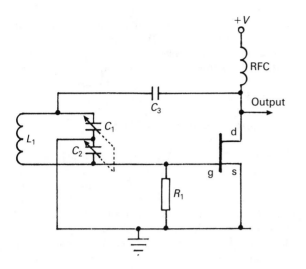

Figure 1.30 Colpitts oscillator.

resonant frequency is controlled by L_1 and C_1 in series with C_2. As connected, this provides zero loop phase shift marginally above the theoretical resonant frequency. The circuit is described as being shunt fed; the resonant circuit being in parallel with the active device relative to the dc power supply. The twin gang tuning capacitors with the grounded vanes ensure that hand capacity effects are minimal and they act as low impedance bypasses to earth for any harmonics, thus providing an output waveform of high purity.

Franklin oscillator. This oscillator, shown in Fig. 1.31, uses the total phase shift of two amplifiers to ensure

positive feedback. The tuned circuit of L_1 and C_1 forms a shunt load on one of the two stages and because of the high gain available, it is possible to use low values for the coupling capacitors C_2 and C_3. This minimises the loading on the tuned circuit so that there is a high loaded Q factor to ensure good waveform purity and a high degree of frequency stability.

Hartley oscillator. Figure 1.32 depicts an oscillator that has a number of features that are common to the Colpitts circuit. In this case, the centre point of the tuned circuit consisting of $L_1 C_1$ is made by centre tapping the inductance. As shown, the circuit is again shunt fed, but it is also possible to feed the dc power via the centre tap, giving rise to a series fed oscillator. If variable tuning is provided by C_1, then hand capacity effects can become a problem by changing the effective capacity during the tuning operation.

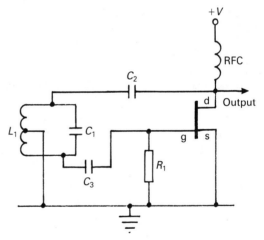

Figure 1.32 Hartley oscillator.

Meissner oscillator. In the circuit shown in Fig 1.33, the tuned circuit L_1 and C_1 is dc isolated from the amplifier, inductive coupling being provided from collector to emitter via L_2 and L_3. Because there is negligible loading on L_1 and C_1 the circuit has a high Q factor and oscillates at the resonant frequency. The output winding should be coupled to the tuned circuit, otherwise stray reactances can be introduced that can give rise to parasitic oscillations.

Negative resistance oscillators

Tunnel diode oscillator. The tunnel or Esaki diode (named after the inventor) is a PN junction formed from germanium or gallium arsenide that is heavily doped

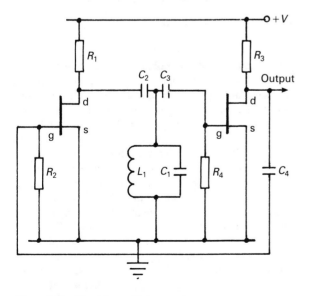

Figure 1.31 Franklin oscillator.

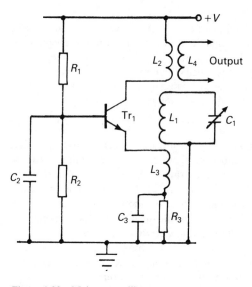

Figure 1.33   Meissner oscillator.

with impurities. This gives the device a forward biased characteristic as shown in Fig. 1.34(a). Between the peak (V_P) and the valley (V_V) voltages it exhibits negative resistance (R_N). If a suitably biased diode is placed in parallel with a tuned circuit with loss resistance R_L, then, provided that the magnitude of R_N is greater than

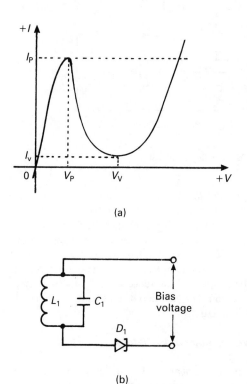

(a)

(b)

Figure 1.34   Esaki or tunnel diode oscillator.

R_L, sustained oscillation will occur at a resonant frequency well into the UHF range.

The oscillator's disadvantages of low voltage swing and poor isolation between input and output are offset by its simplicity, low cost, low noise and low power requirement.

Resistance–capacitance oscillators

Parallel tee oscillator. A parallel tee filter network that has the component value ratios $R_1 = R_2 = 2R_3$ and $C_1 = C_2 = C_3/2$, produces maximum attenuation and $180°$ phase shift at a frequency $f_o = 1/(2\pi R_1 C_1)$ Hz or $1/(2\pi R_3 C_3)$ Hz. If this is combined with an inverting amplifier with sufficient gain, as shown in Fig. 1.35, then the circuit will oscillate at f_o. The filter attenuation at harmonically related frequencies is such that these are effectively rejected. The purity of waveform is such that this type is well suited for use as a signal source for testing high quality audio systems. For variable frequency operation the main difficulty arises when trying to maintain the ratio-tracking between the three resistors or three capacitors.

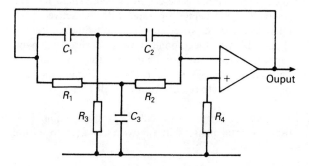

Figure 1.35   Parallel tee oscillator.

Phase shift oscillator. For the circuit shown in Fig. 1.36, if $R_1 = R_2 = R_3$ and $C_1 = C_2 = C_3$, there will be a frequency at which the phase shift through this network is $180°$. The arrangement shown is described as a phase advance network and this produces an oscillation frequency $f_o = 1/(2\pi\sqrt{6}RC)$ Hz. If the resistors and capacitors are interchanged to form a phase retard network, the same effect occurs but at the new frequency $f_o = \sqrt{6}/(2\pi RC)$ Hz. To overcome the attenuation in either network, the amplifier must have a voltage gain greater than 29. For a bi-polar transistor, h_{fe} must be greater than 44.5.

Quadrature oscillator. Figure 1.37 shows how two active first-order filters/integrators in cascade can be used to produce a sinusoidal oscillator with two quadrature outputs. R_4 provides overall positive feed-

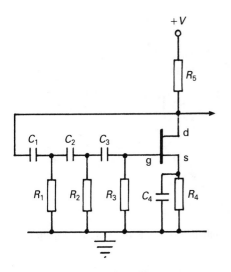

Figure 1.36 Phase shift oscillator.

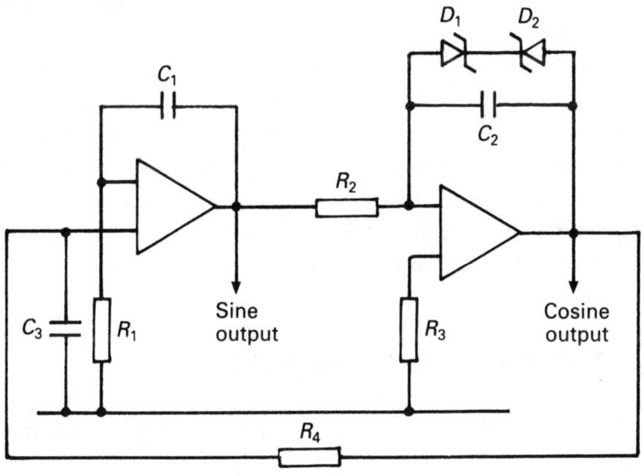

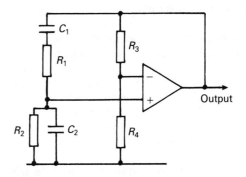

Figure 1.38 Wien bridge oscillator.

Amps the maximum frequency would be slew rate limited. For a practical variable frequency oscillator, where either C_1 and C_2 or R_1 and R_2 would be ganged, the limiting factor is more likely to be due to the minimum value that can be achieved in the variable components.

Surface acoustic wave oscillator. The surface acoustic wave (SAW) filter is effectively a delay line that indirectly produces a relative phase shift. Provided that the total loop phase shift through the filter, sustaining amplifier and the variable element are multiples of 2π radians, positive feedback results if the loop gain is greater than about 8 dB, the typical attenuation of a SAW filter. The variable element may be a varactor diode that can be electronically tuned. This then allows the oscillator to be directly frequency modulated. The basic principle of this concept is shown in Fig. 1.39. Although the circuit has a lower Q factor than a crystal type, the SAW oscillator has a higher power handling capability, good frequency stability and low noise. The low frequency limitation is imposed by the long delays that would be needed, while the upper frequency is set by the accuracy of the technique used to produce the filter. Typical applications would therefore range from about 20 MHz to 2 GHz.

Figure 1.37 Quadrature oscillator.

back and provided that $R_1 = R_2 = R$ and $C_1 = C_2 = C$, then the frequency of oscillations $f_o = 1/(2\pi RC)$ Hz. Zener diodes D_1 and D_2 are used as amplitude stabilisers.

Wien bridge oscillator. The oscillator circuit of Fig. 1.38 uses the reactive arms of the Wien bridge, a circuit that can be used to measure the value and loss resistance of a capacitor. If $R_1 = R_2 = R$ and $C_1 = C_2 = C$, there will be one frequency at which the phase shifts across the series CR limb will be exactly opposite that across the parallel limb. If this zero phase shift is combined in a feedback loop with the zero phase shift of a non-inverting Op Amp, oscillations will result if the loop gain is higher than the attenuation of $1/3$. The frequency of oscillations will be given by $f_o = 1/(2\pi CR)$ Hz. The gain requirement of $(1 + R_3/R_4)$ thus requires that $R_3 > 2R_4$. For Op

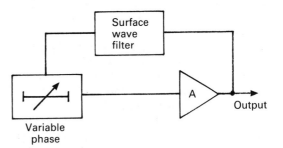

Figure 1.39 Surface wave filter oscillator.

Temperature compensation

MCXO (mathematically compensated crystal oscillator). The temperature coefficient of a particular crystal can be computer-analysed and the correction values for different temperatures stored in a digital memory. If a microprocessor is incorporated in the oscillator system and provided with this information, then by monitoring the operating temperature, the processor can exert control over the frequency stability.

OCXO (oven compensated crystal oscillator). In this arrangement, the crystal is actually mounted in a glass envelope that contains a thermostatically controlled heater element, thus maintaining the crystal at a constant temperature and stabilising the operating frequency.

TCXO (temperature compensated crystal oscillator). This technique utilises a thermistor with the required temperature coefficient to stabilise the amplifier. It produces a lower stability than either of the above, but is cheaper.

UCXO (uncompensated crystal oscillator). For a reasonable degree of stability this concept relies on selecting a crystal with a suitable cut.

1.6 Parametric amplifiers

Whereas the conventional amplifier draws energy from a dc power supply to provide signal amplification, the parametric amplifier draws energy from an ac source known as the pump circuit. This energy is then used to modify a voltage-dependent parameter of the active device, which is usually a varactor diode. In spite of this complexity, these devices provide a valuable gain, typically about 20 dB, with very low noise at UHF and microwave frequencies.

The general principle is indicated in Fig. 1.40, where it is seen that the signal, pump and idler circuits are effectively coupled in parallel by a varactor diode. The pump oscillator works at a frequency above the signal, while the idler circuit is tuned to the difference between these two. Because the idler circuit is not directly connected to a load or source, it controls the relative voltage and current phases in the circuit.

The operation resembles that of the diode mixer in that $f_1 + f_2 = f_3$, except that in this case, it is the self-capacitance of the diode rather than its non-linear resistance property that is important. The pump oscillator adds charge to the diode capacitance while the voltage across it is high and extracts charge when the voltage is low. The signal source has the opposite effect,

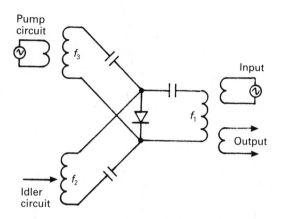

Figure 1.40   Configuration of a basic parametric amplifier.

adding charge while the voltage is low and extracting it when the voltage is high. In this way, energy is transferred from the pump into the signal circuit where it is extracted.

Theoretically the stage gain is given by the ratio, pump frequency/signal frequency, but this is somewhat reduced due to the small losses in the tuned circuits. Because these devices are used at UHF and microwave frequencies, the tuned circuits are usually constructed from resonant lines or tuned cavities, with the source and load being coupled by microwave structures.

1.7 Phase locked loop (PLL)

This is a feedback system in which a voltage controlled oscillator (VCO) is forced to track or follow the frequency of an input signal. The basic arrangement of the PLL is shown in Fig. 1.41, where the phase comparator is used to generate an output signal that is proportional to the difference between the phases of the two input signals, f_s and f_0. This output is a complex wave containing sum and difference frequencies. The sum component is removed by the low-pass filter to

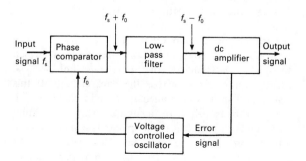

Figure 1.41   Basic phase locked loop system.

leave the difference frequency, which is a slowly varying dc level representing the error between the two inputs. After amplification this is applied to the VCO as a control voltage to correct its frequency. Thus when the loop is locked, the error signal is zero and $f_o = f_s$, but with a phase difference of $90°$. The device in integrated circuit (IC) form finds applications in such areas as demodulators, AM and FM, frequency tracking, synthesis, and multiplication.

Capture or pull-in range or detection bandwidth. This is the frequency range centred around f_o for which the circuit will detect an input signal and snap into lock. The input signal frequency has to approach the filter bandedge frequency before such a condition can arise and is, to a lesser extent, dependent upon the circuit loop gain.

Lock, tracking or hold range. This is the frequency range centred around f_o for which the VCO remains in synchronism with the input signal. This is always greater than the capture range and is independent of the filter bandwidth. Lock only fails when the VCO control device reaches the limit of its characteristic.

Static loop phase error. The theoretical phase shift of the PLL is $\pm 90°$. However due to internal offsets this value can differ by up to about $\pm 5°$.

Frequency synthesis. If a divide by N frequency divider is placed between the VCO and phase comparator, the circuit will lock when the VCO is running at N times the reference frequency, but it will still have the same stability and drift characteristics as the reference source. This principle is depicted in Fig. 1.42. By using a programmable divider, a wide range of highly stable discrete frequencies can be produced from a single reference. The lock condition can also be achieved at harmonics of the reference source. If the Mth harmonic is chosen, then the VCO will be running at f_o/M. By combining these two techniques and defining N and M, it is possible to synthesise fractional values of the reference frequency.

1.8 Power amplifiers

Such amplifiers are designed to deliver significant power, usually into low impedance loads, and can function in any classification (see Section 1.1). By definition, these devices need to be power driven so that the driving stages themselves must deliver a small amount of power. As with all amplifiers, the source of the output signal is the dc power supply and, in the interests of energy conservation, considerable attention

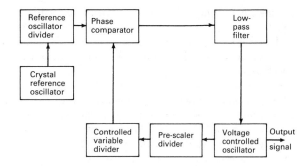

Figure 1.42 Frequency synthesis system.

is focused upon this energy conversion process. Ignoring the small level of the input signal power, the stage efficiency can be expressed as:

(ac signal power delivered to load/dc power supplied)
$\times 100\%$

The total dc power supplied (P_S) is also equal to the sum of the signal power output (P_o) and the power dissipated as heat (P_D), so that the efficiency can also be stated as:

$$(P_o/(P_o + P_D)) \times 100\%$$

The peak power output is dependent upon the power supply voltage (V_S), the amplifier peak current characteristic (I_P) and the load impedance. The optimum value of the load impedance is being given by V_S/I_P.

For reactive loads, the power output is reduced to $VI \cos \theta$, where $\cos \theta$ is the power factor for the load.

If the power dissipation and hence the operating temperature is likely to be exceeded, then the amplifier will need to be derated in the manner shown in Fig. 1.43.

As power amplifiers are necessarily large-signal amplifiers, the effects of non-linearity and distortion

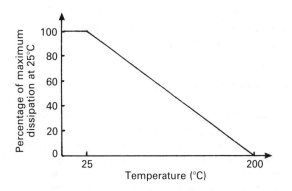

Figure 1.43 Temperature/power derating curve.

become of paramount importance. Negative feedback (NFB) is commonly used to minimise these effects. At high power levels, excess voltage and current protection will be needed, as will the stabilisation of the dc power and bias supplies.

For power outputs up to about 1 kW, the choice between field effect devices and bi-polar transistors is a compromise. Above this level, thermionic valves are commonly used. In general, FETs can operate with a dc voltage about twice that for the bi-polar device. Hence for the same power output they pass less current and dissipate less heat. However, the bi-polar device generally has a lower output impedance.

Directly coupled loads are avoided because the quiescent current represents a power loss. AC coupling using either a capacitor or a transformer is therefore used, the latter having the additional advantage of impedance matching between the amplifier and load.

Power amplifiers may be coupled in parallel to increase the output capability. This technique is commonly used for RF power amplifiers of solid-state transmitters, these being constructed in a modular manner to provide a wide range of power outputs.

Darlington amplifier. A particularly useful bi-polar device is the compound arrangement of the Darlington amplifier shown in Fig. 1.44(a). This consists of a low power device directly coupled to the larger power amplifier, so that the circuit appears at the output terminals as a single transistor of very high gain. Such devices are available in integrated form or can be configured from discrete transistors. Values of dc current gain H_{FE} of around 2000 can easily be achieved.

VMOS FET devices. This structure, shown in Fig. 1.44(b), was developed in order to obtain low output impedance from FET devices. Due to the V- or U-shaped groove that is formed, the conducting path that is induced by a gate voltage is vertical rather than horizontal as is the case in the planar FET. This results in a very short conduction channel of relatively large cross-sectional area, a feature that ensures a low impedance. In addition the drain contact has a large surface area to permit the easy removal of excess heat.

1.9 Push–pull amplifiers

Power output can also be increased by using push–pull operation, where two active devices are series-coupled to amplify the signal alternately in anti-phase. If the non-linear characteristics of these amplifiers are assumed to obey a power series, then this leads to a cancellation of the dc and even-order harmonic components in the output signal. Third harmonic distortion will therefore

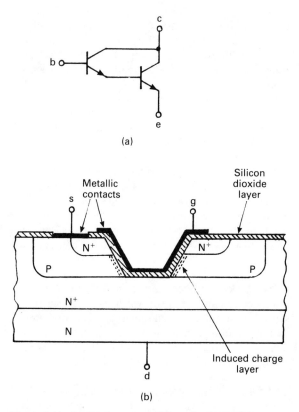

Figure 1.44 (a) Darlington amplifier pair; (b) VMOS semiconductor structure.

predominate. As the amplitude of the power series components decrease monotonically, this method will produce much less distortion than parallel operation. Push–pull operation therefore produces more than twice as much output power as does a single stage, for the same level of distortion. In addition, if a transformer output coupling is used, the cancellation of the dc term means that the risk of magnetic saturation is much reduced, further reducing the distortion that this effect would introduce.

The very low quiescent currents of Class AB and Class B push–pull working are well suited to intermittent operation, particularly when being used with battery or solar panel powered equipment. However, for high power requirements, the large swings in amplifier current can enforce the use of stabilised power supplies.

The output transformer can be dispensed with if a complementary push–pull stage is used. The emitter follower version shown in Fig. 1.45(a) can provide a very low output impedance. The circuit consists of a PNP/NPN transistor pair (N-channel/P-channel FETs can also be used), chosen to have the same parameters and driven with the inputs in parallel. This is possible because of the complementary nature of the conduction

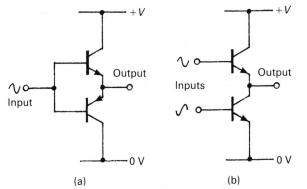

Figure 1.45 Basic push–pull amplifier stages: (a) complementary push–pull; (b) quasi-complementary push–pull.

of the two devices. A variation of this circuit shown in Fig. 1.45(b) may be used. This gives a quasi-complementary push–pull effect and uses two identical devices. Because the emitter and collector currents are large and practically equal, a similar effect occurs when the two bases are driven in anti-phase. Negative feedback is used in both types of circuit to minimise the harmonic distortion that is produced. Assuming a power series for the non-linearity, and a sinusoidal input, the output currents in the two amplifiers are:

$$i_1 = I_c + A_o + A_1 \cos \omega t + A_2 \cos 2\omega t + A_3 \cos 3\omega t ...$$

and, as the devices are driven in anti-phase,

$$i_2 = I_c + A_o + A_1 \cos(\omega t + \pi) + A_2 \cos 2(\omega t + \pi) \\ + A_3 \cos 3(\omega t + \pi)...$$

Now $\cos(\omega t + \pi) = -\cos \omega t$, $\cos 2(\omega t + \pi) = \cos 2\omega t$, $\cos 3(\omega t + \pi) = -\cos 3\omega t$, etc. The output signal current is now

$$i_1 - i_2 = 2A_1 \cos \omega t + 2A_3 \cos 3\omega t ...$$

where i_1 and i_2 are the individual signal currents,
I_c and A_o are dc components, and
A represents the coefficients of the distorting effects.
The signal current output is thus shown to be twice that of a single stage, with only half the number of harmonic components and the dc component being self-cancelling.

1.10 Stability and frequency response

Bode plots

Figure 1.46 shows the gain and phase response of an amplifier plotted against normalised frequency. The bandwidth is defined as the frequency space between the

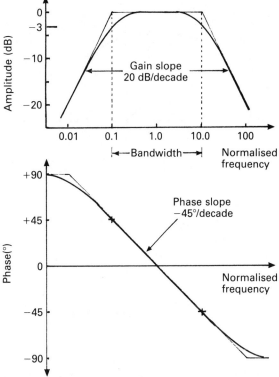

Figure 1.46 Bode plot: gain, phase response versus frequency.

two $-3\,\text{dB}$ or half-power points. It will also be noted that the phase shift ranges over $\pm 45°$ within this bandwidth, relative to the centre frequency. This pair of diagrams is referred to as a 'Bode plot'. Also shown, to display the gain/phase slope characteristics, are the 'piece-wise' linear approximations. If two such amplifiers were connected in cascade and NFB applied over the two stages, the phase shift at any frequency would be twice that shown. It can now happen that at some frequency the loop phase shift might be 180° more or less than intended, transforming NFB into positive. Provided that the loop gain was high enough at this frequency, then the amplifier would become unstable and oscillate.

Nyquist criteria

The criteria for oscillations to occur are that the loop gain A should be at least equal to unity and the loop phase shift be zero, i.e. $\beta A = 1 \angle 0°$. The magnitude and angle for βA can be plotted on a Nyquist diagram as depicted in Fig. 1.47 for all real frequencies. Provided that the locus of βA does not include the point 1,0 then the amplifier will be stable. Figure 1.47(a) thus depicts an unstable amplifier that will oscillate. If the magnitude of βA were reduced so that the point 1,0 moved to the

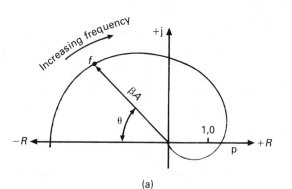

(a)

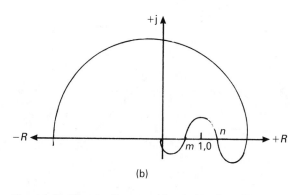

(b)

Figure 1.47   Nyquist diagrams: (a) unstable; (b) conditionally stable.

region p, the amplifier would become 'unconditionally stable'.

In Fig. 1.47(b), the point 1,0 lies outside the locus, but if βA is reduced at some frequency 'n' or increased at another frequency 'm', it would become unstable. Such an amplifier is thus described as being 'conditionally stable'.

1.11 Miscellaneous terms

Bootstrapping. In theory, the emitter/source follower amplifier has a unity gain and infinite input impedance. However, in practice, the use of a bias network significantly modifies these parameters. For the basic circuit, the input impedance is approximately the parallel combination of R_1, R_2 and the base spreading resistance. This produces a relatively low value that consequently reduces the stage gain. If the circuit is modified as shown in Fig. 1.48, the input impedance is effectively increased by the value of R_3, restoring the gain to near unity. The capacitor C_1 has a low reactance at the lowest operating frequency so that the bottom end of R_3 is effectively coupled to the output terminal. Since the top end is connected to the base, there is now a constant voltage developed across

R_3. Thus as the input signal voltage rises, so does the output voltage which appears to be pulled up by the voltage across the *bootstrap* resistor R_3.

Bridge amplifier. A differential amplifier added to the output of a Wheatstone or similar bridge circuit to act as a signal conditioner.

Corner frequencies. The intersections of the straight line approximations to gain and phase (see Fig. 1.46). These correspond to the $-3\,\text{dB}$ or *break frequencies*.

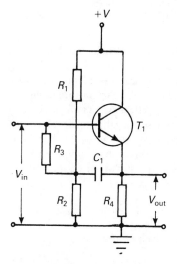

Figure 1.48   Bootstrap amplifier.

Early effect. The width of the space charge region within the base of a junction transistor is reverse voltage dependent, decreasing with an increase in voltage. This modulation of the effective base width is referred to as the Early effect. It has the effect of making the transistor gain, signal amplitude dependent. In particular, large signal levels can lead to *punch through* when the emitter and collector regions become short circuited.

Repeater. The name given to an amplifier used in telephone and cable distribution systems.

Thermal run-away. Semiconductor devices have a negative temperature coefficient of resistance, which leads to an increase in leakage current with a rise in temperature. The effect is cumulative so that unless action is taken to control this, it will result in self-destruction.

Video amplifier. A wide band pulse amplifier whose

frequency response extends from dc up to several tens of MHz.

1.12 Useful references

Abuel Ma'Atti M.T. and Humood N.A. (1986) Current conveyors and sine wave oscillators. *Electronics and Wireless World*, March, p. 282.

Baxendall P.J. (1978) Audio and power amplifier design. *Wireless World*. January, p. 53, March, p. 41, May, p. 83.

Clayton G.B. (1992) *Operational Amplifiers*, 3rd edition, Butterworth-Heinemann, Oxford.

Clayton G.B. (1983) *Operational Amplifier Experimental Manual*, Butterworth-Heinemann, Oxford.

Connor F.R. (1986) *Introductory Topics in Electronics and Telecommunications*, Vol. 2, Networks, Edward Arnold.

Gardener F.M. (1979) *Phase Lock Techniques*, John Wiley, Chichester.

Graeme J.G., Huelsman L.P. and Tobey G.E. (Burr-Brown) (1971) *Operational Amplifiers, Design and Applications*, McGraw-Hill, New York.

Millman J. (1979) *Microelectronics. Digital and Analog Circuits and Systems*, McGraw-Hill, New York.

Millman J. and Halkias C.C. (1972) *Integrated Electronics. Analog and Digital Circuits and Systems*, McGraw-Hill, New York.

Wilson B. (1986) Using current conveyors. *Electronics and Wireless World*, April, p. 28.

2 Antennas or aerials

Antennas operate as transducers interfacing the energy propagating in a waveguide or cable system to the radiant or Maxwellian energy in free space. The roles and properties of transmitting and receiving antennas are the dual of each other in respect of power gain, directivity, equivalent noise temperature and efficiency, which can be expressed as follows:

If a transmitter, matched to antenna 1, excites a given power into the input of a receiver, matched to antenna 2, then the same power will be excited into the receiver, if the transmitter and receiver are interchanged, provided that the matching conditions are maintained.

This feature is particularly valuable as it allows the same antenna to be used alternatively for transmission and reception, as in the case of a radar system. Additionally the antenna may be designed to optimise the energy in or from some given direction in preference to any other directions. Thus the antenna now has an effective gain in this direction.

2.1 Useful theoretical concepts

Hertzian dipole or doublet. A hypothetical current driven antenna that, in terms of wavelength, is very short. It consists of a pair of equal but opposite charges, either electric or electromagnetic, separated by an infinitely small distance.

It forms a valuable mathematical model from which the theoretical properties of many other antennas can be calculated. It can be shown from electromagnetic wave theory that such a dipole has a gain relative to an isotropic source of 1.5 or approximately 1.76 dBi. Knowing this, it is possible to calculate the universal antenna constant.

Isotropic radiator. A hypothetical device whose energy radiation pattern is a perfect sphere. It thus behaves similarly to an isolated electrical point charge such as an electron. The RMS value of the electric field strength (E) at a distance d from this charge can be calculated as

$E = (30P)^{1/2}/d$ volts/metre, where P is the power of the source.

In the same way, the field strength at the same distance from a standard half-wave ($\lambda/2$) dipole can be calculated or measured. The gain of the dipole is therefore the ratio of these two field strengths. It can be shown that a standard dipole has gain of approximately 2.14 dBi (relative to the isotropic radiator).

Poynting vector. A travelling electromagnetic wave carries energy with it that can be calculated from the intensities of the electric (E) and electromagnetic (H) fields. The Poynting theorem states that the power flow in a plane normal to the wave motion is given by the product of the instantaneous values for E and H and the angle θ between them. Thus the power density is given by $P_A = EH \sin \theta$ watts/metre^2, or using the notation of vector algebra, $\boldsymbol{P} = \boldsymbol{E} \times \boldsymbol{H}$.

2.2 Antenna properties

Bandwidth. Most antennas have characteristics that include inductance, capacitance and resistance and this may lead to a resonant frequency. Unlike tuned circuits, the bandwidth of an antenna is usually described in terms of the frequency range over which its important parameters remain within some stated bounds. While aperiodic or untuned antennas may have broadband characteristics that cover a range of upper to lower frequency of about 40:1, the bandwidth of resonant antennas may only cover a range of ±5% of the midband frequency.

Directivity. This is the ability of an antenna to concentrate radiated energy in a preferred direction, in the transmit mode, or to reject signals that are received off-axis to the normal or *bore-sight* of the antenna. This gives rise to an amplifying effect to signals that are received from the preferred direction. The directivity of the radiated energy in space is given formally by:

$$D(\theta, \phi) = \frac{4\pi(\text{Power radiated per unit solid angle in direction } \theta, \phi)}{\text{Total power radiated by antenna}}$$

This directivity feature is depicted in Fig 2.1(a), which shows a polar plot of the radiation/reception pattern for one type of antenna. Figure 2.1(b) shows a similar feature for another antenna, but in this case, plotted in a linear fashion. In both cases the diagrams show the major lobe together with the secondary side lobes. The angle α represents the beamwidth of the antenna, being the angle subtended by the axis and the $-3\,$dB points.

Figure 2.1(b) also shows a curve that can be used to describe the limiting level of side lobe response for certain cases. This is typically defined as $(a-b\log\theta)\,$dB, where a and b are definable constants.

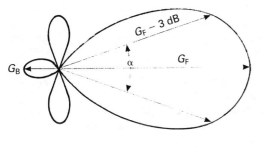

(a)

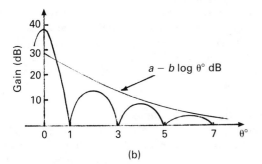

(b)

Figure 2.1 (a) Polar plot of antenna response; (b) linear plot of antenna response.

Efficiency. This parameter may be quoted in one of two ways. In the transmit mode, this is known as the *radiation efficiency* η, given by:

$$\eta = \frac{\text{Total power radiated by antenna}}{\text{Total power accepted by antenna from source}}$$

$$= G(\theta,\phi)/D(\theta,\phi)$$

Alternatively η is described as the *aperture efficiency*, where

$$\eta = \frac{\text{Effective area of antenna}}{\text{Physical area of antenna}}$$

For parabolic reflector types, η varies between 50% and 75%.

Equivalent noise temperature. This parameter is most important in the receive mode and is related to the *antenna radiation resistance*. Just like a physical resistor, this component generates a thermal noise that degrades the S/N ratio.

Far field or Fraunhofer region. A region of the electromagnetic radiation field, some distance from the antenna, where the field distribution is unaffected by the presence of the antenna structure. Also, the electric and electromagnetic field components are mutually at right angles to the direction of propagation and the amplitude of these components is decreasing as the square of the distance from the antenna. If the antenna has a major dimension D, that is many wavelengths (λ) long, the far field can usually be considered to extend beyond a distance of $2D^2/\lambda$.

Near field region. Figure 2.2 shows two elements of this region.

(i) The reactive or Rayleigh region immediately surrounding the antenna, which may extend to about $\sqrt{0.4D^3/\lambda}$. The dominant features of the electromagnetic field are those due to the near-oscillatory action represented by the transfer of energy between the antenna and the transmission medium.

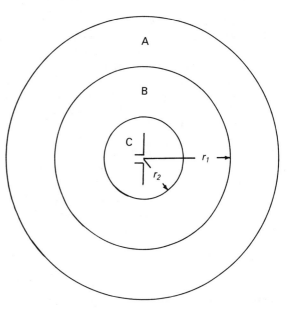

Figure 2.2 Field regions for an antenna: A = far field or Fraunhofer region; B = radiating near field or Fresnel region; C = reactive near field or Rayleigh region.

(ii) The radiating or Fresnel region. If the antenna has a major dimension D that is large compared with the wavelength (λ), then this transition region will exist between the reactive and far field regions.

Phase function of antenna. A function giving the phase of either the electric or electromagnetic field, produced by an antenna, relative to a sphere of fixed radius centred on the antenna. Because of its well-behaved properties, the far field is normally implied.

Polarisation of antenna. The angle of polarisation for the antenna is taken from the plane occupied by the electric (E) field of the radiation. Thus the polarisation may be either vertical or horizontal for linear polarised waves. An antenna aligned to maximise the response to a wanted or co-polar signal will have minimum response to signals of the opposite or cross-polar polarisation, the discrimination against cross-polar signals being described by the isolation ratio or the polarisation loss factor.

Power gain. This is formally defined by the ratio:

$$G(\theta, \phi) = \frac{4\pi(\text{Power radiated per unit solid angle in direction } \theta, \phi)}{\text{Total power accepted by antenna from source}}$$

or alternatively by:

$$G(\theta, \phi) = \frac{\text{Power radiated by practical antenna in preferred direction}}{\text{Power radiated in same direction from an isotropic source}}$$

with both antennas supplied with the same power.

The gain for a parabolic dish antenna is given by $G = 4\pi A_e/\lambda^2 = 4\pi\eta A/\lambda^2$ where A and A_e are the physical and effective areas, respectively.

The front-to-back ratio $= (G_F - G_B)\,\text{dB}$ can be derived from Fig. 2.1(a). This shows the property that describes how the antenna discriminates against signals arriving 180° off bore-sight.

Partial gain of antenna. Reflector antennas, in particular, may be designed to handle two orthogonal polarisations simultaneously. The absolute gain for such an antenna is then the sum of the gains for each polarisation. For antennas designed for a single polarisation the partial gain = total gain.

Radiation resistance/input impedance. Each antenna has a unique value of characteristic resistance R, which must be matched to the feeder cable for maximum transfer of power. R is the voltage/current ratio on the antenna and is not a physical resistance, but more an equivalent one. If the antenna current is I amps, then the radiated power will be I^2R watts. If the antenna were replaced with a physical resistance of the same value, then this would dissipate the same amount of power but in the form of heat. The radiation resistance for a Hertzian dipole is given by $R = 80\pi^2(l/\lambda)^2$ ohms, where l is the length of the dipole.

2.3 Classification of antennas

Any metallic structure, given the right conditions, will receive, re-radiate or radiate electromagnetic energy. There are thus a very wide range of structures that are suitable for use as antennas. Standing waves are the result of a mis-match in the loading of a transmission line. Under such conditions not all of the energy sent to the line by the generator is dissipated in the load, the surplus energy being reflected back along the line towards the generator. Standing waves are then formed due to the vector (phasor) addition of the forward and reflected energies.

Antennas may be broadly classified by the distribution of the energy on the structure, and more particularly classified by the frequency ranges for which that structure is commonly used.

Standing wave antennas
Since these structures support standing waves they are often described as *resonant* antennas and, as such, typically have a bandwidth of about ±10% of the resonant frequency. They are commonly based upon dipoles of lengths that are integer multiples of a half wavelength ($\lambda/2$).

Travelling wave antennas
These are usually long wire antennas, often varying from one to several wavelengths. They are non-resonant and, to avoid standing waves, the end remote from the generator is terminated in a matched power absorbing resistance. Although this wastes up to 3 dB of output power, the antenna is simple and can be used for a wide range of frequencies.

Travelling waves can be classified as *slow waves* where the phase velocity v_P is less than the velocity of light in free space, c, or as *fast waves* when v_P is greater than c. $v_P = \omega/k$, where $\omega = 2\pi \times$ frequency and k is the phase constant.

Some standard antenna structures classified by frequency range

Up to 300 kHz (very low and low frequency, VLF and LF). Due to the long wavelengths involved, it is difficult to achieve vertical antennas of even quarter wavelength structures. For example, an antenna for say 100 kHz (3000 metres) would require a $\lambda/4$ structure about 750 metres high. For this reason, transmitting antenna masts are often about 100 metres high and used to support 'inverted L' or 'T' antennas. The effect of the additional top element adds *top capacitance*, which increases the antenna current to improve the radiating efficiency. Such antennas have little directivity and the radiation at these frequencies is chiefly *ground wave* (see also Propagation, p. 245). Because of the relatively small bandwidth available for only a few channels, there is no great need to be able to retune transmitters to new frequencies.

Beverage antenna. This is a *long wire* antenna, at least a wavelength long, consisting of either a single wire or a parallel pair supported about 10 metres above and parallel to the earth surface. As the parallel pair should be separated by $\lambda/2$, the structure occupies a large ground area. The system is terminated at each end in a resistance equal to the *characteristic impedance* of the antenna with one resistance providing a load across which the received signal is developed. Because of the power loss across the terminating resistances, this type is not often used for transmission purposes. However, the simplicity makes it valuable for receiving long wave signals over a wide range of frequencies. The directivity is *end fire* (see Coplanar arrays, p. 41), with a good response to signals reflected down from the *ionosphere*.

300 kHz to 3 MHz (medium frequency, MF). This part of the frequency spectrum is used chiefly for the national medium wave broadcasting services. In general this requires an all-round coverage from any given transmitter. At these higher frequencies, a vertical antenna structure can meet this requirement. For a 1 MHz transmission, the $\lambda/2$ mast is now only 150 metres high and if fed as a dipole will provide all-round coverage. Alternatively a quarter-wave-high mast can be used to support conductors arranged as a *folded dipole* (see Half-wave dipoles, p. 31).

3 MHz to 30 MHz (high frequency, HF). At these frequencies, antennas that are multiples of $\lambda/2$ high become feasible.

Standing wave types

Franklin antenna. This structure consists of three end-fed $\lambda/2$ dipoles coupled in series as shown in Fig. 2.3. The inter-dipole coupling device is either a special choke with distributed inductance and capacitance as shown or more commonly a short-circuit $\lambda/4$ section of transmission line. Both methods ensure that the standing wave current pattern on the three elements is series aiding. The folded version of the Franklin antenna is also shown. The overlapping of the three sections reduces the

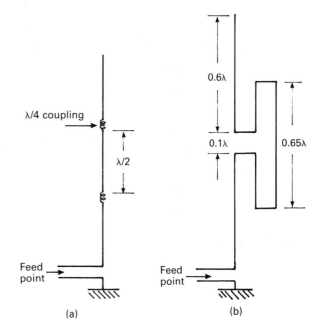

Figure 2.3 The Franklin antenna.

total height of the structure without significantly affecting the gain. Since the total radiation is the sum of that from each of the three elements, the gain is that of three collinear dipoles, with all-round coverage.

Horizontal $\lambda/2$ dipole. The dipole elements may either be supported on masts vertically, which gives all-round coverage, or horizontally. The conductors may be folded to increase the impedance (see Half-wave dipoles, p. 31) to better match the feeder impedance. If the horizontal dipole is suspended about $\lambda/8$ to $\lambda/4$ above a good earth plane, the maximum gain can approach 6 dB. The bandwidth is typically about $\pm5\%$ of the resonant frequency.

Lazy-H antenna. This structure, which is suitable for both transmission and reception, consists of two full-wave dipoles as shown in Fig. 2.4. The dipoles are spaced by $\lambda/2$ with the feeds transposed and driven by open-spaced feeders. Each dipole has a characteristic impedance of about 2000 ohms, so that the total antenna load becomes 1000 ohms. This value can be reduced by adding a reflector curtain of conductors which is placed between $\lambda/8$ and $\lambda/4$ behind the dipole structure. The spacing between the curtain wires should be about $\lambda/10$. This antenna has a typical forward gain of up to 10 dB.

Sterba array. This antenna, shown in Fig. 2.5(a), has similar properties to the Franklin antenna. The cross-coupling used between the vertical sections ensures that the radiation from these is in phase. As shown, the

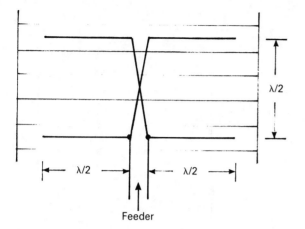

Figure 2.4   Lazy-H antenna.

structure produces vertical polarisation, but the antenna can be turned on its side to produce horizontal polarisation. The antenna can be made to produce a unidirectional beam either with a reflector curtain similar to the lazy-H, or by adding a second identical structure $\lambda/4$ behind but fed with current 90° out of phase. For comparison, a pair of parallel-coupled Franklin antennas is shown in Fig. 2.5(b). This configuration also gives a unidirectional beam.

Wyndom antenna. This is a half-wave dipole that is suspended horizontally between two masts and fed typically offset from centre by about $\lambda/12$, this being an impedance matching technique (see also Delta match, p. 32). The antenna is fed with a single cable, the return line being connected to a good earth point.

Zeppelin antenna. This is another half-wave dipole that is also end fed, but with a pair of open wire feeders, the second being terminated at a null point in the supporting

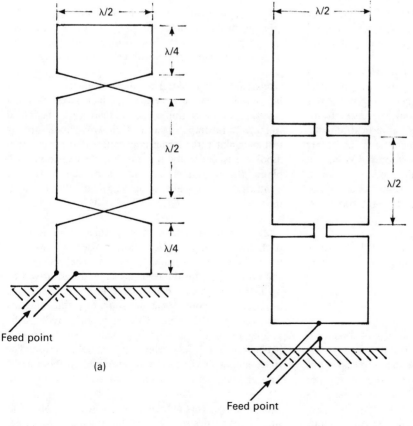

Figure 2.5   HF transmitter arrays: (a) Sterba curtain; (b) pair of parallel Franklin antennas.

structure. The feeders are normally tuned to give the conjugate of the antenna impedance.

Travelling wave types

There are many variations of V-type structures used to give directivity to long wire antennas. But only two classical examples will be considered here.

Inverted V. The structure shown in Fig. 2.6(a) has two arms that are each several wavelengths long and behave as long wire antennas with the main lobes of radiation/ reception being along the directions shown. By adjusting the height of the central mast, the radiation from the main lobes of each arm becomes in phase and additive, increasing the gain in the direction shown. The antenna impedance is of the order of 400 ohms and this is the value of the terminating resistance needed to avoid standing waves. When used as a transmitting antenna, the grounded feed terminal and the terminating resistance are usually connected to earth mats (see p. 46) to improve the earthing and hence the total radiation.

Rhombic antenna. From Fig. 2.6(b) it can be seen that this type is similar to two inverted V antennas in parallel but as this antenna is horizontally mounted and each leg is usually 4λ long, it requires a fairly large ground area. The antenna impedance is in the order of 600 to

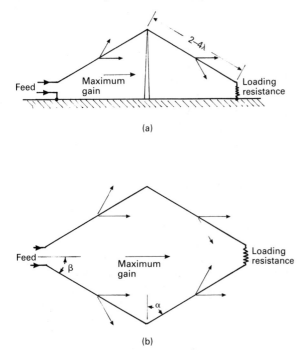

(a)

(b)

Figure 2.6 Travelling wave antennas: (a) inverted vee; (b) rhombic.

800 ohms and this is the value required for the terminating resistance. It is a broadband antenna and covers a frequency range of 2:1, with each leg being cut to an optimum frequency that is the geometric mean of the band edge frequencies. The two angles α and β are made such that the phase difference from the mid-points of two series legs are $n\pi$ radians greater than for the direct path length, where n is some odd integer value. Under this condition, the radiation/reception due to the main lobes become additive for the optimum frequency. Typically the gain is flat over a 2:1 frequency range within ± 3 dB.

When used for transmission, the terminating resistance can provide a problem. The antenna is then often constructed from iron or similar wire that has a high attenuation at radio frequencies. This causes the heat to be dissipated continually along the structure.

30 MHz to 1 GHz (very high to ultra high frequency, VHF/UHF). Over this frequency range many of the antennas used for both transmission and reception are based on the dipole or monopole structure. Resonant antennas in this range have relatively small dimensions, can be constructed from aluminium or copper tube for light weight and have well-defined characteristics. Above about 300 MHz, the length to diameter ratio of the elements can be reduced to improve the bandwidth. At these frequencies the feeder attenuation has to be taken into consideration. Ideally the dipole, which has a radiation resistance of 73 ohms, should be fed with a balanced type of cable (see Transmission lines and waveguides, p. 381). Other features being equal, this type has a greater attenuation than co-axial cable which has a characteristic impedance of 50 or 75 ohms. This latter is therefore very often used, in conjunction with a *balun* (see p. 382).

Polarisation of the electromagnetic wave is controlled by the plane occupied by the dipole. A vertical mounted dipole generates a vertical electric field and this is used to define vertical polarisation. The occupancy of the band can be extended by *frequency re-use* (see Propagation, p. 242), where alternative vertical, horizontal or circular polarisation of the same frequency can be used. At the high frequency end of this band, the $\lambda/4$ monopole is very short and often referred to as a *stub antenna*.

Basic half-wave dipole

For a practical antenna, the dipole will be cut slightly shorter than $\lambda/2$, typically about 0.46λ, due to *end effects* (see Effective length, p. 46). Figure 2.7(a) shows the voltage and current distribution along a dipole when energised under resonant conditions. This shows quite clearly how the centre point is at a low impedance (73 ohms).

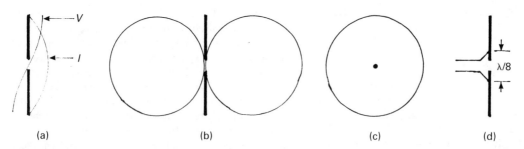

Figure 2.7   Half-wave dipoles: (a) voltage/current distribution; (b) vertical polar diagram; (c) horizontal polar diagram; (d) delta-match.

Figure 2.7 (b and c) show the broadside and plan view radiation patterns, respectively. When it is more convenient to feed the dipole using 600 ohms balanced twin feeder, the *delta match* technique shown in Fig. 2.7(d) can be used.

Dipole arrays
The single-element antenna has little gain and in order to increase this, several identical devices can be coupled in parallel, simultaneously driven, to form an array. The total radiation is now increased by an *array factor*.

Array factor. The ratio of the radiated field strengths in a given direction due to the array and the single element, with each system radiating the same total power.

Broadside array. Figure 2.8 (a and b) show two ways of producing an array factor. If *n* co-phased dipoles are placed *d* metres apart and in-line broadside, the total gain will be *n* times the gain of a single dipole. The beamwidth of the main lobe will be approximately λ/nd rad.

Collinear array. *n* co-phased dipoles may be mounted end-on with their centres *d* metres apart as shown in Fig. 2.8(a). The gain is again *n* times that of a single dipole and the main lobe beamwidth approximately equal to λ/nd rad.

M × N array. The arrangements shown in Fig. 2.8 can be combined to form a structure of *M* vertical and *N* horizontal dipoles. The total gain of this array becomes *MN* times that of a single dipole and the beamwidth approximately λ/MNd rad.

Dipole variants. The gain of the dipole can be increased by adding so-called *parasitic elements* in front and/or behind the main element as shown in Fig. 2.9(a). The element in front is called a *director* while the one behind is called a *reflector*. Both of the parasitic elements are almost $\lambda/2$ dipoles without any loads and as such re-radiate all the energy received, either from the dipole as

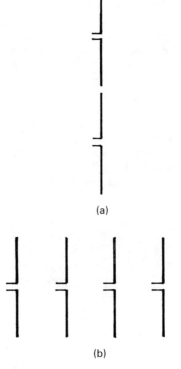

Figure 2.8   Half-wave dipoles: (a) collinear; (b) broadside.

a transmitting device, or the electromagnetic wave as a receiving antenna. By adjusting the lengths and spacing between the elements, the total radiation generated or received is additive, thus giving the structure a gain relative to the dipole. Typically the reflector is cut to $\lambda/2$, while the dipole is made about 0.46λ and the directors about 0.43λ. The spacing, reflector to dipole, is typically 0.15λ to 0.25λ, while the dipole to director spacing is about 0.1λ. This type of structure can be extended to as many as 18 elements before it becomes unwieldy. Typical approximate gain figures for these

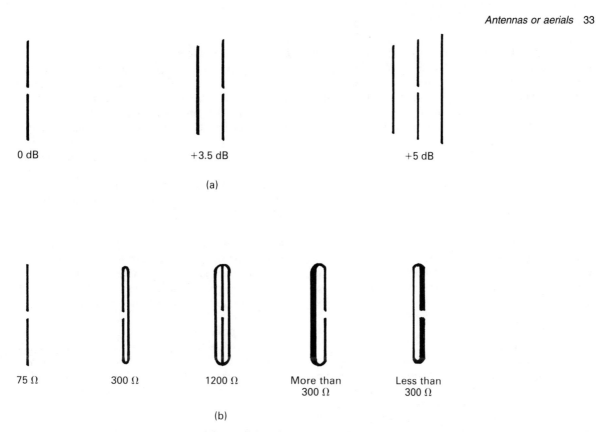

(a)

75 Ω   300 Ω   1200 Ω   More than   Less than
                                            300 Ω   300 Ω

(b)

Figure 2.9 Half-wave dipoles: (a) gain relative to dipole; (b) dipole impedance.

structures, which are known as Yagi–Uda arrays, are as follows:

    5/6 elements – 10 dB,
    9/10 elements – 13 dB,
    18 elements – 16 dB.

The addition of parasitic elements to the dipole reduces the feed-point impedance and reduces the bandwidth but increases the gain. However, the disadvantages can be countered by *folding* the dipole as shown in Fig. 2.9(b). A single fold increases the impedance by a factor of 4, while a third fold increases it by 4^2 or 16. These values can be modified by changing the length-to-diameter ratio as indicated in the last two diagrams. While folding the dipole has no effect on gain, it increases the feed-point impedance and the bandwidth.

Helical beam antenna

The helical beam antenna shown in Fig. 2.10 produces circular polarisation (see Propagation, p. 243) and is particularly suitable for use at frequencies up to about 600 MHz. A helix for left-hand circular (LHC) polarisation is shown together with the important dimensions that affect the characteristics. The feed-point impedance is low and suitable for use with

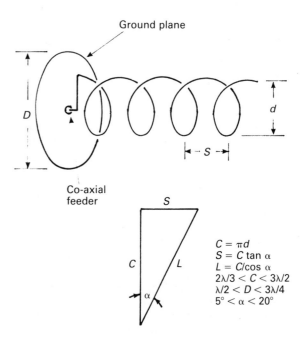

$C = \pi d$
$S = C \tan \alpha$
$L = C/\cos \alpha$
$2\lambda/3 < C < 3\lambda/2$
$\lambda/2 < D < 3\lambda/4$
$5° < \alpha < 20°$

Figure 2.10 Helical beam antenna and its important dimensions.

co-axial cable. The antenna has a good bandwidth and low voltage standing wave ratio (VSWR) (see also Transmission lines, p. 377) and small physical size. When the circumference is about one wavelength, the radiation is almost entirely in front of the ground plane. The directivity is quoted empirically as

$$D = 15n(S/\lambda)(C/\lambda)^2$$

where $C = \pi d$ the circumference,
    d = diameter,
    $S = C \tan \alpha$
    α = helix angle,
    n = number of turns,
    λ = operational wavelength.
For the most commonly used values of $\alpha = 12°$, $C = \lambda$, the forward gain for n between 3 and 25 turns is approximately $G\,\mathrm{dB} = 7 + 2\sqrt{n}$.

Log-periodic antenna

There are many different types of structure that have a performance that is a periodic function of the logarithm of the operating frequency. These vary from plates or slots that have a logarithmic spiral form, to structures similar to those shown in Fig. 2.11, and all are intended for use with linear polarised waves. For any given frequency only a part of the structure is effective, the active part varying with frequency. Thus these antennas have a very wide bandwidth that is only limited by its physical size.

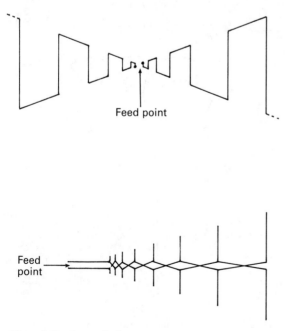

Feed point

Feed point

Figure 2.11 Log-periodic antennas.

Monopole or unipole

A single $\lambda/4$ or $\lambda/2$ element that is mounted perpendicular to a good reflecting ground surface will radiate energy directly and by reflection as indicated in Fig. 2.12(a). The ground reflection may be considered to be due to an image antenna below the ground helping to generate the main lobe radiation as shown. In plan view the monopole antenna produces all-round coverage. This concept can be used at VHF/UHF with the metallic roof of a vehicle acting as the ground plane. The feed-point impedance of the monopole is only about 38 ohms and this can cause matching problems.

Sleeved monopole. By enclosing part of the monopole within a metallic sleeve as shown in Fig. 2.12(b), the feed-point impedance is raised to a level that allows direct coupling with co-axial cable. The sleeve acts in the manner of a transmission line transformer.

Sleeved dipole. The dipole should ideally be fed by balanced feeders and the use of a co-axial cable would require the use of a *balun* (see Transmission lines and waveguides, p. 382). Such a device can be provided by sleeving the centre section of the dipole in the manner of the sleeved monopole.

Skirted monopole. A variation of the ground plane and monopole is shown in Fig. 2.12(c), the conical skirt performing in a similar way, but allowing a lower angle of radiation. In situations where the wind loading on the skirt might create mechanical problems, the skirt can be replaced without loss of effectiveness by a wire framework of similar shape.

Quad or cubical quad antenna. This simple two-element structure consists of two square loops of $\lambda/4$ sides. One has a gap to provide a feed point, while the second is a closed loop. The antenna has a cubical form with the first element acting as the radiator/receiver while the other functions as a reflector. The antenna has a useful forward gain with sharp nulls along the line of the loops. It can be used below 30 MHz but its size then tends to make it unwieldy.

Quadrifilar helix antenna. There is a family of these antennas, differing chiefly in the number of turns in the helix. They are designed for circular polarisation and one example is shown in Fig. 2.13. This device has a gain of about 4 dBi but with a beamwidth approaching 180°. The helices consist of thin brass tapes wound on a cylindrical dielectric former, with the lower ends short-circuited and the upper ends parallel connected. The structure forms a balanced circuit with a radiation resistance of 25 ohms. By using a sheathed balun, the antenna can be fed via a co-axial cable.

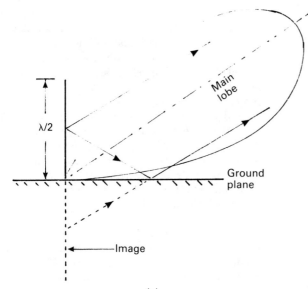

(a)

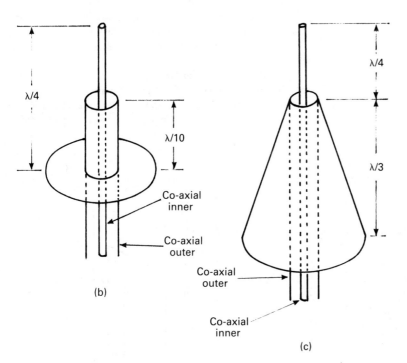

(b)

(c)

Figure 2.12 Monopoles: (a) monopole above conducting plane; (b) sleeved monopole; (c) conical skirt monopole.

Slot antenna. If a rectangular slot of appropriate dimensions is cut in a large sheet of metal and fed in the manner shown in Fig. 2.14(a), an electric field will be set up between the long edges. Currents will then circulate around the slot so that radiation occurs, both in front and behind the slot. Unlike the normal dipole, the electric field is at right angles (normal) to the resonant length so that the radiation has the opposite linear polarity. Figure 2.14(a) shows two ways of feeding

the slot, either with a balanced feeder to match the typical impedance of 500 ohms or by co-axial cable. The second method is equivalent to using a $\lambda/4$ matching transformer between the co-axial cable and the slot. The open slot will radiate energy both in front and behind to provide a dipole-like pattern. If, however, the sheet is rolled into a cylinder with a diameter of $\lambda/4$, the backward radiation will be suppressed and the forward radiation increased to give a forward gain. A typical

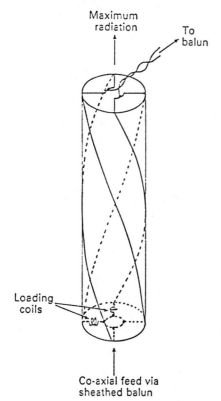

Figure 2.13   Half-turn quadrifilar antenna.

transmitting antenna has a group of slots arranged as shown in Fig. 2.14(b), and this is further extended vertically to provide an array of 8 × 4 slots in total. Due to end effects (see Miscellaneous terms, p. 46) the slots are usually 0.65λ to 0.7λ in length and are spaced 1λ centre to centre, the tube diameter being 0.6λ. Each vertical collinear array of slots has a forward gain of 8 × 1.64 or about 11 dBi relative to an isotropic source. The beamwidth is given approximately by $\lambda/nd = \lambda/8\lambda = \frac{1}{8}$ rad = 7°. The array of slots is co-axially fed using cable lengths that ensure co-phasing. The bandwidth is typically in the order of ±5% of the resonant frequency.

Skeleton slot antenna. For reception purposes, the slot has been used in a modified form. Most of the conducting plane has been removed to leave just a metal frame of suitable dimensions. Both the narrow sides then carry directors and a reflector in the manner of a pair of parallel Yagi arrays, the signal take-off points being provided from the mid-points of each long side. This provides an antenna with little wind loading, and good gain, directivity and bandwidth.

Super turnstile antenna. The basic structure of this antenna is shown in Fig. 2.15 and consists of four metal plates or fins electrically coupled and mounted at 90° to each other on a vertical mast. The width of each vane is

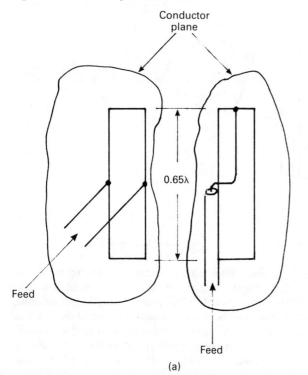

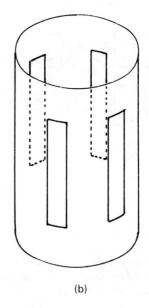

(b)

Figure 2.14   Slot antennas: (a) alternative feed systems; (b) part of slot array.

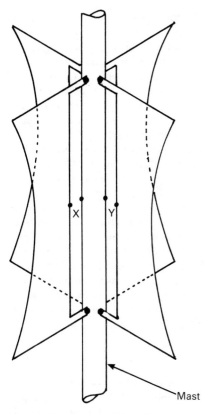

Figure 2.15 The super turnstile antenna.

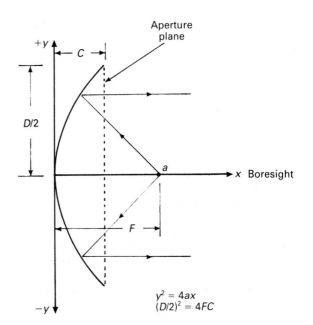

$$y^2 = 4ax$$
$$(D/2)^2 = 4FC$$

Figure 2.16 The parabolic reflector.

typically $\lambda/4$. Slots of about 0.6λ length are left in the structure to provide the feed points X and Y. The feed current to each pair is in quadrature to ensure all-round coverage. For greater gain, several identical structures can be stacked vertically. To reduce the wind loading it is usual to replace the plates with grid or mesh panels.

Turnstile antenna. This is the primitive version of the super turnstile antenna and simply consists of two crossed half-wave dipoles with the same 90° phasing. It forms a very simple but effective receiving antenna.

Yagi–Uda array. See Dipole variants, p. 32.

1 GHz upwards (ultra high and super high frequencies, UHF/SHF). The common feature of antennas designed for these frequencies is the reflector that has dimensions that are large compared with the operating wavelength. These are so shaped as to produce or receive the electromagnetic energy in the form of a narrow beam. One very common form is the paraboloid whose dimensional relationships are shown in Fig. 2.16. The mathematical equation for a parabola is $y^2 = 4ax$. Such a shaped curve has some special

properties. Any ray emanating from point 'a' will reflect off the curve parallel to the x axis. If the curve is rotated around the x axis, the volume so produced is called a *paraboloid* or parabolic dish. If this surface is used as a transmitting device, energy emanating from 'a' will be reflected to form a beam parallel to the x axis. Conversely, energy received along a complementary path will be concentrated at 'a', the focal point. Further, the total distance by any path from the focal point to the aperture plane via reflection is constant. Thus as a transmitting device, any energy leaving 'a' will theoretically pass through the aperture plane completely in phase and any plane wave entering the aperture plane will be focused on point 'a'. It is these properties that are responsible for the forward gain of the device. This shape is the most efficient in terms of narrow beamwidth for smallest area. For a practical dish of diameter D, centre depth C and focal length F, the equivalent mathematical equation become $(D/2)^2 = 4FC$. This can be rearranged as $C = D/(16(F/D))$, showing that the depth of the dish depends on the diameter and the F/D ratio. The dish shape can be completely described from a knowledge of this ratio and either F or D.

If $F/D = 0.25$ the focal point lies on the aperture plane and it becomes difficult to illuminate the whole dish surface without wasting energy. A high F/D ratio results in a flatter dish, which optimises the forward gain, while a lower value gives a better degree of side lobe suppression. Typical practical compromise values therefore lie in the range 0.35 to 0.45.

Because the surface area is very large compared with the operational wavelength, the reflector dimensions are largely responsible for the antenna gain. This is given by:

$$G = \eta\pi^2 D^2/\lambda^2 \text{ or } 4\pi A_e/\lambda^2$$

where η is the aperture efficiency,
  D is the diameter, and
  A_e is the effective area.

Antenna feed configurations

When using a reflector antenna it is important to maximise the radiation/reception efficiency by just illuminating the full area of the dish. The feed system is therefore a critical feature. The simple way involves mounting the electronics unit at the focal or *prime focus* point. However, for large antennas in particular, this makes access for servicing rather difficult. If this unit were mounted at the centre back of the dish the access problem would be removed. There are two basic ways of achieving this situation, either by using *sub-reflector* or *waveguide* feeds.

Cassegrain sub-reflector. This concept is shown in Fig. 2.17(b) where a metallic reflector with a hyperbolic shaped surface is positioned near to the prime focus point that is indicated in Fig. 2.17(a). This then focuses the rays on the centre of the dish.

Gregorian sub-reflector. This sub-reflector surface shape forms part of an ellipse to achieve the same aim. Whichever type is used it is important that the diameter of the sub-reflector should not exceed about 30% of the diameter of the main antenna structure.

Sub-reflector with frequency selective surfaces. At times it is necessary to operate on two different frequencies simultaneously. This could be achieved by using two antenna systems. However, if the sub-reflector is made from a *dichroic* material, which is transparent at a prime focus feed frequency and reflective to a frequency provided from a Cassegrain/Gregorian feed mounted in the conventional position, then the cost of the second antenna would be saved.

Frequency selective surfaces have just these properties. They consist of an array of metallic elements printed over the surface of a suitable dielectric material that has been shaped into a sub-reflector form. Some of the elemental shape patterns that are used are shown in Fig. 2.18. These patches have a self-resonant frequency that depends upon dimensions and this gives rise to these properties. The losses in both the reflective and transmissive modes are typically less than 0.5 dB. In the reflective mode, the patches behave as resonant devices without a load. Any energy thus received will then be re-radiated. In the transmissive mode the patches are off-resonance so have very little effect on the signal energy.

Waveguide feeds. A section of a crook-shaped waveguide can be used to couple signals from a rear-mounted electronics unit to the prime focus feed point. The dish is then illuminated from the open end of the guide. The shape of this guide section gives it its descriptive name, the *J-hook*. An alternative may be found where the open end of the waveguide is split into two branches which turn through 180° so that both ends face the reflector (Cutler feed).

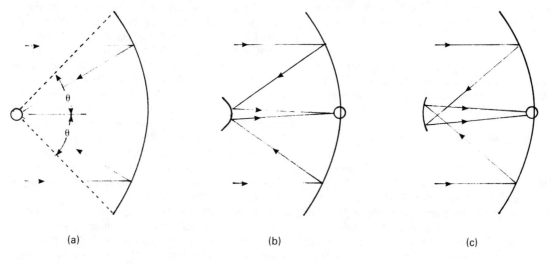

(a)   (b)   (c)

Key: ◯ = Electronics unit

Figure 2.17   Feeding the parabolic reflector antenna: (a) prime focus; (b) Cassegrain sub-reflector; (c) Gregorian sub-reflector.

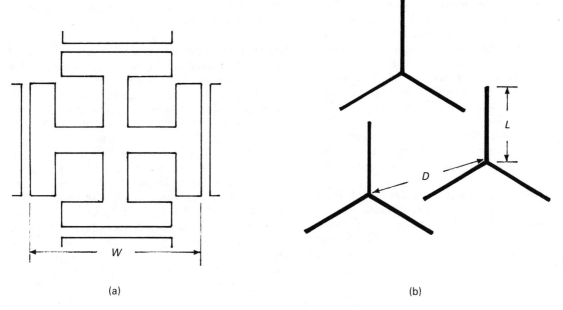

(a)   (b)

Figure 2.18 Some frequency selective surface patterns: (a) Jerusalem cross; (b) tripole.

Feed horns

Simple open-end waveguides provide an inefficient way of coupling energy to free space because of the abrupt impedance mismatch. Matching can be improved by flaring out the sides of the waveguide and various types of feed horn have been designed to solve this problem. A correctly designed feed horn can easily add an extra 2 dB gain to an antenna system.

Pyramidal or tapered horn. This principle, shown in Fig. 2.19, together with the relative dimensions, can be applied to either rectangular or circular waveguides. In true pyramidal form, the waveguide is tapered in both planes. The interior of the flare is sometimes corrugated to improve the rotational symmetry of the radiation energy and to obtain equal beamwidths in the E and H planes.

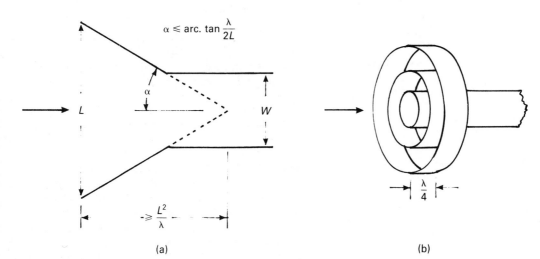

(a)   (b)

Figure 2.19 Feed horns: (a) tapered; (b) scalar.

Scalar feed horn. This device, shown in Fig. 2.19(b), is often used for reception purposes. It has between three and seven concentric rings that behave as quarter-wave choke slots to modify the E field component and so achieve equal response to both the E and H signal components. The device is insensitive to polarisation, gives very low VSWR and good side lobe discrimination. Its use can often add 1 to 2 dB gain to the overall antenna system.

A further significant improvement in illumination efficiency can be achieved by coating the interior surface of a metallic feedhorn with a dielectric material.

Sectoral horn. These horns are only tapered or flared in one dimension and thus have a pair of parallel sides. *E*-plane sectoral horns have the taper applied to the narrow dimension of the waveguide, while *H*-plane sectoral horns are tapered on the opposite sides (see also Waveguides, p. 378).

Hoghorn antenna. This is a specific combination of feed horn and reflector antenna. The reflector surface is a section of a larger paraboloid and this is directly coupled to the waveguide by either a pyramidal or sectoral horn. The concept is shown in Fig. 2.20. The sideplates of the horn reduce the beamwidth of the antenna and so help reduce interference. This type of

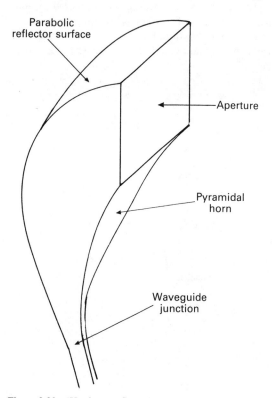

Figure 2.20   'Hoghorn reflector' antenna.

Parabolic
reflector surface

Aperture

Pyramidal
horn

Waveguide
junction

structure is often used on terrestrial microwave communications links.

Illumination distribution and efficiency
An important feature of the gain of a reflector antenna system is the manner in which the surface is *illuminated* with energy in the transmission mode. The *illumination efficiency* is quoted as the percentage of the total radiated energy that falls on the reflector surface. Therefore to maximise this parameter, radiant energy should not be lost beyond the edge. Such a condition not only wastes energy, but also radiates interfering signals behind the antenna. From the reception point of view, this also creates additional sources of interference. The most effective illumination technique involves tapering the energy distribution across the reflector diameter. Often it is arranged so that the energy at the reflector edges is about -10 dB relative to that at the centre. This distribution effect is dependent upon both the feed horn and the dish F/D ratio.

Lens antennas
Lens antennas respond to electromagnetic waves in a similar manner to an optical lens and light. That is, they convert a spherical wave with diverging energy into a plane wave with directivity to provide a gain parameter. These devices are designed around two basic concepts.

Dielectric lens. The basic principles can be explained with the aid of Fig. 2.21(a). Any ray emanating from the focal point 'O' should pass through air and the lens in the same time, so that all the energy radiated from the aperture plane YY is in phase. In a complementary manner, energy received along the x axis will be focused at point 'O'.

If the velocity of the wave energy in air and the lens is c and v, respectively, then

$$OA/c + AB/v = OD/c + CD/v$$

If the refractive index $n = c/v$, then it can be shown that $r = ((n-1)L)/(n \cos \theta - 1)$, this being the equation for a *hyperbola*.

As the dielectric constant of the materials used for such a lens is relatively independent of frequency, these antennas have a wide bandwidth.

Metal plate lens. Electromagnetic waves travelling in a waveguide have a greater *phase velocity* than those travelling through free space (see Transmission lines and waveguides, p. 378). Figure 2.21(b) shows how this feature can be used to provide a lens effect using a set of parallel plates. The plates act as waveguides to the energy and by suitably designing the path lengths inside the structure, a condition similar to that in the dielectric

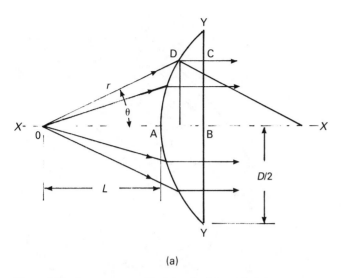

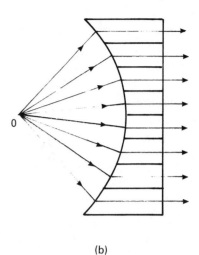

(a)   (b)

Figure 2.21 Lens antennas: (a) dielectric; (b) metal plate.

lens can be achieved. Since the refractive index in this case is less than unity, the equation for the shape becomes $r = L(1 - n)/(1 - n \cos \theta)$. Again a hyperbolic curve but with a complementary shape. Metal plate lens antennas have a relatively narrower bandwidth than dielectric lens types.

Stepped lens. Simple lenses have to be many wavelengths thick and so tend to be heavy and bulky. The problem can be minimised for both dielectric and parallel plate types, by introducing steps into the surface in the form of a series of concentric rings, each ring having a hyperbolic cross-section and the steps being arranged in integer multiples of a wavelength. For lens antennas with p steps and refractive indices n, the approximate bandwidth for each type is $25/(p - 1)\%$ (dielectric) and $25n/(pn + 1)\%$ (metal plate). The technique of stepping is often referred to as *zoning*.

Losses due to feed configuration

Blocking, masking or shadow losses. Whichever method is used to feed the reflector, there will be some structure mounted close to the prime focus point. This structure and its supporting struts will block some of the energy and throw a *shadow* on the dish surface to reduce the illumination efficiency. The effect is shown in Fig. 2.22(a).

Scattering losses. The structure mentioned above will also cause some scattering of both transmitted and received signals to give a further loss. This effect also tends to degrade the side lobe rejection and cross-polarization properties of the antenna.

Spillover losses. Energy radiated from the focal point that does not fall upon the reflector surface is wasted energy and so represents a signal loss (see Illumination distribution, p. 40).

Offset feed antenna. The effect of blocking losses can be minimised by using an offset feed arrangement as shown in Fig. 2.22(b). The reflector surface is part of a larger paraboloid and so the focal point now falls outside of the actual aperture plane. Prime focus or subreflector feeds can be used with the same effect. The diagram also shows how the *boresight* is offset from the normal of the apparent aperture plane by an angle of about 26° to 28°. Although the *effective aperture* is less than the actual value, these reflectors can provide an extra 1–2 dB gain because of the improved efficiency.

2.4 Coplanar arrays

If a number of identical radiators are equally spaced, arranged in line and then fed from the same current source, the radiating fields from each element will combine vectorially to give gain and directivity to the array of elements.

The magnitude of the resulting total electric field strength E_R is given by

$$E_R = E_o[(\sqrt{n} \sin(n(\theta \pm \alpha)/2))/n \sin((\theta \pm \alpha)/2)]$$

where E_o is the radiation from a single element,
 n is the number of elements,
 θ is the phase difference in current between each element and depends upon the separation distance d and λ,
 α is an additional current phase difference.

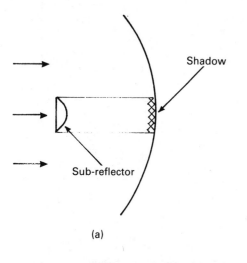

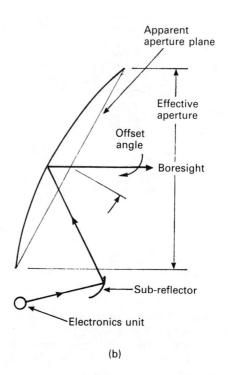

Figure 2.22   (a) Blocking losses. (b) Offset feed antenna.

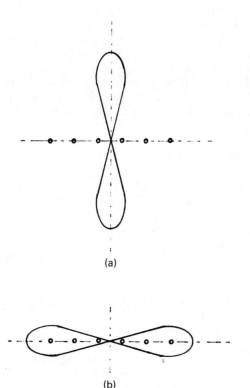

Figure 2.23   Coplanar arrays: (a) broadside; (b) end fire.

Broadside array. If α in the equation above is zero, then the main lobe of the radiation is normal or *broadside* to the line of the array. This is shown in Fig. 2.23(a).

End fire array. If α in the above equation is equal to π rad, then the main lobe will be in line with the array line as indicated in Fig. 2.23(b).

Phased or scanning array. It follows from the above, that if the phase excitation between the elements of the array is modified by a variable angle α, then it will be possible to introduce steerable directivity into the array. This is usually done by adding ferrite phase shifters between each element. The magnetic field associated with these and hence the current phase can be varied by changing the current flowing through the control winding.

Planar arrays

These antennas designed for use at microwaves resemble a double-sided printed circuit board. One side of the substrate carries an etched pattern of microstrip, while the other is left completely metallic to act as a ground plane. Many various track patterns have been used and Fig. 2.24 shows two of the most popular. Gains up to 35 dBi at up to 85 GHz can be achieved fairly easily. By rearranging the feed points to the array, it is possible to introduce some steerable directivity into the radiation pattern.

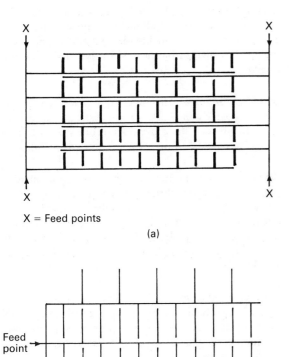

X = Feed points

(a)

Feed
point →

(b)

Figure 2.24 Microwave planar antenna arrays: (a) 'steerable' by feeding at one corner 'X'; (b) single section of multiple array.

Suppressed or conformal antenna. The planar antenna concept adapts well to mobile operation for both ground vehicles and aircraft. In these cases, the metallic skin of the vehicle acts as a ground plane. The flat surface can easily be modified to conform with the shape of the vehicle, in which case they are often described as *conformal* antennas.

2.5 Fresnel zone plate antennas

This antenna is based on the Fresnel principle of a series of concentric rings, alternately transparent and opaque and deposited on a transparent surface. This has the ability to focus electromagnetic waves on to a point behind the surface. The alternate zones are formed by screen printing on to a clear plastic layer or flat glass sheet using reflective or absorbing inks. Silver or graphite loaded ink is used for the plastic surface and metallic oxides can be used on glass. This last form leaves the structure completely transparent to visible light. If the dielectric layer is backed by a metallic surface acting as a reflecting ground plane, the focal point switches to being a mirror image of the transmissive version. This has the effect of increasing the gain of the reflector by almost 3 dB. The principle of both modes is shown in Fig. 2.25. Furthermore, if the rings are made elliptical, the focal point shifts so that electromagnetic waves approaching from a suitable angle *see* concentric circles, the secant of this angle being proportional to the ratio semi-major axis/ semi-minor axis of the ellipsoid. This feature is therefore used to give the reflector surface a *squint*.

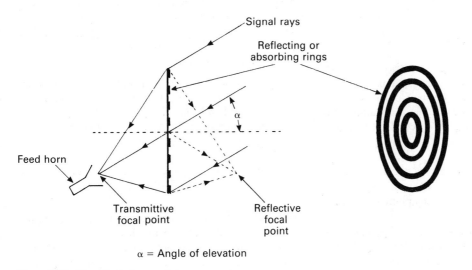

α = Angle of elevation

Figure 2.25 The elliptical zone plate.

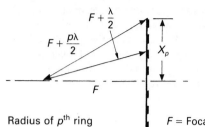

Radius of pth ring

$$X_p = \sqrt{p\lambda F + \left(\frac{p\lambda}{2}\right)^2}$$

F = Focal length
λ = Wavelength
p = Number of ring
X_p = Radius of ring

Figure 2.26 Signal focusing and dimensions of concentric rings.

Figure 2.26 shows the relationship between the wavelength-dependent dimensions.

The gain of the reflector is approximately proportional to the number of rings and the beamwidth is inversely proportional to the area. The bandwidth is practically equal to $2\omega_0/N$ where ω_0 represents the midband frequency and N is the number of rings. For a typical reflector, the bandwidth is about $0.15\omega_0$. Working at 11 GHz, a 90 cm diameter zone plate will give good television reception from a signal source of about 50 dBW. Comparative gain values in dBi are shown in the following table:

Number of zones	Theoretical gain	Practical gain
1	6.0	4.5
2	11.8	9.0
3	15.2	11.5

2.6 Miscellaneous terms

Adaptive antenna array. An antenna array in which the received signal is continually monitored in respect of interference (usually adjacent or co-channel). Its directional characteristics are then automatically adjusted to null out the interference. Such a concept often employs computer control of a planar type antenna.

Antenna feed-point impedance. The complex impedance at the feed point of the antenna, the resistive element of which is the antenna feed-point resistance.

Antenna feed-point resistance. This resistance value accounts for the power radiated from the antenna. It is equal to the power radiated from the antenna divided by the RMS current at the feed point.

Antenna resistance. This value takes into account the total power radiated and the loss dissipation of the antenna system. It is equal to the total power supplied to the antenna divided by the RMS value of the antenna current.

Aperiodic antenna. Theoretically a non-resonant antenna whose characteristics are not frequency dependent. In practice these devices are often described as being *wideband* (see Log-periodic antennas, p. 34).

Aperture. For a parabolic reflector or a horn antenna, this is the dimension of the open mouth and represents a surface over which it is possible to calculate the radiation pattern. For a series of *n* stacked transmitting elements such as dipoles or slots, the vertical aperture is usually defined as *n* times the element spacing in wavelengths.

Aperture illumination. The distribution of the electromagnetic energy across the aperture. In the interests of antenna efficiency (see Antenna properties, p. 26), this distribution may be arranged to be *tapered* or *gabled* in *contour* to minimise the spillover losses.

Azimuth. The compass bearing of the main beam of an antenna, relative to true north as defined by the fixed stars in space.

Azimuth/elevation mounting (az/el). An antenna mounting structure with two degrees of freedom of adjustment or movement so that the main beam of the antenna can be steered to follow a signal source.

Beamwidth. Apart from the conventional definition of the beamwidth of the antenna main lobe, defined as the angle subtended by the two −3 dB or half-power points (HPBW), there are three other parameters that may be quoted in the literature.

   (i) First null beamwidth (FNBW), being the angle subtended by the total extent of the main lobe.
   (ii) First side-lobe beamwidth (FSLBW), which is the angle subtended by the nulls between the first two side-lobes,
   (iii) First side-lobe maximum to maximum (FSLMM), being the angle subtended by the two peaks in the first two side-lobes.

Boresight. The axis of the main radiation lobe usually defined for a reflector antenna. If this is obtained from design calculation or by optical means it is described as the *reference boresight*. The boresight defined by actual measurement is then known as the *electrical boresight*.

Any angular difference between these two values is described as a *boresight error*.

Combiner. A device that allows the outputs of two or more transmitters to be coupled simultaneously to one antenna, without unwanted interaction (see also Diplexer, p. 45). A combiner may also be used to provide the same function in a CATV distribution system.

Community antenna television (CATV). A cable system for distributing television signals received from a single antenna located at an optimum location.

Contoured beam antenna. An antenna designed so that its beam illuminates a specific area to a given power flux density (see also Footprint, p. 46).

Coplanar twin loop antenna. A structure consisting of two concentric circular elements, each with a small gap for the insertion of a tuning capacitor. The impedance of the inner loop is very low while that of the outer is much higher. Both loops are tuned to the same frequency and the coupling between them is adjusted by the use of a variable resistor in the outer loop. By tuning both loops and the coupling, it is possible to obtain a cardioid response with a null of up to 20 dB.

Counterpoise (see also Earth mat, p. 46). For operation at frequencies below about 30 MHz, a good high conductivity earth is almost as important as a good antenna. This is particularly so for antennas that are less than a quarter wavelength in height. Where this is missing and an *earth mat* cannot be provided, a *counterpoise* consisting of one or more radial wires mounted close to the ground and insulated from it, can provide a good substitute for a direct earth.

Cross-polar response. In order to almost double the occupancy of the wavebands from VHF through to microwave frequencies, it is common to use *orthogonal polarisations,* either vertical/horizontal or left-hand/right-hand circular simultaneously. In order to minimise the cross-talk between transmissions using the alternative polarisation, it is important that the antennas used should have a good *cross-polar isolation* to give satisfactory rejection from this form of interference. For reflector antennas, the cross-polar rejection along the boresight is usually high, with the main problem coming from signals that are received via the minor lobes close to boresight. The feed system forms an important feature of the antenna performance in this respect.

Cutler feed. An antenna waveguide feed that passes through the vertex of the reflector to provide a feed at the prime focus point, the waveguide being split at the

end and turned through 180° to face the reflector surface.

Cymo-motive force (cmf). A parameter defined for a lossless propagation path which is the product of electric field strength and distance from a transmitting antenna in a stated direction. At a distance of 1 km, the value is numerically equal to the field strength in mV/m.

Declination angle. The additional angular offset to the *inclination angle* of a *polar mount* system required to ensure that a parabolic reflector antenna can accurately track satellites in the geostationary orbit.

Depolariser. A device constructed from a dielectric material that is placed either in a waveguide, or the path of an electromagnetic wave, in order to modify the plane of polarisation (see also Propagation, p. 242).

Despun antenna (see Satellite systems, p. 269). In a spin stabilised communications satellite, the antenna systems have to be continually readjusted to maintain the satellite to earth links. This is achieved by mounting these on a platform that is separate from the revolving section, that is, only the satellite casing is spun leaving the antennas pointing in a fixed direction.

Diplexer. A three-port filter device that allows two receivers or transmitters to be connected to a single antenna without unwanted interaction.

Director. The parasitic elements of a Yagi–Uda array that is mounted in front of the dipole to increase the strength of the forward radiation or reception.

Dish. The jargon term often used to describe a parabolic reflector.

Diversity reception antennas. The use of a number of antennas in parallel and designed to improve the quality of service under conditions of various forms of fading.

Driven element. The primary element in a compound antenna to which the power for transmission is fed or the element that provides the feed to a receiver.

Dual polarised antenna. An antenna designed to receive or radiate the orthogonal polarisations.

Duplexer. A waveguide device designed to allow an antenna to be used for both transmission and reception simultaneously.

E-plane lens (see Lens antenna, p. 40). A structure consisting of a number of parallel metallic plates aligned parallel to the electric field vector to provide convergence of the waves.

E-plane sectoral horn (see Sectoral horn, p. 40). A tapered waveguide feed that has the parallel sides normal to the electric field vector.

Earth mat. An array of metallic conductors buried beneath an HF antenna and used to improve the conductivity of the natural earth.

Effective aperture (see Antenna properties, p. 40).

Effective electrical length. This is the physical length of an antenna element modified by a factor due to *end effects.* Even in free space, the electrical length is slightly greater than the physical length. Theoretically it is only in free space and with an infinitely thin element that the two dimensions are equal. The difference is due to the fact that the velocity of the waves is slightly less on the antenna than in free space, so that the antenna wavelength is longer by about 2% to 8% on average. The difference tends to increase with an increase in the diameter of the element. Below about 30 MHz an empirical formula $L_e = 291/f$ may be used, where L_e = effective electrical length in wavelengths, and f = frequency in MHz.

Effective height (of MF/HF antennas). It is sometimes convenient to assess the efficiency of a vertical receiving antenna in terms of its effective height H_e, given by the ratio of voltage V delivered by the antenna to the field strength E at the receiving site.

Effective isotropic radiated power (EIRP). The input power to an antenna times the gain relative to an isotropic source. This may also be similarly quoted as effective radiated power (ERP) but relative to a standard dipole.

Elevation. For an azimuth/elevation mounting. The vertical angle between the boresight or line-of-sight path to the object satellite and the tangent surface to the earth at a ground station site.

Figure of merit; G/T ratio (see also Satellite systems, p. 272). It is important to maintain a high signal to noise ratio in any communications system and the limiting feature is often the input stages to the receiver. For a satellite receiver, this is very largely governed by the antenna in use, the gain to noise temperature ratio being a valuable parameter.

Footprint. Jargon term referring to the coverage area on earth of a satellite communications link.

Front to back ratio. The ratio of the gains of a directional antenna at the maximum response from the main lobe to that obtained when the antenna is rotated through 180°.

H-plane lens. See Lens antenna, p. 40.

H-plane sectoral horn. A rectangular feed horn that is flared in one dimension only and operated so that the electric field vector is parallel to the parallel sides.

Height gain. A measure of the gain that can be achieved by mounting an antenna above the ground. It is the ratio of field strength achieved using a given mode of propagation at a given height relative to the field strength obtained at ground level.

Inclination. The term usually used to describe the elevation of the antenna when used in conjunction with a polar mount.

Intrinsic impedance of antenna. The theoretical or ideal input impedance. The elements are considered to be lossless, uniform dimensions and used with a perfect earth or ground plane.

J-hook. The descriptive name for a length of waveguide with one end turned through 180°. This passes through the reflector vertex to illuminate the surface from an electronics unit mounted behind the structure.

Leaky feeder/leaky wave antenna. A length of waveguide with a longitudinal slot or row of holes, or co-axial cable with purposely poor shielding, can lose energy continuously along its length. This provides an effective way of transmitting signals within a tunnel or underground system. These are often described as *fast* or *guided wave structures.*

Linear array. An antenna structure in which all the elements are positioned in a straight line.

Look angles. Jargon for the terms azimuth and elevation.

Low noise temperature antenna. Highly directional antennas used for satellite communications are designed to have a very small side-lobe response. Very little noise is then introduced from terrestrial sources while the antenna main lobe points above the horizon.

Luneberg lens antenna. The spherical structure of this lens is formed from material of varying relative permittivity (ϵ_r). The variation of permittivity should obey the formula $\epsilon_r = 2 - (r/r_o)^2$, where r is the radius of a particular shell and r_o is the outer radius. ϵ_r thus varies from 1 at the surface, to 2 at the centre. A practical structure can be produced by making the sphere from a series of concentric shells, each of different permittivity to approximate the design formula. Increasing the number of shells, increases the bandwidth of the lens, which can easily be as great as 20 GHz. Any plane wave that is incident upon the spherical surface becomes focused on to a point on the opposite side. Conversely, any energy applied at this point will propagate from the opposite side as a parallel beam. By modifying the structure, it is possible to arrange for the focal point to lie outside the surface of the sphere. It then becomes possible to introduce several feed points spaced around it. If these are each mounted on a robotically controlled arm, the lens can function on several frequencies simultaneously, without mutual interaction.

Metre-ampere. A parameter used for regulatory purposes to assess the range of ship's radio installations. It applies to linear vertical antennas that are less than a quarter wavelength in height and is the product of maximum height above sea level and the RMS drive current at the antenna feed point.

MUSA array (multiple unit steerable array). A combination of a line of rhombic antennas, the receiving outputs of which are combined via adjustable phase delays. Phase delay variation then generates a steerable effect of the polar response in the vertical plane that contains the centre line of the rhombics.

Near-vertical incidence radiation antenna. An HF antenna designed so that the major radiation produces a near-vertical pattern. A technique that is particularly useful for single hop, medium distance communications links.

Null steerable antenna (see Adaptive arrays, p. 44). An antenna that has a near omni-directional radiation/reception pattern of directivity but includes a null which is electrically steerable.

Partial antenna gain. The absolute gain of an antenna is the sum of the gains for two orthogonal polarisations. In the case of an antenna designed for a single polarisation, the partial gain is equal to the total gain. This may also be expressed as the ratio of the radiation intensity obtained for a given polarisation and at a given distance to that which would be obtained if the same energy had been radiated from an isotropic source.

Passive reflector. A simple reflector used to change the direction of radiation from a microwave beam. For example, a reflecting surface mounted on a hill top and so positioned as to direct the energy down onto a valley receiving site.

Passive repeater. A passive reflector system constructed from two reflectors that are simply coupled together with a short length of waveguide. The first reflector acts as a receiver while the second transmits but in a different direction.

Phase displacement antenna. An antenna constructed from a driven element and a group of reflectors, the secondary radiation from which produces an antenna with directivity. The Yagi–Uda array is a member of this family.

Phase function of antenna. A mathematical function relating to the phase of either the electric or the electromagnetic fields produced by the antenna. It is based on a spherical surface with the antenna at its centre. The far field region is normally considered because of its well-behaved characteristics.

Phased array (see Adaptive arrays, p. 44). An antenna with a steerable radiation/reception pattern produced by varying the relative phases of the currents in the various elements from which it is constructed.

Pyramidal horn. A waveguide feedhorn in which both opposite faces are tapered.

Radiation function. A mathematical function that expresses the far field radiation in terms of three-dimensional co-ordinates, incorporating the directivity, absolute gain and radiation intensity or cymo-motive force.

Radiation intensity. The density of the radiation in the far field region in terms of the power radiated per unit solid angle or watts/steradian. The average radiation intensity (in all directions) is then the total power radiated by the antenna divided by 4π steradians. For polarised waves, the total radiation intensity is equal to the sum of the intensities of the orthogonal radiations.

Reactive reflector or reflective array antenna. Often constructed from a cluster of waveguide sections, each containing a phase shifter and terminated in a short-circuit to provide reflections. When used in conjunction with a parabolic reflector, the adjustment of the phase shifters can be used to generate the desired directivity pattern. Often used with satellite systems to provide a contoured *footprint*.

Reflector. One or more conductors or a conductive surface, placed behind a driven element so that the direct radiation and the re-radiation from the reflectors add to give directivity.

Reflector element. The element placed behind the dipole of a Yagi–Uda array to increase forward gain.

Reflector surface errors. For a parabolic reflector, the radiation efficiency depends upon all the energy leaving the aperture plane being in phase. Any discrepancies in the shape of the surface will cause scattering of the energy and consequently reduced forward radiation. In practice, provided that the surface errors are dimensionally small compared with the wavelength of the signal in use, the loss will be negligible. Thus for higher frequency operation it is important that surface blemishes are small.

Reversible antenna array. A directional antenna whose main lobe can be rotated through $180°$ simply by changing the feed.

Rotating field antenna. If two identical antennas are mounted at right angles to each other on the same axis and driven by phase quadrature signals, the directivity will be omni-directional and the phase function in any given direction will be proportional to the angle of the given direction.

Scale modelling. Because the product of frequency and wavelength is a constant (2.997925×10^8 m/s, the velocity of electromagnetic waves in free space), it is possible to scale-model large antennas at the design stage. The scaling used represents a reduced wavelength so that if the frequency of energisation is increased by the inverse factor, the experimental results have an acceptable degree of accuracy.

Scattering factor. All antennas, particularly those with a significant mis-match which results in a reactive impedance, suffer from re-radiation to some extent. This form of re-radiation can be quantified by the ratio of the total power that would have to be radiated by an isotropic source to produce the same degree of re-radiation intensity; to the power flux density of the incident electromagnetic wave.

Screening. In order to minimise the radiation or reception of interfering signals, it is possible to introduce screening at the antenna site. This can be achieved by the use of metallic shielding or by judicious use of hills or mounds of earth. This feature is particularly important at satellite ground stations.

Secondary antenna properties. These include such parameters as blocking or shadowing losses, cross-polar response, equivalent noise temperature (see Noise, p. 220), pointing accuracy (see Satellite systems, p. 270), side-lobe response, voltage standing wave ratio (VSWR) (see Transmission lines and waveguides, p. 378), and secondary radiation (see also Scattering factor). When an antenna element is more or less than half a wavelength long, it will have a reactive impedance. If it is then exposed to electromagnetic radiation, the mis-match will result in re-radiation. In antennas such as a Yagi–Uda array, this re-radiation is intentional and it is designed to aid the primary radiation and so give the antenna its directive properties.

Series loaded antenna. If an antenna has a reactive impedance, i.e. it is off-resonance, it is usual to add some series reactance as a cancelling element to the feed point. This results in a modified current distribution on the antenna and an improved radiation efficiency.

Service area. This is a parameter that varies with the type of transmission. In general, it can be expressed by the signal to noise ratio, which for an analogue system provides a level of interference that is perceptible but not annoying, and for a digital system, the level that causes the maximum acceptable error rate.

Shielded loop antenna. An antenna often used for direction finding which consists of one or more turns of wire enclosed by a tubular shielding sleeve. This contains a gap to avoid producing a short-circuit turn effect.

Single polarised antenna. An antenna designed to respond to or radiate a single polarisation. In a practical situation, this often only holds good for a limited angle within the main beam.

Sleeve dipole/monopole (see Dipole variants, p. 34). A dipole/monopole whose feed section is enclosed in a short co-axial sleeve to improve the impedance matching and hence the radiation efficiency.

Specific cymo-motive force. The cymo-motive force produced in a stated direction when the antenna is energised by 1 kW of power.

Spill-over factor or loss (see also Radiation efficiency). The ratio of the power radiated by a reflector to the total power provided from the generator source.

Squarial antenna. British Satellite Broadcasting (BSB), now BSkyB, proposed the use of a *squarial* for the *direct to home* (DTH) satellite broadcasting service provided from a high power satellite (approximately 60 dBW).

This 40 or 50 cm diamond-shaped square unit provides enough gain to produce excellent picture quality even when working with a C/N ratio as low as 11 dB. The planar-based antenna is designed around a structure of either 144 or 256 resonant cavities separated by 0.9λ, each provided with a separate probe. These are arranged to feed an embedded low noise block-convertor (LNB) via a combining network, which introduces unwanted attenuation. This has to be countered by using a very low loss dielectric material for the substrate.

Squint angle (see Planar array, p. 42). The difference in angle between the actual and the expected radiation, taking into consideration the geometry of the antenna. Certain adaptive arrays can be made to *squint* by tuning to give steerable characteristics.

Steerable beam antenna (see Adaptive arrays, p. 44). An antenna whose main beam can be directed in various directions either by an electrical or mechanical drive system.

Surface wave antenna (see End fire array, p. 41). An antenna whose radiation can be considered as being due to an electromagnetic wave travelling along its surface or some imaginary surface associated with the antenna.

Tilt angle (see also Boresight, p. 44). An alternative expression for boresight error. The angular difference between the major radiation axis and the reference axis (see also Offset fed antennas, p. 41).

Universal antenna constant. Consider two antennas R metres apart, directed towards each other, and the gains and areas of each are G_1, A_1 and G_2, A_2, respectively. A power of P watts is transmitted from antenna 1 towards antenna 2 so that the power flux density at 2 is $PG_1/4\pi R^2$ watts/m^2. The total power received at 2 is therefore $PG_1A_2/4\pi R^2$ watts. If the transmitter and receiver are now interchanged

without changing any other parameter, the total power received at 1 is $PG_2A_1/4\pi R^2$ watts. By reciprocity, $PG_1A_2/4\pi R^2 = PG_2A_1/4\pi R^2$ so that $G_1/A_1 = G_2/A_2$, a constant.

Since the gain of a Hertzian dipole can be determined, the gain of all others can be derived. The universal constant is thus $G = 4\pi A_e/\lambda^2$, where A_e is the effective area and λ is the operational wavelength.

Zoned antenna. For metal plate and dielectric lenses, the thickness of the structure can create mechanical problems. Therefore it is common to reduce this by constructing the lens in a series of concentric steps. The general shape of each step is parabolic and the steps are one wavelength deep so that the reflections off the surface are in phase. Although this reduces the thickness of the lens, it also reduces the bandwidth of the antenna.

2.7 Useful references

Balanis C.A. (1982) *Antenna Theory: Analysis and Design*, Harper and Row, New York.

Connor F.R. (1985) *Introductory Topics in Electronics and Telecommunications*, Vol 4, *Antennas*, Edward Arnold, London.

Glazier E.V.D. and Lamont H.R.L. (1966) *The Services Textbook of Radio*, Vol 5, *Transmission and Propagation*, H.M.S.O., London.

Kraus J.D. (1984) *Electromagnetics*, 3rd Edition, McGraw-Hill, New York.

Lewis G.E. (1988) *Communication Services via Satellites*, BSP Professional Books, London.

Mitchell M.A. and Sandford J.R. (1988) Design and analysis of multi-shell spherical microwave lens antennas. *ANTEM '88, Symposium on Antenna Technology and Applied Electromagnetics*, August.

Mitchell M.A. and Sandford J.R. (1989) Luneberg SP lens revival. *Electronics and Wireless World*, May 1989.

Rudge A.W., Milne K., Oliver A., Knight P. (1982) *Handbook of Antenna Design*, Vols. 1 and 2, Peter Peregrinus, London.

3 Audio signal processing

The quality of audio communications is chiefly governed by the originating sound level, system distortion and background noise levels (see also Noise, p. 219, and TV sound channels, p. 361). For telephony, there are the added problems associated with echoes and round-trip delays. The primary aim of all audio processing is therefore to maximise the system signal to noise (S/N) ratio. This, like power levels, is usually quoted in decibels (dB) and is a convenient unit because the human ear can just about recognise a change in sound level of 1 dB. Since noise in all its forms is a destroyer of information and proportional to bandwidth, it is important to ensure that the system bandwidth is restricted to that which is just necessary to convey the information. All analogue processing amplifies both the input signal and the noise and then adds further distortion and noise to degrade the input S/N ratio.

Round-trip delays in speech channels should not exceed about 250 ms otherwise the context of the conversation can be lost. Echo effects simply provide a further addition to the distortion and background noise levels.

The bandwidth of 300 to 3400 Hz is adequate to provide readily recognisable speech quality for both male and female voices at S/N ratios less than about 30 dB. For high quality monophonic (mono) broadcast music channels, the bandwidth needs to be around 20 Hz to 15 kHz, with a S/N ratio of at least 50 dB. For stereophonic (stereo) broadcasting, the bandwidth allocation needs to be twice this. Since modern music tends to have a greater high frequency content, many audio systems provide for a bandwidth extending from around 15 Hz to 20 kHz.

3.1 Analogue signal processing

Apart from using companding and pre-emphasis/de-emphasis systems to combat the effects of noise, waveshape compression is also possible. This operates by controlling the loudness and raising the level of the low amplitude components in a complex wave. A recent development operates in two stages by using an instantaneous waveshape compressor and a loudness controller. The latter controls the mean level of the signal by using the peak amplitudes obtained between successive zero crossing points. This value is then used to control the system gain which changes in a step-like manner at the zero crossing points.

Further developments make use of the natural *masking effect* of the human ear. It has been shown that the audio frequency range can be sub-divided into about 24 bands, within which signal components with a larger amplitude conceal or mask lower level components, including noise, in the adjacent bands.

Whereas analogue signals degrade gracefully with a decrease in S/N ratio, digitally processed signals maintain their high quality down to a very low level and then suddenly crash.

3.2 Digital signal processing (DSP)

Of the many advantages gained through using digital processing, the most important include robustness under noisy conditions, wide dynamic range, ability to provide exact copies, quality unaffected by the transmission or storage medium and a very low level of distortion.

With DSP the high quality requirement is achieved by using a suitable sampling frequency and number of bits per sample. Thus for an upper frequency of 15 kHz the sampling frequency must be at least 30 kHz. At 8 bits per sample, this requires a bit rate of at least 240 kbit/s per channel, which obviously occupies a wide bandwidth. Thus a number of bit rate reduction techniques are available that compensate for this restriction.

Although the system S/N ratio is often quoted, the energy per bit (e/b) ratio, which degrades with an increase in bit rate, tends to be the more important parameter.

If a delay can be tolerated, then many digital techniques can be employed to improve the audio quality. For example, in addition to the conventional forward error control (FEC), error concealment can be implemented. If a word is detected as being in error, then three possibilities exist:

- Discount the word and replace with a zero value.
- Repeat the previous word.
- Interpolate between two unerrored words.

If such a delay data stream must be synchronised with another, such as a video signal, then the latter will have to be delayed by the same period.

3.2.1 Bit rate reduction techniques

APT-X100. This system (devised by Audio Processing Technology Ltd), which is not based on the audible masking effect but can be used with it, divides an audio channel signal into four sub-bands. Bit rate reduction is achieved by using linear predictive and adaptive differential pulse code modulation (ADPCM). The audio sampling rate is either 32 kHz or 48 kHz, with 4 bits per sample, which yields data rates of 128 Kbit/s or 192 Kbit/s.

ATRAC (adaptive transform acoustic coding). This system, devised by Sony Inc., uses the masking effect together with a transform coding to provide a bit rate reduction in the order of 5:1. The original signal is sampled at 44.1 kHz, with 16-bit samples, and then divided into time segments of up to 11.6 ms, depending upon the signal dynamics. The digital bit stream is then split into frequency sub-bands by employing a modified discrete cosine transform (DCT). Since the human ear is most sensitive around 4 kHz, many component frequencies on either side will be masked and need not be coded.

Dolby systems. Dolby Inc. have been responsible for a great deal of audio research and development associated with noise reduction systems and stereo and surround sound systems. The AC-1 system uses adaptive delta modulation (ADM) with a bit rate of about 200 kbit/s and is described under Section 37.5.1.

System AC-2 uses a sampling frequency of 48 kHz and a data rate of 128 Kbit/s per channel. It utilises a low bit rate transform coding technique that operates with a delay in the order of 55 ms.

System AC-3 is designed for use with cinema film and employs five channels for surround sound. Digital bit rate reduction is achieved by making use of the signal similarities between channels.

MASCAM (masking-pattern adaptive sub-band coding and multiplex). This system is adapted to make use of the audio masking effect of the human ear. Each high quality audio channel is sampled at 32 kHz and quantised to 8 bits per sample to produce a gross bit rate of 256 Kbit/s. The signal is split into 24 sub-bands by using quadrature mirror filters, and further divided into 4 ms time frames. Each frame carries an information block containing control bits, scale factors and sub-band data. The last contains information about the bit allocation in individual sub-bands which is particularly important within the frequency range of about 1 to 3 kHz. Channel coding involves the use of punctured convolutional codes which not only adds error protection, but also provides a degree of bit rate reduction. The final transmission rate is 136 Kbit/s of which 24 Kbit/s is allocated to scale factors and error control.

MUSICAM (masking pattern universal sub-band integrated coding and multiplexing). This system also utilises the aural masking effect and provides for a range of processing delays with bit rates ranging from 192 Kbit/s down to 32 Kbit/s, the higher bit rate reductions being achieved by an increased processing delay. The system divides each audio channel into 32 sub-bands and the encoder scans each band and codes only those samples that are above the masking threshold. If any bit periods are unused, these can be dynamically allocated to other sub-bands. Huffman coding is then employed so that long runs of 1s and 0s can be more efficiently encoded.

PASC (precision adaptive sub-band coding). This has many of the characteristics of MUSICAM but with interleaved coding. The data stream is then organised into an 8-bit parallel format for recording purposes.

3.2.2 8 to 14 modulation (EFM)

In any synchronous digital system, the decoder clock rate must be accurately synchronised to the transitions in the data stream. If this contains long runs of 1s or 0s, then the lack of transitions will allow the clock frequency to drift and introduce bit errors. EFM is introduced to reduce the maximum run length for two bytes from 14 to 10.

An 8-bit code group produces 256 unique code words, a number that rises to 16 384 for a 14-bit code. Of these, only the words that contain 10 or less 0s and with at least two 0s between successive 1s are selected. A further 21 words with the longest run lengths are discarded to leave 256, 14-bit code words.

A ROM look-up table at both the encoder and decoder is then used to produce a one-to-one relationship between the 8- and 14-bit code words.

3.3 Miscellaneous terms

Dynamic range. The range of signal levels between the lowest and highest permitted amplitudes. In practice, the lowest level signal is that which appears just above the noise floor. The highest amplitude is that at which unacceptable distortion sets in. The dynamic range is thus practically equal to the S/N ratio.

Gain riding. In any system where the several parts of a complex signal must be processed separately, some form of channel balance and gain control must be employed. Such cases include surround sound and stereo systems. The gain of each channel can be controlled by a voltage controlled amplifier (VCA) which is in turn controlled from a monitor circuit that samples the signal levels in each channel. Then the system balance depends upon the control exerted by the smallest and largest signal level present in any one channel to maintain an overall balance.

Haas effect. A member of a cinema audience viewing the screen from one side of the auditorium will hear the output from the nearest loudspeaker first, followed by a distorted and reflected sound from the others later. The same effect also occurs, but to a lesser extent, with stereo sound television. It is therefore most important that the signals produced by two-channel stereo systems provide a phantom sound image near the middle of the sound field. The effect in cinemas is largely avoided by using surround sound systems.

VOGAD (voice-operated gain adjusting device). A voice-operated compressor circuit that is designed to provide a near-constant level of output signal from a range of input amplitudes. Such a circuit has a fast attack time with a relatively slow release time to avoid excessive volume compression at the system output.

3.4 Useful references

Rumsey F. (1991) *Digital Audio Operations*, Butterworth–Heinemann, Oxford.

Rumsey F. and Watkinson J. (1993) *Digital Interface Handbook*, Butterworth–Heinemann, Oxford.

Sinclair I.R. (1989) *Audio Electronics Reference Book*, Butterworth–Heinemann, Oxford.

Thomas L.D. (1993) Putting punch into voice communications. *Electronics and Wireless World*, January, pp. 14–20.

Wech Ch. and Theile G. (1988) Digital audio broadcasting. IEE Pub. No 293. *Proceedings of the International Broadcasting Convention*, pp. 360–363.

4 Bar code technology

The use of bar coding principles is progressively spreading throughout the communications industry. The applications range from component identification, product serial numbers, books and service manuals, the control of library systems for video cassettes and the programming of video recorders, through to personnel security badges. With minor variations for in-house applications, all the code formats in use are based on the European and Japanese Article Numbering (EAN/JAN) Associations and the American Universal Product Code (UPC) standards.

The code patterns are formed as a series of black (or coloured) and white bars, read sequentially from left to right and bounded by two blank areas referred to as *quiet zones*, as shown in Fig. 4.1. For two-level codes the wide to narrow spacing ratio is either 2:1 or 2.5:1. For the four-level codes, the bars and spaces are either 1, 2, 3, or 4 times the width of the narrow unit. The code patterns vary from 8 to 14 characters in length.

The responsibility for the issue of the unique numbers is delegated by EAN/JAN or UPC to the various country Article Numbering Associations (ANA).

The overall size of the pattern depends upon the sensitivity of the reader device that will be used, and the typical minimum width feature is in the order of 0.13 mm.

The security feature can be enhanced by printing the codes with infra-red reflective ink on a black background. When viewed under normal lighting, the pattern then appears completely blank.

The reading of a bar code is achieved either by scanning the surface with red laser beams or flooding the area with light from a red LED. A photo-detector, typically a charge coupled device, then senses the reflected light pattern.

4.1 Common code formats

Codabar

This is a two-level, seven-element, mainly numeric code that includes four start/stop characters (A, B, C, D) as indicated in Table 4.1. A binary 0 or 1 is coded as a narrow or wide bar or space, respectively, with elements

The four main symbols to represent the different numbering systems

EAN-13, representing an EAN-13 number.

EAN-8 representing the 8-digit short number.

ITF ('Interleaved Two of Five'), representing a 14-digit number

054 12345 67890 8

UPC, representing the Universal Product Code (UPC) 12-digit number used in North America.

Figure 4.1 Bar code formats (courtesy of EAN Association).

being separated by a narrow space. Each message code is delimited by start and stop characters, e.g. B123456B. Because only 20 out of the total 128 possible code words are used, there is no great need for error protection.

ABC codabar (American Blood Commission) or NW7 is widely used throughout the blood transfusion service.

Table 4.1 *Codabar*

0	0000011	..	0001100
1	0000110	$	0011000
2	0001001	:	1000101
3	1100000	/	1010001
4	0010010	=	1010100
5	1000010	+	0011010
6	0100001	A	0011010
7	0100100	B	0101001
8	0110000	C	0001011
9	1001000	D	0001110

Code 39 (code 3 of 9)

This coding scheme, which represents 44 alphanumeric and special symbols as shown in Table 4.2, is used extensively outside of the retail industry. It is a nine-element, two-level code using only wide and narrow strips for bars and spaces, with a wide to narrow ratio of 2.5:1. Each character is represented by 5 bars and 4 spaces (three 1s and six 0s) with a narrow space between each character. A wide space or bar represents binary 1, while a narrow symbol represents 0. Since only 44 patterns out of the 512 available are used there is generally no need to add error protection. However, where needed a check sum can be added. This is obtained by adding all the appropriate check values quoted in Table 4.2, dividing by 43 and back coding the remainder. For example, if the remainder is 22, which is M, this would be coded as 101000010. An extended Code 39 is available in which two extra characters are added so that the complete ASCII (American Standard Code for Information Interchange) code set can be implemented.

EAN/JAN 8

This is a four-level numeric only code system as indicated in Table 4.3. The bars or spaces are either 1, 2, 3 or 4 units wide and bars and spaces represent binary 1s and 0s, respectively. Thus 00111 would be encoded as a two-unit wide space followed by a three-unit wide bar. Each character is represented by a 7-bit word with start, centre and stop guard bars. Characters to the left of the centre guard bars are encoded from either of the left-hand (LH) columns of Table 4.3, while those from the right of centre are encoded from the right-hand (RH)

Table 4.2 *Code 39*

Character	Binary code	Check value
0	000110100	0
1	100100001	1
2	001100001	2
3	101100000	3
4	000110001	4
5	100110000	5
6	001110000	6
7	000100101	7
8	100100100	8
9	001100100	9
A	100001001	10
B	001001001	11
C	101001000	12
D	000011001	13
E	100011000	14
F	001011000	15
G	000001101	16
H	100001100	17
I	001001100	18
J	000011100	19
K	100000011	20
L	001000011	21
M	101000010	22
N	000010011	23
O	100010010	24
P	001010010	25
Q	000000111	26
R	100000110	27
S	001000110	28
T	000010110	29
U	110000001	30
V	011000001	31
W	111000000	32
X	010010001	33
Y	110010000	34
Z	011010000	35
"	010000101	36
=	110000100	37
Space	011000100	38
*	010010100	
$	010101000	39
/	010100010	40
+	010001010	41
%	000101010	42

column. The difference between the two LH columns is that LHA and LHB are encoded with odd and even parity, respectively. The total pattern would be encoded as LH guard bar 101, followed by two flag characters and the first two data characters, encoded as LHA. The

Table 4.3 *EAN/JAN 8*

Number	LHA	LHB	RH
0	0001101	0100111	1110010
1	0011001	0110011	1100110
2	0010011	0011011	1101100
3	0111101	0100001	1000010
4	0100011	0011101	1011100
5	0110001	0111001	1001110
6	0101111	0000101	1010000
7	0111011	0010001	1000100
8	0110111	0001001	1001000
9	0001011	0010111	1110100

Table 4.4 *Two-digit supplement code*

A/A	A/B	B/A	B/B
00	01	02	03
04	05	06	07
08	09	10	11
12	13	14	15
16	17	18	19
20	21	22	23
24	25	26	27
28	29	30	31
32	33	34	35
36	37	38	39
40	41	42	43
44	45	46	47
48	49	50	51
52	53	54	55
56	57	58	59
60	61	62	63
64	65	66	67
68	69	70	71
72	73	74	75
76	77	78	79
80	81	82	83
84	85	86	87
88	89	90	91
92	93	94	95
96	97	98	99

centre guard pattern 01010, the last three data characters and the check character encoded as RH followed by the right-hand guard bar 101.

To calculate the check sum for the number 12345 prefixed with the flag 60 (6012345), add all the characters in the odd number position $(5+3+1+6=15)$ and multiply this by 3 (45). Add all the characters in the even number position $(4+2+0=6)$ and add the sub-totals $(45+6=51)$. The check sum is then the smallest digit which must be added to obtain a multiple of 10 (9). The coded pattern would thus be, 60123459.

EAN/JAN 13

This system is based on EAN/JAN 8, with the same four-level code, guard patterns and check sum procedure, but extended to include 10 data characters. In general, the first two digits identify the country of origin. Digits 3 to 7 identify the manufacturer or code user, digits 8 to 12 provide the product serial code and digit 13 provides the check sum.

EAN/JAN supplementary codes

Two supplementary codes for additional information are provided within this standard.

Two-digit supplement. These two digits are encoded as follows; left-hand guard bars 1011, first data character, separator 01 and second data character. The data characters are encoded from LHA and LHB of Table 4.3 according to the position in Table 4.4.

Five-digit supplement. The five digits are encoded using Table 4.3 and Table 4.5 as follows:

Left-hand guard bars 1011, followed by the five characters encoded from LHA or LHB and separated alternately by either 1 or 01. Each data character is selected from LHA or LHB (Table 4.3) according to Table 4.5 and the check sum, which is calculated for 12345 as follows:

Table 4.5 *Five-digit supplement code*

Number	\multicolumn{5}{c}{Data digits}				
	1	2	3	4	5
0	B	B	A	A	A
*1	B	A	B	A	A*
2	B	A	A	B	A
3	B	A	A	A	B
4	A	B	B	A	A
5	A	A	B	B	A
6	A	A	A	B	B
7	A	B	A	B	A
8	A	B	A	A	B
9	A	A	B	A	B

Add all the odd number position digits $(1+3+5=9)$. Multiply by 3 (27). Add all the even number position digits $(2+4=6)$. Multiply by 9 (54). Add the two sub-totals $27+54=81$. Divide Mod 10, so that the check sum is 1.

Referring to Table 4.5, the data characters would be encoded LHB, LHA, LHB, LHA, LHA.

Interleaved 2 of 5 (ITF). This is another two-level, numeric only coding scheme. Narrow and wide bars or spaces are used to represent binary 0s and 1s respectively, with a wide to narrow ratio of about 2.5:1. Each bar code pattern is contained within a start code of 0000 and a stop code of 100. The data symbols are encoded from Table 4.6. The first data symbol is encoded in the five bars following the start code and the second symbol is interleaved into the first five spaces. Due to the interleaving, no character separators are needed and because each symbol consists of two wide and three narrow units, the code pattern is largely self-checking. A check sum may be appended to the rightmost position and is calculated as in the following example:

67543210. Add all the odd numbered position digits $(0 + 2 + 4 + 7 = 13)$ and multiply by 3 $(13 \times 3 = 39)$. Add all the even numbered position digits $(1 + 3 + 5 + 6 = 15)$. Add the two sub-totals $(15 + 39 = 54)$; the check sum is then the smallest digit that needs to be added to make this total a multiple of 10. The check sum is thus 6, so that the pattern becomes 675432106, but since this contains an odd number of digits, a non-significant leading zero is appended so that interleaving can be applied. The final coded number is thus 0675432106.

Universal product codes (UPC)

These are very similar to the EAN/JAN formats. UPC A is a 12-digit code similar to EAN 13 but with a shorter

Table 4.6 *Interleaved 2 of 5 coding*

Number	Binary code
0	00110
1	10001
2	01001
3	11000
4	00101
5	10100
6	01100
7	00011
8	10010
9	01010

left-hand guard zone and is split into a number of specific applications. 0 = basic codes, 1 = no specified use, 2 = random weight items, 3 = National Health items, 4 = non-food items, 5 = used for special items. UPC E is a short 6-digit code for small items.

4.2 Useful references

Article Numbering & Symbol Marking Operating Manual.
Tradacoms-Manual of Standards for Electronic Data Exchange.
Both are published by Article Numbering Association Ltd, Catherine Street, London.

5 *Codes and coding formats*

With all forms of signal transmission, it is important to minimise the effects of noise, which is essentially an information destroyer. With analogue systems, this effect can be quantified by the signal to noise voltage, current or power ratio (see Noise, p. 219). In digital systems where the information is transmitted in binary digits or bits, of fundamental On/Off, mark/space or 1/0 elements, a similar concept can be applied. In this case the parameter is measured in terms of energy/bit per watt of noise power/hertz (E_b/N_o). If the noise power becomes comparable with the energy in each bit, then *bit errors* are produced. Thus the degradation of digital signal to noise ratio leads to a *bit error rate*. The Morse Code is one of the earliest forms of digital signalling. All the characters of an alphabet are represented by On/Off voltage or current pulses of either short or long duration or dots and dashes. The coding periods are based on the dot duration, with a dash being equal in length to three dots, the spacing between each element, character and word being one, three and five dots, respectively. Such a code format is really only suitable for use by human operators, the variable durations being difficult to accommodate in electronics high speed communications.

5.1 Primary codes and pulse shapes

The bit error rate can be minimised by using pulses of maximum width and/or amplitude, the obvious choice being a square shape. However, this introduces a number of problems. To pass a square wave, a transmission channel requires a wide bandwidth. To retain a good approximation to a square wave requires that the channel bandwidth should extend up to at least the 13th harmonic of the fundamental frequency. In any case the transmission of such pulses through a typical channel will produce *dispersion* or pulse spreading, which can lead to an increase in *inter-symbol interference* (see Basic digital processing, p. 75), and an increase in bit error rate. Increased pulse width reduces the signalling speed and an increase in pulse amplitude

introduces further problems. The final pulse shape is thus a compromise. One very popular pulse shape is described as a *raised cosine*. This is chosen because 50% of the pulse energy is contained within a bandwidth of half the bit rate, thus reducing the bandwidth requirement and going some way towards maximising the pulse energy.

5.2 Secondary codes and formats

A code format is an unambiguous set of rules that defines the way in which binary digits can be used to represent alphabetic, numeric, graphic and control character symbols (see American Standard Code for Information Interchange (ASCII) and Extended Binary Coded Decimal Interchange Code (EBCDIC)), the level of the code being identified by the number of bits per symbol used.

Shannon's equation for a communication channel capacity in bits/second (b/s) (see Information theory, p. 143), states that capacity $C = B\log_2(1 + S/N)$ b/s, where B is the bandwidth in Hz and S/N is the signal to noise ratio. Since the bit error rate is dependent upon the S/N ratio, this shows that bandwidth and the S/N ratio can be balanced to maximise the channel capacity for an acceptable bit error rate suitable for a particular service.

To take advantage of this trade-off, binary codes are designed by inserting extra bits into the data system in a regular and controlled way. Some of the methods used are depicted in Fig. 5.1. The general aim is to minimise the number of similar consecutive bits and balance the number of 1s and 0s in the message stream. The greater number of signal transitions is used to improve the locking of the receiver clock and so reduce bit errors. The balance of 1s and 0s produces a signal without a dc component in its power spectrum which allows ac coupling to be used in the receiver and reduces its low frequency response requirement. The commonly adopted codes are generated and decoded using dedicated integrated circuits (ICs) and a *non-return-to-zero* (NRZ) basic code, which is shown in Fig. 5.1(a), where a '1' is signified by a full width pulse and a '0' by no pulse. A further variant of this is shown in Fig. 5.1(b), where '1' is signified by a signal transition at the bit cell centre and a '0' by no transition. Further variants of the NRZ code which have the same characteristics include inversions of these two. The *return-to-zero* format where a '1' is represented by a half-width pulse and a '0' by a negative pulse, finds little use because its reduced width pulses represent an energy/bit penalty.

Figure 5.2 shows comparatively how the power in the spectra of some common code formats is distributed. Coding rules of the 'nBmB' type are often used. These convert *n* bits/symbol of original code into *m* bits/

symbol for transmission $(m > n)$, ensuring that on average there is an equality in the number of 1s and 0s, any excess of 1s or 0s being described as a *disparity*. The additional bits introduced represent a *redundancy* which slows the rate of transmission, but this can be turned to advantage in error control (see Error control, p. 104). The more efficient formats use the least redundancy and in general there is a trade-off between the complexity of balancing the number of consecutive similar bits and the added redundancy. The increase in bit rate that this introduces represents a power penalty. For example, using a 3B4B format involves a transmission rate increase of 4:3 or a power increase of about 1.25 dB. A simple form of error monitoring can take the form of counting the running average of the number of

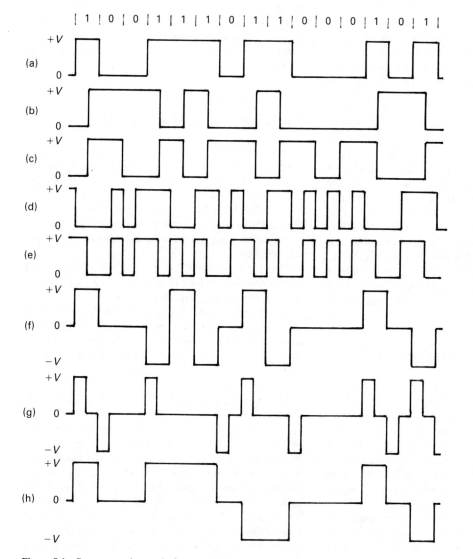

Figure 5.1   Some secondary code formats: (a) Non-return to zero (NRZ); (b) NRZ-M version; (c) Miller; (d) Code mark inversion (CMI); (e) Manchester II (bi-phase-L); (f) Alternate mark inversion (AMI); (g) Dicode; (h) Duo-binary.

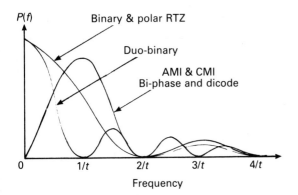

Figure 5.2 Comparison of power spectra for some secondary code formats.

1s and 0s. Any variation outside some predetermined bound then indicates an error situation.

Rules for forming the codes shown in Fig. 5.1

Alternate mark inversion (AMI). This is a three-level (ternary) code where a 0 is represented by no transition while a 1 is represented by alternate positive and negative pulses.

Code mark inversion (CMI). This is a two-level code where a signal transition is introduced at the bit cell centre for a 0, so that 0 is represented by 01. There is no transition for a 1 which is represented alternately by 00 and 11 as indicated in Fig. 5.1(d).

Dicode. As shown in Fig. 5.1(g), this is another ternary format. The half-width pulses are formed from the *derivative* of the original signal. Thus positive and negative transitions generate positive or negative pulses in the derived format with the same energy/bit penalty as return-to-zero codes.

Duo-binary. This ternary format has several variants. That shown in Fig. 5.1(h) is generated from the original NRZ format by the transform 0 is represented by no transition and 1 by alternate positive and negative pulses according to the rules:

positive if the 1 is preceded by an even number of 0s, and reversed if the 1 is preceded by an odd number of 0s.

On average this new code has no dc component and a bit rate that is half that of the original signal.

Manchester code. This is a *bi-phase* format in which each bit in the original signal is represented by two bits

in the derived format. The basic rule for the transform is that 0 is represented by 01 and 1 by 10. This ensures that there is never more than two identical bits in series. One variant of this format uses the opposite of this transform, while two further variants adopt their inverses.

Miller code. This is a popular format for use with magnetic storage media. A '1' in the original format is represented by a transition at bit cell centre and a '0' by no transition, with the following exception: after two consecutive zeros an additional transition is introduced at the end of the first zero cell. A variant of this format utilises a transition for each 0 and no transition for a 1 with the opposite exception.

Miller² code. In the Miller format, a dc component is produced whenever an even number of 1s occur between two 0s. The Miller² format, which is very similar, avoids this by omitting the last previous 1 transition, whenever this condition arises.

5.3 Miscellaneous terms

Alphanumeric character. A symbol from the set that contains characters from an alphabet, a numeral subset and punctuation marks, all of which can be represented by a binary code.

American Symbolic Code for Information Interchange (ASCII). A seven-bit binary code that is commonly used to establish communications between one device that operates in pure binary numbers and another that functions with displayable characters, i.e., a computer to printer type of communication. The seven bits produce a $2^7 = 128$-character set which provides for *device control* characters as shown in the first two columns of Table 5.1, together with punctuation and mathematical signs, and both upper and lower case alphabetical characters. The matrix also shows the simplicity with which bit 5 switches the codes between the upper and lower symbols of the latter sub-set. With minor variations, this code is equivalent to the International ISO-7 Code. In normal applications an extra eighth bit is added as a *parity bit* (see Error control, p. 104).

Anisochronous signal (non-synchronous). A digital signal where the duration between the transitions that represents each bit is variable.

Asynchronous character. A binary character that is used in asynchronous transmission systems, which contains bits of equal time duration. The character bit stream contains *start* and *stop* bits to define its length.

Table 5.1   *ASCII code*

b6				0	0	0	0	1	1	1	1
b5				0	0	1	1	0	0	1	1
b4				0	1	0	1	0	1	0	1
b3	b2	b1	b0								
↓	↓	↓	↓								
0	0	0	0	NUL	DLE	SP	0	@	P		p
0	0	0	1	SOH	DC1	!	1	A	Q	a	q
0	0	1	0	STX	DC2	"	2	B	R	b	r
0	0	1	1	ETX	DC3	#	3	C	S	c	s
0	1	0	0	EOT	DC4	$	4	D	T	d	t
0	1	0	1	ENQ	NAK	%	5	E	U	e	u
0	1	1	0	ACK	SYN	&	6	F	V	f	v
0	1	1	1	BEL	ETB	'	7	G	W	g	w
1	0	0	0	BS	CAN	(	8	H	X	h	x
1	0	0	1	HT	EM	)	9	I	Y	i	y
1	0	1	0	LF	SUB	*	:	J	Z	j	z
1	0	1	1	VT	ESC	+	;	K	[	k	{
1	1	0	0	FF	FS	,	<	L	\	l	--
1	1	0	1	CR	GS	-	=	M	]	m	}
1	1	1	0	SO	RS	.	>	N	↑	n	~
1	1	1	1	SI	US	/	?	O	←	o	DEL

(b6 = MSB; b0 = LSB)

Baud. A unit of signalling speed that represents the number of discrete signal events occurring per unit time. Only when each signal event represents just one bit condition, as in a binary system, are the baud and the bit rates identical.

Baudot code. A five-bit code that represents 64 alphanumeric characters and is used in teleprinter communication systems. Each character is preceded by a start bit and followed by a stop bit period of 1.42 bits. Named after Emile Baudot, a pioneer of printing telegraphy.

Binary coded decimal (BCD). The decimal numbers 0 to 9 represented by the four-bit binary numbers from 0000 to 1010.

Binary digit or bit. The smallest element of information in a binary system. Each bit thus represents a choice between one of two conditions.

Binary numeration or weighting. In a binary number, the bits are normally arranged in a sequence from right to left, in *weighting* of powers of 2, i.e. $---2^3, 2^2, 2^1, 2^0, 2^{-1}$, etc. The rightmost bit has the smallest weighting and thus is known as the least significant bit (LSB). The leftmost bit is then the most significant bit (MSB). Thus the actual value of each bit in the number depends upon its position in the sequence, or its weighting.

Bi-phase coding. The generic title for codes of the Manchester type where each bit in the original code is replaced by two bits in the derived format.

Bi-polar coding with zero suppression. With formats such as alternate mark inversion (AMI), a sequence that contains a long string of zeros will cause the receiver clock to become unsynchronised leading to an increase in the bit error rate. Formats such as BnZS are designed to introduce code violations that will break up such long strings (see also BnZS codes, below, and High density bi-polar codes, p. 62).

BnZS codes. In order to combat the problem just described, this format introduces violations of the alternate ones rule. Typical codes operate for *n* equal to 3, 6 or 8. Thus whenever a string of three, six or eight zeros are encountered, bipolar pulses that violate the rule are introduced. When the receiver decoder finds two adjacent pulses of the same polarity, it recognises that violations have occurred and takes the necessary corrective action (see also HDB3 code, p. 63).

Byte. A group of eight bits that are used to represent alphanumeric characters in a digital system.

Character. An element of a code set.

Code. A unique and unambiguous set of rules that are used to relate a character set to the binary digits that represent it.

Code efficiency. The ratio of the number of bits in the original code to the number in the final format after the addition of redundant bits. For an nBmB code this will be $n/m \times 100\%$ (see p. 58).

Code level. The number of binary digits used to represent each character in the code set. For example, the ASCII and ISO-7 codes have a level of 7.

Code rate. A measure of the reduction in information rate due to the addition of redundant bits.

Control character. A non-printing member of a character set which is devised simply for the control of a peripheral component in a digital system.

Delay modulation. An alternative term for Miller coding (see Secondary code formats, p. 57).

Dibit. A group of two bits which represent the four possible states 00, 01, 10, and 11 (see p. 59).

Dicode. A three-level or ternary code derived from a binary representation. Each positive and negative going transition in the original format produces a positive or negative pulse in the derived format (see also Secondary code formats, p. 59). If the final code pulses are full width, then this version is sometimes described as a *twinned binary code.*

Differential encoding. The generic group that contains the code mark inversion format. In general, a transition at bit cell centre is used to represent a logic 0 with no change in level for a logic 1. For synchronous operation, the original signal can be recovered from the transmitted sequence by sampling and testing to check where transitions have occurred, a process known as *differential detection.*

Digital sum. This parameter can be used to estimate the level of dc component in coded signal. If positive and negative pulses are given the values of $+1$ and -1, and no pulse a zero value, then for zero dc component, the running digital sum over some particular period should be zero. Codes such as the 4B3T (4 binary to 3 ternary) and MS43 have been devised to closely achieve this situation.

Dipulse. A pulse of one polarity followed immediately by one of the opposite polarity.

Dipulse code. A ternary code (three-level) where the zero signal level is used to represent one binary value and a dipulse of one bit period is used to represent the complementary value.

Disparity. The difference between the number of 1s and 0s in a binary word. If there is an excess of 1s, then this value is positive, if not, then it is zero or negative.

Duobinary code. A ternary coding that is generated by precoding the serial data stream. In the precode, each 0 is represented by a transition at bit cell centre and a 1 by no transition. This signal is then passed through a low-pass analogue filter with a cut-off frequency equal to half the Nyquist value. This generates an analogue type of signal with reduced bandwidth. In the processing, opposite sense pulses tend to self-cancel to produce a zero level, while adjacent 1s produce positive and negative peaks. The original information can be recovered from the received signal by full-wave rectification and then slicing at the half-signal amplitude (see Television signal processing systems, p. 353).

Error correction/detection codes (see Error control, p. 104). Redundant bits are added to a message stream in a controlled way to signal the validity of that message at the receiver.

Excess 3 code (XS3 code). This is an example of a non-weighted binary decimal code. The binary representation of each decimal number n is given by $n + 3$. The chief advantage of this format is that when complemented the code becomes the 9s complement of the original number and is thus useful where subtractions are required.

Extended Binary Coded Decimal Interchange Code (EBCDIC). This is an eight-bit binary code that provides for 256 different characters. With respect to the ASCII code, this provides more control codes plus some graphics symbols.

Five unit code. This is a binary, equal length code that provides initially for 32 characters. This is virtually doubled by including case shift control codes to switch between alphabetic and numeric symbols. These control codes are referred to as *letter* and *figure shift*, respectively. It is commonly used with teleprinter systems, when the code characters are delineated by 1 start pulse and 1.5 stop pulses, making it more correctly a 7.5-bit code.

Folded or symmetrical binary code. A technique whereby the most significant bit in a binary number is used to represent polarity. By convention a leading 0 or 1 signifies a positive or negative number, respectively.

Four binary three ternary code (4B3T). The serial binary message system is divided into groups of four bits and these are then converted into a ternary (three-level) representation according to Table 5.2. The first six of the binary codes are always represented by the same patterns but the remaining ten can be represented by one of the two alphabets depending upon the running digital sum. Because of the disparities between the two alphabets this format has some valuable properties.

(a) Because of the near-zero disparity, the signal will have practically no dc component.
(b) If the digital sum exceeds some bound, then errors have been generated in the message.
(c) The maximum number of consecutive zeros is limited to four and this helps to avoid recovery clock errors in the receiver.
(d) The transposition to ternary code reduces the bit rate to 75% of the original value, thus reducing the bandwidth required to transmit the signal.

Gray code. This is described as a *unit distance, cyclic code* as each binary number that represents the decimal values changes only by one bit at a time as shown:

Decimal	0	1	2	3	4	5	6	7
Gray	0000	0001	0011	0010	0110	0111	0101	0100

Decimal	8	9	10	11	12	13	14	15
Gray	1100	1101	1111	1110	1010	1011	1001	1000

The table repeats every 8 or 16 bits as necessary. Since only one bit in a number changes at a step, any double changes mean that an error has occurred.

Gray/XS3 code. This is sometimes used with telemetry monitoring systems to combine the error checking properties of the Gray code with the complementing advantage of XS3.

Hexadecimal. A digital counting system to base 16 (2^4). The sequence of decimal number equivalents from 0 to 15 is given by the direct mapping for 0 to 9, plus the upper case letters A to F representing 10 to 15. Any 8-bit byte can therefore be represented by just two characters.

High density bipolar codes. A development of the BnZS and AMI code formats that is designed to combat the problem generated by long strings of zeros. When

Table 5.2   *4B3T code alphabets*

				Alphabets						
					1			2		
0	0	0	0	+	0	−	+	0	−	
0	0	0	1	−	+	0	−	+	0	
0	0	1	0	0	−	+	0	−	+	Balanced words with
0	0	1	1	+	−	0	+	−	0	zero disparity
0	1	0	0	0	+	−	0	+	−	
0	1	0	1	−	0	+	−	0	+	
0	1	1	0	0	0	+	0	0	−	
0	1	1	1	0	+	0	0	−	0	
1	0	0	0	+	0	0	−	0	0	
1	0	0	1	+	+	−	−	−	+	
1	0	1	0	+	−	+	−	+	−	Inverse pairs with
1	0	1	1	−	+	+	+	−	−	equal and opposite
1	1	0	0	0	+	+	0	−	−	disparity
1	1	0	1	+	0	+	−	0	−	
1	1	1	0	+	+	0	−	−	0	
1	1	1	1	+	+	+	−	−	−	

+ = positive pulse, − = negative pulse, 0 = zero level.
Alphabet 1 used when the running digital sum at the end of the previous word has the value 0, −1 or −2.
Alphabet 2 used if the running digital sum is 1, 2 or 3.

such is located in a message stream, some zeros are replaced by code rule violations (see HDB3 code, below).

HDB3 code. This is a pseudo-ternary format that processes a string of zeros found in a code according to the following rules:

(a) The three states or signal levels are represented by $\pm V$ for a 1 and 0 V for 0.
(b) The 0s in the binary system are coded as 0s in the HDB3 code, except for a string of four zeros.
(c) The 1s in the binary stream are coded alternatively as $+V$ or $-V$, except for code violations when a string of zeros is involved.
(d) For a string of four zeros the following rules apply:
   (i) the first zero is coded as a zero if the preceding 1 has a polarity opposite to the previous violation and is not a violation in itself; if it is, then the new violation has the opposite polarity to the last;
   (ii) the second and third zeros are coded as zeros;
   (iii) the last zero is always coded as a 1 with a polarity that obeys the alternation rule.

Line codes. Codes devised according to specific rules for transmission over copper cables or optical fibres. The rules are such that the new format provides a signal with parameters that meet such criteria as spectral bandwidth, dc component and low or high frequency content to match the properties of the transmission media.

Mark (see also Space, p. 64). A telegraphy term used to represent a '1' in binary coding formats.

Modified AMI code. A signal coded according to AMI rules but including code violations according to a subset of rules (see, for example, BnZS codes, p. 60 and HDB3 code, above).

MS43 code. A 4-binary to 3-ternary code transformation similar to the 4B3T but having three alphabets selectable according to the running digital sum. Most of this code's energy is concentrated in the mid-range of frequencies. Under certain conditions MS43 thus has a better frequency spectral distribution than 4B3T.

n-ary digital signal. A signal that can assume any one of n amplitudes or discrete states. This signal contains the maximum information if it is represented by an n-ary digit. If an n-ary signal, by coding, produces less than n states for transmission, it is no longer n-ary (see Pseudo n-ary code, p. 64).

nBmB code format. A coding technique that converts n bits of original code into m bits for transmission ($n < m$), the aim being to achieve zero disparity (see also 4B3T code, p. 59).

Nibble. A group of four bits or half a byte. A group of bits that can be represented by one hexadecimal character.

Non-return-to-zero inverted code (NRZI). Also referred to as *invert on coding*. A coding technique where a signal transition at bit cell centre is used to represent 0 and no change of level represents 1 (see also NRZ, p. 63).

Non-weighted binary code. The binary coded decimal code format is a weighted code because of its orderly arrangement of each bit in powers of two. The XS3 code on the other hand has no such ordering and is thus a *non-weighted* code.

Octal code. A digital counting system to base 8 (2^3). Each octal value is therefore represented by three bits, the decimal numbers 0 to 7 being used to represent the corresponding characters in the decimal or denary scale.

Octet. A group of eight bits or one byte.

Packet. A block of bits arranged in a defined format and containing both control and data fields.

Pair selected ternary. A technique that divides the inputs binary stream into pairs of bits and then recodes each pair with a pair of ternary digits. Two of the pairs are represented by a zero disparity code pair while the other two pairs are represented by opposite disparity codes. This results in the minimum overall disparity.

Paired disparity code. A family of codes in which each input bit of the original code may be replaced by one of two signal levels of opposite polarity. These are used to minimise the disparity of the new code (see AMI, p. 59).

Partial response coding. The serial binary bit stream is passed through an analogue low-pass filter with a cut-off frequency less than the bit rate. This removes the high frequency components from the signal and produces a multi-level signal with a narrower bandwidth (see Duobinary coding, p. 61).

Polar signal. A signal format that contains pulses of one polarity only.

Polybinary coding. The coding process converts the two-level binary signal into a multi-level code. All even number levels are then used to represent one binary value and the odd levels its complement. Duobinary coding is the lowest order of this group.

Polybipolar coding. A similar technique to polybinary coding except that an odd number of levels is generated which are balanced about the zero level to produce a format with no dc component.

Pseudo n-ary code. A coding technique whereby an *n*-ary input signal produces output code at the same symbol rate. For example, a pseudo ternary signal derived from a line code.

Pseudo-ternary code. The general class of codes where three signal levels are used to represent binary information. The derived format has the same bandwidth and transmits at the same rate as the original code. By comparison, true ternary data conveys $\log_2 3$ Shannons of information (see AMI, p. 59, HDBP coding, p. 62 and Information theory, p. 143).

Quaternary signal. A signal having four distinct levels or states.

Redundancy. The inclusion of extra bits which are transmitted but carry no additional information. These are used only to detect or correct errors that have been introduced in the channel.

Redundant n-ary signals. Each digital signal element can occupy one of *n* discrete levels, and on average, the equivalent binary information is less than $\log_2 n$, which is the theoretical maximum. The redundancy in a code is usually quoted as $(1 - r_e/r_d \log_2 n) \times 100\%$, where r_d is the symbol rate and r_e the equivalent bit rate. AMI and 4B3T codes are members of this class.

Reflected binary code. See Gray code, p. 62.

Space. The telegraph code equivalent of binary 0 (see also Mark, p. 63).

Symmetrical or folded binary code. A code used in pulse code modulation systems whereby the leading bit is used to signify the polarity of the value represented by the remainder of the code bits (see also Folded/symmetrical binary code, p. 62).

Ternary signal. One having three discrete levels or states.

Transliteration. The translation of characters from one alphabet into a second.

Twinned binary code. The generic term for the coding methods that low-pass filter a binary data stream to reduce the bandwidth and create a multi-level signal. A pseudo-ternary signal can be produced by delaying the input by one bit period and then either adding or subtracting this from the original signal. The technique thus converts a binary signal into a pseudo-ternary one by a digital integration process (see also Duobinary code, p. 59).

Two out of five code. A binary coded decimal system in which each decimal character is represented by five bits. The code usually uses two 1s and three 0s, although the inverse may be used.

Unit disparity code. A binary code in which the disparity is ±1.

Variable length coding. A pseudo-ternary signal produced from an input binary stream to form code words of several different lengths. The length of each code word has to be proportional to the number of bits that it represents because the data must normally be transmitted at a constant rate.

VL43 code. This is a variable length code that converts blocks of four or eight bits into three or six ternary digits. Like the MS43 format, this is a state-dependent code but with a lower variation in digital sum with a greater number of signal transitions. These latter ease synchronism of the receiver data clock.

Zero disparity code. A binary code where each word consists of an even number of bits and an equal number of 1s and 0s.

Zero suppression. In a digital code, a long string of zeros leads to an increase in the bit error rate because of synchronism failure. This can be countered by the addition of extra 1s that form a code rule violation. Provided that the receiver can recognise these as violations, then corrective action can be taken (see BnZS codes, p. 60 and HDB3 code, p. 63).

2B1Q line code. Each two bits of binary code is replaced by a single quaternary or four-level bit (see Phase shift keying, p. 190 and Digital modulation, p. 189). This technique reduces both the bandwidth requirement and the transmission rate by a factor of two but at the expense of a S/N ratio penalty.

8 to 9 bit transcoding. By adding an extra bit to each byte in a controlled manner, 512 possible code words are available, from which only 256 are selected in a 1 to 1 relationship with the input byte. The encoded words are chosen using a read only memory (ROM) look-up table so that there is a maximum number of signal transitions in each word, to aid receiver clock synchronism and thus reduce errors. This concept can be further extended as in 8 to 14 bit modulation (see Audio signal processing, p. 51).

5.4 Useful references

Connor F.R. (1975) *Signals. Introductory Topics in Electronics and Telecommunications*, Vol. 1, Edward Arnold, London.

Lewis G.E. (1988) *Communications Services via Satellite*, Blackwell Scientific Publications, London.

6 Computers in communications

The term, *Command, Control, and Communications* (C³), which is often associated with military communications, epitomises the applications of computers and micro-processors in telecommunications. After less than 50 years of development of the electronic computer, telecommunications channels are no longer provided just for computer-to-computer communications. The ubiquitous microprocessor now finds applications within many of the traditional user terminals. Systems such as the telephone now contain such a device to control and manage dialling and message handling, while larger computers can control the whole system.

Today there is a hierarchy of computers available to manage the data processing for industry, commerce and the home. The *microprocessor* is found at the lowest level; this device is an integrated circuit (IC), and on its own has limited computing ability. However, when combined with some *memory* and external communicating circuits as shown in Fig. 6.1, it becomes a *microcomputer*. At this level the machine is sometimes described as a *personal* or *home computer*, and is capable of such a powerful function as desktop publishing

(DTP). The *minicomputer* is a larger machine capable of providing sufficient power to manage process control systems and handle all the data processing for a fairly large establishment. These are often found in research laboratories. Mainframe computers are, by contrast, large machines that are often found in large corporations such as banks, where it is necessary to provide computing power for many sites.

For most of their development period, computers have been constructed according to the Von Neuman architecture, the term architecture referring to the abstract structure of the system that is driven by a fixed set of instructions. The concept is named after John Van Neuman whose work in the 1940s had such a lasting influence on computer development. This concept is based on *stored program control* (the program being stored within the computer), with the answers being arrived at by a series of small uniquely defined steps. During the last decade there has been such a massive improvement in the technologies of architectures, electronic components and programming languages, that Von Neuman machines are now capable of

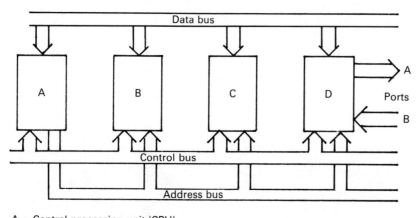

A = Central processing unit (CPU)
B = Read only memory (ROM)
C = Random access memory (RAM)
D = Input/output device (I/O)

Figure 6.1 Basic microcomputer system.

vast amounts of data processing at high speed. Even though such developments have taken place, there are now many applications where Von Neuman machines are too slow and as a consequence, *parallel processing* has become the important feature of new computers. One such important concept is referred to as the *Harvard architecture* where data and instructions are moved around simultaneously or *concurrently*, at the same time as the computer is carrying out some other function. This parallelism results in a very significant increase in computing power.

The minimum computer system shown in Fig. 6.1 consists of a microprocessor or central processing unit to handle all the computing functions, together with memories (see also Memories, p. 164) to hold the instructions and data, and a device to manage communications with the outside world. The *read only memory* holds the instructions for the basic operation of the microprocessor, while the *random access memory* holds the data and user programs. The *input/output* device is used for communications purposes with the user terminals. Communication between the elements of this basic system is via three sets of parallel conductors, known as *buses*, one each to carry data and control signals and the third to address the memory locations where data and instructions are to be found. Timing signals for all the operations are provided from an oscillator or *clock* that is usually included within the microprocessor.

Much jargon is used within the computer industry, including the terms *hardware*, which describes the physical components that make up the system, and *software*, which refers to the intangible elements of the system like the computer language and the programs. Often the computer instructions or programs are contained within an item of hardware, this combination is then referred to as *firmware*. The human operators are often referred to as *liveware*.

6.1 Hardware

Bit slice technology. This originally applied to the concept of constructing general purpose processors from 4-bit integrated circuits or chips. By using parallel connection, word lengths of any desired value could be achieved. For example, a 16-bit arithmetic and logic unit (ALU) could be built up from four chips, the data being effectively sliced horizontally into groups of four bits. Since the system contained the minimum of microcode and with control being largely by hardwired logic, these systems had a very fast response time. The terminology now tends to be used rather inaccurately, referring to any device that can provide processing functions with external control, but is incapable of standing alone.

Some examples from typical computation and control applications include floating point processors, multipliers, sequencers, direct memory access controllers, etc.

Cache memory (see also Memories, p. 169). The time taken to move instructions and data between the central processor unit (CPU) and the memory forms a significant overhead that slows the rate of processing. The cache memory technique that provides a small fast memory section, usually on the same chip as the CPU, considerably increases the throughput of data by allowing parallel, concurrent operation. The cache takes the form of a number of registers, some *local* for specific tasks, others *global* for general use and separated into *instructions* and *data* applications. Instructions can be pre-fetched from memory in blocks and held in the cache for future use, all this while the processor is still operating on previous data. As the most recently calculated data is most likely to be required first, this is temporarily held in the data cache prior to being finally stored in the main memory.

Chip. A descriptive jargon term that refers to integrated circuits (see also Semiconductor devices and technology, p. 284).

Coprocessor (see Multiprocessor, p. 68). A second processor that is coupled via an interface so that it operates in parallel with the main processor. Such an extension provides for additional operations and memory control and considerably increases the processing speed at relatively little extra expense.

Enhanced instruction set computer (EISC). The concept of designing a processor architecture to suit some particular tasks and using some specific high level language, together with a large and complex instruction set, can result in extremely fast processing (see also Reduced instruction set computer, p. 68). A system that employs object-oriented program structures, which have long word lengths, requires large amounts of memory, but if used with a large set of very high level operation codes that can perform complex functions for each code, then the improvement in speed becomes very significant. A typical word length may be 5 bytes, with 4 bytes for data and 8 bits for tag purposes, to identify the type of data, i.e. an integer, a character or some other data type. The objects of the program are accessed by a unique label rather than the address that is common in Von Neuman machines; it is therefore immaterial whether the object is in RAM or on disk. Whenever an op-code is encountered, the microcode translates this directly into hardware control actions so that complex instructions can be carried out in a single step (see also Object-oriented programs, p. 70).

Front-end/back-end processor. In any one data processing operation, large mainframe computers are often employed for a wide range of applications. For some tasks, the input data may need to be subjected to statistical analysis before processing in the main program. In a similar way, the data output from the mainframe may need to be further processed before it can be conveniently used in the outside world. By arranging for a small desktop type of computer in a peripheral mode, to handle this pre- or post-processing, the mainframe can be released to carry out its other tasks. The front-end or back-end processor is thus a useful technique to expand the workload of a large computer system.

Harvard architecture. Describes the techniques that have been devised to overcome the relatively slow serial processing that is the feature of Von Neuman machines. The concept provides for parallel and concurrent operation. Data and instructions can be fetched simultaneously in one clock cycle due to the use of separate channels dedicated to these functions. Cache memory is used, and main memory devices have dual ports so that data accesses can be made during instruction fetches. The concept also uses *pipelining* (see Multiprocessors, below) which further increases data throughput.

Multiprocessors. In this technique the term processor means any device that is capable of reading an instruction and then executing it, while multiprocessors refer to two or more processors coupled together to improve the throughput of data. True multiprocessors are contained within the same housing and are tightly coupled, as opposed to distributed computer systems (see also Networks, p. 204). There are two basic approaches to this concept: one is based on a shared dual port memory and the other is based on a common bus system. The various processors in the system all cooperate to solve a particular problem or control a specific operation and do not work on a number of unrelated tasks. The processors communicate with each other and share the same data and hardware facilities. Multiprocessors are used because a number of low-cost devices operating in parallel can often perform a task more efficiently, economically and faster than a single high-cost, high-speed processor. In addition, the use of parallel processors often increases the system reliability and allows it to degrade gracefully, rather than crash, as in the case of a single-device system. Multiprocessor systems are very flexible; some are even redefinable under software control. A good example of multiprocessing is in air traffic control. The radar system continually receives echoes from a target, each being a function of distance and bearing. Due to noise there is

an uncertainty about each signal. The computer has to calculate the target's current position from constantly varying data and then estimate the track with a view to forecasting any conflicts of airspace. Such a system requires very considerable computing power and is ideal for multiprocessing. One processor can calculate the current track while a second can be used to predict the future position. Multiprocessor operation can be classified by the multiplicity of instructions and data as follows:

(a) Single instruction/single data stream. This is the trivial case of a single processor.
(b) Single instruction/multiple data stream. The instructions are performed sequentially on parallel data and this is suitable for calculations on a vast amount of data.
(c) Multiple instructions/single data stream. This forms the basis of the pipeline processor where multiple operations can be carried out on a single data stream. The pipeline consists of a series of registers through which the data flows while it is being modified by consecutive instructions. Processing is therefore managed in a production line manner.
(d) Multiple instructions/multiple date stream. Each processor receives its own data and instructions so that it can perform a sub-section of the main task.

Register. An electronic circuit that is capable of temporarily holding a limited number of bits of data. Often in groups of 8, 16 or 32 bits (see also Memories, p. 164).

RISC processor. The term RISC is an acronym of reduced instruction set computer, as opposed to the corresponding term complex instruction set computer (CISC) which was coined to differentiate between the two concepts. Throughout most of the period of computer development, an increase in computer power has been achieved by increasing the number of operations that it could perform together with increasing its clock frequency. As the complexity of the operations increased, so did the time taken for the processor to search through its operation code list that was stored in microcode (see also Programming languages, p. 70) in its memory. Research has shown that the computer uses its operation code set almost on an 80/20 basis. That is, for 80% of its calculations it uses only 20% of its instruction set. Thus the RISC chips developed along with the Harvard architecture, allowing the computations to be carried out within the registers, with processing requiring the minimum number of operation codes, each of which is directly executable without the need for decoding. Complex instruction can then be carried out, either by being built up from a series of simple operations, or from decoding logic that is hardwired into the silicon of the chip. By using a pipeline that is maintained full by on-chip data and instruction caches, most operations are carried out in a single clock cycle so that the RISC processor can run 2 to 5 times faster than the CISC chip. A memory management unit (MMU) chip is required to look after the transfers between the processor and the main memory. The concept was developed chiefly for high level language applications and this in turn has produced a demand for optimised *compiler* languages (see also Programming languages, p. 70).

Scalable processor architecture (SPARC). This terminology due to Sun Microsystems Inc, USA, emphasises the flexibility that allows a RISC chip to be used in both large and small computers. The term *scalable* refers to the small dimensions of the conductors on the chip (line lengths). The operating speed of a chip increases almost in inverse proportion to its conductor lengths. Thus as semiconductor fabrication technology improves, the chip can be shrunk to proportionally smaller dimensions and hence run at a higher speed. Thus by choosing a suitable *floorplan* or layout, the original design does not have to be changed to increase speed as technology develops.

Scoreboard. For RISC-type processors, the speed of operation depends upon making maximum use of the registers and if these can be kept to a minimum, the chip can be made smaller and consequently faster. The scoreboard is a technique that can be used to ensure that any given register is only used as long as its data is needed. If a bit in the scoreboard that corresponds to a particular register is set, then that data is needed, otherwise the register is available for more data.

Tag. Usually applied to cache memory systems, and tagging implies that a particular word carries one bit that identifies that word with a particular address.

Transputer. The transputer as developed by Inmos Ltd, UK, is a computer on a single chip, containing processor, random access memory, communications ports, programmable interface and clock. The transputer and its concurrent programming language, Occam, were developed in parallel, specifically to be used in multiprocessor parallel arrays for very high speed processing, concurrent programming being required because large programs can be split into a number of tasks which can operate independently. For the transputer, the *kernel* of Occam is implemented   in hardware to increase speed. Each I/O port is implemented as an autonomous device so that it can perform communications with a peripheral device as

a background task while the processor carries on with its foreground task.

6.2 Software

Benchmark programs. These are test programs designed to evaluate and compare the speed performance of various computer systems and architectures. Common, but not entirely satisfactory, are the rates at which certain operations can be carried out for a given clock frequency. These parameters are stated in mega-instructions per second (MIPS) which states the rate at which instructions are carried out, or mega-flops per second (MFLOPS) which refers to the rate at which mathematical *floating point* operations can be processed. The major problem with both MIPS and MFLOPS is that they are dependent upon the number of clock cycles required to process an instruction and this can vary between different processors. Typical early benchmarks included programs to test load an ASCII file of specified length, search a file of fixed length to find a specified record, sort an unindexed file into some prearranged order or to extract all the prime numbers below some upper value. Over the years the benchmarks have been considerably refined and currently a number of pseudo-standard programs are available. These include:

(a) *Ackerman.* A short recursive program that is suitable for initial evaluation of architectures that use the cache technique.
(b) *Dhrystones.* Devised by Wiecker to exercise the system utility programs. It provides a standard mix of load, store and branch instructions that tests the ability to manipulate strings of characters. The parameter is stated in Dhrystones/second.
(c) *Puzzle.* This is a lengthy benchmark that evaluates the system by solving a three-dimensional matrix problem. It provides a run on a good mix of instructions and procedures.
(d) *Sieve.* This is a small program devised to find all the prime numbers below some stated value n. It is based on the *sieve of Eratosthenes* and provides a recursive search of all the numbers less than n to eliminate the non-prime numbers from the list.
(e) *Whetstones.* Devised by Whetstone as a program to evaluate the operational speed of the processor. It consists of a mix of floating point, integer and data processing instructions that is representative of scientific programs. The parameter is stated in Whetstones/second.

Bugs and debugging. In all except simple programs, some operational errors are likely to be found when the program is run. These errors are referred to as *bugs* and the process of eliminating them is known as *debugging*.

Hierarchy of programming languages

Machine code. The fundamental code for operation of the processor is in terms of the binary values 0 and 1, a concept known as *machine code*. To program a computer in this manner is not only tedious and time consuming but is also error prone.

Assembler language. In order to minimise these problems, an *assembler* language is used whereby each instruction is written in a mnemonic form that helps to describe the function. For example, 'LDA' might be used to represent the operation 'load register A'. The computer then holds an assembler program that converts each mnemonic code into a machine code on a one-to-one basis. That is,

1 assembler instruction = 1 machine code instruction

Whichever of these two languages is used, the programmer must have an extensive knowledge of the way in which the computer actually processes the data and instructions internally. *High level* languages have been developed in order to overcome this particular problem, the programs being written in a language closer to the programmer's normal language.

Compiler. A *compiler* is a program written in a suitable language and used to convert a *source program* into an equivalent *object program*. In the past, compilers have been written in the assembly language of the processor but this is changing. Compilers written in a high level language are easier to read, easier to debug and consequently reduce the programming time. Compiler–compiler languages that are a sub-set of *translater programs* are written in high level languages that are specifically designed for compiler writing. In application, a compiler analyses and decomposes the source program then synthesises the object program. To do this some compilers need to make several passes through the original code, thus slowing down the process.

Interpreter. An interpreter for a high level language accepts a source program and then executes it. The difference between this and a compiler is that the interpreter does not generate on object program but executes the source program itself. A pure interpreter analyses a source program statement each time it is used in order to find out how to perform the instruction.

Translator. This is a program that converts a source program into an object program. If the source language is high level and object language assembler or machine code, then the translator is known as a compiler.

Instruction set/operation code. The group of fundamental operations that each processor is capable of performing is described as its instruction set and this varies from processor to processor. Each instruction in the set is represented by a unique machine or operation code (op code).

Linker. A short program designed to link together the several parts or modules of a larger program.

Macro. A program sub-routine written in assembler language.

Meta-assembler. For processors that have been designed to optimise the assembler efficiency, such as RISC machines, part of the assembler has already been inbuilt on the chip. The assembler programs that are used with these, then follow on behind, i.e. *meta*.

Microcode/microprogram. The internal programs that are used to define the processor's instruction set are known as its microprogram and this is stored within its permanent memory as microcode. Such machines are described as being microprogrammable, as opposed to those where the programs are hardwired into the silicon of the processor.

Object-oriented programs (OOPs). Conventional programs are structured around collections of data that are processed by algorithms and procedures, producing more data. OOPs structures are very different, viewing the information more as a series of objects, each self-contained and capable of communicating with other objects via messages. For example, an object in a computer aided design (CAD) program might be 'circuit board'. This would contain data about dimensions, shape, component description, costs, etc., together with the rules that describe how each is related. A message input might be 'print PLAN view' or 'print interconnections'. Thus messages can initiate whole sequences of actions. For this reason, object-oriented programs require large memory capacity. This concept is particularly well suited to image processing.

Operating system. The fundamental programs that are supplied with the computer, to enable it to carry out the basic functions such as start up, load and run user programs, are known as the operating system.

Pseudo-code. The first step in writing a program, be it in machine code or a high level language, is to produce an algorithm. This is a series of steps that would produce a solution to the particular problem. In all except fairly trivial cases, the conversion of the algorithm into the appropriate language can be time consuming. A useful half-way stage is to convert to pseudo-code, which looks like a computer program but is only an aid in the production of the final program.

Software engineering. This concept has developed into a discipline because of the need to ensure high standards within the software section of the industry. Because of the extensive use of computing, system failures can be catastrophic. The application and adoption of formal and systematic procedures for program development, maintenance and documentation is therefore essential to ensure the integrity of both the software programs and consequently the processed data. Additionally, the conflicting requirements of producing efficient, cost-effective programs on an acceptable time scale, plus legislation regarding product liability, demand a rigorous approach to validation and documentation. Because of the large number of discrete paths through large programs, some of which can be totally unexpected, it is important to probe deeply with suitable test data using standard techniques, to achieve the high level of user confidence that is essential in communications and control applications.

Structured programs. These are constructed in a modular manner, with each module capable of being run and tested independently. Effectively these are a series of sub-programs or sub-routines linked together to form a much larger program.

6.3 Viruses

A virus is a jargon term used to describe the action of a small program illicitly entered into a computer operating system with the illegal intention of corrupting a database. The illegal code infects a system by replicating itself. The program is usually cleverly buried within the operating codes so that it is difficult to detect until the system fails. These problems spread from program to program, from disk to disk, or even across a network. It is thought that there are a few thousand such viruses in circulation, varying from the amusing but annoying, to the absolutely disastrous. The following are just two examples of these parasites:

The *Trojan horse* represents a range of programs that provide a function other than that claimed. For example, a program might be described as a data compressor, when in fact it compresses all the data in a file to zero and thus erases it.

The *logic bomb* describes a range of virus that has a delayed action which is triggered by some specific condition. For example, a program may load all zeros to corrupt a data file when the computer system is operated on Friday the 13th.

Anti-viral injection techniques are available to counter many of the viruses, but too often only after the virus has struck. Programs have been devised to detect other programs that attempt either to self-load permanently into the operating system or to replicate itself. When a viral condition is detected, the search program then freezes further computer activity. If used in conjunction with a range of password levels, this gives a good degree of security. A typical password set might include:

Level 1 to load the main operating system;
Level 2 to gain access to the working system;
Level 3 to gain access to individual programs or data-bases.

In addition, a physical system key can be used to blank out displays and to disable printers, document readers and communications ports.

6.4 Miscellaneous terms

Address. Each memory location and peripheral or terminal device must be assigned a unique identification label to allow the system controller to recognise the source or destination of the information being processed. This label is known as an address and each memory location and peripheral is connected to a number of parallel conductors known as address lines which together form the address bus.

Automatic test equipment (ATE). A concept originally devised to use a computer control system to continually monitor and log the test data from a production test environment. This has been extended to include communications system whereby a live operating system is continually monitored in real time, to report its density of activity and operational status.

Bit orientated. A communications protocol in which the control information is coded within fields that may be as short as one bit.

BRISC (bipolar or BiCMOS reduced instruction set computer). These are high speed processors based on the RISC chip technology and fabricated using either bipolar or BiCMOS semiconductor techniques. They are designed to function with clock rates in excess of 100 MHz and with operation rates in excess of 60 MIPS.

Buffer. A temporary storage device that is used to compensate for the difference in timing or rate of data flow within a system.

Bug induced errors. Such errors arise from a less than perfect understanding of the tasks and purposes for which the software is intended. These can be categorised as follows:

(a) Programs behave in a manner that is unexpected by the user but the result is not catastrophic.
(b) Similar to (a) but is unexpected by the software provider.
(c) The system fails in a catastrophic manner so that the computer fails to complete its task.
(d) A catastrophic failure that results in the complete loss of data.
(e) A catastrophic failure that results in physical danger to the persons involved in the controlled system.

Bus. A parallel data, address or control path that can be shared by many devices in a computer communications system.

Character orientated. A communications protocol in which the control information is coded within fields that have a minimum length of one character or byte.

Clock. The oscillator that generates the timing signals and pulses within a computer system.

Clock skew. The difference in arrival times of the clock synchronising pulses at different devices due to unequal signal path lengths. This causes switching errors that may be countered either by adding delays to the faster clock paths, or by transmitting the clock pulses earlier over the longer paths. The latter technique usually requires that several versions of the master clock pulses are generated with different timings.

Closed architecture. A system that is only compatible with the hardware and software of a particular manufacturer (see also Open system interconnection, p. 73).

Computer aided design (CAD). The application of a computer to solve problems of design. Because by its nature the computer is particularly efficient where repetitive tasks are involved, this concept enables the user to see the effects of various design changes very rapidly.

Computer aided engineering (CAE). The generic term for the application of computers in all aspects of engineering.

Computer aided manufacturing (CAM). The use of a computer control system in the manufacturing process.

Computer integrated manufacturing (CIM). The application of computer control to design, manufacturing, data logging and communications.

Control character. A non-printed/displayed character that is used specifically for system control.

CRISP. An acronym for complex reduced instruction set processor. This hybrid of the complex and reduced instruction set processors has been derived due to the development of a 'million transistors on a chip' concept. By using on-chip cache memory for both data and instructions, most operations are carried out in fewer machine cycles so that processing has the speed of the RISC concept. At the same time, this allows the use of the very wide range of software that is available for CISC machines to provide a more flexible system.

Daisy chain. The serial connection of devices within a network (see also Networks, p. 204).

Data communications. The generic term used to describe the collection, processing, transmission and use of data in an electronic form.

Data integrity. A measure of the accuracy and the absence of errors in the data stream.

Deadlock or deadly embrace. The jargon phrase often used in multiprocessing to describe the situation where two tasks are unable to proceed because each task holds some resource that the other requires to complete its task. Two processors A and B both require the use of resources X and Y. If A seizes X and Y before B, there is no problem as B simply waits until A releases the resources. If however A seizes X at the same time as B seizes Y, there is deadlock. A is waiting for B to release Y so that it can complete its task, but B cannot do this because it still needs X to complete its own task. The problem arises because of a hardware or software design fault.

Diagnostics. Programs or procedures used by the ATE system to test a terminal, communications link or similar system.

Distributed architecture. A communications system where the transmission medium is shared between several services.

Distributed computing. The term applied to describe the use of computing services closer to the operator, i.e.

workstations and personal computers on each desktop (see Networks, p. 204).

Distributed processing. A technique that allows separate computers to share the work on the same applications programs.

Driver. The software program that is used to control or drive an input/output device or port.

Electronic design automation (EDA). The generic title for computer aided design, engineering and manufacturing.

Highway addressable remote transducer (HART). The facility of using several remote transducers to provide input to a system but over a single bus system.

Host computer. The central computer of a communications system that provides the primary data processing, such as system management, computing, database control, etc.

Hypertext. A term coined to describe a technique for database searching. Humans communicate by speech in a serial form because we only have a single set of vocal organs. We also read and write in a serial fashion, probably due to the way in which printing was developed from speech sounds. However, human thought is not sequential. Hypertext is a software development that allows a database to be scanned in a more human manner through many parallel paths simultaneously.

Interface. The boundary between two parts of a system across which data signals have to pass. Because of physical differences between the two parts, the signals and connections for both sides of the interconnecting device, the interface unit, must be precisely specified.

Latency. The time taken for a device to respond to a request for some action.

Minimum instruction set computer (MISC). A RISC processor that typically operates with just nine instructions. The small size allows this chip to be fabricated using GaAs (gallium arsenide) technology so that it is capable of running at up to 225 MIPS. Designed as a 32-bit device, software is used to break down the microcode of other processors into the nine instructions, so that the chip can emulate other processors. The device uses a 129-bit long instruction word, the first half of which instructs the microprocessor how to handle the second half, which consists of two operands.

Multiplexer. A device that divides a data channel into several parts for independent use.

Multiplexing. The process of combining several signal channels into one composite stream or dividing a transmission medium for simultaneous use by several signals (see also Telephony and associated systems, p. 327).

Multi-tasking. The ability of a system to carry out several tasks simultaneously. The ability to run more than one applications program at once or the ability to process data at the same time as communicating with a peripheral device.

Multi-threading. Multi-tasking but applied to a single applications program. For example, reading input data while carrying out calculations or updating back-up store while data is being output to a terminal.

On-line processing. A method of processing whereby data is input from its source, processed and then directly applied as output to the point where it is to be used to control some feature of an environment.

Open system interconnection (OSI). Standardised protocols for the exchange of information between terminals, computers and devices of varying characteristics (see also Networks, p. 204).

Polling. The procedure by which the controlling computer in a system interrogates its various peripheral devices in a regular manner to test if each has a data transfer need.

Port. A point in a computer system that connects to the outside world. A point that provides either input or output facility.

Port concentrator. A multiplexer device that allows several ports to communicate with the outside world over a common transmission medium.

Protocol. A definitive set of rules that are designed to ensure accurate communications between two or more elements of a system.

Protocol convertor. A device that converts the protocol of one system into that of another to allow cross-communications between the two.

Real-time system. A computer system in which a service is available instantly and on demand.

Smart card. A plastic credit card sized device in which some simple computing facilities are embedded. Each card contains a microprocessor, some fixed or read only memory, some variable or random access memory and a portable database. Can be used as a means of payment (credit card), control of telephone (phone card) or as a simple electronic security key.

Smart system. A computer-controlled system that has the ability to learn. It therefore possesses a degree of pseudo-intelligence.

Static analysis. This is a software tool that is designed to expose any errors in programs. A network model of the program is generated to show the data flow of processing. Such a device can identify redundant coding and exposes program areas which once entered have no escape route (black holes). The process, which is fairly expensive, requires that programs should be written in a highly structured manner. However, it is useful in providing 'proof of correctness', particularly when designing systems on a chip (application specific ICs).

Statistical Multiplexer. A multiplexer that divides a common channel facility between a number of devices whose total data rate is greater than that of the channel capacity. It does this on the basis that not all devices wish to transmit simultaneously. The capacity, thus freed, is available for other users.

Telemetry. The transmission of analogue data from a remote site using some form of digital coding usually in real time.

Tempest. The jargon term used to describe the electromagnetic radiation from a computer system, particularly its visual display unit (VDU). The importance of this is not so much the interference that can be generated, but the fact that this radiation can be received, recorded and redisplayed on a second computer system in an illicit and illegal manner.

Timesharing. Computer operations that allow several terminals to use the facilities in sequence. The high speed of processing makes it appear that all are using the services simultaneously.

Transaction processing. A real-time system where the data processing of all the tasks is carried as received.

Very long instruction word (VLIW). The concatenation of several instructions into a multi-byte word. This technique is useful in multi-processor systems, as a management unit can be used to separate out the individual instructions and direct these to the appropriate processor.

Wait states. For high speed data processing a high clock rate is necessary and fast random access memories (RAM) tend to be expensive. In cases where much of the processing will not involve memory read/write operations, less costly slower RAMs may be used with a high clock rate system. When RAM accesses are needed, the processor is then made to idle in a *wait state* for several clock cycles to give the memory time to respond. A *no wait state* memory is thus one that has an access time compatible with the memory address clock cycle time.

6.5 Useful references

Milne A.C. (1987) *Interfacing Standards for Computers*, I.E.E.I.E. (London) Monograph.

Price Waterhouse (1990) *The Complete Virus Handbook*, Pitmans, London.

Woolfe R. (1980) *Videotex. The New Television/Telephone Information Services*, Heyden and Sons Ltd, London.

7 Digital communication systems

Wideband communications applications have traditionally used analogue signal processing, primarily for reasons of bandwidth conservation and the fact that the technology is matured and well understood. However, each application tends to be unique in certain ways. When such systems are concentrated into integrated circuits (ICs) these devices become specialised, relatively few are made, and so their costs are higher than they would be if mass produced. When the analogue signals are converted into digital form for processing and then back into analogue again for output, the only dedicated ICs are those associated with the interfaces between the two types of signal. The digital signal processing (DSP) region of the system then uses standard digital components that are mass produced so that the system becomes more cost-effective.

The increased transmission bandwidth required for digital signals may be available over certain types of link, when this penalty can be offset by the considerable advantages of digital processing. The system becomes more flexible, different systems can be integrated and computer control introduced. Such a technique leads to the concept of an *integrated services digital network* (ISDN) where each service can be accommodated with equal performance. The principal benefits of digital processing can be summarised as follows:

(1) It is more appropriate for linking devices that operate in the digital mode.
(2) It can provide a significantly higher transmission speed than can usually be achieved with analogue processing.
(3) It provides for improved transmission quality in noisy environments. The noise component can be reduced using signal regenerators and error detection/correction techniques.
(4) It is more compatible with the digital switching techniques used to control distribution and is a natural technique to use for systems involving an optical fibre link.

(5) Encryption/decryption can easily be adopted for data security.
(6) Where necessary, signal compression/bit rate reduction techniques can be employed to minimise the bandwidth requirement.
(7) For many applications, time division multiple access (TDMA) can be used more effectively than frequency division multiple access (FDMA) that is common for analogue transmission systems (see also Networks, p. 204 and Satellite systems, p. 267).
(8) For systems involving reception and retransmission, signal regeneration can be used at the intermediate stage to improve signal quality.

Figure 7.1 (a) and (b) clearly show the noise advantage that digital signals possess. By a process of slicing or sampling, the noise component can be completely removed. Figure 7.1(c) compares the behaviour of analogue and digital signals under noisy conditions. As the input signal to noise ratio degrades, the analogue signal quality may be said to degrade gracefully, while the digital signal is relatively unaffected until it suddenly crashes.

7.1 Basic digital processes

Figure 7.2 shows the principle of converting an analogue signal into a digital format, the process being described as *sampling and quantisation*. The analogue signal is evaluated rapidly (sampled) at very precise intervals of time, to measure its amplitudes. As only discrete values are applicable in a digital system, each of the sampled values is allocated the integer value that represents each level (quantised). The waveform shown in Fig. 7.2 would thus be represented by the sequence, 101,110,110,110,110,101,100,010,001,001,001,010. . . . It will be noted that each of the eight levels can be coded with just three bits. The general rule for binary coding is $M = 2^n$, where M is the number of discrete levels and n is the number of bits per sample.

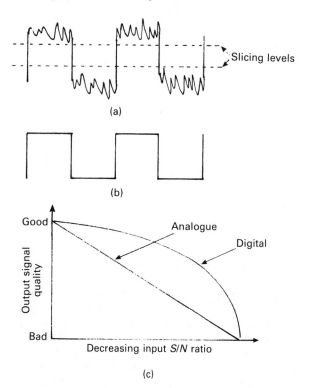

(a)

(b)

(c)

Figure 7.1   Digital signals in noise environments: (a), (b) regeneration of noisy digital signal; (c) comparison of signal behaviour in the presence of noise.

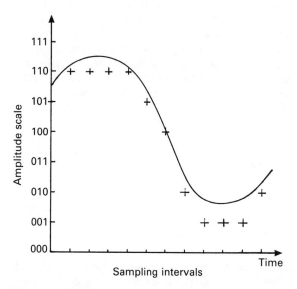

Figure 7.2   Sampling and quantising an analogue signal.

It can be shown by Nyquist's theorem (see also Signals, p. 300) that provided a complex analogue signal is sampled at a rate at least twice that of its highest frequency component, then the original signal can be recovered from these samples without error. In practice

any signal that is reconstructed from such samples will suffer from the approximations that have been made during the quantisation process. Such errors produce *quantisation noise*. There are two ways of quoting the signal-to-quantisation-noise ratio (SQNR) depending on the nature of the analogue signal. If this is a dc signal such as that found in video, then the peak signal to RMS noise ratio of $(10.8 + 6n)$ dB is applicable. In the more general case of ac signals, the value quoted is that of the RMS signal to RMS noise which is calculated from $(1.76 + 6n)$ dB. In each case n is the number of bits per sample. Reference to Fig. 7.2 shows that quantisation noise can be reduced to any low level desired simply by increasing the sampling rate and/or the number of allowable levels, the cost of this being an increase in the bandwidth of the digital signal which can be calculated from $2nf_m$, where f_m is the maximum component frequency and n is the number of bits per sample.

The process of sampling produces a frequency spectrum similar to amplitude modulation (see also Modulation, p. 186), but with an infinite range of harmonics as represented by Fig. 7.3. The receiver demodulator circuit must include a low-pass filter to separate the baseband component from the harmonics. If the sampling frequency is not high enough or the filter cut-off not sharp enough, interference from the first lower sideband will result. This effect, known as *aliasing*, due to use of a sampling frequency below the Nyquist frequency, is shown in Fig. 7.3.

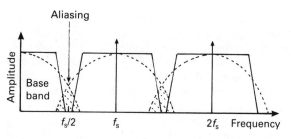

Figure 7.3   Spectrum of sampled signal and aliasing.

Quantising ac signals

The quantising process described above is satisfactory for signals, such as video, that have a large dc component. For ac signals such as audio, an alternative approach is needed. One method involves adding a constant to each sampled value, using the *offset-binary* technique. But in certain cases, such as with audio mixers, where signals from different sources have to be added together, the sum can overflow or exceed allowable peak values. The commonly adopted solution involves the use of the two's complement way of representing a binary number. By convention a leading zero indicates that the remaining code represents a positive number, while a leading one signifies a negative quantity. The two's complement is formed by inverting each bit in turn and then adding 1 to the value. Thus the two's complement of $01010101 = 10101010 + 1 = 10101011$. An alternative short cut involves copying each bit in turn from the right-hand side up to and including the first 1. Then invert all the other bits in turn. When recovering the analogue signal from a two's complement sample, the excess 1 should be removed before inversion. However, in practice, the failure to do so causes such a small error that it is often neglected.

Distortion. Figure 7.4 shows a single rectangular pulse of amplitude V and time duration t, together with its transformation into the frequency domain. The pulse is defined by

$$f(t) = V \text{ for } |t| < t/2, \text{ and}$$

$$f(t) = 0 \text{ for } |t| > t/2$$

while the transformation is given by $F(\omega) = (Vt \sin(\omega t/2))/\omega t/2)$, where $wt = 2\pi f$ the angular velocity (see also Signals, Fourier transform, p. 304). When such a signal is passed through a practical transmission system the pulse will spread in time, giving rise to *pulse dispersion*. This effect is shown in Fig. 7.5 for a serial bit stream where the overlapping of the pulse edges gives rise to intersymbol interference (ISI), which causes decoding errors at a receiver.

Analogue to digital conversion processes (A/D)

Algorithmic A/D convertor. An N-bit convertor consists of N cells, each constructed from a current mirror amplifier and a comparator on a single integrated

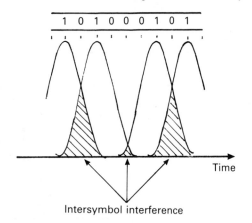

Figure 7.5 Intersymbol interference in a digital bit stream.

circuit. The device is thus current driven. Two inputs have to be provided: the analogue input to be quantised I_{in} and a current reference I_{ref}. Each cell provides a digital and an analogue output. The amplifiers double the analogue input to $2I_{in}$ and this is compared with I_{ref}. If $2I_{in} > I_{ref}$ the digital output goes to logic 1 and the analogue output becomes $2I_{in} - I_{ref}$. This difference is applied as the analogue input to the next cell. If $2I_{in} < I_{ref}$, the digital output goes to logic 0 and $2I_{in}$ becomes the analogue input to the next cell. The digital output from the converter is provided by the logic outputs of the sequence of cells. The technique is very similar to successive approximation, but because the circuits contain no capacitors, operational amplifiers or control logic, these devices are fast, very small and ideal for IC construction.

Counting or voltage to time A/D convertor (Fig. 7.6) The clock circuit provides a source of precisely timed pulses that are counted in a digital fashion, the count being set to zero by a clear or start signal. The binary count provides the digital output and is also used to provide a comparison signal after being converted into analogue form. As long as this value is below the level of the analogue input, the AND gate is enabled so that the count can accumulate. As soon as the comparator inverting input exceeds the analogue input, its output goes negative, the AND gate is disabled and the count ceases. The value held in the binary counter can then be output as the digital representation of the input analogue signal.

Dual slope or ratiometric A/D convertor. In the method depicted in Fig. 7.7, the sequence starts with the analogue input signal applied via the switch to an integrator. As soon as the integrator output starts to rise, the gate opens and clock pulses are passed to the counter, which has been preset to some value and now

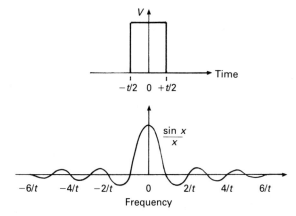

Figure 7.4 A rectangular pulse and its spectrum.

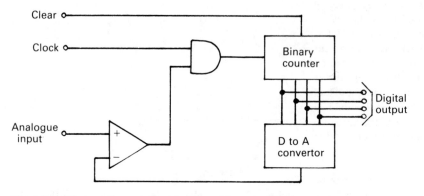

Figure 7.6   The counting or voltage to time analogue to digital convertor.

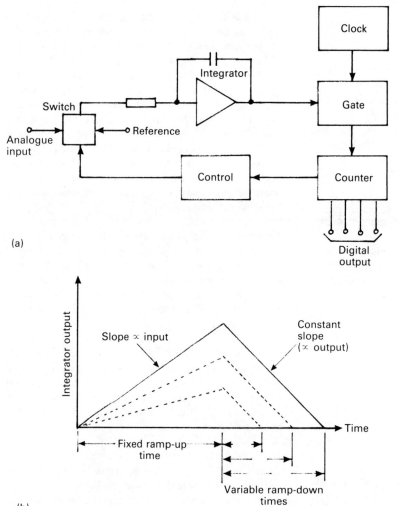

(a)

(b)

Figure 7.7   (a) Dual slope analogue to digital convertor; (b) characteristic of dual slope analogue to digital convertor.

starts to count down. When the count reaches zero, a signal is passed to control and the integrator input now switches to a reference voltage of opposite polarity. This causes the capacitor to discharge, but at a constant current, the counter counting upwards during this period. The time taken to discharge the capacitor is proportional to the level of input voltage originally applied. Thus the count at this point provides a digital equivalent of the input signal. From Fig. 7.7(b), it will be seen that the *ramp-up* period is fixed so that the accumulated charge is proportional to the level of input signal. As the discharge period occurs at a constant current, the *ramp-down* time must be proportional to the charge on the capacitor and hence the level of the input signal. Clock frequency drift is relatively unimportant in this system, as the voltage levels are dependent more upon the number clock pulse than their rate.

Flash convertor. This aptly named very fast technique is shown in Fig. 7.8. The analogue input is applied simultaneously to a number of parallel comparators. For M levels this system requires $M - 1$ comparators and b bits to encode the digital output ($M = 2^b$). The second input to each comparator is obtained via a potential divider network so that appropriate multiples of $1/M$th of the reference voltage are obtained. The parallel outputs from the comparators are then encoded to provide the digital equivalent of the input signal. The half-flash conversion process provides a convenient way to double the number of output bits. The input is first quantised to provide the eight most significant bits (MSB) and this is then converted back into analogue form to be subtracted from the original signal. When this difference is converted into digital form, it represents the eight least significant bits (LSB).

Single slope or voltage to frequency A/D convertor. In this system shown in Fig. 7.9, clock pulses pass through the gate circuit to trigger a pulse generator, the pulse widths being defined by these pulses and their amplitudes controlled by the reference voltage. Thus, when filtered, these pulses provide a dc input to the comparator that is proportional to the amplitude of the reference. The analogue input is compared with this level to provide a drive for the integrator and as this output rises the gate opens to allow the pulse generator to be triggered by the clock. At the appropriate times the pulse generator also resets the integrator. The feedback system balances the number of generated pulses against the level of input signal. When these pulses are digitally counted, they represent the digital conversion of the input signal. The circuit is also self-compensating for clock frequency drift. If the clock frequency rises, the comparator input will tend to fall so that the count is

performed in a shorter period of time. The number of pulses being counted are thus unchanged.

Successive approximation A/D convertor. In this system, which is shown in Fig. 7.10, a digital divider, the D/A convertor, presents accurately known increments of the reference level to one input of the comparator, these increments being offered in a fixed sequence, largest first and decreasing in binary proportion. If any step is offered and it produces a running total that is larger than the analogue input, this value is rejected and the next one tried. The successive approximation register thus accumulates a total in binary form that represents the analogue input. As the sequence ends, the counter stage commands the register to output its contents. A start pulse is used to start conversion and a delay circuit is used to provide automatic recounting, these pulses being used to open the clock gate to provide the necessary drive to the counter circuits. Often the successive approximation register outputs provide the digital signal in serial form and some indication of status, i.e. counting is in process.

Companding (see also Noise, p. 223)

(see also Noise, p. 223)

Reference to Fig. 7.2 shows that linear quantisation produces proportionally more error on smaller signals. Additionally, large amplitude signals are better able to mask the effects of noise. This imbalance can be improved by using a nonlinear form of quantisation that acts as level compression during encoding and level expansion during decoding, the technique being referred to as *companding*. Figure 7.11 shows this concept applied to an overall transmission system where the degrees of compression and expansion are complementary. In both theory and practice, a logarithmic companding law would be suitable and fairly easy to produce. However a true, logarithmic function has a discontinuity at zero and therefore cannot be used for companding. In practical circuits, a modified transfer characteristic with the form $\log(1 + x)$ is used. Two common companding schemes are in use: the A-law devised by the European CCITT (Comité Consultatif Internationale Télégraphie et Téléphonie) and μ-law, devised by Bell as follows:

$$y = 0.18(1 + \ln(A \mid x \mid))\mathrm{Sgn}(x) \qquad \text{for } 1/A \le x \le 1$$

$$y = 0.18(A \mid x \mid)\mathrm{Sgn}(x) \qquad \text{for } 0 \ge x \le 1/A$$

where $A = 87.6$ and

$$y = \ln(1 + \mu \mid x \mid)\mathrm{Sgn}(x) \qquad \text{where } \mu = 255$$

Thus both laws cater fully for ac signals.

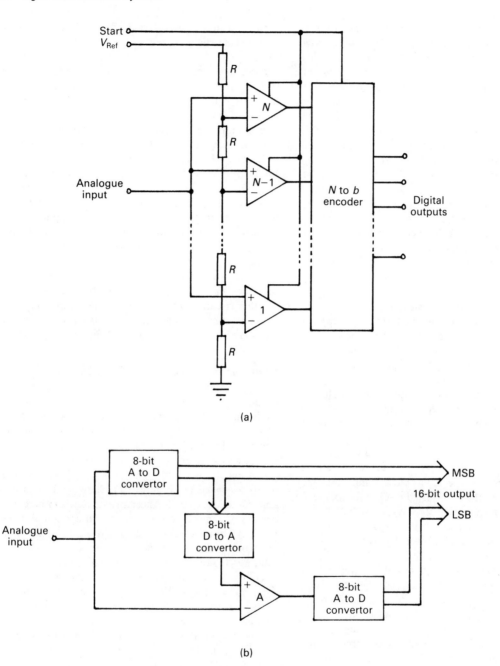

(a)

(b)

Figure 7.8   (a) A '*b*-bit' flash convertor; (b) a half-flash convertor.

The system shown in Fig. 7.11 utilises the same type of companding D/A convertor at each end of the chain. At the encoding stage, the analogue signal is fed to a comparator which uses a compressed version of it as a reference. The serial output data therefore represents a digitally compressed version of the input. At the decoder, the received signal is first reconverted into parallel format before being expanded.

Digital to analogue conversion (D/A)

Binary weighted resistor network. A series of parallel transmission gates provide inputs to an inverting operational amplifier (Op Amp) that functions as a summing amplifier. Reference to Fig. 7.12 shows that these gates switch the inputs to either earth potential or to a stable negative reference voltage, $-V_{Ref}$. A binary

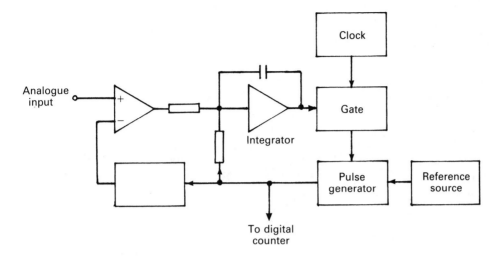

Figure 7.9 A single slope or voltage to frequency analogue to digital convertor.

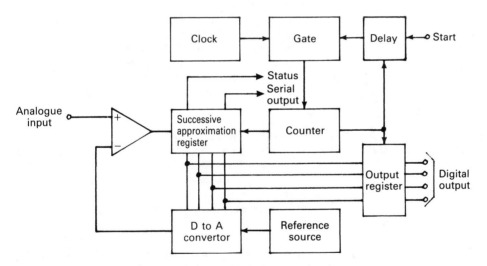

Figure 7.10 Analogue to digital conversion by successive approximation.

'0' on any data line connects that input to 0 volts, while a '1' sets the input to $-V$ volts. The resistor network is chosen in binary proportions as shown, so that the gain of the Op Amp can vary as the series $1/2$, $1/4$, $1/8$, $1/16$, etc. The negative reference is chosen so that the analogue output V_{out} is positive. Thus the analogue output signal is:

$$V_{out} = V_{Ref}(a_1/2 + a_2/4 + a_3/8 + \ldots)$$

where a_1, a_2, etc., are set to 1 or 0 according to the bits in the input binary word. A low-pass filter is required to remove the quantisation steps that are present at the Op Amp output. The chief disadvantages of this system lie in the resistor values, which must be very accurate, stable and track each other as the operating temperature varies. Also if there are a large number of bits in each binary word, some resistor values become very large and this adds to the stability problem as well as tending to slow the rate of conversion.

Multiplying D/A convertors. From the above it will be seen that the D/A convertor output is the product of the reference voltage and the binary word and always less than the reference voltage. If the fixed reference is replaced with an analogue signal, the device becomes a *multiplying* D/A convertor or *programmable attenuator*, with the degree of attenuation being controlled by the digital word. This concept of using one dc and one ac voltage is referred to as a two-quadrant multiplier. The

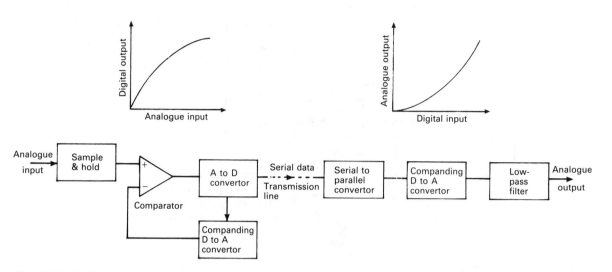

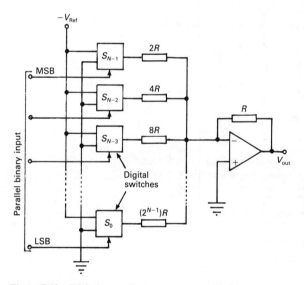

Figure 7.11   Nonlinear companding system and its characteristics.

Figure 7.12   Digital to analogue convertor using binary weighted resistors.

combination of a bipolar binary signal with an ac reference forms a four-quadrant multiplier.

R/2R ladder network D/A convertor. Each leg of the ladder network shown in Fig. 7.13 can be connected either to ground or the reference voltage, by the transmission gates which are controlled by each bit of the input binary word. The network is a current splitting device that due to the $R/2R$ ratio of the resistors produces a sequence of binary ratios. Analysis of the circuit shows that a resistance of $2R$ is seen from any node, looking left, right or towards the switches. At

node N the Op Amp gain $= -3/2$; voltage due to V_{Ref} at $N = -V_{Ref}/3$, so that $V_{Out} = -V_{Ref}/3 \times -3/2 = V_{Ref}/2$. Similarly at node $N-1$, $V_{Out} = V_{Ref}/4$, etc., again producing the ratio, 1/2, 1/4, 1/8, etc., with the Op Amp acting as a summing amplifier. Temperature stability with this network is easily achieved, as only two resistor values in a 2:1 ratio are needed for accuracy.

Counting and dividing

The bistable or flip-flop circuit forms the basis of all counters. Since this circuit is capable of being set to 1 or reset to 0, it forms a 1-bit memory cell. A series of such cells are commonly formed on an integrated circuit to form an $n \times 1$ array (see also Shift register, p. 83). For counter applications the cells are formed on the IC in a serial chain, with the output of one providing the input to the next. The data is then transferred from cell to cell on the action of a clock pulse. As every cell goes through one cycle of transitions every two input pulses, the device acts in a divide-by-two fashion. Apart from the direct action of counting or dividing/scaling electrical pulses, the electronic counter finds many other applications. With suitable interfaces and transducers, the device can be used indirectly for measuring such parameters as frequency, time, distance, velocity, acceleration, etc.

Asynchronous counters. In these counters, only the first flip-flop in the chain is clocked and synchronised to the pulse stream. The rest of the chain is driven by the output from the previous stage.

Divide by 2N−1. While a series of N flip-flop circuits will divide by $2N$, by combining a feedback loop with an

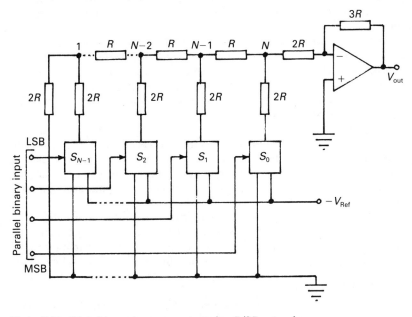

Figure 7.13 Digital to analogue convertor using $R/2R$ networks.

exclusive-OR (Ex–OR) gate, as shown in Fig. 7.14, division can be extended to cover all integer ratios.

Ring counters. A ring counter is formed by taking the output from the last flip-flop and feeding it into the input of the first. Such a construction is often referred to as a *circulating memory* because if all cells except one are set to 0 (the exception is thus set to 1), on a series of clock pulses the 1 will circulate around the chain. For an N-stage counter, the 1 will complete one cycle every N pulses. As a counter, the count is taken by noting which cell is set to 1.

Ripple counter. As an N-stage ring counter can divide only by N it does not make efficient use of flip-flop circuits. N devices can define 2^N different states, and thus can, if correctly coupled, divide by 2^N. The flip-flops are again connected in series but the parallel outputs of each now form a binary word that represents the count. The device is so named because the bits of the count ripple through the cells.

Twisted ring counter (also known as Moebius or Johnson counter). By inverting the output state of the feedback pulse in a ring counter, the circuit will now divide by $2N$. Thus five flip-flops can provide a decade counter.

Synchronous counters. Asynchronous counters are relatively slow due to the *carry propagation delay*. This is defined as the time taken for the counter to complete

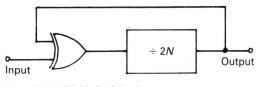

Figure 7.14 Divide by $2N - 1$.

its response to an input pulse. The more cells that have to change state simultaneously, then the greater this lag becomes. A synchronous counter is formed by arranging for all the cell clock inputs to be connected in parallel; thus all cells change state in synchronism with the clock signal.

Up–down counter. By inverting the signal feed polarity between each stage of a counter, the count can be made to run backwards from some preset value towards zero. Thus by arranging some switching logic between stages, the counter can be controlled to work in either direction.

Registers

Registers, which are constructed from a series of bistable or flip-flip circuits, form $n \times 1$ arrays of single-word memory cells. Because they can be configured in several ways, the data can be shifted in or out in a serial or parallel form, thus allowing these simple devices to perform many valuable digital processing functions. The general arrangement is shown in Fig. 7.15(a).

Barrel shifters. These consist of a series of parallel shift registers, arranged to transfer data in a parallel format and in a pipeline-like manner, as indicated in Fig. 7.15(b). Each data transfer may be direct to the corresponding bit cell in the next register or may be shifted to the left or right during this process. As each bit position represents a binary weighting, a shift of 1 bit to left or right is equivalent to multiplication or division by 2, respectively. When five parallel registers are used these are arranged either to load the next register direct or to provide a left or right shift of 1, 2, 3 or 4 bits. Each transfer can represent a scaling factor (multiplication or division) of 2^0, 2^1, 2^2, 2^3 or 2^4. Then by the use of appropriate control signals to obtain direct or shift transfers, scaling factors of all integer values from 2^0 to 2^{10} can be achieved.

Parallel in–parallel out shift register (PIPO). Shifting a digital signal through such a register provides a 1-bit delay for the whole byte. During this period the byte may be tested or compared to check for some feature (see also Error control, p. 104).

Parallel in–serial out shift register (PISO). This performs the task of converting a parallel or spatial code into a serial or temporal one, i.e. a data format changer.

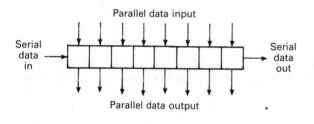

(a)

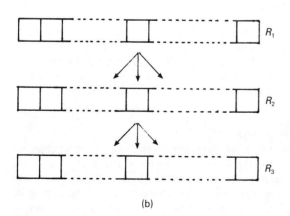

(b)

Figure 7.15 Shift registers: (a) general-purpose register; (b) barrel shifter.

Serial in–parallel out shift register (SIPO). This performs the inverse format conversion of the above.

Serial in–serial out shift register (SISO). The registers perform a delaying function allowing the data to be loaded at one code rate and unloaded at another. Provided that there is a time gap between bytes, then the output rate can be either faster or slower than that during input.

Sequence generators. There are two basic ways of using shift registers to generate specific binary bit sequences. Type 1 uses simple shift register techniques, while type 2 utilises a complex feedback arrangement.

Type 1. A shift register that has its output returned to the input is described as a *re-entrant shift register*. If such a device is loaded with a particular bit pattern, then this will continually cycle with each clock pulse. When required, the sequence can easily be changed.

Type 2. This is often described as a *linear feedback shift register* or *pseudo-random binary sequence* (**PRBS**) generator. For a series of binary bits to be in random order, each symbol must occur by chance and not be dependent upon any previous bit. Over a long period of time, the number of occurrences of ones (n_1) and zeros (n_0) should be equal. In addition, runs of two, three or more similar bits should be equiprobable. Sequences that very nearly meet this criterion can be generated by the arrangement shown in Fig. 7.16, where the logic states of the switches control the feedback paths through modulo-2 adders (exclusive OR gates), placed between the serial shift register input and output. The state of the switches are set according to a *characteristic polynomial*, to close just one of the switches S_1 to S_{n-1}. Assuming a 4-bit register with the switches set, $S_1 = 1$ and the other set to 0, then, irrespective of the initial states of the

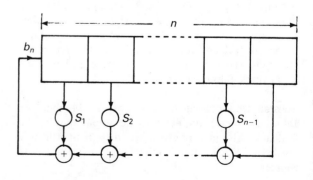

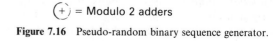 = Modulo 2 adders

Figure 7.16 Pseudo-random binary sequence generator.

Table 7.1 *Register binary pattern*

State	r_1	r_2	r_3	r_4	r_b
1	1	1	1	1	0
2	0	1	1	1	1
3	1	0	1	1	0
4	0	1	0	1	1
5	1	0	1	0	1
6	1	1	0	1	0
7	0	1	1	0	0
8	0	0	1	1	1
9	1	0	0	1	0
10	0	1	0	0	0
11	0	0	1	0	0
12	0	0	0	1	1
13	1	0	0	0	1
14	1	1	0	0	1
15	1	1	1	0	1

register cells, the binary pattern shown in Table 7.1 will be produced. The bit pattern b_n is the required sequence which is generated on a periodic and cyclic basis. Changing the characteristic polynomial will produce a new sequence.

There are 2^n possible states for the shift register cells, but the all-zero combination is not valid as this would cause the generator to stop. Therefore the length of a sequence is $2^n - 1$, the period of repetition being independent of the initial conditions.

It can be shown for these sequences that $n_1 = 2^{n-1}$ and $n_0 = 2^{n-1} - 1$, thus if n is large these sequences have near-random properties. A maximal length sequence is usually described as an *m*-sequence and finds many uses in communications systems.

Because of the pseudo-random properties, PRBSs have a number of uses:

(1) as a repeatable noise source for testing digital systems;
(2) to add redundancy to a transmitted data stream, by coding logic 1 as the *m*-sequence and logic 0 as its inverse;

(3) added to a data stream, the *m*-sequence can act as a key in a controlled access system (see also Encryption/decryption, p. 100).

Universal shift register. By adding suitable control logic to the basic shift register, it is possible, using a 2-bit code, to be able to make the shift register function in any one of the four configurations.

7.2 Digital signal processing (DSP)

Although the concept of processing signals in a digital fashion is old, the term *digital signal processing* has relatively new connotations. DSP implies the use of computing power in the use, transmission and management of communications. The basic principles are shown in Fig. 7.17, where the analogue signal is first low-pass filtered using a filter with a cut-off frequency equal to half the sampling frequency (to ensure that there are no components present that will cause aliasing), and then converted into a digital format. At this stage the signal can be processed, transmitted over a network or used in a control environment, where the advantages of noise reduction, error correction, etc., can lead to improved signal quality. Thus most of the temperature drift and noise problems associated with analogue signals can be practically eliminated by using digital processing. At a suitable stage, the digital signal can then be converted back into an analogue form.

Very large scale integrated (VLSI) circuits are available to perform these functions and these are often described as DSPs. These are basically very sophisticated microprocessors whose design has been optimised for signal processing. DSPs have a very high sampling rate and can execute complex processing algorithms in real time. Speeds as high as 100 MFlops can be achieved, at relatively low cost and with low power consumption.

DSPs are designed around an enhanced Harvard architecture, that includes on-chip random access memory (RAM) and read only memory (ROM), and

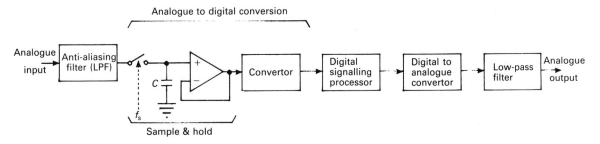

Figure 7.17 Principle of digital signal processing.

use multiplying accumulators that are capable of adding and multiplying simultaneously in one instruction cycle.

The applications for DSP range from missile guidance and radar, through automobile engine control, telecommunications, speech recognition and synthesis, to intelligent toys. Some operations, such as video image and graphics processing, would not be possible in the analogue domain.

7.3 Echo cancelling/suppression

Echoes are produced when a signal arrives at a receiver after travelling over propagation paths of different distances, or energy is induced from one line into another. For digital full duplex operation over a normal two-wire telephone circuit, it is possible for the transmitting modem (modulator/demodulator, see also Networks, p. 204) to receive echoes of its own signals down the return line due to inductive effects or mismatching. Echo cancelling techniques have been developed to minimise this effect by comparing the received signals with the last ones transmitted. When echoes are detected, a replica of the echo is produced and a signal is generated to add in anti-phase to cancel the echo, thus ensuring that only the signals arriving from the distant modem are processed.

7.4 Integrated services digital network (ISDN) (see also Telephony and associated systems, p. 327)

An ISDN is a communications network that has evolved because of the problems associated with analogue telephony systems. The concept recognises the considerable advantages that can be gained by changing to a system that will allow the end-to-end transfer of information in a digital manner. Once such a transition has been made, the advantages gained include:

(1) Greater reliability due to the use of digital integrated circuits.
(2) Faster access speed due to the introduction of *dual tone multi frequency* (DTMF) dialling instead of pulse dialling.
(3) Allows for computer access to the system which gives rise to the concept of *computer-integrated telephony* (CIT), where the power of the computer can be combined with the communications power of the ISDN.
(4) By introducing CIT, many new services such as video, data and group 4 high speed facsimile can be introduced into the telephony system in a truly integrated manner.

For reasons of economy, the service has to be compatible with the current analogue systems and introduced in an evolutionary manner. Thus voice signals are processed in analogue to digital and digital to analogue converters to be carried over the new system by *pulse code modulation* (PCM) (see also Digital pulse code modulation, p. 91).

7.4.1 ISDN access

Two classes of service are offered. A basic rate access (BRA) which most subscribers will use, known as 2B + D, and consisting of two 64 Kbit/s voice and data channels, plus one 16 Kbit/s digital signalling channel making a total of 144 Kbit/s. The former is circuit switched, while the latter is packet switched. The second, more costly service is known as primary rate access (PRA) and consists of 30B + D channel groupings in Europe (CCITT standard, 32 channel multiplex) and 23B + D (Bell standard, 24 channel multiplex) in North America, giving total signalling rates of 2.048 Mbit/s (E1/T2) and 1.544 Mbit/s (T1), respectively (see also Pulse code modulation frame structures, p. 91). Thus the D channel forms a *common signalling channel* using the CCITT Common Channel Signalling System C7. The domestic and small business users will use BRA, while large users with Private Automatic Branch Exchanges (PABX) will use PRA.

7.4.2 ISDN D channel

The D channel is used in the following manner with the calling party transmitting the called number plus further information. If just a voice connection between two digital telephones is required, then only one 64 Kbit/s B channel will be needed. However, if voice plus data is required, then two B channels must be engaged from end to end. The calling party thus negotiates the necessary channel capacity using time sharing of the D channel. When not needed for call routing control, the D channel can be used for packet switched data or network monitoring. The D channel signalling protocol uses LAP D (line access procedure D) which consists of layers 1 to 3 of the Open Systems Interconnections (OSI) reference model (see also Networks, p. 214).

7.4.3 ISDN D channel packets

D channel access to the packet switched network is obtained via a *packet assembler/dissembler* (PAD) which is positioned between the ISDN and the packet switched public data network (PSPDN). For the packet switched

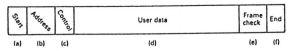

Figure 7.18 Typical packet structure.

system, the data is organised into blocks or packets in the manner shown in Fig. 7.18, where:

(a) Start flag indicates the beginning of the packet and can contain a *preamble* sequence of bits designed to aid synchronisation.
(b) Each packet must contain the address of the originator and the destination.
(c) Control section holds the packet sequence number that is needed to reassemble the message in the correct order if it exceeds one packet in length.
(d) The data bits.
(e) A frame check can be used to provide error protection and correction (see also Error control, p. 104).
(f) An end flag to signify the last bit in the packet.

When the user has obtained access to the network using the appropriate indentification, he specifies a destination address and then inputs the message. The PAD breaks this up into suitable blocks to form the packets and automatically adds the controlling information before transmission to the network. Should the receiving end detect an error, then a *repeat transmission* request will be generated.

7.4.4 ISDN system and interfaces

Figure 7.19 illustrates the digital data flow through the system. The subscriber's end terminates in two types of terminal equipment (TE): TE1 equipment that is ISDN compatible and TE2 devices that are not. These latter are connected to the system through terminal adaptors (TA) that provide bit rate adaptation plus the functions of a network terminator 2 (NT2). The two types of network terminators have the following functions:

NT1 provides the physical and electromagnetic termination;
NT2 manages the protocol functions, switching concentration, maintenance, etc. Each NT2 can support up to eight TEs.

Signals are then routed through the ISDN via Exchange terminators (ET), switch points and the packet handler (PAD) to the ISDN access unit ports (AU) of the packet switched public data network (PSPDN).

R interface. This provides system matching for non-ISDN compatible terminal equipments.

S interface. This is a four-wire unit that provides the bidirectional basic rate service for up to eight items of terminal equipment. Each B channel can be used by only one TE at a time, while the D channel is shared by all TEs on a packet-by-packet basis. This interface also allows for ISDN testing.

T interface. This interface ensures that all the input signals to the NT1 terminators have the same standard.

U interface. There is still some flexibility in the specification for this interface which is basically a two-wire connection between the subscriber's network terminator and the exchange. It provides a basic rate access of 2B + D, two 64 Kbit/s plus 16 Kbit/s, for up to 10 km. The specified line codes are 2B1Q or 4B3T (see also Codes and coding formats, p. 57) to reduce the bit rate by 50% or 25%, respectively. This interface allows system parametric measurements to be made.

7.5 Miscellaneous terms

Bit-oriented protocol. Describes one method of transmitting serial data in a packet format. This includes such protocols as HDLC and its derivatives, SDLC, LAPB and LAPD.

Bit stuffing (see also Justification and rate adapter, p. 89). The technique of adding extra bits to a data stream in order to make small adjustments to the bit rate.

Broadband-ISDN. An ISDN service that provides for videoconferencing, facsimile, community antenna TV, videophone, etc. The system bit rate is commonly accepted to exceed 140 Mbit/s.

Character-oriented. Describes the protocol in which control information is coded into character length fields.

Circuit switching or network. A technique where the physical circuits are completed from end to end via switches.

Cluster. A group of terminal devices connected to a channel at a single point or node.

Concentrator (see also Statistical multiplexer, p. 90). A device that can divide a data channel into several other channels to share its facilities usually at a slower data rate. Channel space is allocated dynamically between the

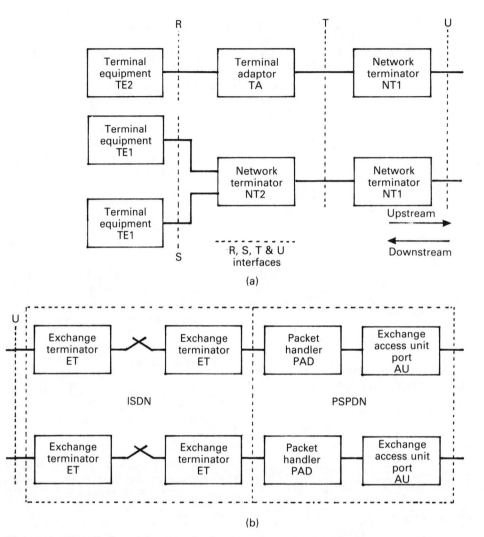

Figure 7.19   ISDN implementation: (a) subscriber's premises equipment; (b) exchange equipment.

users according to demand to increase the throughput of data on the original channel.

Data communications equipment (DCE). A device such as a modem that is used to establish, maintain and terminate a data transmission.

Data terminating equipment (DTE). The equipment connected to a channel that acts as a data source, data sink, or both. For example, a keyboard or printer, respectively.

Digital access signalling system (DASS). A message-based signalling system that is in accord with the Open Systems Interconnect (OSI) standard.

Digital filling (see Bit stuffing, p. 87).

Digital private network signalling system (DPNSS). A message-based signalling system devised to allow manufacturers of PABXs to implement varying levels of supplementary services in a compatible manner.

Downstream. That part of a system that is closer to the user than some other section, e.g. the telephone unit is downstream from the exchange.

Dual tone multi-frequency dialling (DTMF). A dialling technique that uses push-buttons and cross-point switches to select two frequencies, one below 1 kHz and one above, according to the matrix of Table 7.2.

Electronic mail (E-mail). A service between subscribers to provide for the exchange of messages via a public or private data communications system.

Table 7.2 *DTMF frequency matrix (all frequencies in Hz)*

	1209	1336	1477
697	1	2	3
770	4	5	6
852	7	8	9
941	*	0	#

Field. A group of bits used to describe a specific characteristic of part of a data packet, e.g. the address field.

Fractional E1/T1 services. These services have been made available in order to provide for on-demand, flexible bandwidth operations. In Europe, this is provided by blocks of $n \times 64$ Kbit/s and in the USA by $n \times 56$ Kbit/s in the E1 (2.048 Mbit/s) and T1 (1.544 Mbit/s) basic channels, respectively. These blocks of bandwidth are available for wide area point-to-point, or point-to-multipoint digital communications services. The basic 30 channel (E1) or 24 channel (T1) circuits are thus divided into multiples of 64 or 56 Kbit/s, so that the assigned data service uses only as many of the available time slots as necessary. This arrangement allows users to tailor their bandwidth and hence operating costs to meet a specific need. E1 to T1 inter-working, and vice versa, can be accommodated as with the conventional PCM approach, making the interface between the two rates transparent to the user.

Frame structure (see Digital pulse code modulation, p. 91)

Full duplex channel. A channel that is capable of transmitting data in both directions simultaneously.

H-Channels. Apart from the provision of contiguous blocks of $p \times 64$ Kbit/s for general broadband use, the Consultative Committee for International Telephony and Telegraphy (CCITT) has defined a set of wide bandwidth channels suitable for use with videoconferencing codecs and other similar systems. The spectrum for these will be allocated in blocks of 384 Kbit/s, 1.920 Mbit/s or higher if necessary.

High level data link control (HDLC). An international standard protocol of frame structures designed for data communications (see also SDLC, p. 90).

Homomorphic signal processing. Mathematically a morphism is the mapping of a function from one domain on to another via an operation. That is $f(a_1) * f(a_2) = f(a_1 o a_2)$, for all a and where $*$ and o are two different operators. In the field of DSP, this is processing approached either through multiplication in the frequency domain or convolution in the time domain to achieve the same effect.

House-keeping bit. An additional digit that is inserted into a data stream to enable the equipment to perform some particular function within the system.

Integrated Services Private Automatic Branch Exchange (ISPABX). A customer-based element of the ISDN that reflects the functions of the public ISDN and is capable of handling a range of signalling protocols.

Justification (see Bit stuffing, p. 87 and Pulse code modulation, p. 91).

Leased, dedication or private line. A line reserved for the exclusive use of the lessee, without interexchange switching nodes. The leased line may be point-to-point or point-to-multipoint.

Line access procedure B (LAPB). A variation of the HDLC protocol used with packet switched networks. It provides the procedure for the establishment, maintenance and termination of communications between data terminal equipment and data communications equipment.

Line access procedure D (LAPD). A bit-oriented packet protocol used for D channel transmissions. It is a variation of LAPB that supports multiple simultaneous connections.

Line concentrator. A multiplexing switching stage in a local exchange. Incoming data from a large number of subscribers is concentrated into fewer lines at the output.

Message switching. Also known as the *store and forward* concept. The technique minimises the system idle time by the sequential transmission of stored messages.

Modem (modulator/demodulator) (see Networks, p. 204).

Multiplexing. The technique of combining several communications channels into one larger channel (see also Time, frequency and code division multiplex, p. 269 and Statistical multiplexer, p. 90).

Narrow band. The term usually used to describe the bandwidth of the 2B + D channels of 144 Kbit/s.

Node. A point in a system where two or more lines converge or diverge. Usually at a switching point.

Octet. A packet switched network term to describe eight bits. This may not be the same as a byte.

Open Systems Interconnection (OSI) (see also Networks, p. 214). The standardised procedures for the interchange of information via different types of networks and communications equipment.

Packet. A block of data with a precisely defined set of fields. These include addresses, control information and error control bits. A basic unit of data transfer in an ISDN system.

Packet assembler/dissembler (PAD). A device that converts the terminals data flow to and from the packets. The PAD also handles all aspects of call set-up and control.

Packet switching. A technique for switching packets of data in a network, with route and destination being controlled by information within each packet. As each packet contains a serial number code, the receiving terminal can restructure the complete message in the correct sequence.

Protocol. The precise set of rules that govern the actions of communicating equipment in a system. The nature of low level functions of packet protocol is real-time bit or character-oriented, often under hardware control. The higher level of functions are usually packet-oriented and control is exerted via software.

Protocol convertor. A device that converts one communications protocol into another.

Rate adapter. A device within a terminal adapter that converts an asynchronous data stream into a 64 Kbit/s B channel stream by adding extra bits to pad out the bit rate.

Statistical multiplexer. A technique that makes efficient use of a common communications channel by allocating time slots only to those terminals that require service (see also Telephony and associated systems, p. 327).

Stored program control. The application of a digital computer to the control of a telephone-type network. As all the information that is required to process the calls is held in the computer programs, it is a simple matter to reconfigure the system for future expansion (see also Telephony and associated systems, p. 327).

Synchronous data link control (SDLC). A system protocol similar to HDLC.

X25/X31. The CCITT standards that are used to define the interfaces for a packet switched public and private data network.

7.6 Useful references

Brewster R.L. (1992) *I.S.D.N.*, Chapman & Hall, London.
Smouts M. (1992) *Packet Switching Evolution from Narrowband to Broadband I.S.D.N*, Artech House, London.
Truelove J.E. (1992) *A Guide to Fractional T1*, Artech House, London.

8 Digital pulse code modulation

The sampling and quantisation process described in Section 7.1 forms the basis of pulse code modulation (PCM). While the following information refers to the system as used for international voice telephony channels, the concept can be extended for use with any other form of analogue signal (see also Television signal processing and systems, p. 343). The 8-bit binary coding in sign and magnitude format represents the baseband signal. This may be converted into a ternary or quaternary form to minimise bandwidth and may then be modulated on to a carrier frequency (see also Modulation, p. 190) for transmission either over a cable or radio network.

The advantages gained from using PCM can be summarised as follows:

(a) As the signals are in digital form, they can be regenerated at intervals along the transmission path to minimise the effects of noise.
(b) For a given distance, the spacing between regenerators can be greater than that required for line amplifiers in an equivalent analogue system.
(c) Digital signal processing is highly reliable and cost effective.
(d) If necessary, the digital signals can easily be computer processed and the system computer controlled.
(e) The effects of noise and interference are minimal.

As with the ISDN concept that has developed from PCM systems, the information is transmitted in bursts and it is therefore necessary that transmitters and receivers are maintained in synchronism, this being achieved by including a common signalling control within the bit stream. The basic parameters of the systems are listed in Table 8.1.

8.1 Frame time multiplexing

As indicated by Table 8.1, there are two systems in use. These are the North American Bell and the European

Table 8.1 *Parameters of pulse code modulation*

| | System | |
Attribute	US (Bell)	Europe (CEPT)
Companding law	μ Law	A Law
Sample rate	8 kHz	8 kHz
Quantisation levels	255	255
Number of bits per time slot	8	8
Number of time slots per frame	24	32
Number of bits per frame	193	256
Frame period	125 μs	125 μs
Bit rate	1.544 Mbit/s	2.048 Mbit/s
Frame alignment signal	First bit in each frame	Time slot 0 in each frame
Control signalling	Time slot 24	Time slot 16

Both systems agree with CCITT Recommendation G 711.
See text for variations in frame alignment and control signals.

CEPT (Conference of European Posts and Telecommunications) systems and the differences are relatively minor. Interfaces are in operation that allow cross-working.

Figure 8.1 shows the CEPT frame format. The basic element is the 8-bit word or time slot and 32, 8-bit groups are formed into a frame of 125 μs duration. The bit rate is thus: $8 \times 32 = 256$ bits per 125 μs or 2.048 Mbit/s. Sixteen frames are then time multiplexed into a multiframe. A frame alignment word is transmitted in time slot 0 (TS0), while TS16 is reserved for control signalling. This arrangement is thus very similar to the 30B + D ISDN system, where synchronism is controlled from channel 0 and the D channel included in slot 16.

Figure 8.2 shows the similarities and differences involved with the US (Bell) frame multiplex. Again the

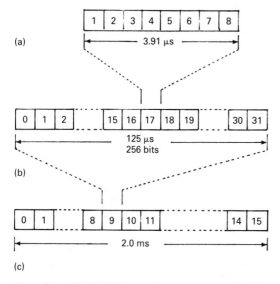

(a)

(b)

(c)

Figure 8.1 CEPT PCM frame format structure: (a) 8-bit time slot; (b) frame; (c) multiframe.

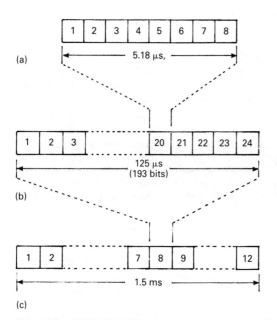

(a)

(b)

(c)

Figure 8.2 US (Bell) PCM frame format structure: (a) 8-bit time slot; (b) frame; (c) superframe.

basic element is the 8-bit word and these are formed into groups of 24 to form a frame of 125 μs duration. An additional bit precedes time slot 1 to act as frame alignment, so that there are 193 bits in each frame. The bit rate is thus 193/125 μs = 1.544 Mbit/s. Each frame is further multiplexed into either 12-frame superframes or 24-frame extended superframes. Control signalling is included in word 24 of each frame and frames 6, 12, 18, and 24 of the superframe or extended superframe.

Table 8.2 *Multiplex hierarchy*

Level	Transmission rate (Mbit/s)	Number of channels
1	2.048	30
2	8.448	120
3	34.368	480
4	139.264*	1920
5	565	7680

*Nominally 140 Mbit/s.
The actual transmission bit rates are multiples of 64 Kbit/s to maintain the 8-bit basic structure.

8.2 Hierarchy of PCM multiplexing

Higher levels of multiplexing are in use as shown in Table 8.2, where it will be noticed that, while the number of telephone channels increases by a factor of 4, the bit rates increase by slightly more. This is because, as the structure expands, relatively more control and synchronising bits become necessary.

8.3 Justification or bit/pulse stuffing

This is a particularly flexible and adaptable process that allows both synchronous and non-synchronous signals to be combined in a multiplex for transmission over a single channel. This is achieved by making small adjustments to the bit rates in the tributary channels to match that of the multiplex.

Positive justification. The total bit rate of the multiplex is higher than necessary to contain all the information bits. This allows dedicated time slots to be available for bit stuffing to pad out the rates of the slower channels. These justification slots can be used either for justification or for information bits. Therefore two condition control signals need to be included to signal to the receiver demultiplexer that it can ignore the non-information bits. For example, it will be noticed that with the level 2 PCM multiplex, 8.448 Mbit/s is provided for four 2.048 Mbit/s channels. There are thus four 64 Kbit/s channels available for justification purposes.

Negative justification. If the transmission multiplex has less capacity than is needed for total bit rates of the tributary channels, there will not be enough time slots for all the information bits. The excess bits can be extracted from the input signal to be transmitted separately along with the control time slots because not all these bit positions are used for signalling.

Positive–negative justification. These two basic concepts can be combined so that bits can be added or deleted in a controlled manner.

Positive–zero–negative justification. This further extension of the justification technique automatically allows for the case where the bit rates of the tributary channels exactly match that of the transmission multiplex.

Justification control signal. This control signal provides status information regarding the bit in the justifiable time slot. An error in this bit can lead to serious loss of frame synchronisation. To minimise this possibility, three identical bits are used to allow a majority logic decision at the receiver and, to avoid the problem of burst errors, these bits are distributed throughout the frame.

Justification jitter. This represents an irregularity in the timing of the digital signal due to the insertion and deletion of the stuffed bits and this can lead to decision errors at the demultiplexer.

Justification or stuffing ratio. A parameter that is used to measure the usage of justification and is simply the ratio of actual justification rate : maximum available justification rate.

Justification service digit. A bit that forms part of the justification control signal.

8.4 Signalling protocol

The CCITT common channel signalling system No. 7 is used in the control of the PCM system. This operates on a packet basis which allows for bits to be allocated for flag, acknowledgement, addressing and error checking. This system is flexible in application as the data field in each packet can be of variable length.

8.5 Speech digit signalling/bit stealing

In the PCM system the time slots are used mainly for the transmission of encoded speech signals. Periodically it is possible to steal the eighth bit in a time slot, or the complete sixth frame to augment the channel signalling system.

8.6 Data rates

Commonly the data rates of the two systems are referred to as T1 for 1.544 Mbit/s and E1 or T2 for 2.048 Mbit/s.

8.7 Miscellaneous terms

Adaptive differential pulse code modulation (ADPCM). Both delta modulation (DM) and differential pulse code modulation (DPCM) systems can give better results in terms of S/N ratio if the step size is made to vary in accordance with the time varying amplitude of the input analogue signal. Such adaptive circuits behave in a companded manner (see also Television signal processing and systems, p. 362)

Anisochronous or asynchronous transmission. A transmission system in which the durations between signal transitions are variable. In such a case, the characters are delineated by start and stop bits.

Bisynchronous transmission. A communications protocol in which a defined set of characters is used to control the synchronism of the data stream.

Burst isochronous transmission. A method that is used when the actual channel bit rate is higher than that required for the input data stream. The input stream is therefore interrupted at intervals and the accumulated data then sent in bursts, the bursts being separated by a time interval that is an integer multiple of the bit periods.

Byte serial transmission. Successive bytes are transmitted in their natural sequence either as a serial data stream over a single channel or in parallel over a bus system.

Circuit switched network. A connection between two or more terminals that is made on demand and used exclusively for the duration of the call.

Codec. A device that is capable of handling transmitted and received signals, i.e. a coder and decoder in one unit.

Data source and sink. The terms used to describe the generating and receiving terminals respectively.

Delay modulation. A binary encoding concept in which a one is signified by a signal transition 0–1 or 1–0 at the bit cell centre and zero is signalled by no transition unless it is followed by another zero. In this case a transition occurs at the trailing edge of the bit cell (see also Codes and coding formats, p. 57).

Delta modulation (DM). The level of the sampled analogue signal is coded by one bit, positive or negative, based upon whether a signal sample is greater or less than the previous value. Because only one bit per sample needs to be transmitted, the sampling rate can be significantly increased to reduce the quantisation noise, simplify the anti-aliasing filter and at the same time reduce the bandwidth required for transmission.

Demand assignment (DA). A circuit that is made available and allocated to a user on demand and when actually required for traffic (see also Networks, p. 206 and Satellite systems, SPADE, p. 282).

Demux. Jargon term for demultiplexer.

Differential modulation. A binary coding system whereby a zero is signalled by a transition at the bit cell centre while a one is signalled by no change of level. Even if this signal becomes inverted there is no loss of information. In synchronous systems the information is recovered by sampling the signal and comparing the polarity with adjacent samples to detect a change of level, i.e. to detect zero values.

Differential pulse code modulation (DPCM). If the pattern of an analogue signal is known up to some point in time, then because of the strong correlation between successive values, it is possible to draw certain inferences about the future behaviour. This process is based on a predictive or extrapolation technique. In a predictive coding system, the signal is sampled at regular intervals and, as each sample time approaches, the probable value of the next sample is predicted. Only the difference between this prediction and the actual value is coded (as for PCM) and transmitted (see also Television signal processing and systems, p. 357).

Digital speech interpolation (DSI). In time division multiplex channels, it is possible to use the gaps that occur naturally in speech for other communication users that are waiting to transmit data. DSI finds applications over satellite links and with the ISDN systems.

Digital switching. The switching of digital telephony signals through an exchange involves two distinct operations. In the first case it is required to connect the incoming line to the outgoing one and this is described as *space switching* because the action is performed by a matrix of cross-point switches. The second action involves transferring the data stream into a new time slot and this requires that the *time switches*

must contain a memory element. The technique has two variants, space–time–space and time–space–time switching. The control of the digital switching network can be carried out by a central computer using the *stored program control* (SPC) concept.

Dither. When an analogue signal is processed in the digital domain, the sampling process adds quantisation noise that is proportional to the step size. By adding a pseudo-random signal to the quantiser input and subtracting it at the output, the random variation reduces the coarseness of the steps. For example, if a signal is quantised to 6 bits, and then 2 bits dither are added, the overall noise effect will be close to 8-bit quantising, but without the extra bandwidth required for the transmission of 8 bits.

Duplex. A system capable of communicating in two directions simultaneously.

Dynamiciser. The term sometimes used to describe a device that can convert digital signals in parallel format into a serial bit stream.

Echoplex. A signalling method whereby the integrity of the received data can be checked by returning the characters to the sending terminal. An all correct transmission would be acknowledged by a simple ACK signal.

Electronic mail (E-mail). A messaging service provided for subscribers over a public or private data network.

Glitch. A jargon term used to describe an extraneous bit that has been introduced into a bit stream usually by a noise source.

Half-duplex network. A system that allows two-way communications but not simultaneously.

Heterochronous. A term used to describe digital signals whose transitions do not occur synchronously.

Heterogeneous multiplex. A multiplex of digital signals where the tributary channels have differing bit rates.

Homogeneous multiplex. A multiplex of digital signals where the tributary channels all have the same bit rate.

House-keeping bit. A service bit that is added to a digital signal to enable the system to provide a range of services under the control of such overhead digits.

Isochronous transmission. A transmission where all signals are of equal duration and are transmitted in a continual stream.

Line concentrator. A device used in a local exchange to multiplex a large number of incoming lines into fewer outgoing lines feeding the switching stages. The smaller number of lines is then allocated according to the traffic loading.

Loopback, loopback testing. A digital diagnostic technique that is used to test a network from end to end. The network integrity and its parameters can be evaluated by comparing the signal that is received back at the transmitting terminal with that which was originally transmitted.

Mesochronous signal. Digital signal whose transition instants occur at an average rate, this small variation of timing varying between specified limits.

Modem. Term used to describe a modulator/demodulator that forms the interface between analogue terminals and a digital network (see also Networks, p. 204).

Muldex or muldem. A jargon term used to describe a device that contains both a digital multiplexer and a demultiplexer.

m-Sequence. A pseudo-random binary sequence of maximal length (see also Digital communication systems, p. 84).

Nyquist bandwidth. The bandwidth of a digital signal is half that of the numerical value of the bit rate. For n bits/s, Nyquist's bandwidth is $n/2$ Hz.

Nyquist interval. The reciprocal of the Nyquist rate (see also Modulation and demodulation, p. 186).

Nyquist rate. This defines the maximum rate at which information can be transmitted over a bandwidth limited channel (see also Modulation and demodulation, p. 186).

Plesiochronous transmission. A signal multiplex situation where the system clocks may run at slightly different rates. For example, in the PCM hierarchy, the basic level is a multiplex of 32, 64 Kbit/s channels, with synchronous clock signals. Four of these 2.048 Mbit/s blocks, each with slightly different clock rates, are then multiplexed to a second-order level of 8.448 Mbit/s. Padding bits can then be added at each

multiplex stage to compensate for the small discrepancy in clock frequencies.

Quantised feedback. Whenever there is a large drift in the running digital sum, there is also a drift in the system low frequency response, which introduces an ac component into the signal and can result in transmission errors. This can be countered by using quantised feedback in the manner depicted in Fig. 8.3.

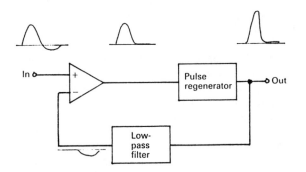

Figure 8.3 Quantised feedback.

Repeater. The generic title given to line amplifiers sited at intervals throughout a network in order to maintain a high S/N ratio. In the equivalent digital network, such devices are usually described as regenerators.

Segmented encoding law. The approximation of the smooth curve required for the companding laws by a series of linear functions.

Software multiplexing. The use of a digital computer to separate the channels of a time division multiplex system by recognising the signal formats from the separate sources. This simplifies the hardware requirement and produces a more flexible system.

Space–time–space switching. In this digital switching concept, the memory element to manage the time slot interchange is interposed between input and output matrices of cross-point switches.

Spacial code. A term sometimes used to describe code patterns that are in parallel format.

Statistical multiplexer. A time division multiplex device that makes efficient use of a network by allocating time slots to only those terminals with data to transmit. If the total channel capacity tends to overflow, the surplus data is stored in memory to await an available time slot.

Telecommunications. The technique of transmitting information from point to point by electronic means over cables or optical fibres or by radio.

Teleinformatics/telematics. A generic term to describe the non-voice services provided by telecommunication services. These cover such methods as facsimile, telex, teletex, and videotex.

Telemetry. The transmission of coded real-time analogue information from remote sensors to a controlling terminal.

Teletex. The enhanced telex type service that allows subscribers to exchange information on a memory-to-memory basis over a telecommunications network. Available on a 24 h basis and provides for the transmission of still images.

Teletext. A television broadcast service whereby data representing graphics and text is transmitted in a coded manner, within the frame blanking period. The signal therefore forms a time multiplex with the normal television image (see also Videotext, p. 383).

Temporal code. A term sometimes used to describe serial code, i.e. code arranged in time.

Time assignment speech interpolation (TASI). This is a technique that increases the traffic capacity of a multi-channel network by allocating a channel to a user only when needed. Since normal speech contains silent periods for up to 60% of the time, channels can be reallocated to other users during these gaps.

Time sharing. A term applied usually to a distributed computing system whereby several terminals communicate with the main processor over a network. By giving each terminal access to the system in short bursts and in a cyclic manner, all users have access to its computing power in turn.

Time–space–time switching. In this variant of digital switching there are time slot memories associated with both the input and output trunk cables. A cross-point switching matrix is then interposed between the two time slot memory banks to reroute the signals through the exchange.

Videography. The generic term used to describe the teletext and viewdata services.

Viewdata. The graphics/text service provided for the television user but transmitted over the telephone network (see also Videotext, p. 383).

X-on/X-off. The communications control characters that are transmitted to instruct a terminal to commence or stop transmission.

8.8 Useful references

Baines R. (1988) Enhanced-instruction set processor. *Electronics and Wireless World*, November, p. 1111.
Halliwell B.J. (ed.) (1974) *Advanced Communications Systems*, STC Monograph, Butterworth-Heinemann, Oxford.
Millman J. (1983) *Microelectronics. Digital and Analog Circuits and Systems*, McGraw-Hill, I.S.I. London.
Wade J.G. (1987) *Signal Coding and Processing*, Ellis Horwood, Chichester.

9 Electromagnetic compatibility/ interference (EMC/EMI)

Electromagnetic compatibility is defined as the ability of a device, equipment or system to function satisfactorily in its electromagnetic environment without introducing intolerable electromagnetic intereference to any other system, while at the same time, its own performance will not be impaired by interference from other sources (see also Noise, p. 219). Since interference is a destroyer of information, it is often a limiting factor in a communications system. For convenience, interference can be divided into two categories, natural and man-made. The former often results in electrostatic discharge (ESD), while the latter usually results in power supply line surges. Natural sources of interference include ionospheric storms and lightning. Man-made interference commonly results from high current switching operations and radio frequency generating equipment, including industrial, scientific and medical (ISM) apparatus. With the expansion in the use of portable digital communications equipment such as computers and cordless telephones, and the introduction of information technology (IT) systems into offices and the domestic environment, the sources of interference are growing rapidly.

A direct lightning strike is not necessary to cause havoc in a communications system. Even a cloud-to-cloud discharge can easily set up an electric field of the order of 50 volts per metre, at a distance of 1 kilometre. Such a situation could induce a voltage as high as 50 kV on a kilometre of exposed power line or telephone cable. Fortunately, such induced surges are usually limited to very short periods of time.

The action of a person walking across a carpet or even sitting in a chair can generate very high electrostatic charges, depending upon clothing and the level of humidity. This *triboelectricity* due to rubbing dissimilar materials together can often be the source of computing disasters. A typical body capacitance can range between 100 and 250 pF and acquire a charge of 0.1 to 5 μC. With a human body discharge resistance of around 150 to 1500 ohms and taking midrange values of 150 pF and 3 μC, this can produce a voltage of around 13.3 kV ($Q = CV$) with an energy of about 13.3 mJ ($E = 0.5CV^2$). (Because of the $Q = CV$ relationship, any movement such as lifting the feet off the floor which lowers the body capacitance will automatically increase the voltage level.) If this is discharged through a typical body resistance in from 0.5 to 20 ns, the peak currents can be in the order of 1 to 50 A.

Even if exposures to such discharges do not prove to be immediately destructive, they can introduce a latency failure mechanism because these effects are cumulative.

A triboelectric series lists the materials ranging from air, hands, asbestos, at the most positive, through the metals, to silicon and Teflon as the most negative. The further apart in this table, then the higher will be the electric charge between two materials when rubbed together.

9.1 Protection and protective devices

Electromagnetic interference can enter communications equipment either by direct radiation or by conduction. The effects of the former can be reduced by screening or shielding, while reduction of the latter requires the use of suitable filters. For very sensitive cases, such as encountered in testing, it will be necessary to enclose the equipment in a *Faraday cage* to exclude the interference. This is a region that is enclosed by well-earthed electromagnetic screens and with the inputs and outputs, including any power supplies, extremely well filtered.

MOSFET, CMOS and Group III/V (GaAs, etc.) semiconductor devices are easily damaged by electrostatic discharges. Because these effects are cumulative, a discharge of as little as 50 V can cause component degradation, which might lead to an early failure. It is therefore important that such sensitive components are only serviced in a workshop area where static electricity can be controlled. The basic workstation should provide for the operator, work bench, floor, test equipment and device under service to be at the same electrical potential. The operator should be connected to the bench via a wrist strap, that allows any static charge to leak to earth via a 1 Mohm resistor, and should wear an anti-static smock. An ionised air ventilation system can be used to advantage. Large quantities of negatively and positively charged air molecules will quickly neutralise any static charges.

There are two common approaches to preventing catastrophic failures due to supply line surges. These involve the parallel connection of components that act as limiters or clamps to prevent line voltages exceeding a nominal value, and the use of *crowbars* that place an almost short circuit across the lines in the event of a major discharge.

Gas discharge tubes. These devices act as clamps and consist of a glass envelope that contains two electrodes and a mixture of gases. Different gas combinations produce different breakdown voltages, so that these can provide for a range of applications from about 100 V, to several kilovolts. They function by reason of an arc discharge, operate very quickly and can handle large currents.

Silicon controlled switches/thyristors. These are avalanche devices that operate very rapidly to provide a crowbar action. They are available in a wide range of voltages, from about 10 V to a few kilovolts. One major problem with crowbars is that the fast response can lead to a reflection of the incident energy back into the circuit.

Varistors. These are voltage-dependent resistors that are made from a variety of metal oxides to provide a range of operating voltages. They have a very nonlinear voltage current characteristic that provides a clamping action to any over-voltage condition.

Most systems will include hybrid protection, using a combination of clamp, series limiter and crowbar. For low voltage communication systems, such as computer networks, where the interstage wiring can act as an aerial, it may be necessary to use power conditioner circuits to prevent mains-borne interference from corrupting the data. In addition, it might be necessary to provide some form of *uninterruptable power supply* (UPS) to allow for the complete loss of power for short periods (see also Power supplies, p. 239).

9.2 Standards and guidelines

While the British Standards Institution (BSI) provides some standards for EMC/EMI that include domestic, industrial and medical applications, the commonly quoted European organisations are CISPR (Comité International Spécial des Perturbations Radioélectriques) which is a committee of the IEC (International Electrotechnical Commission) and VDE (Verband Deutsche Elektrotechniker) (see also Standards organisations, p. 317).

9.3 Techniques of testing

The management of ESD/EMI protection is a relatively young and rapidly growing discipline due to the wide use of integrated circuits (ICs) that are sensitive to such interference. International standards have been laid down that govern not only the measurement of the problem, but also state the standards required for equipment suitable to establish the degree of immunity of a system. The tests for system sensitivity must mimic the conditions found in the real world and must be precisely repeatable for production control purposes. For this reason, the IEC has defined two tests to evaluate a system's ability to withstand ESD/EMI. These use either *air discharge* or *current injection* in a strictly controlled way. At first sight, it might appear that the polarity of test waveforms is unimportant. However, reverse bias applied to many transistor junctions is more destructive than forward bias. Test waveforms are specified for peak voltage or current.

Type 1: Impulsive. This waveform is unipolar and has exponential rise and decay and refers to either a current or a voltage. Apart from the amplitude, the wave is specified by two numbers, such as 8×20 or 8/20, which represents a double exponential pulse with $8\,\mu s$ rise time and duration $20\,\mu s$, the rise time being defined as the time taken for the pulse to rise from 10% to 90% amplitude, the duration being the standard half amplitude value.

Type 2: Damped oscillatory. This wave has a cosine characteristic with a fast rise to the first peak. This ringing waveform is characterised by two numbers, such as $0.5\,\mu s/100\,kHz$ (rise time and frequency). It is discharged into a low Q circuit so that it decays quickly.

9.4 Miscellaneous terms

Air discharge. A high level of electric charge first generates an ionisation effect in the space surrounding it. If the charge is high enough, this can create a conducting path to some earthed body. If such a path exists, then an arc forms to collapse the electric charge and generate a magnetic field that causes radiation.

Bonding. The grounding or earthing of a system's metalwork to provide a very low impedance discharge path to earth. If this path is not short and direct it can act like an aerial to pick up further interference. If the link becomes corroded, then the grounding line can actually reradiate the energy instead of shunting it to earth.

Burst interference. Interference that occurs repeatedly in bursts with short enough duration to corrupt a sequence of binary digits.

Corona discharge. The luminous discharge which appears around a charged body or conductor, due to the ionisation of the surrounding gas or air. This results in the total charge being distributed throughout a larger volume of space to lower the potential gradient. The potential gradient has to exceed a critical value but is not high enough to promote an arc discharge to some nearby earthed body.

Electromagnetic pulse (EMP) (nuclear EMP, NEMP). Surges of energy result from the redistribution of electromagnetic energy. This can occur from lightning or the switching of highly inductive loads. The nuclear EMP is a very short transient that follows a high altitude or distant nuclear explosion. This leads to the use of *radiation hardened* equipment in certain critical communication systems.

Electrostatic discharge remnant. The proportion of the electrostatic charge that leaks through the protective mechanisms and into a system.

Hard failure. The results of levels of ESD/EMI that produce an equipment failure (i.e. hardware failure).

Harmful interference. Defined by international standards as the level of interference that endangers the operation of radio, navigation and safety services, or seriously degrades a communications service that is operating according to the Radio Regulations.

Nuclear electromagnetic pulse (NEMP). See EMP, p. 99.

Permissible interference. A level of interference that does not allow a system performance to fall below some specified level of integrity.

Precipitation static. Interference due to the charges that build up on a support structure due to rain, sleet, ice or snow.

Protection ratio. The minimum ratio between the wanted signal power and that of the unwanted interference under specified conditions of reception, that is deemed necessary, to achieve a specified standard of system performance.

Radiation hardened. Systems and components that have been specifically screened and shielded in manufacture, to withstand the effects of a nuclear electromagnetic pulse.

Secondary radiation. Reradiation of interference that occurs due to oxidation and other imperfections in the continuity of a grounding or earthing system.

Soft failure. The results of ESD/EMI that are insufficient to cause system hardware failure, but give rise to data errors (i.e. software failure).

Tempest. The jargon term used to describe the compromising radiation of signals from a system. In certain cases this radiation may be regarded as interference, but more importantly it is a breach of security (see also Encryption and decryption, p. 193).

Weighting. The artificial adjustment of measured values to allow for their significance in a particular application. For example, the measured levels of noise are often scaled or weighted to account for the nuisance value on a more subjective basis.

9.5 Useful references

CCITT Vol. IX, Rec.K.17, 1977, Protection: Tests on power-fed repeaters using solid state devices in order to check the arrangements for protection from external influence.

Electrostatic Discharge, Protection Handbook, 2nd edition, KeyTek Instrument Corp. USA, 1986.

European Computer Manufacturers Association (ECMA), Electrostatic discharge susceptibility, TR/23, September 1984.

Freeman E.R. (1983) *Electromagnetic Compatibility Design Guide*, Artech House, Boston.

Stoll D. *EMC*, Elitera Verlag, Berlin.

Surge Protection Test Handbook, 2nd edition, KeyTek Instrument Corp. USA, 1986.

Violette (1987) *Electromagnetic Compatibility Handbook*, Van Nostrand Reinhold, New York.

Williams T. (1993) *EMC for Product Designers*, Focal Press (Butterworth-Heinemann), Oxford.

10 Encryption and decryption

The terms *scrambling* and *encryption* tend to be used synonymously (see also Television signal processing and systems, p. 366). In this text, however, the term scrambling will be taken to mean the rearrangement or *transposition* of the order of the original information, while encryption will imply that the original information (often referred to as *plain* or *clear text*) has been replaced by the *substitution* of an alternative code pattern (known as *cypher* or *encrypted text*). Scrambling alone is not considered to be secure, because a study of the signal pattern can lead to the design of a suitable descrambler.

The encryption operation is quite simple if the signal is the binary electronic form. When a second binary sequence is added to the first, using Modulo-2 arithmetic, the resulting bit stream carries no obvious information. The original information can be recovered at a receiver by carrying out a complementary operation. The rules for Modulo-2 arithmetic can be stated as follows:

$$
\begin{array}{ll}
0 + 0 = 0 & 0 - 0 = 0 \\
0 + 1 = 1 \quad \text{and} & 0 - 1 = -1 \\
1 + 0 = 1 & 1 - 0 = 1 \\
1 + 1 = 0 \text{ carry } 1 & 1 - 1 = 0
\end{array}
$$

If the 'carry' and minus sign are ignored, it will be seen that both operations produce the same result and can be carried out using an Exclusive OR (Ex.OR) logic operation. For example, it is required to secretly transmit the binary character 10001110 and an 8-bit key, 10101010 is chosen for encryption. The transmission/reception process then becomes:

Character to send	10001110
Key	10101010
Sum Mod-2	00100100 (this is transmitted)
Key	10101010
Sum Mod-2	10001110 (the original character)

Thus encryption and decryption are the same operation and, in the general case, the keys can be produced by a pseudo-random binary signal (PRBS) generator. Such keys have several advantages, including:

- they are practically random and easy to generate and change;
- the longer the key, then the more difficult becomes unauthorised decryption.

The important rules for any encryption system can be stated as follows:

(1) The number of keys should be very large to prevent a pirate from testing all possible keys in turn.
(2) Any fixed encryption operation should be very complex, making it impossible to deduce the operation from a few plain text/cypher text pairs.
(3) If security is to be based on secret information, then this must be created after the system is built and, if subsequently revealed, it should not jeopardise the security of the entire system.

The operation of this process can be expressed mathematically as follows:

The operation of the encryption function E on plain text characters P, using key K, results in cypher text C. Thus the encryption process is given by $E(K).P = C$.

Now the decryption function $D(K)$ is the inverse of $E(K)$ or $E(K)^{-1}$. Therefore decryption is given by $D(K).C = P$.

As most of the encryption/decryption processes are carried out using integrated circuits, the operation of such systems is completely transparent to the user. In addition, there is no need to disguise the system structure, because the security lies in the strength of the key schemes.

10.1 Key systems

One-key system. The one-key system or *one-time pad*, is very secure as long as the key remains secret and this is also the weakness. The key has to be transmitted in some way to all authorised users before the message.

Although this increases the time delay, the key can easily fall into the wrong hands.

Two-key system (also known as a public key system). In this system, one key is made public for encryption and decryption, while the second is kept secret and is used as a *modifier*. This key can be used to rearrange the logic used to set up the PRBS generator and is therefore sometimes known as the *keystream*.

Three-key system. This further variant of the public key system, uses three keys, two secret ones, primary and secondary, both user programmable, which can be stored in a digital memory. The third non-secret key, which acts as a modifier, can be generated as a new PRBS at the start of each transmission. This technique provides a very high degree of security and is at the same time very flexible (see also Television signal processing and systems, p. 366).

10.2 Data Encryption Standard (DES) system

This Federal Information Bureau Data Encryption Standard, which is fully described in the reference, translates blocks of 64 bits of plain text into similar sized blocks of cypher text, using 56-bit keys. Each plain text block is divided into left (L) and right (R) groups, each of 32 bits and then processed through the alogorithm in the manner shown in Fig. 10.1(a). Successive R groups are combined with successive keys using a very complex function f. Each R group is then added modulo-2 to the corresponding L group, throughout the 16 stages according to the following formulae:

$$L_J = R_{J-1}$$

$$R_J = f(R_{J-1} - K_J) \oplus L_{J-1}$$

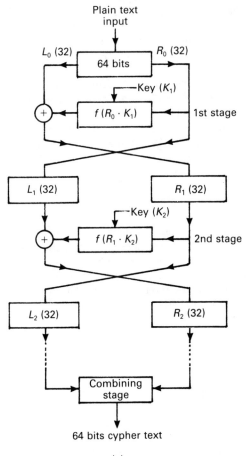

(a)

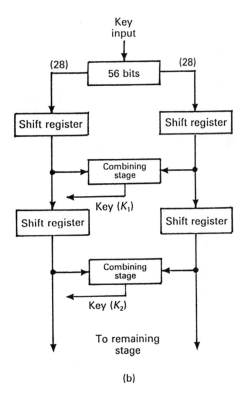

(b)

Figure 10.1 (a) Operation of DES algorithm; (b) key generator.

After processing using the 16 keys, the L and R groups are recombined in the reverse order (R, L).

The key generator stage is shown in Fig. 10.1(b) where the initial 56-bit group is divided into 28-bit sub-groups. These are increased to 32-bit groups by the addition of parity bits and then processed through the shift registers. A new key is formed by combining the sub-groups after one or two left shifts. After 16 keys have been generated, each key input is ready to repeat.

The function f accepts 32-bit key inputs which are then expanded into 48 bits via a look-up table. The addition modulo-2 of R_J and K_J produces another 48-bit group and this is compressed into 32 bits via a further look-up table. This 32 bits is then added modulo-2 to the previous L group, to provide a new R group.

By using the DES alogorithm repeatedly in overlapping blocks, it is possible to encrypt plain text blocks that are very much longer than 64 bits.

10.3 Diffie Hellman algorithm

This shared key system requires that each end of the communications link shall contain some computing power to calculate the shared *secret key S*. The system contains two numbers, P which is a prime and X, both of which may be public.

The two communicators, A and B, both choose random values (a) and (b), respectively, that lie between 0 and $P - 1$. The secure link is established as described by the four following stages:

A computes $X^a \bmod P$ which is sent to B
B computes $X^b \bmod P$ which is sent to A
A then computes $S = (X^b \bmod P)^a \bmod P = X^{ab} \bmod P$
B then computes $S = (X^a \bmod P)^b \bmod P = X^{ab} \bmod P$.

Both A and B now know the secret key S that is used for encryption.

Provided that a, b and P are large, it is extremely difficult to find the secret numbers by taking logarithms, even though $X^a \bmod P$ and $X^b \bmod P$ have been openly transmitted.

10.4 Rivest Shamir and Adleman (RSA) public key system

This secure system relies upon the choice of two large prime numbers P and Q chosen at random and used to calculate a modulus N from the product PQ. Since it is not possible to factorise such large numbers, mod N can be made public without jeopardising P and Q. The encryption key (E) and decryption key (D) are then evaluated from the formula:

$$ED \bmod (P - 1)(Q - 1) = 1$$

The cypher text C is then obtained from the plain text M by

$$C = M^E \bmod N$$

The message is recovered from the received text by using

$$C^D \bmod N = M$$

For example, if $P = 3$, $Q = 11$, then $N = 33$ and $(P - 1)(Q - 1) = 20$; thus when $ED \bmod 20 = 1$, $E = 7$ and $D = 3$ are possible values. If $M = 4$, then $C = 4^7 \bmod 33 = 16$ and this is transmitted. For decryption, $16^3 \bmod 33 = 4$, the original message.

10.5 Security of encryption systems

DES. Based on the 56-bit key, the possibility of a pirate deciphering the code first time is 1 in 2^{56}. Alternatively $2^{56} = 7.206 \times 10^{16}$, so that a pirate making one attempt every nanosecond, would on average take about 3.6×10^{16} ns or more than 1.140 years to decipher the code.

RSA. A pirate attack on this scheme requires the factoring of 'mod N' to discover the two primes P and Q or taking the discrete logarithms of $C = M^E \bmod N$. Both operations are equally complex and require the same number of steps. The fastest known algorithm that will achieve this is quoted as having the following number of steps S:

$$S = \exp(\ln(n) \ln(\ln(n)))^{1/2}$$

where $n =$ modulus value. Assuming that a computer is available that will perform the steps at the rate of one million per second, then for 200 binary digits which give a modulus $n(2^{200})$ of 1.607×10^{60}, the time taken to break this code will be about 2.625 days. The problem is very similar for the Diffie Hellman scheme.

10.6 Voice systems

Within ISDN systems it is possible to obtain security of voice channels using any of the schemes previously described. Because of the bandwidth restriction (300–3400 Hz) of the voice-grade analogue telephone network, security of communications is more difficult. However two basic techniques are possible, using either *spectral* or *temporal* transpositions. In the former method a range of contiguous bandpass filters separate the audio signal into sub-bands. These are then either inverted, or translated to new frequencies and then reassembled into a new analogue signal, still within the same frequency range, but now completely unintelligible. The intended receiving terminal must, of course, be equipped with a complementary filter/processor unit. For the temporal transposition technique, the audio signal is converted into digital format using delta modulation (see also Modulation and demodulation, p. 190). The data stream is then divided into blocks of about 100 ms and these are then time reversed before being added together for conversion back to analogue form. This 'new' audio signal is completely unintelligible and can be transmitted over a voice network or radio link (cordless telephone). For a further level of security, each block can be sub-divided into segments which may be shuffled and then relocated within the original block period. The shuffling sequence can be controlled either from a read only memory at each pair of terminals, or by a pseudo-random generator. A telephone so equipped will operate in the scrambled mode only with a similarly equipped receiving terminal, but in the normal mode can communicate with any telephone. The synchronising of two scrambled telephones is carried out using a handshaking technique to agree the scrambling sequence before the start of voice transmissions.

10.7 Miscellaneous terms

Cryptoanalyst. A code breaker. A person whose task is to analyse a code sequence in an attempt to discover the coding techniques in use. Often in a piratical or illegal sense.

Cryptographer. A person whose task is to devise secure encrypting strategies.

Key distribution systems. Encryption systems that depend upon the distribution of one or more keys in a manner that maintains system security.

Keystream. The series of random digits that form one or more of the controlled access system keys.

Masking. A term often used to describe a one-key system or one-time pad.

Multiple level security (access). Access control to a computer or communications system according to the security status of the user, and the degree of classification of the facility required. Different users thus have different levels of access to a system.

One-time pad. See One-key systems, p. 100.

One-way function. These mathematical operations that are used to generate encrypted text are the ultimate aim of the cryptographer. The function chosen, although reversible, operates easily when all the keys are known, but in the absence of these, the decrypting operation becomes practically impossible.

Pseudo-random digit generator (PRDG). A pseudo-random binary sequence generator.

Public key systems. A multiple key system in which one or more of the keys is not kept secret, i.e. public knowledge.

Tempest. The jargon term used to describe the emission of compromising signals from a system, installation or building. Since networks of cables can act as aerials to radiate electromagnetic signals, which can be picked up by suitably equipped receivers, such leaks can be an embarrassment. Hence the importance of a secure encrypting system.

10.8 Useful references

Coates R.F.W. (1985) *Modern Communication Systems*, 2nd edition, Macmillan, London.
Denning D. (1983) *Cryptography and Data Security*, Addison Wesley, Reading, MA.
Lewis G.E. (1992) *Communication Services via Satellite*, 2nd edition, Butterworth-Heinemann, Oxford.
Mason A.G. (1984) Proposal for a DBS over-air addressed conditional access system. IBA Experimental Report 132/84, Independent Broadcasting Authority, UK.
NBS (1977) DES Algorithm, FIBS:PUB 46. National Bureau of Standards, Data Encryption Standards, Washington DC.

11 Error control

The technique of forward error control (FEC) is so called because the means of detecting or correcting bit errors in a digital system during transmission over a noisy channel is contained within the transmitted message stream. This is achieved by the addition of extra redundant bits which, when suitably processed, are capable of identifying errors. The alternative technique involves scanning the received bit stream to make a running check of the disparity (see also Codes and coding formats, p. 57) and if an error is detected, automatically generate a request for a repeat transmission of the bytes in error (Van Duuren's ARQ, automatic request repeat). Either method makes additional demands upon the spectrum: FEC requires additional time or bandwidth to include the extra bits, while ARQ requires a free return channel. Since the rate at which bit errors are generated is dependent upon the system signal to noise power ratio, FEC which can correct errors can be particularly useful where a system tends to be power rather than bandwidth limited (see also Satellite systems, p. 269).

There are three prime causes of bit errors and two types of error. White (or Gaussian) noise produces errors that are completely uncorrelated and thus random in occurrence, while impulsive noise or slipped bit stream synchronism will generate errors that occur in groups or bursts.

There are three classes of error that need to be considered:

(1) detectable and correctable;
(2) detectable but not correctable; and
(3) undetectable and hence uncorrectable.

For any errors detected under Class 2, the concept of error concealment can then be applied as follows:

(a) ignore the error and treat it as a zero level;
(b) repeat the last known correct value;
(c) interpolate between two known correct values.

Forward error control thus considerably enhances the robustness of any digital communications system in a noisy environment, giving a significant advantage over a similar analogue system.

The ASCII code (American Standard Code for Information Interchange) is a commonly used method for representing alphanumeric characters in a digital system. This 7-bit code allows for $2^7 = 128$ different alphabetic, numeric and control characters. The most commonly used digital word length is eight bits or one byte, therefore there is space for one extra redundant bit in each code pattern.

11.1 Even and odd parity

A single-error detection (SED) code of n binary digits is produced by placing $n - 1$ information or message bits in the first $n - 1$ bit positions of each word. The nth position is then filled with a 0 or 1 (the parity bit), so that the entire code word (or code vector) contains an even number of 1s. If such a code word is received over a noisy link and found to contain an odd number of 1s, then an error must have occurred. Alternatively, a system might use *odd parity*, where the nth bit is such that the code word will contain an odd number of 1s. In either case, a parity check at the receiver will detect when an odd number of errors has occurred. The effects of all even numbers of errors is self-cancelling, so that these will pass undetected. The even or odd parity bits can be generated or checked, using Exclusive OR or Exclusive NOR logic, respectively.

Such an arrangement of bits is described as an (n, k) code, n bits long and containing k bits of information. It thus follows that there are $n - k = c$ parity or protection bits in the code word. The set of 2^k possible code words is described as a *block code*. It may also be described as a *linear code* if the set of code words form a sub-space (part) of the vector-space (entirety) of all possible code words.

11.2 Hamming codes

Error correcting codes have been devised and named after R.W. Hamming, the originator of much of the early work on error control. For code words of length n bits, it is possible to define n ways in which a single bit error can arise. Including the possibility of a correct word, there must be $n + 1$ different code patterns to be recognised. Hamming showed that the number of parity bits c needed for a single error correcting code was given by

$$c \geq \log_2(n+1) \quad \text{or} \quad 2^c \geq (n+1)$$

Hamming also defined the case where $2^c = (n+1)$ as being a *perfect code*.

Figure 11.1(a) shows how the message is expanded with redundant check bits. These are usually interleaved with the message bits and placed in positions 2^0, 2^1, 2^2, etc., in the encoded pattern. The mechanics for the encoding/decoding process can be explained using Table 11.1 for a (7,4) block code (block length $n = 7$, message length $k = 4$ and parity bits $c = 3$). The message to be transmitted is (0011) and these bits are placed in the 3rd, 5th, 6th and 7th positions, respectively. The three parity checks are carried out to determine the values to be placed in positions 1, 2, and 4. In general, the xth parity check bit is given by the sum modulo-2 of all the information bits in the positions where there is a 1 in the xth binary position number. The transmitted code word thus becomes 1000011. If this is now received over a noisy link as 1000001, there is an error in the 6th (110) position. The receiver decoder then performs the same three parity checks and generates the following results:

1st check 0
2nd check 1
3rd check 1

The reverse of this series, which is called the *syndrome* (syndrome being a medical term for the symptoms of a

Table 11.1 *Parity checking*

Bit position	1	2	3	4	5	6	7
(Binary)	001 (P)	010 (P)	011 (M)	100 (P)	101 (M)	110 (M)	111 (M)
Check 1	*		*		*		*
Check 2		*	*			*	*
Check 3				*	*	*	*
Message			0		0	1	1
Parity bits transmitted	1	0		0			
Code word received	1	0	0	0	0	1	1
Code word recheck of parity	1 0	0 1	0	0 1	0	0 (reverse order)	1

Syndrome $= 110 = 6$.
Error in position 6. Invert bit 6 to correct.

disease) points to an error in bit-6 position. Now that the error has been pointed out, it can be corrected by simply inverting the bit 6. An all-correct transmission would have produced an all-zero syndrome.

By adding an overall parity (O/P) check bit as shown in Fig. 11.1(b), the single-error correction capability is extended to double-error detection, the error patterns indicating the following conditions:

(1) No errors; zero syndrome and overall parity satisfied.
(2) Single correctable error; non-zero syndrome and overall parity fails.
(3) Double errors, non-correctable; non-zero syndrome and overall parity satisfied.

To further quantify the error patterns, Hamming also defined the *distance* of a code as being the number of bit positions in which one valid code word differs from another valid one.

The *minimum distance d* between any valid code words is the minimum number of bit changes necessary to convert one code word into another. For the correction of all combinations of t or fewer errors, the minimum distance must be at least $(2t + 1)$, thus ensuring that if t errors occur then the received code word will be nearer to the one transmitted than any other word (i.e. $d \geq 2t + 1$). This concept leads to a simple method of error correction. If a code word is detected as being in error, the decoder checks the distance between this and all the valid code words and selects the one with the minimum number of errors.

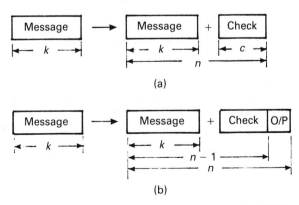

Figure 11.1 Hamming codes: (a) single error correcting (SEC); (b) single error correcting/double error detecting (DED).

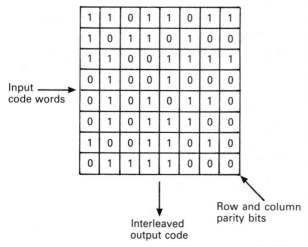

Input code words

Interleaved
output code

Row and column
parity bits

Figure 11.2 Interleaved coding.

11.3 Interleaved or interlaced codes

This is a simple but powerful way of dealing with both random and burst errors. Any cyclically coded (n, k) set of vectors can be used to generate a new $(\alpha n, \alpha k)$ code, by loading the coded vectors from the original code into a matrix of n columns and α rows and then transmitting the bits column-by-column in the manner shown in Fig. 11.2. The transmission sequence is then an interleaved or interlaced code, with an interleaving factor of α. If a burst of errors α or less in length occurs, there will only be one error in each affected word of the original code. As a lower bound, if the original code corrects t or fewer errors, the interleaved code will correct any combination of t bursts each of length α or less. Because the original code was cyclic (see Polynomial codes, p. 106), so will be the interleaved one. If the original generator polynomial had been $G(X)$, the generator of the new code would be $G(X^{\alpha})$. Therefore the interleaved code can also be encoded and decoded in a similar manner to cyclic codes.

11.4 Polynomial or cyclic codes

Cyclic codes are a sub-class of linear block codes that are suitable for combating burst errors. These have a form such that if a code vector 0110 is a valid code word, then so are all the cyclic translates such as 1100, 1001, 0011, etc., which are formed by shifting the sequence one bit at a time to left or right. The integrated circuits that are used as encoding/decoding devices are based on

feedback shift registers. A code that contains four bits of information represents 16 unique valid code words. If this is then incorporated into a (7,4) cyclic code, then there will be a total of 128 different code patterns of which only 16 form valid words. This large degree of redundancy makes it easy to develop an error checking system.

The *cyclic redundancy checking* (CRC) error detection concept is most effectively described in algebraic form using modulo-2 arithmetic. The basic rules being $1 + 1 = 0$ (ignore carry) and $0 - 1 = 1$ (ignore − sign). Thus addition and subtraction produce the same results.

The message and check bits are expressed by the use of polynomials in terms of a dummy variable X, the lowest order term X^0 representing the least significant bit (LSB) and the highest order term X^n the most significant bit (MSB). There are thus $n + 1$ bits in each message. The coefficients of the terms of the polynomials indicate whether a particular bit is set to 0 or 1. For example, the 5-bit message stream 11010 would be represented by $1.X^4 + 1.X^3 + 0.X^2 + 1.X^1 + 0.X^0$ or more simply as $X^4 + X^3 + X$. The data stream is written with the MSB on the left when this is transmitted first. The degree of a polynomial is the power of the highest order term, in this example 4.

To generate the code for transmission, three polynomials are used: the message polynomial $k(P)$, a generator polynomial $G(P)$, which is selected from a group of *primitive polynomials* to produce the desired characteristics of block length and error detection/correction capability, and a parity check polynomial $c(P)$. As for block codes, $n = k + c$ gives the coded word length.

During encoding, $k(P)$ is loaded into a shift register and then multiplied by X^c to move the message c bits to the left, to make room for c parity bits. This is then divided by the generator polynomial to produce a remainder that forms the parity check polynomial $c(P)$, which is then loaded into the remaining shift register cells. Thus:

$$(k(P).X^{n-k})/G(P) = Q(P) + c(P)$$

where $Q(P)$ is a quotient polynomial.

The following example shows how the code vectors are produced using a (7,4) cyclic code $(k = 4, c = 3)$ using $X^3 + X + 1$ as the generator polynomial:

$$\text{message code} = 1101, k(P) = X^3 + X^2 + 1$$

$$k(P).X^c = (X^3 + X^2 + 1).X^3, = X^6 + X^5 + X^3$$

Now divide by $G(P)$ and note that $+1 = -1$:

$$X^3 + X + 1 \overline{)\ X^6 + X^5 + X^3}\ (X^3 + X^2 + X + 1$$
$$\underline{X^6 + X^4 + X^3}$$
$$X^5 + X^4$$
$$\underline{X^5 + X^3 + X^2}$$
$$X^4 + X^3 + X^2$$
$$\underline{X^4 + X^2 + X}$$
$$X^3 + X$$
$$\underline{X^3 + X + 1}$$
$$1 = \text{Remainder}$$

Therefore $X^6 + X^5 + X^3 + 1$ divides exactly by $G(P)$, and so forms the polynomial for transmission, 1101001, the first four bits being the original $k(P)$ (1101) and the remaining three bits $c(P)$ (001). The full list of the $2^k = 16$ code words are shown in Table 11.2.

Table 11.2 *Polynomials*

Message	Parity
0000	000
0001	011
0010	110
0011	101
0100	111
0101	100
0110	001
0111	010
1000	101
1001	110
1010	011
1011	000
1100	010
1101	001
1110	100
1111	111

If a code word $T(P)$ is transmitted, and received without error, $T(P)$ divides exactly by $G(P)$ to leave a zero remainder. The last c bits are then stripped off to leave the original message code. If, however, an error occurs, then division by $G(P)$ leaves a remainder polynomial that forms a syndrome. There is a one-to-one relationship between this and the error pattern, so that correctable errors can be inverted by the error correcting logic within the decoder.

The effectiveness of these codes depends largely on the generator polynomial. For a polynomial of degree n, the decoder is generally capable of detecting error bursts of n or less bits (some errors outside of this bound may be detectable), and an odd number of random errors. A further advantage where speed is less important is that CRC can be operated with microprocessor-based coding, where the system characteristics are reprogrammable.

Bose–Chaudhuri–Hocquenghem (BCH) codes

This subset of cyclic codes is widely used for the control of random errors. For any positive integers m and t, such that $t \leq 2^{m-1}$, codes exist with the following parameters:

Block length	$n = 2^m - 1$
Parity bits	$c = n - k \leq mt$
Minimum distance	$d \geq 2t + 1$

which are capable of correcting t or less errors in a block of $n = 2^m - 1$ bits. The generator polynomials must be at least of degree mt. In general, the number of check bits c is almost mt, but if t is less than 5, then $c = mt$. Values for n, k, and t are tabulated in the references for various polynomials. For example, when $m = 5$, $n = 31$, $k = 26$, $c = 5$; t is equal to 1, but for $m = 5$, $n = 31$, $k = 11$, $c = 20$; t rises to 5.

Fire codes

This class of burst error correcting cyclic code, which was discovered by P. Fire, is systematically constructed for correcting or detecting a single burst of errors in a block of n digits, with small redundancy $(n - k)$. If $P(X)$ is an irreducible polynomial of degree m and e is the smallest integer so that $X^e + 1$ is divisible by $P(X)$, then an l-burst error correcting Fire code is generated by the polynomial $G(X) = P(X)(1 + X^{2l-1})$, where $l \leq m$ and $2l - 1$ is not exactly divisible by e. The number of parity bits is $m + 2l - 1$ and the code length is given by $n = \text{LCM}(e, 2l - 1)$, i.e., the lowest common multiple of e and $2l - 1$. A particularly popular Fire code that is capable of correcting any burst of length l or less and also detecting any burst of length $d \geq l$ is generated by the polynomial $G(X) = P(X)(1 + X^c)$ where $c \geq l + d - 1$ and the length $n = \text{LCM}(e, c)$.

Golay codes

Of the limited members of this sub-set of cyclic codes, the (23,12) version using 11 parity bits is very popular. This is capable of correcting any combination of three random errors, including a burst of three errors in a block of 23 bits. BCH codes are based on the generator polynomials:

$$G_1(P) = X^{11} + X^{10} + X^6 + X^5 + X^4 + X^2 + 1$$
$$G_2(P) = X^{11} + X^9 + X^7 + X^6 + X^5 + X + 1$$

both of which are factors of $X^{23} + 1$. Encoding and decoding is effected using integrated circuits based on 11-bit feedback shift registers.

Reed–Solomon codes

These are a sub-set of the BCH codes, which correct any combination of t or fewer errors and require no more than $2t$ parity bits. The parameters are $n - k = c = 2t$ and $d = 2t + 1$. As with all cyclic codes, a generator polynomial is used for encoding, and the decoder again produces a syndrome that identifies the error pattern.

An extension of the interleaving technique is sometimes used in special cases. This involves the generation of two Reed–Solomon codes from the data, and then cross-interleaving the coded bit patterns before transmission. Provided that the encoder and decoder are synchronised, relatively very long burst errors become correctable (see also Two-dimensional coding, p. 109).

11.5 Recurrent or convolution codes

These are so called because the encoding technique involves the convolution (see Signals, convolution, p 304) of the message stream with a generator matrix (also referred to as Hargelbarger codes after D.W. Hargelbarger who carried out much of the early work on convolution codes). They provide a most effective way of dealing with burst errors. Unlike cyclic codes, the parity bits within a block of convolution code check the message bits in previous blocks as well. Information about one particular bit therefore appears in several places, thus improving the error correction properties. The range of blocks over which the check bits are operative is defined as the *constraint length m* for the code (a block code might then be described as a convolution code for $m = 1$), the parity bits formed on the current block being dependent upon the message bits in the previous $m - 1$ blocks. The transmitted code structure, shown in Fig. 11.3, has the following parameters:

Block length	$= n$
Message bits per block	$= k$
Parity bits per block	$= c$
Constraint length n_o	$= m$ blocks $= mn$ bits
Message bits within n_o	$= k_o = mk$ bits
Parity bits within n_o	$= c_o = mc$ bits
Typical code rate	$= \frac{1}{2}$ or $\frac{1}{3}$

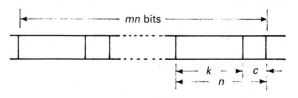

Figure 11.3   Convolution coding structure.

For these codes, n and k are normally small integers, often less than 5.

The encoder consists essentially of feedback shift registers and Ex.OR logic. In the Hargelbarger codes, alternate bits in the transmitted data stream are parity bits that have been formed by the addition modulo-2 from the xth and yth previous message bits. For example, with the code where $x = 14$ and $y = 7$ (i.e. one parity bit per message bit), bursts of up to six consecutive bit errors can be corrected, provided that there are at least 19 error-free bits in the previous signal stream. For decoding, there are two strategies that can be used. Logic or algebraic decoding is restricted to the constraint length and this uses the same circuit as the encoder plus the necessary logic to invert any correctable bits in error. The alternative is *probabilistic* or one of the *maximum likelihood* techniques which requires storage for more than the constraint length.

11.6 Specific decoding techniques

Maximum likelihood, minimum distance or correlation decoding

This technique operates on the basis of determining the cross-correlation between the received signal and each member of the code word set, the code word that generates the maximum energy being the one that exhibits minimum distance or bit errors. Since binary signals are discrete, this process is effectively one of summation, rather than the performance of the correlation integral (see Signal, correlation functions, p. 305).

Permutation decoding. At the decoder, the information and parity bits are separated. The parity bits are then recomputed and the two sets compared. If there is one or less differences, then the message bits can be accepted. If more than one difference is detected an error in the message has been found. This is corrected by shifting the bits in the received code word one step at a time through the shift register, computing the parity bits at each step, thus checking all possible permutations of parity bits. This testing stops as soon as agreement between recomputed and original parity is achieved.

Threshold decoding. The decoder holds, within a memory, a look-up table of all the correctable error patterns and their corresponding syndromes. The larger the table, then the greater the error correcting capability. This allows indentification of the bits in error, then by using suitable logic circuits to invert these, the errors are corrected. Beyond the threshold of this table are the non-correctable errors, of which some may still be detectable.

Viterbi algorithm and decoder. This maximum like-lihood decoding algorithm devised by A.J. Vitberbi in 1967, together with the development of Viterbi decoder integrated circuits that make the concept transparent to the user, has done much to make convolution codes both practical and popular. As an example, a typical IC can handle raw data rates in excess of 1 Mbit/s, and using a code rate of $\frac{1}{2}$ can achieve the same bit error rate as for uncoded data but with about 5 dB worse signal to noise ratio. The technique uses a search tree or trellis structure as shown in Fig. 11.4 and the decoder continually calculates the Hamming distance between received and valid code words within the constraint length. This is checked at each node of the tree and if an error path is detected, the decoder backtracks to the previous node and takes the alternative path.

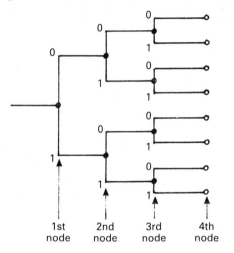

Figure 11.4 Decoding search tree.

11.7 Two-dimensional coding

Two-dimensional Reed–Solomon codes have been devised so that an inner code can be used to correct burst errors of a few bytes length, while the outer code can correct very long bursts. In one such code, the data is organised into a matrix of 600 bytes ×30 rows. The first or outer code is then used to add two check bytes to each of the columns to create 32 rows. Each row is then divided into 10 blocks of 60 bytes and the second or inner code is used to add four check bytes to each block of 60 bytes, to give an array of 640 bytes ×32 rows. The inner code is able to correct all single-byte errors and tag all errors that are longer than this. The outer code can correct two inner code blocks of 60 bytes each, provided that they are tagged when passed to the decoder. It can therefore correct 120 bytes. However, since the inner code is interleaved to a depth of 10 bytes, the outer code,

when using the inner code tags, can correct an error up to 1200 bytes long.

11.8 Miscellaneous terms

AUTOSPEC. An acronym of 'automatic single-path error correction' and relates to a coding format devised for use with the 5-bit 'Baudot' teleprinter code. This is a 10-bit block code that consists of the standard 5-bit alphanumeric codes plus five parity bits. The latter are formed modulo-2, from the information bits four at a time. The first parity bit is formed after dropping the first message bit, the second parity bit after dropping the second message bit, etc. The parity bits are then appended or concatenated on to the Baudot code.

Block code. A code format of fixed length $n = k + c$, where k and c are the number of message and parity bits, respectively.

Block systematic code. A block code where all the message bits appear unchanged in the code word but not necessarily in their natural order. The parity bits may thus be interleaved with the message bits.

CCITT V41 code. A (256,240) cyclic code that uses the generator polynomial of $X^{16} + X^{12} + X^5 + 1$ and was devised for those situations whereby a long message stream could be formed by the concatenation of a number of shorter elements. The parity bits then refer to several characters transmitted in sequence to reduce the redundancy overhead.

Code rate. The ratio of the number of message bits to the total number transmitted or k/n. The code rate is equal to (1 − redundancy) (see Code redundancy, below).

Code redundancy. The ratio of the number of parity bits included for error control, to the total number transmitted, or c/n. It follows that, since $n = k + c$, the code redundancy is equal to (1 − code rate).

Error extension or multiplication. When using equipment such as data scramblers or code converters, an input error of one bit can give rise to more than one error at the output. For example, with the 4B3T coding, a 1-bit error in the ternary word can create up to four errors in the decoder output. The effect is particularly troublesome in units where the error forms part of a feedback signal.

Error rate or bit error rate (BER). The ratio of the number of bits received in error to the total number of bits transmitted. Usually expressed in the form, 1 in 10^x, or as 10^{-x}.

Error spread. Due to error extension, a single random bit error can give rise to a burst of errors.

Fano's algorithm. All the messages allowed in the system are replaced by a variable length code that depends upon the probability of a message being transmitted. The probability of each message is calculated and the messages arranged in descending order of probability. The list is then divided into two groups of approximately equal total probability; the upper and lower parts are allocated the value 1 and 0, respectively. This process is continued until all the messages have been assigned a unique code sequence. The principle is explained with the aid of Fig. 11.5.

Hard errors (see also Soft errors, below). This refers to data errors that cannot be corrected no matter how many times the data is re-read.

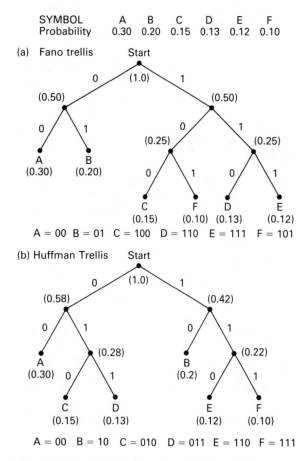

Figure 11.5 Fano/Huffman encoding algorithms. (a) Fano trellis, (b) Huffman trellis.

Huffman's algorithm. As with the Fano algorithm, the messages are again arranged in descending order of probability, but in this case, the two lower probabilities are combined to produce a new value. The lowest message is then designated 0 and the upper group 1. The upper group is then treated as a new list and the division repeated. Continuing this process, results in the upper or most likely messages are allocated the shortest code words (see also Facsimile, p. 114, and Fig. 11.5.).

Iterated coding. Successive message words are loaded into a matrix and then row and column parity bits calculated and added to the matrix in a manner similar to interleaved coding. The derived code words together with the parity bits are then transmitted in the same order. The technique allows any single-bit error to be corrected, because any such error occurs at the intersection of row and column parity checking.

Metric. A mathematical function related to the separation of two points in space, but in terms of coding, this concept is extended to the Hamming distance between two code words. In probabilistic decoding, the decoder has to make *hard decisions* based on distance (see Soft decision decoding, below).

Punctured codes. At higher data rates, Viterbi/Trellis decoding becomes more complex and hence more costly. *Punctured* convolution codes can achieve higher data rates without this penalty. The technique operates by selectively and periodically removing bits from the encoder output to effectively raise the code rate.

In practice, the output from an adaptive convolutional encoder is passed through an analogue gate circuit in blocks of parallel bits. The switching action of the gates is controlled by the *puncture matrix*, which defines the characteristics of the specific error protection for the bits in an information block.

The puncture matrix has to be available at the receiver decoder either as a hardware element or in software. The decoder then treats the punctured code in the same way as erased bits so that these do not affect the decoder decision making.

Residual or undetected error rate. The ratio of the number of bits received in error that escape the detection/correction system. In certain cases the ratio may be extended to include characters or blocks of data.

Run length coding. In certain operations such as image processing, the digital code remains unchanged for relatively long periods of time. When such states are detected, it is only necessary to transmit the actual value once, together with the length of time that this persists, i.e. a value followed by a stop symbol.

Soft decision decoding. This technique allows the metric to be modified in response to some other changing system parameter such as signal to noise ratio. It allows the decoding strategy to be varied to suit varying channel conditions.

Soft errors (see also Hard errors, p. 110). This refers to data errors that can be corrected if the data is read again.

Wagner coding. Under certain conditions, a single-parity-bit code can provide error *correction*. If a byte is received with incorrect parity and it is clear that one particular bit was received under abnormal conditions, then there is a high probability that the inversion of this bit will correct the error and restore parity.

11.9 Useful references

Coates R.F.W. (1985) *Modern Communication Systems*, 2nd edition, Macmillan, London.

Connor F.R. (1972) *Introductory Topic in Electronics and Telecommunication, 'Signals'*, Edward Arnold, London.

Hagenauer J. (1987) Rate compatible punctured convolutional codes, *Proc. ICC '87*, June, pp. 1032–1036.

Hamming R.W. (1950) Error detecting and error correcting codes, *Bell System Technical Journal*, **26**(2) , pp. 147–60.

Lewis G.E. (1992) *Communication Services via Satellite*, 2nd edition, Butterworth-Heinemann, Oxford.

Peterson W.W. and Weldon E.J. (1980) *Error Correcting Codes*, 2nd edition, Massachusetts Institute of Technology, USA.

Shu Lin (1970) *An Introduction to Error-Correcting Codes*, Prentice Hall, Englewood Cliffs, NJ.

Shu Lin and Costello D.J. (1983) *Fundamentals and Applications of Error Control Coding*, Prentice Hall, Englewood Cliffs, NJ.

12 Facsimile (Fax) systems

These systems, originally developed to transmit photographic and documentary information over standard telephone cables, are now used over radio and satellite links (Docfax) as well. The wide variety of services available range from the transmission of simple letters, to weather and geological mappings and the transmission of complete pages of a newspaper for remote printing (Pagefax).

The information in a document to be transmitted usually lies in dark markings on a light background. This can be analysed by segmenting the document into elemental areas small enough to resolve the finest detail needed. To achieve this, the document is scanned sequentially by a light beam, in a series of very narrow strips. The magnitude of the reflected light from each *picture element* (or *pel* or *pixel*) is then used to generate an electrical signal. In the earlier systems, this was accomplished by using a photocell/electron multiplier tube, but this has been superseded by an array of semiconductor devices, either photodiodes or charge-coupled image sensors (see also Semiconductor devices and technology, p. 292). Since Fax service terminals are transceivers, this signal can be used to construct an accurate facsimile of the original document, either locally or at a distance.

It is not always necessary for the facsimile to be the same size as the original and certain terminals have the facility to enlarge or reduce. However, if distortion in the document is to be avoided, the page aspect ratio should remain constant. The compatibility of aspect ratios is reflected in the *factor* or *index of cooperation* (FOC or IOC), which is based on the ratio of scan line length to vertical scanning density or line width. FOC = effective scan length × vertical scan density (effective scan length = actual length + 5% to accommodate a phasing signal). Since the IOC ratio was defined for the original drum scanned systems, IOC = FOC/π.

Provided that the terminals working together have the same ratio, the document shape will be retained, even though the actual size may be different.

12.1 Facsimile terminal (description and operation)

Generating the Fax signal. In the early and now obsolete terminals, the document to be transmitted was wrapped around a drum, and while this rotated, an optical system driven by a lead screw scanned the document to generate an analogue electrical signal that represented the varying shades of a *grey scale* for a single line. The drum system gave way to a *flat-bed* technique, where the document was scanned by moving it over an illuminated window with the optical system producing a narrow strip of reflected light that represented a single line. This was then scanned by a rotating helical slit to provide the light drive to the photocell, which again produced an analogue signal.

The current techniques again involve moving the document over an illuminated window, but with reflected light being focused on to an array of up to 5000 semiconductor photodiodes. This effectively produces a digital representation of the dark and light areas, but a whole line at a time. This signal can be temporarily stored in a semiconductor memory. Since each pixel signal can be generated in about 2 μs and the line signal can only be changed and transmitted in about 140 μs, there is ample time for signal coding and companding.

Operation of the terminal. After the signal has been generated, it is modulated on to a suitable sub-carrier and filtered to ensure that its bandwidth lies within the spectrum of the telephone channel (300 to 3400 Hz). The actual method of modulation used depends upon the level of technology used (see Table 12.1). A line amplifier is then used to achieve the correct signal level and impedance match for the transmission medium. The receiver section of the terminal is practically complementary, the received signal first being amplified and filtered to remove any line noise. Demodulation restores the baseband signal, which is then passed to a *marking amplifier* which drives the facsimile producing mechanism.

Table 12.1 *Standard FAX groups*

	Group 1	Group 2	Group 3	Group 4
Generic title	Low speed	Medium speed	High speed	Ultra high speed
Transmission speed (A4 document)	6 min	3 min	1 min	2–4 s
Modulation	FM	AM/PM-VSB	DPSK	Digital (64 Kbit/s)
Carrier frequency	1700 Hz ± 400 Hz	2100 Hz	1800 Hz	
White level	1500 Hz	Max. Amp.		
Resolution — Vertical	3.85 1/mm	3.85 1/mm	3.85 1/mm	7.7 1/mm
		5.3 1/mm	7.7 1/mm	15 1/mm
Resolution — Horizontal		5.3 pel/mm	8 pel/mm	16 pel/mm
Image signals	Analogue		Digital	Digital
Handshake signals	Audio tones		300 b/s FSK (CCITT V21)	ISDN compatible (CCITT V29/33)
Redundancy reduction	None	None recommended	READ	Modified READ

Microprocessor control of the Fax terminal can be introduced with advantage. This allows the use of encryption for security reasons and *auto dialling* and *auto answering* for unattended operation.

Recording methods

The actual method used on any terminal depends upon the particular application. For instance, units designed for portable operation introduce limitations that do not exist with office based units.

Electrolytic methods. These use a special paper that is impregnated to render it conductive. Current flows from a scanning stylus, through the paper to a fixed back contact. This causes the paper to discolour in proportion to the magnitude of the current density.

Electrostatic methods. This method, which is popular in office photocopiers, may also be found in Fax machines. Styli attached to a flexible belt are driven across standard paper to leave a charged pattern on it. The styli voltage varies between +600 V for black and −450 V for white. The charged paper passes over a magnetic roller bristling with magnetised iron particles which hold a black toner powder in the interstices of the bristles. The powder is attracted to the positively charged areas and repelled by the negative regions. The deposited powder thus forms a facsimile of the original document. Passing the paper over a heated roller causes the toner to bond permanently to the paper.

Thermal methods. This is a very common method that employs a special thermosensitive paper, which discolours when its temperature exceeds about 70°C, full black being obtained at around 110°C. The paper is passed over an array of minute heating resistive elements that are constructed using thin-film technology. These provide heating elements with very little thermal inertia so that they can respond to the rapidly varying signal. One end of each element is held just below a critical voltage level where zero signal level produces no mark on the paper. This makes maximum use of the signal which can cause the paper temperature to vary between about 60°C and 110°C, to produce a copy with varying tonal shades where these occur in the original. Copying on to standard paper is also carried out using a thermal transfer technique. A special meltable ink film is carried on a base film and this can be transferred to the copy paper by heated thermal heads. Colour printing is also possible using this technique, by transferring coloured inks (yellow, cyan or magenta), successively overlaid, to provide a wide range of colours.

Weatherfax recording systems. Because of the highly automated processes and the large size of print-outs required, two special techniques have been developed for the printing of weather maps. A photographic unit which uses photosensitive paper, stored either in cassettes or as separate sheets, for automatic feeding, provides the facsimile. The exposure light source may be a laser or a high brightness lamp. The unit contains facilities to complete the photographic development, and this follows on automatically after exposure. This method has the advantage of high grade reproduction with a grey scale of more than 16 levels. However, since the process is a wet one, the equipment is not suitable for portable operation and the cost of the unit and the paper is high. A dry silver process is also available and this can provide good quality reproduction but with a reduced tonal range. The process uses a laser as the exposure light source and the paper is relatively expensive. Storage temperature of printed documents can also be a problem, as heat causes the facsimile quality to degrade.

Synchronising and handshaking. Handshaking describes the sequence of signals used between communicating terminals prior to the transmission of Fax signals. The procedure is carried out to check the status and mutual compatibility, to issue control commands and, in some cases, to monitor the line conditions. It is a way of identifying each terminal's capability, whereupon one will select a particular mode, which in turn will be acknowledged by the other. Following this procedure, the originating terminal will transmit a start signal, the nature of which depends upon the group (see Table 12.1). This may either be a 30 s white burst or one of two audio tones. A short phasing burst follows to provide synchronism before the Fax signal starts. Stop is indicated either by a loss of carrier for more than 5 s or another tone burst. After the transmission, further handshaking confirms satisfactory reception and indicates if more pages are to be sent.

12.2 Signal compression techniques (see also Image processing, p. 138 and Television signal processing and systems, p. 356)

Reference to Table 12.1 shows that a Group 3 terminal at minimum resolution produces about 30 pel/mm^2. So that for an A4 page 210 mm × 297 mm there will be approximately 1.9 million pixels. If all of these have to be transmitted, then the coding system used must be very efficient to minimise the transmission time and/or the bandwidth required. A study of typical test documents has shown that considerable redundancy exists which can be eliminated by suitable coding:

(a) Many sections of each scan line are continuously white and hold no useful information to print. Omitting these produces a horizontal economy known as one-dimensional coding. This technique provides the least compression but yields the highest error immunity.
(b) On average, each scan line has a high correlation with both of its neighbours. This fact can be exploited by only transmitting information on how the current line differs from the previous one and so produces a vertical economy. The use of both horizontal and vertical coding (two-dimensional) is referred to as *relative element address designate*

(READ) coding and can reduce the transmission time by a factor of 10 or more. Because one error can propagate through several lines, the error immunity is rather lower than for one-dimensional coding, but this is offset by the considerably higher compression ratio.

One-dimensional coding. Statistical histograms of black and white pixel runs obtained from test pages show that black runs have a peak probability of approximately 0.15 for runs of 2 to 4 pixels, with a spread extending beyond 100. The corresponding values for white runs have a flatter distribution, peaking at about 2 to 9 pixels with a probability of about 0.05, but with a spread extending beyond 1000. Run length coding is therefore most suitable for handling the redundancy, which is achieved using a *modified* Huffman code (see Error control, Huffman code, p. 110). The code format is modified to handle the double-sided distribution of the message probabilities, both black and white, and still retain the feature of fewest bits of code for the most probable message or run length.

Two-dimensional coding. The statistics also reveal the line-to-line correlation for transitions between black and white and vice versa. About 50% of the transitions on any line fall directly beneath a similar transition on the line above. Further, 25% of the transitions occur within one pixel of a similar transition on the line above. Therefore about 75% of all transitions occur within ±1 pixel of a similar transition. Modified READ coding takes both the vertical and horizontal redundancy into consideration by storing the previous line as a reference and continually referring back to it as the system encodes and decodes the next scan line.

12.3 Standard groups

Table 12.1 lists the important parameters and characteristics of the four standard groups.

Group 1. These terminals are now virtually obsolete. Frequency modulation (FM) of the sub-carrier was used to minimise the effects of line noise but the technique proved to be too slow for modern applications.

Group 2. Amplitude modulation (AM) was introduced to increase the operational speed. A normal double-sideband AM signal can be filtered to remove part of one sideband, leaving a *vestige* of it to provide a *vestigial sideband signal* (VSB). This reduces the AM bandwidth and still leaves a signal that can be processed in a normal AM manner. The use of VSB saves about 30% to 40% on bandwidth to support an increase in transmission speed. A further bandwidth saving was effected by introducing a *phase modulation* (PM) component. Assume that white represents a positive voltage and black zero volts. A series of transitions between black and white thus represent a particular frequency. If alternate white peaks are now inverted to produce a negative voltage (see also Codes and code formats, p. 57, Alternate mark inversion, p. 58), the fundamental frequency is halved. When this signal is used to modulate the sub-carrier, the negative peaks produce a carrier phase reversal, producing a white signal without any change of frequency of amplitude. This form of modulation, known as AM/PM-VSB, has a reduced bandwidth such that the transmission speed can be doubled. While these terminals are on the obsolescent list, they are still capable of being worked with most current Group 3 terminals.

Group 3. Currently these terminals provide the major part of the Fax service, because the operating speed is just about the maximum that voice grade telephone lines can support. They are normally made so that they can communicate with Group 2 machines but at the Group 2 speed, the handshaking signals being used to set up such a link. As *differential quadrature phase shift keying* (DQPSK) (see also Modulation, digital, p. 190) is used, each sub-carrier phase represents two bits, thus halving the bandwidth and allowing for data rates of up to 9.6 Kbit/s to be employed over a standard voice grade telephone line. Because of the digital nature of the signals involved, Group 3 working can easily be implemented on personal computers.

Group 4. Like the ISDN systems that this group is designed to be compatible with, this standard is still developing and is still not fully implemented.

12.4 Miscellaneous terms

Auto-answer. A facility that allows a Fax terminal to answer a call while unattended.

Auto-dial. A facility that allows a Fax terminal to originate a call even when unattended. This allows the transmission of documents automatically during off-peak rate periods.

Broadcast. The facility to transmit to a number of receiving terminals simultaneously.

Dial-up. The means of setting up a communications link over the Public Switched Telephone Network (PSTN) by dialling a telephone number.

Duplex/half duplex. A duplex Fax terminal can send one document while simultaneously receiving a different one. Half duplex terminals can only send or receive at different times.

Grey scale. The number of distinguishable grey tones between black and white.

Huffman code (see Table 12.2). The messages are arranged in descending order of probability as shown.

Table 12.2 *Basic Huffman coding structure*

Message	Probability							
m_1	0.42	0.42	0.42	0.42	0.42	0.42	0.58 ⌐1	1.0
m_2	0.20	0.20	0.20	0.20	0.23	0.35 ⌐1	0.42 ⌐0	
m_3	0.12	0.12	0.12	0.15	0.20 ⌐1	0.23 ⌐0		
m_4	0.11	0.11	0.11	0.12 ⌐1	0.15 ⌐0			
m_5	0.06	0.06	0.09 ⌐1	0.11 ⌐0				
m_6	0.04	0.05 ⌐1	0.06 ⌐0					
m_7	0.03 ⌐1	0.04 ⌐0						
m_8	0.02 ⌐0							

Message codes
m_1 0
m_2 111
m_3 101
m_4 001
m_5 0011
m_6 01011
m_7 111011
m_8 011011

The two lowest are then combined to provide a new value. The lowest message is then designated 0 and the upper group as 1. The upper group is then treated as a new list and the procedure repeated. Continuing this procedure results in the most probable message being allocated the shortest code, each code word being derived by tracing a path through the structure. This code structure lends itself to modification for the run length coding used for data compression (see One-dimension coding, p. 114). In a practical case, the code word patterns are held in a read only memory look-up table and accessed via an input of run length.

Modem (modulator/demodulator) (see also Networks, p. 204). A device used to convert the data signals into suitable format for transmission over a network. In a complementary sense, a similar device will decode the data at the distant receiver.

Modified modified READ coding. The modified READ coding uses the modified Huffman run length coding for the first line of a page and then compares following lines on a pixel-by-pixel basis to determine the difference to be used as the signal for transmission. By contrast, the modified modified READ (MMR) coding is a pure 2D coding system which can give a greater degree of data compression, assumes that the first line is all white and then derives the differences using this as a reference. However, because an error created on one line propagates throughout the rest of the page, MMR is only used in the relatively error-free environment of leased lines.

Password. A security feature that can be used to enhance the integrity of the system and ensure privacy.

Polling. The ability to request a distant terminal to transmit automatically after checking the password to ensure that both are valid users.

Resolution. The number of scanning elements or pixels used to produce the final image. Horizontal resolution is measured in pels/line, while vertical resolution is defined in scanning lines per millimetre.

Skip-white-spaces. An alternative time saving technique that can be used when no black/white transitions are detected, i.e. in a white area. This technique requires a feedback network in the scanner mechanism and is used successfully in the Pagefax system that can transmit whole pages of newsprint.

Synchronising signals. The start or beginning of a document is signalled by a short *white burst*, while the start of each line is defined by a *phasing burst* of white. The end of the document is indicated by a *stop* signal. These signals all appear in the dead space of the document area.

12.5 Useful references

Muirhead Data Communications Ltd, UK (1986) What is this thing called Fax. Fax Appreciation Course Notes.
Lewis G.E. (1992) *Communication Services via Satellite*, 2nd edition, Butterworth-Heinemann, Oxford.

13 Filters

Filters are linear circuits that are devised to modify the frequency distribution of a signal spectrum and are available to cover the whole of the electromagnetic spectrum, from dc through to light frequencies. They are described as linear, because the principles of superposition and proportionality apply and the excitation (input) and response (output) are related by a linear differential equation. Filters are also *time invariant*, that is, their characteristics do not change with time. A common feature of all filters is that they have at least one pass band with zero attenuation and at least one stop band in which a finite attenuation is produced. This leads to the concepts of *low-pass* (LPF), *high-pass* (HPF), *bandpass* (BPF) and *bandstop* (BSF) filters, terms which describe the filters' frequency characteristics. The slope of the transitions between pass and attenuation bands is a particularly important parameter. It is interesting to compare filters with attenuators and equalisers. Attenuators are devised to produce a fixed degree of attenuation that is frequency invariant, while the attenuation of equalisers varies gradually with a change of frequency. Filters are usually constructed from reactive components only, capacitive and inductive; attenuators are constructed from resistive components only, and equalisers from a combination of both reactive and resistive components. Given an input signal with a spectrum $V_1(\omega)$, any unwanted signal components can be removed using a filter with a *transfer function* $H(\omega)$. The output spectrum is then obtained by multiplying the input spectrum by the transfer function. Thus, the output spectrum $V_2(\omega) = V_1(\omega) H(\omega)$, or $H(\omega) = V_2(\omega)/V_1(\omega)$, the transfer voltage ratio. The concept of signal filtering in the time domain is equivalent to the application of the *convolution* operation (see also Signals, p. 304). The parameters of a filter are usually defined in terms of the amplitude ($A(f)$ or $A(\omega)$) and phase ($\theta(f)$ or $\theta(\omega)$) responses when plotted against frequency. Both can be combined in the transfer function $H(j\omega)$ which can be converted into the complex frequency concept, $s = \sigma + j\omega$. The transfer function $H(s)$ then becomes the ratio of two polynomials; for example $H(s) = 4s/(s^2 + 2s + 2)$.

When this function is factorised, its properties are more clearly revealed by a *pole–zero* diagram representing the steady-state conditions. This can be drawn in the manner shown in Fig. 13.1. Poles and zeros are defined as those values of s that make the denominator and numerator polynomials respectively equal to zero:

$$4s/(s^2 + 2s + 2) = 4(s)/((s + 1 + j)(s + 1 - j))$$

Therefore there is one zero at $s = 0$ and two poles at $s = (-1 + j)$ and $(-1 - j)$. These are plotted on the diagram of Fig. 13.1, which reveals much about the circuit that this represents. For example, a pole or zero at the origin represents an integrator or differentiator,

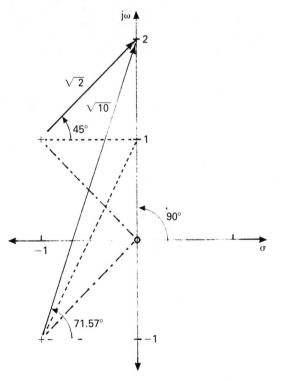

Figure 13.1 Pole–zero diagram for $H(s) = 4s/(s^2 + 2s + 2)$.

respectively, while stability is only certain if there are no poles in the right-hand side of the plane.

Using Fig. 13.1 the magnitudes of the amplitude response $A(\omega)$ and phase response $\theta(\omega)$ can be obtained as follows:

$$A(\omega) = \frac{\prod_{i=0}^{n} \text{vector magnitudes from the zeros to a point on } j\omega \text{ axis}}{\prod_{j=0}^{m} \text{vector magnitudes from the poles to same point on } j\omega \text{ axis}}$$

$$\theta(\omega) = \sum_{t=0}^{n} \text{angles of vectors from zeros to } j(\omega) \text{ axis}$$
$$- \sum_{j=0}^{m} \text{angles of vectors from poles to } j\omega \text{ axis}$$

Thus for $\omega = 2$
$$A(\omega) = 4(2/(\sqrt{2} \times \sqrt{10})) = 1.789,$$
$$\theta(\omega) = 90° - 45° - 71.57° = -26.57°$$

For $\omega = 1$
$$A(\omega) = 4(1/(1 \times \sqrt{5})) = 1.789$$
$$\theta(\omega) = 90° - 0° - 63.44° = 26.56°$$

For $\omega = 0$ (dc)
$$A(\omega) = 4(0/(\sqrt{2} \times \sqrt{2})) = 0$$
$$\theta(\omega) = 90° - (-45°) - 45° = 90°$$

Thus the response curve has a peak at approximately 1.5ω and then decreases to zero as ω approaches infinity. An accurate response curve could easily be calculated using a digital computer with a suitable program.

13.1 Analogue filters

The most common circuit elements of analogue filters are of T or π form, the low- and high-pass versions of which are illustrated in Fig. 13.2. These basic structures, from which more complex filters can be made, are often described as *prototype* filters or *constant k* sections. The constant arises as follows, for either T or π sections: if Z_1 and Z_2, the series and shunt impedances, respectively, are related by $Z_1 Z_2 = R_0{}^2$, where R_0 is a real constant (k) resistance (independent of frequency), R_0 is described as the *design impedance* of the filter. Consider a T and a π network each constructed from the same series and shunt impedances Z_1 and Z_2 with *characteristic impedances* (Z_0) related as follows.

If $Z_{0T} Z_{0\pi} = Z_1 Z_2 = R_0{}^2$, then both are constant k sections. Clearly Z_{0T} and Z_0 will be real or imaginary together and both will change from real to imaginary at the same frequency, so that both have the same cut-off frequency and same pass bands.

13.1.1 Classical filter theory

This theory starts with a filter structure which is analysed, usually assuming perfect components, to obtain the necessary information about its characteristics. The complete study requires a knowledge of the characteristic impedance Z_0 and propagation constant γ, over the frequency range of interest. The characteristic impedance is defined in the same way as for a transmission line; that is, the input impedance of an infinite number of identical sections connected in series,

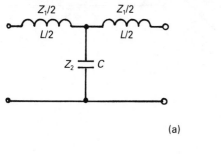
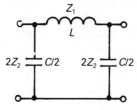

(a)

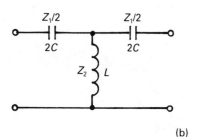
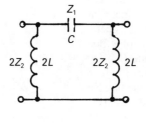

(b)

Figure 13.2 (a) T and π low-pass filters; (b) T and π high-pass filters.

or the input impedance of one section when terminated in that value of impedance. The propagation constant is complex and given by $\gamma = \alpha + j\beta$, where α, the real part, reflects the attenuation and β, the imaginary part, relates to the phase shift through the filter.

The characteristic impedances of the two networks are related as follows:

$$Z_{0T} = \sqrt{Z_1^2/4 + Z_1 Z_2}$$

$$Z_{0\pi} = (Z_1 Z_2)/Z_{0T}$$

The design impedance $R_0 = \sqrt{L/C}$ ohms and the cut-off frequency or $-3\,\mathrm{dB}$ point is given by $f_c = 1/\sqrt{LC}$ Hz. Thus substituting known values for R_0 and f_c allows the values of L and C to be calculated.

Complex filters. Several filter sections can be connected in series to increase the attenuation and reduce the width of the transition region between the pass and attenuation bands.

Half-filter sections. Because the characteristic impedance of the protype network is not constant over the whole of the pass band, matching arrangements need to be made at both input and output. This can be achieved using the half sections depicted in Fig. 13.3, which also

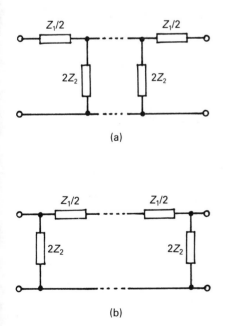

(a)

(b)

Figure 13.3 Derivation of half sections: (a) T network; (b) π network.

shows how these can be produced from the basic T and π circuits.

Lattice filters. Where large degrees of attenuation with very sharp transitions between pass and attenuation bands are needed, circuits of the type shown in Fig. 13.4 can be used. These are commonly constructed from piezo-electric crystals where Z_1 and Z_2 are carefully matched pairs of crystals with the series resonant frequency of one pair equal to the anti-resonant frequency of the other pair. The high Q of the circuit results in a response curve with steep skirts and a narrow pass band. Tyically the bandwidth is in the order of 0.75% of centre frequency.

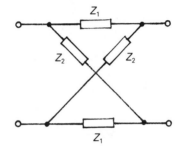

Figure 13.4 Lattice filter.

m-Derived filters. Prototype filters suffer from a variation of both attentuation and characteristic impedance in the pass band, together with a gradual transition between pass and attenuation bands. These problems can be minimised by modifying the circuits in the manner shown in Fig. 13.5. The factor m has a value between 0 and 1 and typically 0.3 to 0.6. The series and shunt impedances are modified by a factor of either m or $(1 - m^2)/4m$ as shown. This diagram also shows the effects of this change on component values. At a certain frequency, the shunt arm in the T network and the series arm in the π network resonate. In one case this acts as a short circuit to signals, while in the second case it produces a very high impedance. In both cases, this results in almost infinite attenuation. The frequency that produces infinite attenuation is defined as f_∞ and has the following relationship with m and the cut-off frequency f_c:

$$\text{LPF}, \; m = \sqrt{(1 - (f_c/f_\infty)^2)}$$

$$\text{HPF}, \; m = \sqrt{(1 - (f_\infty/f_c)^2)}$$

Thus by making m small, f_∞ approaches f_c and increases the steepness of the response curve roll-off.

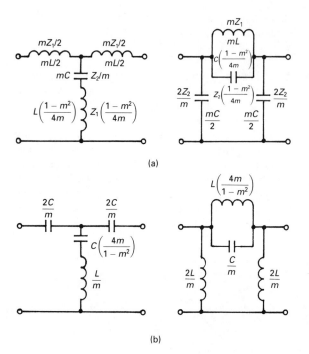

(a)

(b)

Figure 13.5 *m*-Derived filters; (a) low-pass; (b) high-pass.

13.1.2 Modern filter theory

The modern approach to filter design uses a step-by-step synthesis of the transfer function of an approximation to an ideal low pass filter (LPF) with a cut-off frequency ω_c of 1 rad/s and a characteristic impedance of 1 ohm. An ideal LPF has a transfer function of unity over the pass band and this falls sharply to zero at the cut-off frequency to give a square cut-off characteristic, a feature that is not realisable in practice. The approximating function is carefully chosen to provide the required effect. Then by the applications of suitable transforms, the values so obtained are converted into those needed for the actual filter.

The design starts by choosing a suitable transfer function $H(s)$ and then from it calculating the conjugate $H(-s)$. Then $H(s) \times H(-s) = |H(s)|^2$ and this gives the square of the amplitude response. The poles are then determined and a plot reveals much about the filter action.

Design data tables for the most popular approximations are given in many of the references. The transfer functions are given in polynomial form and the order of the polynomial defines the number of poles present in the synthesis and the *order* of the filter. The amplitude responses for a number of LPF of different order using two of the approximations are shown in Fig. 13.6.

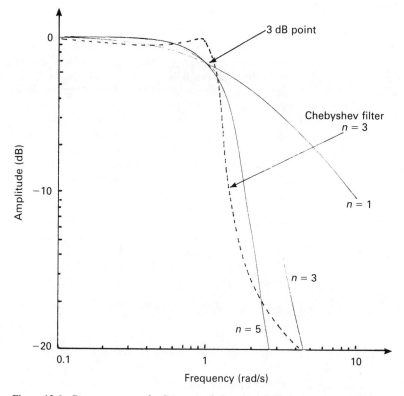

Figure 13.6 Response curves for Butterworth (——) and Chebyshev (- - - -) filters.

Bessel approximation. These filters give maximally flat group delay response over frequency domain pass band and are described by the relationship

$$H(s)_n = B_n(s)/B_n(0)$$

where $B_n(s)$ is the nth order Bessel polynomial defined by the recursive formula

$$B_n(s) = (2n - 1)B_{n-1}(s) + s^2 B_{n-2}(s)$$

where $n - 1$ is the previous term, or alternatively by:

$$B_n(s) = \sum_{i=0}^{n} b_i s^i$$

where each b_i is given by

$$b_i = \frac{(2n - i)!}{2^{n-1}!(n - i)!}$$

Thus

$B_0(s) = 1$, $B_1(s) = s + 1$ and $B_2(s) = s^2 + 3s + 3$.

$B_n(0) = [1 + \overline{2 \times 0}][1 + \overline{2 \times 1}][1 + \overline{2 \times 2}]...[1 + \overline{2 \times (n - 1)}]$.

The slope roll-off of the Bessel LPF is more gradual than that of the Butterworth approximation for the same filter order and gives a somewhat wider bandwidth. Beyond its $-3\,\mathrm{dB}$ point, the slope roll-off is approximately the same as Butterworth.

Butterworth approximation. These filters give a maximally flat amplitude response over the pass band, a characteristic that is often described as being monotonic. The amplitude response is given by:

$$A_n(\omega) = \sqrt{1 + (\omega/\omega_c)^{2n}}$$

where n is the filter order and the cut-off frequency ω_c and the $-3\,\mathrm{dB}$ point occur where $\omega = \omega_c$. Figure 13.6 also shows how all of the characteristics pass through this point irrespective of the filter order. The attenuation slope rolls off at the rate of $6n\,\mathrm{dB/octave}$ or $20n\,\mathrm{dB/decade}$.

Chebyshev (Tschebycheff) approximation. These filters have a sharper roll-off than Butterworth filters of the same order, as indicated by Fig. 13.6. For $\omega > 1$, the roll-off is given by, $20 \log \epsilon + 6.02(n - 1) + 20n \log \omega\,\mathrm{dB}$. Thus close to cut-off, the filter has a response at least one order better than the Butterworth filter. But at frequencies well beyond cut-off, the slope is again approximately $6n\,\mathrm{dB/octave}$.

The filters are non-monotonic in the pass band as shown; however, this ripple error, ϵ, can be made negligible by design. The transfer function is obtained as

$$|H(s)|^2 = 1/(1 + \epsilon^2 C_n^2(\omega))$$

The filter design thus depends upon the two variables, n the order and ϵ the permissible ripple amplitude.

The Chebyshev polynomials $C_n(\omega)$ are given by:

$$C_n(\omega) = \cos(n \cos^{-1} \omega) \quad \text{for } |\omega| \leq 1 \text{ and}$$
$$= \cosh(n \cosh^{-1} \omega) \quad \text{for } |\omega| > 1$$

Thus for $n = 0$, $C_0(\omega) = 1$ and for $n = 1$, $C_1(\omega) = \omega$. The higher order polynomials are then more easily obtained from the recursive formula:

$$C_n(\omega) = 2\omega C_{n-1}(\omega) - C_{n-2}(\omega)$$

so that

$$C_2(\omega) = 2\omega(\omega) - 1 = 2\omega^2 - 1$$
$$C_3(\omega) = 2\omega(2\omega^2 - 1) - \omega = 4\omega^3 - 3\omega$$
$$C_4(\omega) = 2\omega(4\omega^3 - 3\omega) - (2\omega^2 - 1) = 8\omega^4 - 8\omega^2 + 1$$

These filters are designed around values for n and ϵ and the normalised ripple amplitude is given by $1 - (1 + \epsilon^2)^{-0.5}$. Tables are published in many of the references for n up to 10 and ϵ values from 0.1 to 1 dB.

Elliptic approximation (also known as Cauer or Zolotarev filters). For a given filter order, these filters have the sharpest roll-off characteristic of all, but with ripple in both attenuation and pass bands. The filters are characterised by both poles and zeros, with the positions of the zeros being related to the elliptic functions of classical field theory, of which only the integral of the first kind:

$$\mu(\phi, k) = \int_0^{\phi} (1 - k^2 \sin^2 x)^{-0.5} dx$$

is commonly used. The parameters ϕ and k are the amplitude and modulus of the integral, respectively, and the integral is real for all real values of ϕ. Solution of this integral yields the transfer function

$$H(s) = \prod_{n=2,4...} \frac{(s^2 + b_{1n} + b_{0n})}{k(s^2 + a_0)}$$

for even n and

$$H(s) = \prod_{n=1,3...} \frac{(s^2 + b_{1n} + b_{0n})(s + c_0)}{k(s^2 + a_0)}$$

for n odd, where a, b and c are constants.

For the LPF shown in Fig. 13.7, this simplifies to $|H(s)|^2 = 1/(1 + \epsilon^2 R_n^2(\omega))$, where $0 < \epsilon < 1$ and $R_n(\omega)$ is an nth-order rational Chebyshev function.

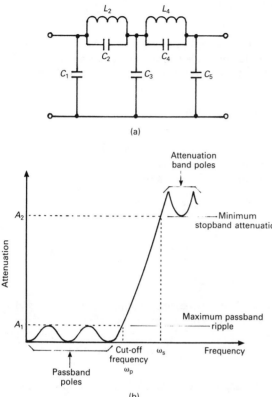

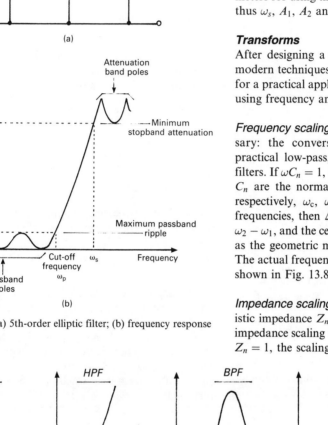

Figure 13.7 (a) 5th-order elliptic filter; (b) frequency response characteristics.

If A_1 and A_2 are the attenuation factors for the pass and stop bands, respectively, then ω_p and ω_s will be corresponding frequencies of interest. If the filter is normalised to $\omega_c = 1$, the maximum pass band ripple attenuation coincides with $\omega_c = \omega_s$. The relevant parameters for using the published tables for filter design are thus ω_s, A_1, A_2 and the filter order n.

Transforms

After designing a normalised low-pass filter using the modern techniques, it is necessary to redefine the circuit for a practical application. This is achieved in two stages using frequency and impedance scaling.

Frequency scaling. Four distinct transforms are necessary: the conversion of the normalised LPF into practical low-pass, high-pass, bandpass and bandstop filters. If $\omega C_n = 1$, the normalised frequency, and L_n and C_n are the normalised inductor and capacitor values, respectively, ω_c, ω_1 and ω_2 are the required cut-off frequencies, then $\Delta\omega$ can be defined as the bandwidth, $\omega_2 - \omega_1$, and the centre or resonant frequency ω_0 defined as the geometric mean of ω_1 and ω_2. Thus $\omega_0^2 = \omega_1\omega_2$. The actual frequency transforms and rescaled values are shown in Fig. 13.8.

Impedance scaling. The normalised value of characteristic impedance Z_n has to be scaled up to Z_0. Thus the impedance scaling factor N_z is equal to Z_0/Z_n and, since $Z_n = 1$, the scaling factor is also Z_0. The denormalised

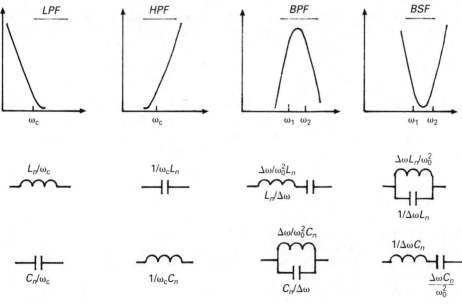

$$\Delta\omega = \omega_2 - \omega_1 \qquad \omega_0^2 = \omega_1\omega_2$$

Figure 13.8 Frequency scaling and transformation.

inductor and capacitor values become L_nN_z and C_n/N_z, respectively.

13.1.3 N-path or comb filters (see also Television signal processing and systems, p. 343)

N-path filters with a periodic response spectrum as shown by Fig. 13.9 are useful for selectively filtering a periodic signal from a background of uncorrelated noise. The path networks have a low-pass characteristic with a cut-off frequency small compared with $1/2T$ and this gives the circuit a response with a series of equally spaced pass bands that form a comb. Using suitable weighting amplifiers this results in an output signal that suffers negligible distortion but with a very significantly reduced noise component. Notch-comb filters with the inverse response are used to remove interference of a periodic nature from a signal that is essentially non-periodic.

13.1.4 Operational amplifier filters (active filters)

The inductors used in electronic circuits tend to be relatively bulky, heavy and generate stray electromagnetic fields. Because they are not perfect reactors, they also dissipate power. Practical operational amplifiers have a very high gain at low frequencies and this falls monotonically to unity at some high frequency, thus they behave as low-pass filters. By adopting resistance–capacity networks to provide phase shift and operational amplifiers to counter the resistive attenuation, a range of active filters become available that overcome these problems.

All pass network. These networks with an all pass characteristic, which are effectively delay equalisers, are often used to correct phase distortion that has been introduced in a transmission system. The transfer function for the circuit shown in Fig. 13.10(b) is given by:

$$(Z_1Z_3 - Z_2Z_4)/(Z_1Z_3 + Z_1Z_2)$$

If $Z_1 = Z_4$ and $Z_3 = R$, this function becomes:

$$(R - jX)/(R + jX)$$

and the response is unity over the frequency range of interest.

Bandpass filter. The multiple feedback circuit shown in Fig. 13.10(a) provides a pseudo-resonant circuit without the use of an inductor. The resonant frequency is given by:

$$\omega_0 = (R'R_3C_1C_2)^{-1/2}$$

where

$$R' = R_1R_2/(R_1 + R_2) \text{ and } Q = \omega_0R_3C_1C_2/(C_1 + C_2).$$

Gyrator circuit. The circuit shown in Fig. 13.11 is a useful element of filter construction. It is a non-reciprocal two-port circuit with an input impedance Z_i proportional to the reciprocal of the load impedance Z_L and $Z_i = R^2/Z_L$, where R is described as the gyration resistance. Thus if the load is a capacitor, the input terminals will appear as an equivalent inductor.

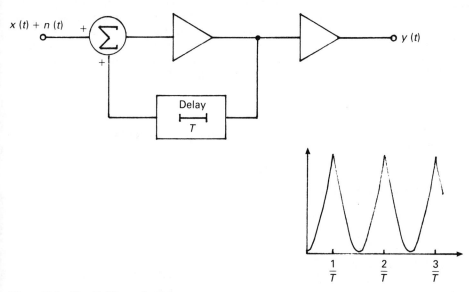

Figure 13.9 N-path filter and spectrum.

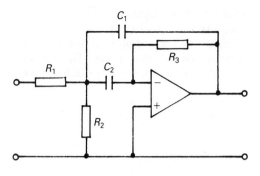

(a)

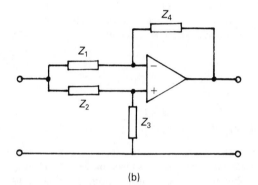

(b)

Figure 13.10   (a) Active bandpass filter; (b) active all-pass network.

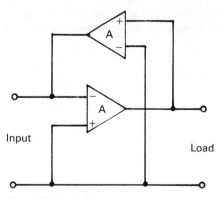

Figure 13.11   Gyrator circuit.

Inductors with values up to 1 henry and Q factors up to 1000 can be produced in this way.

Parallel or twin-T network. This circuit is formed by the parallel combination of low-pass and high-pass T networks which can be combined with a unity gain operational amplifier. By a suitable choice of the resistance and capacity component values a bandpass

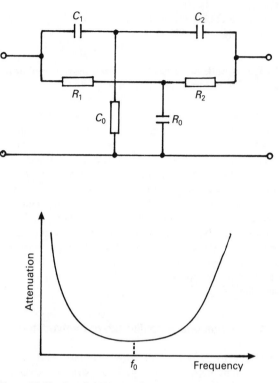

Figure 13.12   Parallel T network.

characteristic can be achieved. The circuit and its response is shown in Fig. 13.12. If $R_1 = R_2 = 2R_0$ and $C_1 = C_2 = C_0/2$, then $\omega_0 = 1/C_0 R_0$ or $f_0 = 1/(2\pi C_0 R_0)$.

Sallen and Key filter. The second-order low-pass filter circuit shown in Fig. 13.13 has the general transfer function $H(s) = G/(s^2 + as + b)$, where a, b and G are constants. For certain values this will give a response that is identical to the Butterworth approximation. Higher order filters are constructed by cascading networks, combined with a first-order network if necessary. If $R_1 = R_2 = R$ and $C_1 = C_2 = C$, the transfer function becomes $H(s) = G/((sRC)^2 + sRC(3 - G) + 1)$, where

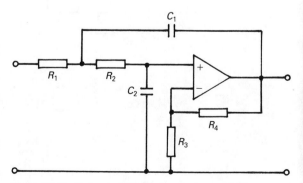

Figure 13.13   Sallen and Key 2nd-order low-pass filter.

$G = 1 + (R_4/R_3)$ is the gain of the non-inverting operational amplifier.

The cut-off frequency is given by $\omega_c = 1/RC$ or $f_c = 1/(2\pi RC)$, while the Q factor is given by $Q = 1/(3 - G)$. If $G > 3$ the circuit would become unstable and hence all Sallen and Key filters are restricted to low values of G and consequently low Q.

The circuit configuration can be changed to provide a high-pass characteristic simply by interchanging R_1 and C_1 and R_2 and C_2.

13.1.5 Surface acoustic wave filters (SAWF)

These devices are constructed on a thin slab of piezo-electric substrate in the manner indicated in Fig. 13.14. The metallic comb-like transducer structures are fabricated using photolithographic techniques and are referred to as *inter-digital transducers* (IDT). When an electrical signal is coupled to the input, this creates a surface wave, which is accompanied by an electric wave, that propagates towards the output. When this passes under the output transducer, the electric field creates the output signal. As the surface wave velocity is in the order of 3000 m/s, relatively long delays can be generated with short path lengths. With the latest technology, operating frequencies above 1.5 GHz are possible.

The optimum electro-acoustic coupling occurs when the IDT fingers are spaced by half a wavelength. By controlling this spacing and overlap during manufacture, asymmetric IDTs can be used to produce a very wide range of characteristics. The resulting filter is thus very much smaller than an equivalent LC device and is particularly compatible with integrated circuits. Although SAWFs can provide sharp cut-off characteristics, they also produce about 20 to 30 dB more insertion loss than an equivalent LC filter. The device depicted in Fig. 13.14 is described as a *transversal filter* where the unconnected multistrip coupler acts as a damper to the propagation of any unwanted bulk waves.

13.1.6 Switched capacitor filters

This integrated circuit approach to relatively low frequency filtering (typically below 50 kHz) is based on the operational amplifier integrator. The devices are usually fabricated for 2nd-order responses which are programmable simply by the addition of a few external resistors. The normal integrator input resistor is replaced by a capacitor and a pair of analogue switches, the latter being driven by an external clock circuit which controls the filter cut-off frequency. The switching action results in the input signal samples being switched to the input capacitor before being transferred to the integrating capacitor. For a cut-off frequency of 50 kHz the clock frequency is typically 1 MHz. The basic 2nd-order device can be programmed to function in the 1st-order mode, so that any desired filter order can be achieved simply by cascading devices. All the standard filter forms (Butterworth, elliptic, etc.) can be programmed and for all possible pass modes. Often up to three pass modes can be achieved simultaneously in a single device. Since the cut-off frequency, filter form and its Q factor are all defined by the clock frequency and a few resistors, a very small filter with little power consumption can be provided. Because of the programmability, these devices find applications in adaptive filtering as well as in anti-aliasing, phase locked loops and digital signal processing.

13.2 Digital filters

The widespread use of personal computers for the processing of statistical data has greatly expanded the concept of digital or number filters. There are many cases where it is necessary to study a continually time varying quantity, say $x(t)$. Instead of recording $x(t)$, it is common to simply record samples that are equally spaced in time, x_n of the function $x(t)$. Rounding-off or approximating is inherent in this process and this gives rise to a quantising error or noise component associated

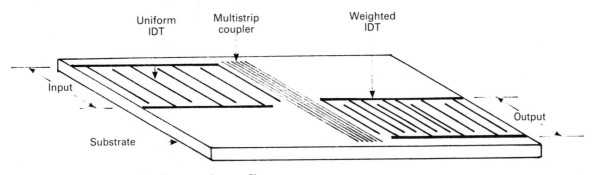

Figure 13.14 Construction of surface acoustic wave filter.

with the recordings (see also Digital communication systems, sampling, p. 75). Digital processing can then be carried out in order to derive an understanding of the effects that gave rise to the observations. Common operations on such data include smoothing, predicting, differentiating, integrating, separating the variability from the general trend of the function and removing the noise of measurement, all of which represent linear applications of filtering. Within the limitation of frequency of operation, digital filters can be produced to process analogue or digital signals, including some filter characteristics that are impracticable in the analogue domain. Due to the rapid acceptance of digital filters, integrated circuits are available that have programmable characteristics. The many advantages of these devices include, freedom from component ageing and drift due to environmental changes, reproducibility and high reliability, tuneable very high order circuits, linear phase response and operation down to very low frequencies. In applications where it is convenient to store the data prior to processing, the work of filtering can be carried out using a standard personal computer. Changes can be made to the filtering by program modification so that the effects of different types of filtering can be clearly seen.

Bilinear transform. The operation of time delay is basic to digital filters and is described by the *unit delay operator* z^{-1}. If the sampling frequency is 10 kHz, then $z^{-1} = 1/10\,000 = 100\,\mu s$.

The transfer functions of classical analogue filters can be written in terms of the Laplace variable s. Delaying a function by time T multiplies the transform by $\exp(-sT)$, the variables being related by $z^{-1} = \exp(-sT)$, where T is the sampling period. An approximation known as the bilinear transform:

$$s = (2/T)(1 - z^{-1})/(1 + z^{-1})$$

is used in the design of digital filters that are based on analogue prototypes. Using this transform, the general transfer function

$$H(s) = (As^2 + Bs + C)/(Ds^2 + Es + 1)$$

transforms to

$$H(z^{-1}) = K(1 + a_0 z^{-1} + a_1 z^{-2})/(1 - b_0 z^{-1} - b_1 z^{-2})$$

from which simultaneous equations can be derived to solve for the constant K and coefficients a and b.

The bilinear transform thus represents a mapping of the frequencies from the s-plane on to corresponding frequencies on the z-plane with the following relationships:

$$\omega_s = 2/T(\tan(\omega_z T/2))$$

or

$$f_s = 1/\pi T(\tan(\pi f_z T))$$

where ω_s, f_s and ω_z, f_z relate to the s and z planes, respectively.

Due to the tangent function in these expressions, there is a nonlinear compression in the frequency scale of the digital filter and this gives rise to an effect described as *frequency warping*.

Frequency transform. Just as in the case of analogous filters designed using the modern techniques, digital filters are designed around basic low-pass versions and then transformed into high-pass or bandpass as required. The complex transforms (listed in Table 13.1) are most easily applied using a suitable computer program.

Impulse invariant transform. As the duration of a large amplitude pulse approaches zero, its bandwidth approaches infinity (see also Signals, p. 300). If such a pulse is applied to the input of a filter, then this is equivalent to simultaneously applying a very wide range of frequencies. The shape of the resulting output pulse therefore conveys information about the frequency response which is described as the impulse response of the filter. If the impulse response of a digital filter is the same as a sampled version of the impulse response of an analogue filter, then the two filters have the same frequency characteristics. Then the transformation between the two is described as an *impulse invariant* one.

13.2.1 Kalman filters

The ability of this system to reduce the effects of noise allows it to be classified as a filter. However, because of its prediction ability it is also described as a system *state estimator*. Such filters are concerned with the processing of discrete sampled data. The concept finds applications in air traffic control radar and inertial navigation systems. Using the air traffic control system as an example, the states of interest would be course, height and speed of an aircraft, all of which need to be correctly assessed. The Kalman filter is then used on a basis of radar measurements and in this sense there is system noise and a degree of uncertainty associated with each measurement. Noise that arises within the system is described as *process* or *system noise*, while the uncertainty of measurement is described as *observation* or *measurement noise*. The latter may arise from such sources as misalignment of antenna or the round-off error associated with computer processing.

Table 13.1 *Corresponding bilinear transforms for derived filters (replace z with the following transformations)*

Parameters used in table:

$$\alpha_1 = \frac{\sin((\omega_c - \omega_H)T/2)}{\sin((\omega_c + \omega_H)T/2)}$$

$$\alpha_2 = -\frac{\cos((\omega_c - \omega_H)T/2)}{\cos((\omega_c + \omega_H)T/2)}$$

$$\beta = \frac{\cos((\omega_U + \omega_L)T/2)}{\cos((\omega_U - \omega_L)T/2)}$$

$$\gamma = \cot((\omega_U - \omega_c)T/2)\tan(\omega_c T/2)$$

$$\delta = \tan((\omega_U - \omega_L)T/2)\tan(\omega_c T/2)$$

where ω_c = cut-off frequency for prototype LPF
  ω_H = cut-off frequency for HPF
  ω_L = lower cut-off frequency
  ω_U = upper cut-off frequency.

Low-pass $\left[\dfrac{z^{-1} - \alpha_1}{1 - \alpha z^{-1}}\right]$

High-pass $-\left[\dfrac{\alpha_2 + z^{-1}}{1 + \alpha_2 z^{-1}}\right]$

Bandpass $-\left[\dfrac{z^{-2} - \left(\dfrac{2\beta\gamma}{\gamma + 1}\right)z^{-1} + \left(\dfrac{\gamma - 1}{\gamma + 1}\right)}{\left(\dfrac{\gamma - 1}{\gamma + 1}\right)z^{-2} - \left(\dfrac{2\beta\gamma}{\gamma + 1}\right)z^{-1} + 1}\right]$

Bandstop $\left[\dfrac{z^{-2} - \left(\dfrac{2\beta}{1 + \delta}\right)z^{-1} + \left(\dfrac{1 - \delta}{1 + \delta}\right)}{\left(\dfrac{1 - \delta}{1 + \delta}\right)z^{-2} - \left(\dfrac{2\beta}{1 + \delta}\right)z^{-1} + 1}\right]$

Unlike the conventional filter, the Kalman filter gain is not defined in terms of a universal stable solution, but in terms of an iterative process. The filter output is a weighted combination of a prediction at that instant, based on an earlier measurement and a noisy observation made at that instant. The weighting depends on whether the prediction or the observation is considered to be the most reliable. The composite output is thus part prediction and part observation in which the total error is less than that in either component taken separately. Figure 13.15 depicts a model of a first-order Kalman filter where the time delay T is one sample period. This feature allows the comparison of each new sample with the previous one in order to update the estimate in a recursive manner. The transfer function is given in terms of two equations:

Measurement equation, $y(t) = Bx(t) + w(t)$
Process equation, $x(t + 1) = Au(t) + Cx(t) + v(t)$

where $u(t)$ is the input, $y(t)$ is an observation at time t, $x(t)$ is the system state at time t, B and C are linear scaling factors and $v(t)$ and $w(t)$ are process and observation noise components, respectively.

13.2.2 Non-recursive or finite impulse response (FIR) filter (also known as transversal filter)

These circuits are designed around shift registers to provide unit delay periods equal to the sampling period and weighting networks to provide feedforward coefficients that lie in the range -1 to $+1$. A 3rd-order system is shown in Fig. 13.16, which contains three delay/weighting networks. An output pulse is the summation of delayed and multiplied versions of an input pulse due to the circuit arrangement. These filters find applications up to at least video frequencies as low-pass devices, interpolators and equalisers, chiefly due to the linear phase response and high level of stability. FIR filters can also be produced

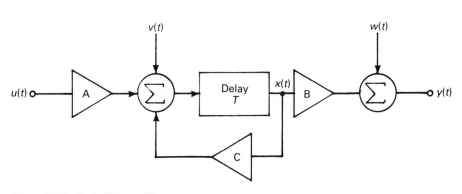

Figure 13.15 Basic Kalman filter.

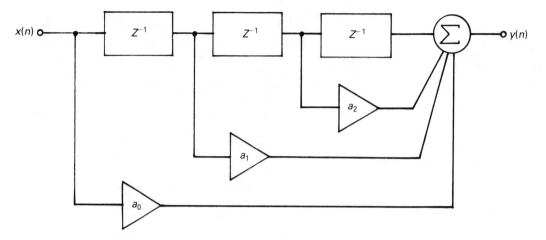

Figure 13.16   3rd-order non-recursive filter.

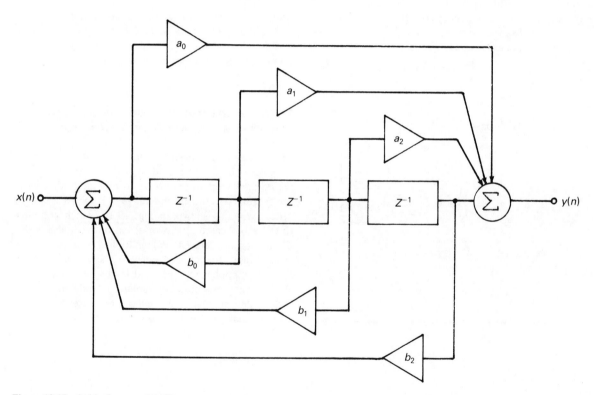

Figure 13.17   3rd-order recursive filter.

using computer software. The design is closely associated with modern analogue filter techniques. Starting with the desired frequency response in the s-plane and then converting to the transfer function $H(z)$ to calculate the weighting coefficients. Conversion from the prototype low-pass filter into the final characteristic can be achieved using the transforms shown in Table 13.1

13.2.3 Recursive filter or infinite impulse response (IIR)

These configurations use both feedforward and feedback to achieve filtering, the basic configuration being shown in Fig. 13.17. Again the delays are produced within shift registers and the weighting coefficients lie in the range -1 to $+1$. Due to the recursive nature of the circuit, an input pulse has the ability to produce an output effect well into the future. Because each delay unit is effectively used twice, the filter has a narrower transition band than an equivalent FIR, but is not necessarily as stable. These filters can be used for real time applications and are found in telephony circuits, in reverberation units in sound studios, as noise reducers and for special video effects. Filter design is again based on analogue concepts, being achieved as a mapping from the *s*-plane on to *z*-plane.

13.2.4 Median filters

The *median* of a sequence of numbers is the middle term when the sequence is arranged in ascending or descending order. Thus for the sequence 1, 9, 7, 6, 2 the median is 6. A median filter has a structure that is similar to the non-recursive or transversal type and is based on a tapped delay line. This passes a sequence of digital samples to a selector switch that gates out the median value. Since this value is one of the original samples, the median filter does not generate any quantisation error as may occur with other digital filters that function by averaging. This advantage is however offset by the introduction of some nonlinear distortion.

Because these filters can be constructed for one, two or three dimensions, they are much used for removing impulsive noise from processed video signals where the pixel values with the lowest noise amplitude are selected.

As configured for 3D filtering, the filter locates the median value for pixels in the horizontal and vertical dimensions and between fields, typically operating over five fields to minimise errors.

13.2.5 Transversal filter or equaliser

This variant of the non-recursive filter illustrated in Fig. 13.18, can be used in the all-pass mode to correct for undesired phase distortion. Because the weighting networks can be programmable, a single unit can be used in many different applications. The main output of the filter is obtained from the central tapping on the delay network with a weighting coefficient of unity. Contributions to the summed output pulse then come from different proportions of the input pulse either delayed or advanced on the main component. By varying the weighting it is possible to obtain the necessary phase correction over the frequency range of interest.

13.2.6 Quadrature mirror filters

These are basically multi-rate digital filters so that the sampling frequency is not constant throughout the system. Figure 13.19 shows the basic principle involved with the system containing *decimators* which *down sample* the signal and *interpolators* that provide complementary *up sampling*.

If a decimator has the digital sequence $X(n)$ as input, it will produce an output $Y(n) = X(Mn)$, where M is an integer that represents the down sampling ratio. If $M = 2$, then $Y(n)$ will consist of every second sample. The decimator thus produces a compression in the time domain. In a practical system, before the signal is decimated, it will first be bandwidth limited to reduce the effects of aliasing using a *decimation filter*. At the receiver, the M-fold interpolator will replace the missing $M - 1$ values between adjacent samples. If $M = 2$, then the interpolator will either average successive values or insert zeros, the effect being a complementary expansion in the time domain.

The concept is closely related to sub-band coding. As indicated in Fig. 13.19 the input sequence $X(n)$ is divided into separate channels using low- and high-pass filters with a response of $H_L(z)$ and $H_H(z)$ before down sampling. This signal is then coded for transmission in any suitable manner, before being processed in a

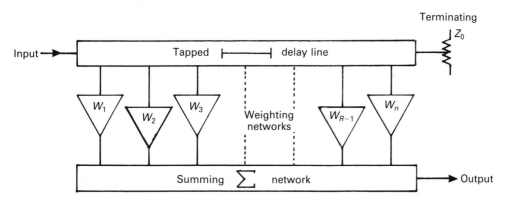

Figure 13.18 Transversal filter.

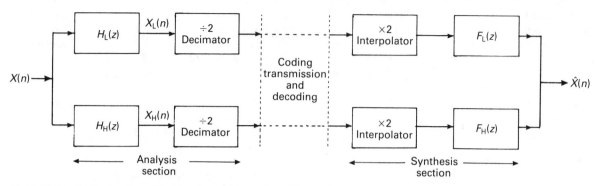

Figure 13.19   Basic operation of 2-channel quadrature mirror filter system.

complementary manner at the receiver. Higher orders of filter can be produced by further sub-dividing each channel with low- and high-pass filters. It will be clear from the comments on sub-band coding, that some of these channels may contain long strings of zeros, thus allowing significant bit rate reductions to be achieved.

The combined response of decimator and interpolator can give rise to aliasing in the frequency domain, but *synthesis filters* with response of $F_L(z)$ and $F_H(z)$ remove this from the respective channels. In a correctly designed system, the recombined output sequence $\hat{X}(n)$ will be a perfect reconstruction of the input $X(n)$, even under noisy transmission conditions.

The name quadrature mirror filter derives from the fact that the response of the filter $H_L(z)$ is a mirror image of that of $H_H(z)$ with respect to the normalised frequency $\pi/2$, which is a quarter of the sampling frequency.

13.2.7 Windows

The frequency characteristic of a digital filter is a periodic function related to the sampling period T which can be expanded as a Fourier series (see Signals, p. 304). The coefficients of this series represent the impulse response of the filter. When, in a practical case, this infinite series has to be truncated to N terms, the sharp

cut-off leads to overshoots and oscillations in the characteristic. This is known as the *Gibbs phenomenon* and the effect can be minimised by multiplying the impulse response by a weighting factor described as a *window function* $w(n)$. This function can easily be incorporated in the design of an FIR. Some of the common window functions are depicted in Fig. 13.20 and these have the mathematical form listed in Table 13.2.

13.3 Optical filters (see also Optical communication devices and systems, p. 225, and Television signal processing and systems, p. 343)

Light is that part of the electromagnetic spectrum that enables the human eye to observe different objects. It is characterised by wavelengths from about 350 to 750 nm and a mean propagation velocity c of 2.997925×10^8 m/s. To the viewer, the variation of wavelength produces the sensation of colours, and when all the colours are present simultaneously, the eye perceives white light. From the special theory of relativity, the velocity of light forms the link between *mass* and *energy* and thus has a special significance. The velocity is also considered to be a limiting value beyond which it is impossible for any body to travel. Because light energy travels with a wave-

Table 13.2   *Typical window functions*

Window	Function
Rectangular	$w(n) = 1 \quad 0 \leq n \leq N - 1$
Triangular or	$w(n) = 2n/(N - 1) \quad 0 \leq n \leq (N - 1)/2$
Bartlett	$w(n) = 2 - ((2n)/(N - 1)) \quad (N - 1)/2 \leq n \leq (N - 1)$
Hanning	$w(n) = 0.5 - 0.5 \cos((2\pi n)/(N - 1)) \quad 0 \leq n \leq N - 1$
Hamming	$w(n) = 0.54 - 0.46 \cos((2\pi n)/(N - 1)) \quad 0 \leq n \leq N - 1$

N is the number of unit delays in each filter.
The Hanning and Hamming windows are very similar, except that the latter sits on a small pedestal.

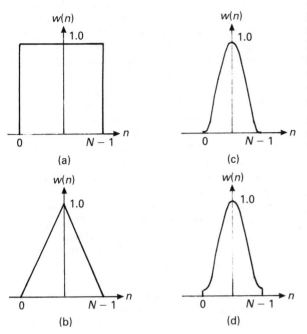

Figure 13.20 Typical windows: (a) rectangular; (b) triangular or Bartlett; (c) Hanning; (d) Hamming.

like motion it is possible to separate out different colours from the spectrum using suitably coloured filters. For example, a red surface or transparency will only reflect or transmit red light, all other colours being suppressed. Typical optical filters are constructed either from coloured gelatine sandwiched between two plates of optical glass, or from optical glass that has been suitably dyed in the mass.

Dispersion. Figure 13.21 indicates the way in which a prism causes light consisting of a mixture of wavelengths to become split into its component colours. Because the light travels at slightly different velocities in different media, this gives rise to a marginal change of wavelength, which in turn causes light of different wavelengths to be bent or refracted by different angles at the interface between the two media. When a light ray passes from a less optical dense medium into a more dense one, the ray always turns towards the normal to the plane of the interface.

Dispersive power. The dispersive power of a medium is defined in terms of light of two specified wavelengths. If n_1 and n_2 are the refractive indices of the medium for the two wavelengths, then the dispersive power for the medium is given by $(n_1 - n_2)/(n - 1)$, where n is the average of the two refractive indices.

Optical density. If one medium has a greater refractive index than another at some particular wavelength, it is said to have a higher optical density at that wavelength.

Refractive index and laws of refraction. These can be defined with the aid of Fig. 13.22 and Snell's laws which state that:

(1) The incident ray, the refracted ray and the normal to the separation interface between two media at the point of incidence all lie in the same plane.
(2) The ratio of the sine of the angle of incidence to the sine of the angle of refraction is constant for any given pair of media.

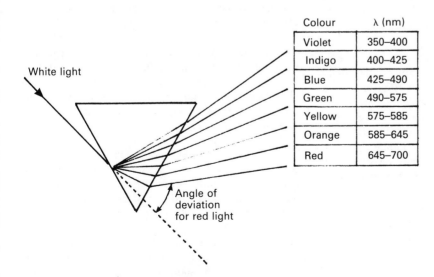

Colour	λ (nm)
Violet	350–400
Indigo	400–425
Blue	425–490
Green	490–575
Yellow	575–585
Orange	585–645
Red	645–700

Figure 13.21 Dispersion of white light.

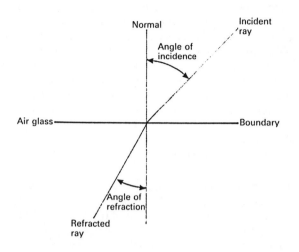

Figure 13.22 Definition of refraction.

When one of the media is a vacuum this ratio is defined as the refractive index for the other medium. (Air is also considered to have approximately the same refractive index as a vacuum.) It can further be shown that the refractive index is also the ratio of the velocity of light propagation in a medium to that in free space.

13.3.1 Terminology of optical filters

Angström unit (Å). 10^{-10} metres and often used in the measurement of light wavelengths.

Colour graduated filter. A filter that is used to accentuate a section of a scene. These are constructed with one half clear and the other based on colour temperature. The transition region may be sharp to produce a hard edge or graduated to provide a soft edge.

Colour temperature. The spectrum of light emitted by a full radiator varies with its temperature. Some light sources have a spectral energy distribution which corresponds to a full radiator at a specific temperature and this can be used to classify that source. The colour temperature in K (Celsius + 273), of a source is thus the equivalent temperature to which a full radiator must be raised to emit matching light. In general, the higher the temperature then the greater the content of blue light.

Colour temperature or compensating filters. These are used to compensate for deficiencies in the colour quality of lighting sources and recording media.

Coral filter. A filter used to enhance the warm red tones or flesh tones in a scene.

Decamired filters (see Mired scale, p. 133). Used for colour balancing because these filters exert the same amount of colour temperature scale change at every point in the scale. There are two ranges in use and these are predominantly red or blue (1.5 decamired = 15 mired).

Dichroic mirrors or filters. These are formed by vacuum deposition of extremely thin layers of materials with alternate high and low refractive indices on to an optical glass plate. The typical materials include zinc sulphide and cryolite. The thickness of these layers is such that the reflection from their surface is in phase with the reflection from the glass surface over some small colour bandwidth. All other colours then pass through the structure with negligible attenuation. The layers are typically less than 0.0005 mm thick and three such layers can reflect about half of the visible spectrum. With 17 layers the reflection bandwidth is around 1 nm or 10 Å

Diffusion filter. A filter that is highly polished on one side and slightly rippled on the other. Light passing through such a filter becomes slightly diffused and this decreases the sharpness of an image.

Filter factor. This defines the light attenuation of a filter and may be expressed in two ways. For example, a filter that reduces the light level by a factor of 2 may be described as ×2 or 0.5. The first value signifies that the light level needs to be doubled to obtain the original level, while the second one signifies that only 50% of the light is transmitted through the filter, reducing the light by a factor of 2.

Fluorescent light filter. A colour temperature filter designed to reduce the effects of blue and ultraviolet light present with fluorescent lighting.

Fog filter. Fog in the atmosphere is formed when small moisture droplets cause diffraction, thus dispersing light. Fog filters are constructed in such a way that glass, gelatine and adhesives components all contribute to diffraction in a similar way. The general effect is to cause light sources to flare.

Glasses used for filter production. Glass is a hard brittle amorphous (non-crystalline) and transparent material commonly made from silica (sand) and metallic compounds. Ordinary soda glass that contains sodium silicate and sodium tetraborate is not transparent enough for use in filters, due to defects in the molecular structure and hydroxyl absorption. Optical glasses include crown glass made from potassium or barium in place of sodium and boric oxide or boric anhydride in place of silica and flint glass that contains lead silicate. Calcium aluminate glass is used for those applications where the pass band must include infra-red through to ultraviolet, as this is the only glass that is free from hydroxyl absorption. Quartz which is the natural crystalline silica (rock crystal, SiO_2) is used for high temperature applications and where the pass band needs to extend from infra-red through to ultraviolet.

Haze filters. Whilst haze has a similar effect to fog, it results from the scattering of light particularly in the blue to ultraviolet region.

Low contrast filter. These devices are constructed from a special material sandwiched between two glass plates. The filler material causes the light to spread from the highlight areas into the shadows. This lowers the overall contrast and mutes the colours to avoid flaring of highlights.

Mired scale. Because of its nonlinear visual effect, the use of the temperature scale in K is not convenient when making filter calculations. By using instead a *micro-reciprocal degree* (Mired) scale, obtained from $10^6(1/K)$, a filter will always shift the mired value of a light source by a fixed amount. A set of eight filters (four pink and four blue) will meet all normal requirements. Using this scale, lowering the temperature increases the mired value and as red light exhibits the lowest temperature this will give a higher mired value than will blue light. Thus the pink mired filters are said to have a positive mired shift. Conversely, the blue range are described as having a negative mired shift (see also Decamired filters, p. 132).

Multi-image lens. While not specifically a filter, this device causes multiple images to be produced within the same field of view. This is achieved by polishing a lens in a multi-faceted manner.

Net filters. These consist of a net material laminated between two plates of optical glass to create a diffusion effect without affecting the dark areas of an image. The density of the filter depends upon the pitch of the net matrix. Colour accentuation can also be achieved by using suitably coloured net materials.

Neutral density filters. These are used to reduce the amount of light available without modifying the tone of colour balance. They are therefore the electric filter equivalent of the all-pass filter. A ×4 or 0.25 ND filter provides a light transmission of 25%.

Polariser filter. These filters are used to reduce the effects of glare produced by ultraviolet light and bright blue skies and are also useful for minimising haze problems. The filter has to be rotated to give the desired effect.

Spot filter. A heavy diffusion filter that has a clear circular spot in the centre. This has the effect of softening the background and making it less distracting.

Star effect filter. These are made from optically clear glass with one surface engraved with a grid pattern to scatter the light to form star shaped bursts. Different star patterns are created by different grid patterns.

Thin polariser filter. These are designed for use with wide angle lens to prevent *vignetting*.

Transmittance or transmission coefficient. Whenever electromagnetic radiation in any form passes through a medium, to a greater or lesser extent, some energy will be absorbed. The ratio of the original intensity to that after passing through unit distance of the medium, is described as the coefficient of transmission.

Vignetting (see Spot filter, above). Only the central area of the image is in sharp focus, the border being gradually diffused.

Woolaston prism. A three-element quartz prismatic structure designed to deflect a light beam in a direction depending upon its polarisation. Vertically and horizontally polarised light will be deflected in opposite directions.

13. 4 Miscellaneous terms

Active filter. A filter constructed around an amplifier, usually an operational amplifier.

Adaptive equaliser. A device used in data transmission lines or modems, that automatically modifies its characteristics to compensate for distortion caused by amplitude, frequency and phase non-linearities and echoes.

Anti-aliasing filter (Nyquist filter) A filter whose cut-off frequency is so chosen that the analogue input signal has no frequency components beyond half the sampling frequency.

Asymmetrical network. A network whose input and output ports have dissimilar characteristics and can therefore not be interchanged without affecting the system response (see Symmetrical network, p. 135).

Causal. A system or network is said to be causal if its response does not anticipate an input signal, i.e. there is no output signal before the input excitation.

Chirp or dispersive filter. An all-pass filter that has a quadrature phase response which can be used to spread digital pulse energy in the time domain. This reduces the peak transmission power requirement but must be used in conjunction with a complementary filter at the receiver.

Comb filter. See N-path filters, p. 123.

Combiner/splitter. A filter network designed to combine several input signals into one output, without mutual interaction at the inputs. Conversely, a device that segments a signal to provide feeds to several outputs, but without mutual interaction between the outputs.

Cosine roll-off filter (see also Full cosine roll-off filter, below). A low-pass filter devised to convert a square pulse into one that minimises the effect of *inter-symbol interference*. The impulse response is then skew symmetric about the half amplitude value and half bit rate f_b. The percentage roll-off is defined by $(f_c - f_b/2)/(f_b/2) \times 100\%$.

Decimation filter. A digital filter for which the term technically implies taking every tenth value, but in practice it is more common to take every other one. It is sometimes convenient to multiplex two 4-bit data streams into one 8-bit channel for processing purposes. A decimation filter would then be used at the post processing stage to recover the processed 4-bit streams. If an analogue signal is sampled at a high rate to retain the high frequency components in the signal and then a decimation filter used to reject alternate samples, this will avoid the aliasing that might have occurred if the sampling rate had been halved in the first place.

Diplexer/triplexer. See Combiner/splitter, above.

Directional coupler. A device with three ports: input, output and tapping port. When a signal flows from input to output a predetermined proportion of the input can be tapped off to provide an extra output. However, in certain conditions signals can flow in the reverse direction, from output to input, when no signal will be tapped off.

Full cosine roll-off filter. A low-pass filter with a response given by $H(\omega) = 0.5(1 + \cos(\pi\omega/2\omega_c))$, where ω_c is the cut-off frequency. The impulse response has zero crossing points at multiples of the Nyquist interval,

the energy in the tails of the response falls off rapidly and the half-amplitude duration is equal to the Nyquist interval. In addition, if such a filtered pulse is sliced at the half-amplitude value, a full-width Nyquist pulse is generated without distortion. This device is thus used to pre-filter a pulse stream in order to reduce the effects of inter-symbol interference. Filters with modified roll-off are sometimes used and these have a cut-off slope defined by $((f_c - f_b/2)/f_b/2) \times 100\%$, where f_c is the cut-off frequency and f_b is the bit rate.

Image impedances. An asymmetrical network has two different image impedances such that when one port is terminated in one impedance, the other value is 'seen' at the other port. In the special case where image impedances are equal, these values are known as the characteristic impedance of the network.

Insertion gain or loss. When a network is connected between a generator and a load then assuming that both impedances are correctly matched, there may be a gain or loss of signal power at the load, depending on the nature of the network. The ratio of the output power to input power, usually quoted in decibels, is described as the system insertion gain or loss, respectively.

Iterative impedances. The input impedances at the two ports of an asymmetrical network are different and are described as iterative impedances. In the special case where iterative impedances are equal, these are known as the characteristic impedance of the network.

Matched filter. A filter used in a digital data decoder prior to the decision detector, to minimise the bit error rate. This is achieved by matching the impulse response of the filter to the input signal characteristics so that the signal to noise ratio is maximised.

Nyquist filter. See anti-aliasing filter, p. 133.

Passive filter. A filter circuit that contains no elements that can introduce energy into it, i.e. contains only inductive, capacitive or resistive components.

Prolate filter. The term prolate refers to a sphere that has been distorted so that the polar axis has been lengthened, resulting in an elliptical shape. A prolate filter is thus a low-pass filter designed around the elliptic function. The circuit has a very good response to a step function.

Rotomode filter. See Microwave devices, p. 172 and Rotomode coupler, p. 179.

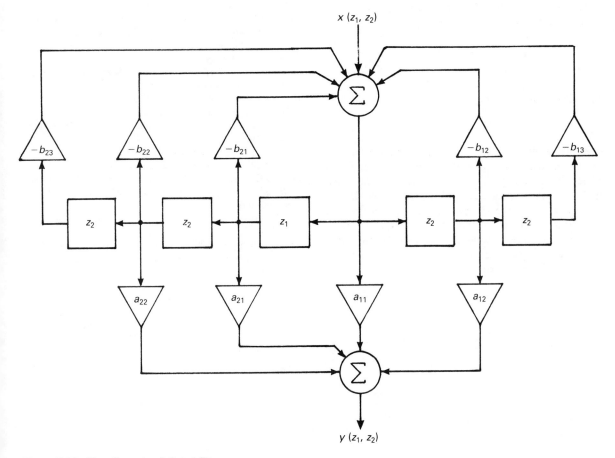

Figure 13.23 Two-dimensional digital filter.

Sensitivity of filter. Resistance/capacity active filters are subject to drift of component and amplifier characteristics with both time and temperature. Such changes can affect the frequency response of the circuit. The effects are usually quantified as the filter *sensitivity* to these variations.

Splitter. See Combiner, p. 134.

Symmetrical network. A network where input and output can be interchanged without its electrical properties being affected. Such a network therefore has a characteristic impedance.

Two-dimensional filters (see also Image processing, p. 141, and Television signal processing and systems, p. 360). Two-dimensional filtering is concerned with the processing of discrete signals that are a function of

two variables such as the elements of an array or matrix. If a digital sequence $(x_{m,n})$ is input to such a filter, then processing with a specific computing algorithm will result in an output of $(y_{m,n})$. Two-dimensional filters are characterised in a similar manner to one-dimensional devices.

The transfer function in the frequency domain is given by:

$$H(z_1, z_2) = z(y_{m,n})/z(x_{m,n})$$

where $z(x_{m,n})$ and $z(y_{m,n})$ are the two-dimensional z transforms of the input and output sequences, respectively. The inverse z transform of $H(z_1, z_2)$ is the impulse response of the two-dimensional filter. The output sequence can also be obtained from the convolution of the input sequence and the impulse response. If the latter contains only a finite number of non-zero samples, the device is described as a *finite impulse response* (FIR), two-dimensional filter. Otherwise it is described as being an *infinite impulse response* (IIR) two-dimensional filter.

The transfer function $H(z_1, z_2)$ is typically given by a polynomial of the form:

$$\frac{(a_{11} + a_{12}z_2 + a_{21}z_1 + a_{22}z_1z_2)}{(1 + b_{12}z_2 + b_{13}z_2{}^2 + b_{21}z_1 + b_{22}z_1z_2 + b_{23}z_1z_2{}^2)}$$

the physical implementation of which is shown in Fig. 13.23.

Two-way filtering. Because recursive digital filters are not symmetrical they introduce a phase change that varies with frequency. If the data is available in stored form, then this can be cancelled by passing the data in one direction and then in the opposite way through the same filter, or through a second oppositely cascaded filter. Since such processing produces a squaring of the magnitude of the transfer function, this must be considered in the design.

13.5 Useful references

Bozic S.M. (1979) *Digital and Kalman Filtering*, Edward Arnold, London.

Connor F.R. (1986) *Introductory Topics in Electronics and Telecommunications*, 2nd edition, Edward Arnold, London.

Cowan C.F.N. and Grant P. M. (1985) *Adaptive Filters*, Prentice Hall, Englewood Cliffs, NJ.

Daniels R.W. (1974) *Approximation Methods for Electronic Filter Design*, McGraw-Hill, New York.

Hilburn J.L. and Johnson D.E. (1973) *Manual of Active Filter Design*, McGraw-Hill, New York.

Kuo F.F. (1976) *Network Analysis and Synthesis*, 2nd edition, John Wiley, New York.

Mazda F. (ed.) (1983) *Electronic Engineers' Reference Book*, 5th edition, Butterworth-Heinemann, London.

Mitra S.K. and Ekstrom M.P. (eds) (1978) *2 Dimensional Digital Signal Processing*, Dowden, Hutchinson & Ross, Stroudsburg, USA.

Terrell T.J. (1983) *Introduction to Digital Filters*. Macmillan, London.

Vaidyanathan P.P. (1987) Quadrature mirror filter banks, M-band extensions and perfect reconstruction techniques. *ASSP Magazine*, **4**, 4–19.

14 Frequency ranges in use

Theoretically all radio frequency signals above dc will radiate electromagnetic energy and thus propagate through space. However, for those frequencies below about 15 kHz the radiation is insignificant. This figure should not be confused with the higher limit of the audibility of sound waves by the human ear. Radio and sound waves propagate via completely different phenomena. Typically then, the part of the electromagnetic spectrum that is suitable for radio communications lies in the range 15 kHz to 300 GHz, although current applications above about 30 GHz are very limited. The ranges and nomenclature of the various sub-bands in use are shown in Table 14.1.

The frequency ranges used for microwave communications are commonly known by the American Radar Engineering Standard (Table 14.2). The alternative classification used in Europe is also shown in Table 14.2.

Table 14.2 *Microwave frequency bands*

	Frequency ranges (GHz)	
Band	American	European
P	0.2–1.0	0.2–0.375
L	1–2	0.375–1.5
S	2–4	1.5–3.75
C	4–8	3.75–6
X	8–12.5	6–11.5
J	–	11.5–18
Ku	12.5–18	–
K	18–26.5	18–30
Ka	26.5–40	–
Q	–	30–47

Table 14.1 *Sub-band ranges*

Frequency	Name	Wavelength
Up to 30 kHz	Very low frequency (VLF)	> 10 km
30 kHz to 300 kHz	Low frequency (LF)	10 km to 1 km
300 kHz to 3 MHz	Medium frequencies (MF)	1 km to 100 m
3 MHz to 30 MHz	High frequencies (HF)	100 m to 10 m
30 MHz to 300 MHz	Very high frequencies (VHF)	10 m to 1 m
300 MHz to 3 GHz	Ultra high frequencies (UHF)	1 m to 10 cm
3 GHz to 30 GHz	Super high frequencies (SHF)	10 cm to 1 cm
30 GHz to 300 GHz	Extra high frequencies (EHF)	1 cm to 1 mm

15 Image processing

An image is defined as the optical counterpart of a physical object produced when light rays pass through a lens or are reflected from a mirror (see also Television signal processing and systems, p. 343). An *image space* is thus an abstract concept formed by a transformation from an *object space* in the real world. However, images can be classified in a wider sense, dependent upon the wavelength of the radiation that is used to produce them:

(1) images obtained from radiation with wavelengths longer than light such as radar mappings;
(2) optical or visible images, including those obtained from ultraviolet or infra-red radiation (infra-red images are often described as *thermic*);
(3) images produced by very short wavelength penetrating radiation such as X-rays.

Each fixed point in an image space can be defined in terms of two-dimensional coordinates (x, y). To this can be added a third (z) dimension to give the effect of perspective, plus movement which is a function of time (t). Thus *real* (live or animated) images are characterised by points in four-dimensional space. All image processing, including cine film, involves the use of sampling, filtering, transforming, etc. Each frame of a cine film is a frozen image of two dimensions representing a scene that varies as a function of time. Thus each frame is a sample of the object space. Image manipulation in this medium is difficult and not easily repeatable, therefore modern image processing is carried out in the electronic digital domain, which is very flexible, repeatable and precise.

Images intended to be displayed on a cathode ray tube (CRT) are conventionally formed on a *raster scan* basis. The object space is scanned in a series of very narrow strips or *lines* to obtain an electrical signal that represents either the varying brightness (luminance) or colour (chrominance) along the line. The total number of such lines needed to span the object space is termed a *frame*. The greater the number of lines per frame, then the better will be the fine detail or *resolution* that can be displayed in the image space. In television-type displays, it is common to use two sub-frames or *fields* to make up the complete raster. These consist of alternate lines from the scanning sequence which are then interleaved or *interlaced* to provide a complete frame, the alternative to interlaced scan being a *progressive* or *sequential* system. If frames of sufficient lines are presented fast enough, then due to the persistence of vision, the line structure on the screen will appear to fill the image space. The number of lines per frame is given by the ratio of line frequency to frame frequency and this can vary from 525 for one television system, to more than 2000 for a high resolution computer graphics system. The smallest detail that can be displayed in such a system depends upon the number of scanning lines and the maximum rate at which the line signal can change. This element of the image space is referred to as a *picture element*, *pixel* or *pel*.

At the basic level, images can be displayed in monochrome (black and white) by simply using the brightness variation of each scanning line (the *luminance signal*). The addition of colour to the images considerably increases the amount of information that has to be processed. This includes *hue* which identifies the colour and *saturation* which describes the degree to which the colour is diluted with white light. Most of the colours that appear in nature can be created by the addition in suitable proportions of red, green and blue light, these being referred to as *primary* colours. White light can be formed by summing suitable levels of red (R), green (G) and blue (B) light. Thus the luminance signal (Y) can be formed from, $Y = R + G + B$, to provide a monochrome signal. An alternative system uses a luminance and two *colour difference* signals Y, R−Y, B−Y, concept. Because the green component is contained in Y, G can be recovered by suitable processing. In either case, three signal components have to be processed.

Image processing systems will be found in many areas of scientific, industrial and graphics art applications, from enhancing vague images transmitted from satellites, robotics, medicine, to automatic inspection and video microscopy.

15.1 Basic image processing

At the simplest level, an image processing system consists of an interface to a video source, such as a camera or video recorder; a host computer accessible frame store; a video output interface to drive the display; and memory look-up tables to handle the processing of the image data before and after it is entered into the frame store.

Typically each pixel value is quantised to 8-bits resolution, giving 256 different grey scale levels. Assuming a very modestly rated system with 1000 pixels per line of 64 μs duration, this represents a bit rate in excess of 15 Mbit/s. With such a high bit rate, the analogue to digital converter in the input interface must be of the flash type. Again, for 1000 pixels per line, a raster of only 625 lines will require more than 5 Mbits of store. Because of the multi-dimensional nature of the images and these high rates, it is common to find parallel or multiprocessor systems in operation. Data compression and efficient coding form important features in the design of such systems.

The video random access memory (VRAM) used in these systems is unique in that they are double or triple ported to allow simultaneous access by both the scanning system and the host computer.

The output interface normally has three ports, red, green and blue, to meet the needs of both colour and monochrome display devices. This typically forms a *colour palette* of 255^3 or 16 581 375 different shades of colour, not all of which will be available at a given instance.

The look-up tables hold the programmable algorithms that are used in the processing of the pixel data before and after it is entered into the frame store. This is particularly valuable as two simple transforms can then be performed in series to provide a very complex manipulation of the image.

15.2 Image analysis (contrast enhancement)

With the pixel values stored in memory, a grey level histogram program can evaluate how many pixels fall within particular bands. This can then be used to determine if the image has a low contrast ratio or dynamic range. If this is detected, then the range of the displayed image can be stretched by equalising the distribution of the grey levels, to enable the eye to see details that would otherwise be missed. As a further aid to resolving vague details, false or pseudo-colours can be assigned to specific pixel values, even in the case of inputs from a monochrome source. An alternative technique employs the *Hough* transform to detect

straight edges in the primary image. This is achieved by dividing the image into parallel lines which are then projected on to a y axis intercept defined by the line slope. When an edge that coincides with one of these lines is detected, the corresponding pixel has a higher value assigned to it.

15.3 Image processing

The principle of all image manipulation is closely allied to the three-dimensional geometrical problems of the ancient cartographers; that of mapping shapes between curved and flat surfaces. All manipulation is then based on the row/column array of pixel data held in the frame store. It is often convenient to consider the unprocessed image to exist on a primary or x, y plane and any transformations of it as mappings on to a secondary u, v plane. Many transformations can be made simply. For example, adding a constant to the pixel addresses produces an image displacement. Varying the constant in a controlled way displaces the image and changes its shape. Loading the frame store in serial form and in rows and reading out in columns, rotates the image by 90°. However, since the pixel data is stored in memory in a matrix form, it is much more convenient to perform mathematical matrix operations on the data. First-order matrix operations using constant coefficients can be used to produce such manipulations as rotation and scaling. The use of second-order matrices with non-constant coefficients can produce motion and *rubber sheet* (warping) effects. Such transforms are used to provide correction for the *keystone* distortion seen on images from the Meteosat and Landsat series of satellites. Digital filters play a significant role in image processing. Decimation filters that drop every nth sample can be used for image reduction, while repeating pixels produces image expansion. The capability to use the pixel samples many times is sometimes described as a *resampling* technique. If an image is suitably processed, it can be given perspective and hence pseudo three-dimensional appearance. Such images can then be rotated to provide visual information about *hidden details*.

The convergence of computing, video, television and associated applications has generated a multi-media concept. This in turn has created a demand for compatible image processing systems. One result of this is the formation of a number of study groups set up by the ISO (International Standards Organisation), CCITT (Comité Consultatif International Téléphonique et Télégraphique), and CCIR (Comité Consultatif International des Radio-communications). From these deliberations, a number of image processing standards have been, and are still being, produced (see also Television signal processing and systems, p. 356).

These groups operate under the acronyms of: JPEG (Joint Photographic Expert Group) for still colour images and MPEG (Motion Picture Expert Group) for moving colour images and *full motion video* (FMV). Phase IV of MPEG is providing a theoretical study into the future image compression needs of the multi-media industry.

15.3.1 Discrete cosine transform coding (see also Television signal processing and systems, p. 350)

This technique, which is commonly used for image signal transmission, is a two-step process. First a linear transform is performed on the original image by splitting the space into $N \times N$ blocks of picture elements or pixels. The amplitude coefficient of each pixel is then mapped onto a transform space for coding and transmission. The discrete cosine transform (DCT) which is related to the discrete and fast Fourier transforms (DFT and FFT) is widely adopted. This method is fast, requires the calculation of coefficients and is compatible with digital signal processing (DSP) ICs. When used with suitable coding the DCT provides for a high degree of image data compression. This transform technique thus finds applications in videophones, tele/video conferencing, colour fax systems, interactive video discs, and high definition colour imaging.

Each block of pixel data is transformed using a two-dimensional matrix operation in the following manner:

$$[T] = [C].[D].[C]^\top$$

where $[T]$ is the transformed block, $[C]$ is the basis of the DCT matrix, $[D]$ is the original data block and $[C]^\top$ is the transpose of $[C]$.

The DCT coefficients are calculated from the relationship,

$$F(u, v) = \frac{4C(u)C(v)}{N^2} \sum_{i-0}^{N-1} \sum_{j=0}^{N-1} f(i,j) \cos A \cos B$$

The inverse transform relationship used at the decoder is given by:

$$f(i,j) = \sum_{u=0}^{N-1} \sum_{v=0}^{N-1} C(u)C(v)F(u, v) \cos A \cos B$$

where

$$A = \frac{(2i + 1)u\pi}{2N} \quad \text{and} \quad B = \frac{(2j + 1)v\pi}{2N}$$

In both cases, i, j and $u, v = 0, 1, 2, ..., N - 1$ (N being the block size), i, j are the spatial coordinates in the original image plane, u, v are the corresponding coordinates in the transform plane, $C(u) = C(v) = 1/\sqrt{2}$ for $u = v = 0$ and $C(u) = C(v) = 1$ for $u = v \neq 0$.

Such transforms are readily performed within dedicated ICs. The high degree of image data compression available arises not from the transform itself, but due to the fact that many of the transformed coefficients are very small or zero, indicating little block-to-block variation. In addition, many of the remaining coefficients can be transmitted with lower precision without significantly affecting the received image quality. Since only a relatively few values need to be transmitted, any form of variable or run length coding can be used to advantage.

Depending upon the system requirements, a standard TV image can be reduced to a bit rate as low 64 Kbit/s, but 2 Mbit/s is necessary to provide an acceptable quality of colour image.

15.3.2 Fractal Transform coding (Fractal Transform is a trade mark of Integrated Systems Ltd)

The term fractal is a derivative of *fractured structure* and was coined to describe the way in which an image may be considered to be composed of many similar shapes and patterns of various sizes. Examples of this are snowflakes or the centre of sun flowers, where quite complex patterns can be defined by an origin, and motion and rotation instructions.

For encoding or compression, an image is divided into small blocks. These are searched for similar patterns that have been moved or rotated in order to determine the precise transforms that are occurring. The basis of this shape relationship is the matrix operation known as an *affine transformation*. An affine transformation is defined by the matrix

$$\begin{bmatrix} x \\ z \end{bmatrix} \rightarrow \begin{bmatrix} a & b \\ c & d \end{bmatrix} \cdot \begin{bmatrix} x \\ y \end{bmatrix} + \begin{bmatrix} e \\ f \end{bmatrix}$$

or

$$\begin{bmatrix} x \rightarrow ax & + & by & + & e \\ y \rightarrow cx & + & dy & + & f \end{bmatrix}$$

where a, b, c, d, e, f are real numbers that can include $\pm r \cos \theta$ and $\pm r \sin \theta$; r represents a scaling factor and θ represents a rotation, relative to the x and y axes.

In operation, this transformation represents a mapping from one shape onto another, point by point. For example, circles to ellipses, squares to parallelograms, right-angled triangles to equilateral triangles, etc. The affine transformation can thus describe a combination of rotation, scaling and translation of coordinate axes in *n*-dimensional space.

The compression/encoding process involves the use of an algorithm that is computationally intensive, which means that moving images cannot be processed in real time. By comparison, decompression/decoding can be performed in real time at 30 Hz frame rate using computer software only. The system is therefore most useful for asymmetrical processes that involve recording.

While it takes several minutes to encode a single colour frame, because video images can be encoded using interframe differences, a video signal can be encoded at an average rate of about 1 frame per second.

The following provides an indication of the degree of compression that can be achieved: a grey scale image can be compressed into about 3 Kbytes, a colour image into about 10 Kbytes, and a video frame using interframe differences into about 800 bytes.

15.3.3 Motion compensation/estimation

A technique commonly applied to the digital representation of images that have been subjected to compression or bit rate reduction. Motion compensation provides a means of improving the performance of any temporal compression, particularly in images where motion is continually taking place (video). In order to apply such compensation, it is first necessary to determine what part of the image is moving and where it is moving to.

There are two common classes of motion estimation algorithm: one which extracts three-dimensional motion information from a sequence of two-dimensional images and the other which estimates the velocities of object on a pixel-by-pixel or area-by-area basis. In the first case it is common to assume that any movement can be described by a translation component, such as rotation or zooming, and a centre about which the translation occurs. This concept works well when attempting to give still images a degree of motion, as when producing cartoons. For video images, the motions are much too complex and so the motion vector concept is normally applied. Differences between the current and previous frames can be used to detect the moving areas and generate *displacement* or *motion vectors*, either on a frame-by-frame, area-by-area, or pixel-by-pixel basis. Ideally, a unique vector would be generated for each pixel, but this would require the use of considerable processing power and memory. Therefore, in practice, the differences and hence the motion vectors are derived

and applied on a *block matching* basis, typically using blocks of 8×8 or 16×16 pixels.

15.4 Overlay techniques

This technique, which allows a section of one image to be overlayed or inset into another, can be performed in either the analogue or digital domains. The final image is thus a composite of two primary sources. Overlaying or *keying* is achieved by placing the insert image in front of a plain background, termed a *matte*. The raster line scanning system for all images is synchronised to the same source so that when the edge of the insert is detected, a keying signal can be generated that switches the appropriate signal to the final display device. If the keying signal is generated from the luminance of the image, as in a monochrome system, brightness variations in the insert can create a false keying action. To avoid this, *chroma-keying* or *colour separation overlay* is used, with the switching action being generated from the colour signals, colour difference, or red, green and blue as appropriate. In this case the matte is chosen as a colour that is least present in the insert, usually blue, to avoid false switching. As this switching action is effectively a convolution between the image and switching signals, aliasing can be a problem. Even if the two image signals are pre-filtered to remove aliasing components, the artifact can be transferred from the keying signal. It is thus important that the sudden keying transition should be softened by filtering or alternative processing.

15.4.1 Text or caption generators

These devices generate electronic versions of text signals directly, without the artwork that has been the traditional method of creating captions. The shapes of each character are stored in a semiconductor memory and these are recalled and scanned whenever required. A keying signal effectively cuts a hole in the main image for each character to be inserted at the appropriate position. Unless this signal is pre-filtered, aliasing can cause the loss of fine detail from the sharp edges of the characters.

15.5 Miscellaneous terms

Anti-aliasing. A circuit attribute, usually of a character generator, that avoids the introduction of aliasing.

Browsing file. When still images are stored in a magnetic memory in digital form, vertical and horizontal decimation filtering can be used to provide image reduction. By suitable scaling, a complete set of representative images can be held at the head of the memory. This can be used as a rapidly accessible browsing file, to enable the user to select the most suitable images from the library.

Contouring. Produces a relief map type of presentation that may be intentional, or unwanted due to using insufficient quantising levels. When intentional, this can be achieved either by discarding some of the low-order bits, or by allocating all pixel values in a particular range to a specified level and displaying all others as normal levels.

Corner pinning. This technique allows each corner on an image to be picked up in turn and placed or pinned in the exact perspective location. The rest of the image is then transformed in a three-dimensional way within this boundary.

Digital chroma-keying. Unlike its analogue equivalent, digital keying can be used to key uniquely from any one of the more than 16 million different colours in the colour palette, allowing the system to be used with subdued colours. This reduces the highlighting effect that can occur when using highly saturated matte colours.

Digitiser. The section of the *frame grabber* that contains circuits for sampling and quantising and maintaining synchronism of the input analogue signals from a video source. The digital outputs, either monochrome or chrominance components, are then loaded into the framestore.

Framestore. A semiconductor memory that is used to store the digital elements of a video signal. Although the term implies that a complete frame will be held in such a memory, it commonly refers to a store that can hold an amount of data varying from a few scanning lines up to a complete frame.

Linear keying. A technique used to minimise the effects of aliasing and provide semi-transparent effects. The overlay of one video signal on to another is performed in a selective manner with the ratio between foreground and background being determined by a linear-scaled control signal.

Mosaic. A two-dimensional process in which selected groups of pixels are averaged to provide *tiles* of uniform brightness and colour. The size of the tiles can be varied to suit the required image definition. A mosaic *window* can be defined so that only selected parts of the image are 'mosaiced'.

Posterise. The selective discarding of colour difference signal bits to reduce the number of colours available. The output then resembles an image produced in poster paints.

Solarisation. These images are characterised in luminance by bright highlights and deep shadows and in chrominance by the lack of pastel shades. This effect is achieved by replacing the linear transfer characteristic of the analogue to digital convertor with one that is S-shaped. This results in expanding one part of the range and compressing the other.

Warping. The process of mapping one image from one shape on to another, the transformation usually revealing more information about the object. Since the image is often distorted, it is sometimes described as a *rubber sheet* process. New addresses for the pixel data have to be calculated using three-dimensional algorithms.

15.6 Useful references

Barnsley M.F. (1988) *Fractals Everywhere*, Academic Press, London.

Clark R.J. (1985) *Transform Coding of Images*, Academic Press, London.

Happe B. (ed.) (1988) *Dictionary of Image Technology*, 2nd edition, Focal Press, Butterworth-Heinemann, Oxford.

Huang T.S. (ed.) (1981) *Topics in Applied Physics*, Vol. 6, *Picture Processing & Digital Filters*, Vols 42/43, *2 Dimensional Digital Signal Processing*, Springer-Verlag, Berlin.

Mandelbrot B. (1982) *The Fractal Geometry of Nature*, W.H. Freeman, CA.

Mitra S.K. and Ekstrom M.P. (eds) (1978) *2 Dimensional Digital Signal Processing*, Dowden, Hutchinson & Ross, Stroudsburg, USA.

16 Information theory

As indicated by the references to this section, the development of a theory of the communication of information owes much to the classical work of Hartley and Shannon. The term communication means the passing or transmitting of data from a source to a receiver. In a human context, this implies the passing of messages constructed from an alphabet of symbols within the constraints of a specific language. Since it produces a worst-case situation, it is convenient to consider that the use of each symbol from the alphabet will be independent and equiprobable. Because the number of possible messages expands in an exponential manner with an increase in the number of symbols available, it is useful to use a logarithmic scale to measure information. The probability of transmitting a particular symbol before it is received is called the *a priori probability* and the probability of having transmitted a particular symbol after it has been received is called the *a posteriori probability*. From these, the information (I) received can be defined as

$$I = \log[a\,posteriori\,\text{probability}/a\,priori\,\text{probability}]\,\text{bits}$$

The greater the number of symbols in the alphabet, then the smaller will be the *a priori* probability and hence the greater the information, thus showing that information increases with uncertainty. (If a particular message to be transmitted was a certainty, then there would be no need to send it.) Noise which is a destroyer of information thus increases the unpredictability. Because information is so related to disorder, it is usually equated with *entropy* of thermodynamics.

16.1 Average information

If an alphabet consists of a finite set of symbols, each with a different probability of occurrence, $p_1, p_2, \ldots p_i, \ldots p_m$, then with no noise present, each transmitted symbol would be received without error. The *a posteriori* probability would then be unity and the *a priori* probability for the ith symbol would be p_i. The *self-information* or *information content per symbol (I)* is then given by:

$$I = \log(1/p_i) = -\log p_i \,\text{bits}$$

For n symbols, the total occurrence of the ith symbol is just np_i, so that the total information received is the summation over all values of i, or

$$\text{total}\,I = \sum_1^i -np_i \log p_i$$

and the average information $H = I/n$ (often described as the source entropy), is given by:

$$H = -\sum_1^i p_i \log p_i \,\text{bits per symbol}$$

showing again, that for certainty, the information content is zero. Strictly, the units for information depend upon the base used for the logarithm as follows:

$\log_2$, binits, bits or Shannons
$\log_{10}$, decits or Hartleys
$\log_e$, natural units or nats

As $\log_2 10 = 3.322$, it follows that 1 Hartley = 3.322 Shannons.

Over a noise-free channel, both the source and received signals have the same entropy, but over a noisy channel the entropy differs because the noise generates further uncertainty.

16.2 Hartley–Shannon law of channel capacity

Shannon uses x and y to define the transmitter output and receiver input, respectively, and then $H'(x)$ and $H'(y)$ to represent the entropy rates of the transmitted and received signals, respectively. Further, $H'_y(x)$ defines the conditional entropy of the transmitted signal, when the received signal is known and $H'_x(y)$ defines the conditional entropy of the received signal, when the transmitted signal is known. Shannon describes the uncertainty $H'_y(x)$ as the *equivocation* and defines the transmission rate (R) as:

$$R = [H'(x) - H'_y(x)] \text{ bits/second}$$

It follows that the maximum channel capacity (C) must be the maximum value of R.

Assuming a binary system and transmission over a noisy channel, then the two signal levels have to be detected, one in the presence of the noise power N, and the other in the presence of the signal plus noise powers $(S + N)$. Shannon uses this to show that the information content H is given by, $H = \log_2 \sqrt{(S + N)/N}$ bits. Shannon then further develops the channel capacity (c) relationship as $C = B \log_2(1 + S/N)$ bits/s, where B is the channel bandwidth in Hz, thus showing that the bandwidth and S/N ratio can be traded to maximise the channel capacity for any acceptable bit error rate suitable for a particular service. For example, it can be shown that a standard speech channel with a bandwidth extending from 300 Hz to 3.4 kHz with a S/N ratio of 30 dB, can support a maximum digital bit rate of almost 31 Kbit/s.

16.3 Redundancy

If a message contains additional bits for reasons of error control, these are described as being redundant to the basic requirement of information transmission. The efficiency of the source coding can then be described in terms of relative entropy. If H is the actual source entropy and the maximum possible value is H_{max}, the redundancy can be defined as, $1 -$ relative entropy, or $R = 1 - (H/H_{max})$.

16.4 Miscellaneous terms

Bayes theorem. It has been shown that noise in a communications channel adds to the uncertainty of the transmission of a particular symbol. This uncertainty can be evaluated using Bayes theorem of conditional probability. This is stated as

$$P(A \mid B) = \frac{P(B \mid A)P(A)}{P(B)}$$

where $P(A)$ and $P(B)$ are the *a priori* probabilities of the transmission of symbols A and B, $P(A \mid B)$ is the conditional and *a posteriori* probability that symbol A was transmitted when symbol B was received, and similarly for $P(B \mid A)$. If A and B are exclusive and independent events, then the joint probability becomes $P(A.B) = P(A).P(B)$.

Gross information content. A measure of the total information in a message, including redundancy, that is required for transmission over a noisy channel, with a specified bit error rate.

Information rate. The product of the average information content per symbol and the number of symbols transmitted per second.

Nett information content. A measure of the essential information in a message that is required for transmission over a noise free channel.

16.5 Useful references

Bell D.A. (1968) *Information Theory and its Engineering Applications*, Pitman, London.

Hartley R.V. (1928) The transmission of information, *Bell System Technical Journal*, **3**, 535.

Proakis J.G. (1989) *Digital Communications*, 2nd edition, McGraw-Hill, New York.

Schwartz M. (1980) *Information Transmission, Modulation and Noise*, 3rd edition, McGraw-Hill, New York.

Shannon C.E. (1948) A mathematical theory of communication, *Bell System Technical Journal*, **27**, 379 and 623.

Shannon C.E. (1949) Communication in the presence of noise, *Proceedings of the Institute of Radio Engineers*, **37**, 10.

17 Logic

The application of mathematical analysis and deductive reasoning to propositions that may be true or false owes much to the work of George Boole in 1847. He formulated a system that could be applied to the relationships between propositions to which only a binary choice of truth existed. The first application of the *Boolean algebra* that derived from this was Shannon's research into the analysis of relay switching circuits in 1938. Although now used extensively in communications, the concept also finds applications in fluidic control systems.

17.1 Boolean logic

This algebra, which behaves mathematically in many ways similar to the algebra of conventional mathematics, allows statements or propositions to be either *true* or *false*; no varying shades of meaning such as *maybe* are permitted. There are three basic elements employed in this system, the AND function or *logical product*, the OR function or *logical sum* and the NOT function which negates a proposition. Because each obeys most of the rules of conventional algebra, they are symbolised by ., + and −, respectively. For applications in the communications field, these elements or *gates* are constructed from electronic circuits, each with one or more inputs and one output. The inputs represent the statements while the output represents the truth or otherwise of the combination. In a binary sense, true = 1 and false = 0. Thus each gate represents a *Boolean function* which can be combined with others if necessary. The circuit symbols most commonly used are shown in Fig. 17.1 which lists both the American National Standard Institute (ANSI) and British Standards Institute (BSI) symbols for these and other derived gates. NAND and NOR gates are effectively NOT AND and NOT OR combinations, respectively. The *Exclusive-OR* (EX–OR) excludes the combined statements A AND B and the EX–NOR is simply its inverse. The behaviour or truth value for such circuit elements can be described in *truth tables*, which for two input devices (for n inputs, there will be 2^n rows in each truth table) are

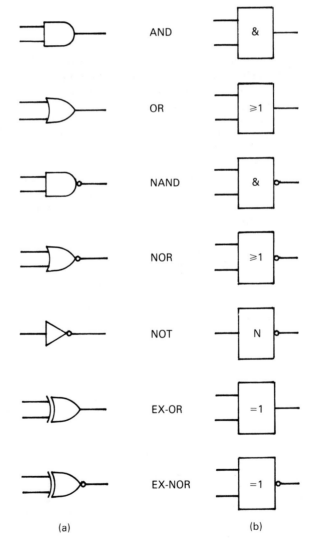

(a)   (b)

Figure 17.1 Logic element symbols: (a) ANSI; (b) BSI.

Table 17.1   *Truth tables*

AND			OR			NOT		NAND			NOR			EX-OR			EX-NOR		
A	B	F	A	B	F	A	F	A	B	F	A	B	F	A	B	F	A	B	F
0	0	0	0	0	0	0	1	0	0	1	0	0	1	0	0	0	0	0	1
0	1	0	0	1	1	1	0	0	1	1	0	1	0	0	1	1	0	1	0
1	0	0	1	0	1			1	0	1	1	0	0	1	0	1	1	0	0
1	1	1	1	1	1			1	1	0	1	1	0	1	1	0	1	1	1

shown in Table 17.1, where the effect of the NOT function can be clearly seen.

The EX–OR gate performs the function of modulo-2 addition, while the EX–NOR gate is often described as a *coincident detector*.

17.2 Combinational logic and circuits

Logic gates may be combined in a wide variety of ways, each combination designed to provide a specific overall function. Thus for any given input bit pattern, the output will always be the same. Figure 17.2 shows how the AND and EX–OR gates can be combined to provide an *adder* circuit. The function of Fig. 17.2(a) can be explained by the truth table shown in Table 17.2.

As this simple circuit does not provide for the possibility of a carry input, it is described as a *half adder*. By combining two half adders with an OR gate, the full function of addition can be provided as shown in Fig. 17.2(b).

Table 17.2   *Truth table for adder circuit*

A	B	Sum	Carry
0	0	0	0
0	1	1	0
1	0	1	0
1	1	0	1

De Morgan's theorem. This particular pair of relationships show the flexibility of elementary logic functions. It can be shown using truth tables and otherwise that a NAND gate can be produced from two inverters (NOT) gates combined with an OR gate as well as by simply inverting the output of an AND gate. A similar relationship exists between NOR and AND gates. In terms of Boolean algebra these can be expressed as:

$$\overline{A + B + C...} = \bar{A}.\bar{B}.\bar{C}...$$

$$\overline{A.B.C...} = \bar{A} + \bar{B} + \bar{C}...$$

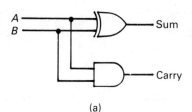

(a)

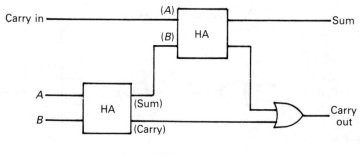

(b)

Figure 17.2   Serial adder circuits: (a) half adder (HA); (b) full adder.

Fundamental laws and identities. Boolean algebra obeys the associative, commutative and distributive laws of conventional algebra.

Associative. $(A + B) + C = A + (B + C)$ and
              $(AB)C = A(BC)$
              (the . sign for AND is often omitted).
Commutative. $A + B = B + A$ and $AB = BA$.
Distributive. $A(B + C) = AB + AC$.

In addition there are a number of very useful identities (see Table 17.3) that can help the logic designer.

Karnaugh maps. This technique, which forms a very powerful tool for minimising complex Boolean expressions, is illustrated in Fig. 17.3. A matrix is first drawn using the leading row and column to represent all the possible values of the variables in the expression to be minimised. In this case A, B and C. For n variables, the matrix will consist of 2^n cells. In this example, the expression to be reduced is $F = \bar{A}\bar{B}C + \bar{A}BC + A\bar{B}C + ABC + \bar{A}B$, shown in the truth table, and the values for $\bar{A}\bar{B}C = \bar{A}BC = A\bar{B}C = ABC = \bar{A}B = 1$ are taken from the table and then transferred to the mapping. It will be noticed that neighbouring-cell values differ from each other by 1 bit and this extends even to the edges of the diagram. The map is filled by placing 1 in the cells represented by $\bar{A}\bar{B}C$, $\bar{A}BC$, $A\bar{B}C$, and $\bar{A}B$. The remaining cells may be filled with 0s. Next, the cells containing 1s are grouped into rectangular blocks containing 2, 4, 8, etc., 1s, including the edges if convenient. In the example shown, one block can be represented by C alone and the other by $\bar{A}B$. Therefore the expression is minimised to $F = \bar{A}B + C$.

Negative and positive logic. Unless otherwise stated, positive logic is normally assumed for both descriptions and circuit diagrams. In this case a logic 1 is represented by the most positive voltage or current level and thus logic 0 is the most negative value. This is shown in Fig. 17.4(a). The alternative negative logic is shown in Fig. 17.4(b). The logic levels for both 1 and 0 are shown with a range of variation, usually to allow for the effects of noise. By changing from positive to negative logic, AND and OR gates become OR and AND gates respectively. Figure 17.5 shows the way in which circuit symbols are modified to cater for signal inversion and logic polarity changes.

A	B	C	F	
0	0	0	0	
0	0	1	1	$\bar{A}\bar{B}C$
0	1	0	0	$\bar{A}B$
0	1	1	1	$\bar{A}BC$
1	0	0	0	
1	0	1	1	$A\bar{B}C$
1	1	0	0	
1	1	1	1	ABC

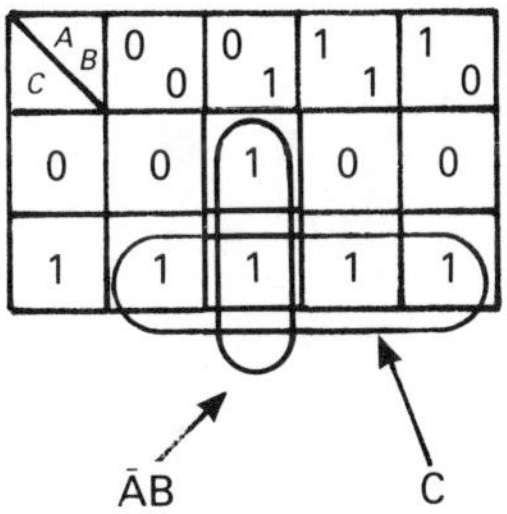

Figure 17.3   Truth table and Karnaugh map for $F = \bar{A}\bar{B}C + \bar{A}BC + A\bar{B}C + ABC + \bar{A}B$.

New standard symbols. Figure 17.6 shows how logic blocks may be combined in a circuit diagram. In Fig. 17.6(a) there are three inputs and one output. There is at least one signal connection between sub-blocks if there is a line perpendicular to the signal flow, the logic connection being indicated by a symbol on the common line. Thus in Fig. 17.6(a) there is no coupling between A, B and C, but all three feed into element D. In Fig. 17.6(b), the upper sub-block is the only non-rectangular element used and is described as a *common control block*. In this case input 'a' combines with inputs b, c and d as AND gates with the following effects:

$$e = a.b \qquad f = a.c \qquad g = a.d$$

Table 17.3   *Useful identities*

$A + 0 = A$	$A.0 = 0$	$A + \bar{A} = 1$	$A + AB = A$
$A + 1 = 1$	$A.1 = A$	$A.\bar{A} = 0$	$A + \bar{A}B = A + B$
$A + A = A$	$A.A = A$	$\bar{\bar{A}} = A$	$(A + B)(A + C) = A + BC$
$A + \bar{A} = 1$	$A.\bar{A} = 0$		

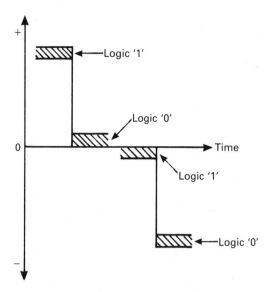

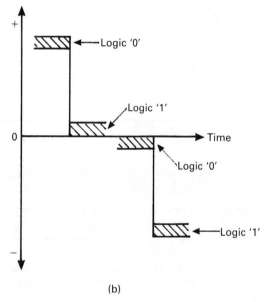

(b)

Figure 17.4   (a) Positive logic levels; (b) negative logic levels.

17.3 Fuzzy logic

This apparent contradiction of words provides a jargon phrase that very aptly describes an important control and processing concept. While conventional logic uses a rigid step-by-step approach, *fuzzy logic* allows the systematic processing of imprecise data using adaptive reasoning, combined with statistical probability to cater for varying levels of uncertainty. The concept was developed around 1965, but has become important because of the increased interest in *artificial intelligence*

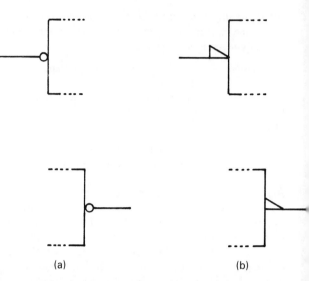

(a) (b)

Figure 17.5   (a) Logical negation at input and output; (b) polarity indicator at input and output signifies negative logic at this point.

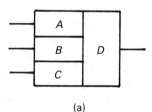

(a)

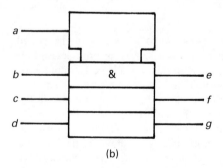

(b)

Figure 17.6   (a) Combination symbol; (b) common control block symbol.

(AI) and *expert systems*. The latter are computer systems that use knowledge plus reasoning procedures to solve problems that are normally the domain of an expert. The concept is becoming increasingly important in the fields of diagnostics and planning. Commercial processors are available that use a fuzzy language whose important elements are statistical *degrees of belief* curves. In this context, belief has a maximum value when the probability of error is zero. This tapers off gradually, rather than abruptly as in conventional logic systems. Data inputs are evaluated against these curves and fuzzy conclusion values are then combined to calculate a numerical output that is typically based upon a *centre of area* of a region of belief.

17.4 Sequential logic

This concept differs from combinational logic, in that the system output state depends not only on the present input, but also on the previous output state. This implies that this form of logic possesses a degree of memory. The basic elements are two-state devices that are described variously as *flip-flops*, *binaries*, *1-bit memories* or *latches*. The circuits provide two complementary outputs, usually defined as Q and $\bar{Q}$. A family of such circuit elements, which are all clock driven, is shown in Fig. 17.7. A complete system normally consists of several flip-flops connected in series and this leads to two possible driving conditions. The system is said to be synchronous when all the clock inputs are driven in parallel. Alternatively, the term asynchronous is used to describe the technique where only the first element is driven by the system clock and the succeeding stages are driven from the previous stage outputs. Flip-flops may either be triggered or clocked on the positive or negative going edge of a signal transition and Fig. 17.8 shows the circuit symbolism for this variation.

D-type flip-flop. This device, shown in Fig. 17.7(a), accepts serial data at the D input and each bit is transferred to the Q output after a clock pulse. (The $\bar{Q}$ output takes up the complementary state.) Thus each input bit is delayed (D) by one clock period. Because of the indeterminate output state at switch-on, this device usually has two further inputs described as *preset* and *clear* that are used to preset the output state before the first clock pulse.

Fan-in. The number of inputs to a given gate.

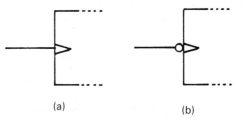

(a)　　　　(b)

Figure 17.8 Dynamic inputs: (a) positive edge trigger; (b) negative edge trigger.

Fan-out. The number of individual inputs that the output of a given gate can drive simultaneously in parallel.

Master–slave flip-flop. This consists essentially of two R–S flip-flops in cascade, with overall feedback from output to input and driven by complementary clock signals. The first stage is described as the *master* and the second as the *slave*. The positive going edge of the clock pulse *enables* the master so that data can be input and the negative going edge enables the slave allowing the data to be transferred to the output section. This technique provides a high degree of isolation between the input and output circuits and avoids the problem of *race conditions*.

J–K flip-flop. This device, shown in Fig. 17.7(b), is universal in that it can be configured to function as either member of the family by using simple additional external components. Additional preset and clear inputs are provided to avoid the indeterminate Q and $\bar{Q}$ states at switch-on and, because all states of output are uniquely defined and controlled, there is no race condition problem. Table 17.4 shows the truth table which describes the operation following the next clock pulse.

R–S flip-flop. This circuit element carries no preset or clear inputs because it is normally driven by complementary R and S inputs which uniquely define an output state as shown by the truth table (Table 17.5). If not so driven, then it is important to prevent the state $R = S = 1$ from occurring.

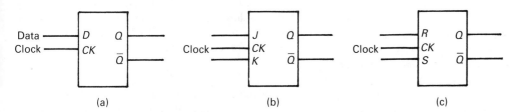

(a)　　　　　　(b)　　　　　　(c)

Figure 17.7 Binary or flip-flop circuit elements: (a) D type; (b) J–K type; (c) R–S type.

Table 17.4   *Truth table for J–K flip-flop*

	J	K	Q	$\bar{Q}$
Hold	0	0	No change	
Reset	0	1	0	1
Set	1	0	1	0
Toggle	1	1	Change state	

Table 17.5   *Truth table for R–S flip-flop*

	R	S	Q	$\bar{Q}$
Hold	0	0	No change	
Set	0	1	1	0
Reset	1	0	0	1
Prohibited	1	1	1	1 (Impossible)

T-flip-flop. This is a special case of the J–K flip-flop where $J = K = 1$. This causes the device to change state on every clock pulse, thus performing a divide-by-two function.

17.5 Miscellaneous terms

Don't care states. When minimising a Boolean expression using a Karnaugh map, it is often found that certain states of a variable can take up a 0 or 1 value without affecting the operation. For example, this can arise in a binary coded decimal system which requires four bits but not all 4-bit patterns can occur. By placing a X in the corresponding cells of the map, these conditions are clearly visible and can be incorporated into any output looping without affecting the operation. This can produce larger groupings that lead to greater simplification of the final expression.

Glue logic. The jargon term used to describe the additional logic components needed to interconnect such devices as a microprocessor or controller to an interface, transducer, etc., to form a system.

Maxterm and minterm. The Boolean expression for any arbitrary logic function can be obtained from its truth table in two forms, either as a *sum of products* or maxterm, or *product of sums* or minterm. Either form can then be minimised using the Karnaugh mapping technique. Consider the truth table shown in Table 17.6 for some arbitrary function *F*.

Table 17.6

A	B	C	F	
0	0	0	0	.. (invert) .. $(A + B + C)$
0	0	1	1	.... $(\bar{A}.\bar{B}.C)$
0	1	0	1	.... $(\bar{A}.B.\bar{C})$
0	1	1	0	.. (invert) .. $(A + \bar{B} + \bar{C})$
1	0	0	0	.. (invert) .. $(\bar{A} + B + C)$
1	0	1	1	.... $(A.\bar{B}.C)$
1	1	0	1	.... $(A.B.\bar{C})$
1	1	1	0	.. (invert).. $(\bar{A} + \bar{B} + \bar{C})$

The maxterm is found by taking all the 0 entries and inverting the variables as shown and combining them in OR fashion. All these are further combined in AND fashion to give maxterm $F = (A + B + C) (A + \bar{B} + \bar{C}) (\bar{A} + B + C) (\bar{A} + \bar{B} + \bar{C})$ (product of sums). In a similar way, the variable values that produce 1 entries are combined first by AND and then by OR to give minterm $F = \bar{A}.\bar{B}.C + \bar{A}.B.\bar{C} + A.\bar{B}.C + A.B.\bar{C}$ (sum of products).

Race conditions. Although abrupt, the digital signals used for logic operation do not change level instantaneously. During the transitions periods the signal levels are thus ill-defined and when such are applied at the inputs of several gates simultaneously, the response can be unpredictable. One or other input thus arrives first and leads to a *race condition* which can have disastrous effects.

Standard product form. When every variable of a Boolean equation is contained in each factor of an expression and all combined by OR in each factor, then the equation is in *standard product form*: for example $(\bar{A} + B + \bar{C}) (A + \bar{B} + C)$ (note that some variables may be negated).

Standard sum form. When every variable in a Boolean equation is contained in every factor of the equation combined by AND, with separate factors combined by OR as follows, then the expression is in *standard sum form*: for example $(\bar{A}B\bar{C}) + (A\bar{B}C)$ (note that negated variables are included).

Tautology. Boolean expressions that are always true no matter what the state of the variables. For example, the identities or theorem values.

Wired logic. It is sometimes necessary to provide a gate that has more inputs than are readily available. The problem can be solved by wiring the necessary number of gates together in parallel and then combining the separate outputs in a suitable manner (see Fan-in, p. 149).

17.6 Useful references

Millman J. and Grabel A. (1987) *Microelectronics*, 2nd edition, McGraw-Hill, New York.

Tokheim R.L. (1988) *Digital Principles*, 2nd edition, McGraw-Hill, New York.

18 Measurement of system parameters

In any communications system the most important parameter is the signal quality or signal to noise ratio (S/N ratio) (see Noise, p. 219). In the general sense, noise is considered to be the summation of all unwanted effects or artifacts superimposed upon the signal due to nonlinearity, in all parts of the link that produce distortion and interference, plus random noise from natural sources. The S/N ratio is thus the power ratio (or in certain cases, the voltage or current ratios) of the wanted signal level to the total noise component. In an analogue system, such effects will distort the signal waveshape and this will be obvious to the ear or eye. Various techniques can be applied that minimise the noise effects while maximising the wanted signal, but ultimately degradation in the S/N ratio sets the threshold of unacceptability. For digital systems a similar problem arises, but it is possible to regenerate the signal waveshape without the noise, but at the expense of bit errors due to wrong decoding decisions.

The concept of measurement/testing of system parameters falls into two broad areas: system design and production, and system service and maintenance. While there are many similarities in these areas, there are also some significant differences. Design and production invariably requires measurements to a very high standard using expensive equipment. For service/maintenance the parameter tolerances can often be relaxed to permit the use of portable test equipment. But in this area, the pressure is to minimise the time (downtime) that a system is out of service. The concept of system *testability* thus becomes an added problem.

Automatic test equipment (ATE). Rapidly changing technology demands increasingly complex and fast test capabilities, both at the production stage and during maintenance operations. The introduction of *application specific integrated circuits* (ASICs), *surface mounting technology* (SMT) and compound *printed circuit boards* (PCBs) has placed further restrictions on the testability of systems and sub-systems. Even for the testing of analogue systems, adapted digital techniques can be used with advantage. A digital oscilloscope under the control of a computer can collect waveforms from various test points or nodes within a system or sub-system. These can then be compared by the computer with waveforms derived from a known good system to provide a *signature profile* of the system under test. Such ATE systems can be operated in a *learn* mode where the test system can be taught to recognise a good system. When operated in the *test* mode, the ATE system then compares each test value with the standard parameters and then uses the computer technology to process this data as required. This concept has a flexibility that allows the test standards to be modified as the product or system develops through its various upgrades.

Bed of nails. The descriptive title given to a test fixture that is used extensively with ATE. It consists of a matrix of spring loaded pins that make contact to the various test nodes of the system under test. Each test fixture is thus unique to each PCB to be tested.

Boundary scan testing. Some system PCBs are so complex that the various test nodes are not readily accessible to a bed of nails. In these cases, a specific area of circuit can be included that is dedicated to testing of the complete assembly. This contains the control logic of a state machine which scans the various testing nodes to verify the status of each section. The test results are held in a memory register which is then interrogated by an external boundary scan control computer.

Current nulling technique. When the current flowing in a part of a circuit is to be measured, it is conventional to break the circuit and insert a current measuring instrument. In many cases the inclusion of this test meter changes the operating conditions of the system. Alternatively a permanent test resistor may be included so that the voltage drop across it may be measured, but in this case the technique adds an extra component that generates heat. To avoid this situation an instrument based on the Hall effect device (see Semiconductor devices and technology, p. 298) can be used. A more elegant method that even allows the current flowing in a track of a PCB to be measured under actual working conditions involves the current nulling technique. The track is probed at two points and the very small voltage drop is measured. This is amplified and used to control a current generator whose output is exactly equal but of opposite polarity to that on the circuit board. This current is applied to the track using a second pair of probes contained within the voltage probes. The level of the mirror current is then displayed on the instrument.

Directed measurement technique. A simple technique where the system output in response to some specified input stimulus is directly measured.

Insertion test signals (see also Television signal processing and systems, p. 372). When either analogue or digital signals have to be transmitted over great distances, either by radio or transmission lines, it is necessary to be able to monitor the signal quality at various points in the link. It is considered uneconomical to dedicate time specifically for such testing. Insertion test signals (ITS) are thus added or multiplexed into the service signal at suitable times to provide for this monitoring. In the television signal, ITS information is added in the field blanking interval and in PCM systems, a digital test code pattern in the form of a pseudo-random binary sequence can be added within the time frame multiplex.

Iterative measurement technique. The technique whereby the system input stimulus is varied until some specified output level is achieved. For example, varying the system input noise level until the output noise level doubles (see Noise measurement, p. 157).

18.1 Analogue systems

The majority of analogue system parameters that have to be measured are associated with signal quality and this has the direct implication of waveform purity.

18.1.1 Cathode ray oscilloscope (CRO) type measurements

The CRO can basically be used in two ways. With the X scale calibrated in time, the instrument provides a window in the *time domain*, so that the time varying amplitudes of signals can be displayed. Direct measurements from the screen are difficult to make, but for production or maintenance testing, where the parameter tolerances provide a Go/No-go state, a *limit mask* may be overlaid on the screen to ensure that the waveform lies between the specified limits. The displayed wave-

forms may be digitised for processing in the digital domain to provide a more accurate assessment. By causing the CRO X axis to be driven by a *sweeping frequency*, the instrument displays a window in the *frequency domain* and is thus described as a *spectrum analyser*. Individual frequency components within some particular bandwidth are then displayed as vertical pulses whose amplitude represents the magnitude of that particular component. Complex signals over a wide range of bandwidths can be examined in this manner. Modern test sets described as *scalar* or *amplitude network analysers* achieve a high degree of function integration. These basically consist of a CRO to display the time or frequency domains. This is controlled by a microprocessor so that the instrument can provide digital processing of the signal, with digital storage and hard copy print-outs. These devices can be used for such measurements as insertion gain or loss, return loss or voltage standing wave ratio (VSWR), gain compression, absolute power, etc. For radio communications networks, a variant of the spectrum analyser is often added to the system to display the frequency components within the intermediate frequency (IF) passband of the receiver, this device being described as a *panoramic display*. The signal components are clearly shown as vertical pulses rising above a background of noise which is often described in the jargon as *grass*. Such a device is useful to distinguish between the wanted signal spectrum and interference. Careful analysis can identify the frequency of such interfering signals. In most of these devices, it is an advantage to be able to display the amplitudes on a logarithmic scale. This not only allows direct measurement in decibels (dB), but is useful to distinguish signals with rapidly changing amplitudes over a wide dynamic range. This is particularly important when measuring the signal strengths over routes where one end is a mobile and subject to fading.

Lissajous figures. The phase difference between two sine waves of the same frequency can be measured using Lissajous figures, a technique that represents the use of a nonlinear timebase. A series of such displays are shown in Fig. 18.1. If two sine waves are input to the X and Y deflection systems and their amplitudes carefully adjusted, the display will vary from a straight line, through a series of ellipses, to a circle in the manner shown in Fig. 18.1(a). As indicated by Fig.18.2 the general phase angle can be obtained from $\sin^{-1} a/b$.

Furthermore, this technique leads to the more complex displays shown in Fig. 18.1(b), when the two waveforms are simple multiples of each other. In general, the ratio of the two frequencies here is given by:

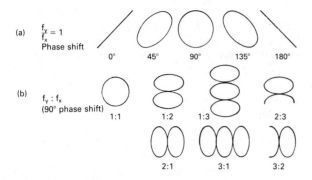

(a) $\frac{f_y}{f_x} = 1$
Phase shift

(b) $f_y : f_x$
(90° phase shift)

Figure 18.1   Lissajous' figures.

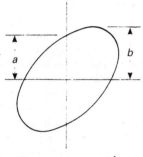

Phase angle $= \sin^{-1}$

Figure 18.2   Lissajous figure for general phase angle.

$$\frac{f_y}{f_x} = \frac{\text{Number of loops horizontally}}{\text{Number of loops vertically}}$$

The Lissajous figure technique can be extended to determine an unknown frequency. This is achieved by making a direct comparison with the output from an accurately calibrated sine wave signal generator. If the unknown frequency is somewhat higher than the standard source generator, the following technique may be used. Use the known frequency to create a Lissajous circle and apply the unknown frequency to the Z input. This produces a set of bright spots on the circle. Adjust the known frequency source to produce a static display and count the dots to obtain the multiplying factor.

Vector network analyser. This extension of the scalar analyser allows for the display of relative phase as well as amplitude. The instrument is interfaced to a personal computer and the digital processing allows not only for direct display on a CRT, but also for storage on disk and hard-copy print-out options. The displays can be configured for logarithmic or linear scaling in either polar or rectangular coordinates using frequency, time or antenna orientation as the second parameter.

18.1.2 Distortions

The general technique used to quantify the effects of distortion is based on separating the frequency components of interest from the distorted waveform, using suitable filters. The amplitudes of these components are then measured and the ratio of unwanted to wanted signal amplitudes used to calculate the level of distortion, either as a percentage or in decibels. In certain cases, these calculated ratios do not equate with the perceived nuisance value of the distortion. The values may then be artificially adjusted or *weighted* to produce a more subjective rating.

Delay distortion. Any arbitrary waveshape can be considered to consist of a fundamental frequency sinewave plus a range of harmonics. If in passing through a network, the relative phase relationship of any of these components is changed, then so is the waveshape. For an undistorted waveshape, the network should have a linear phase/frequency response; that is, the network should generate *n* times the phase shift for the *n*th harmonic as it does for the fundamental frequency.

Deviation distortion (see also Frequency modulation, p. 187). In a frequency modulated (FM) system any nonlinearity in the receiver demodulator will cause distortion of the recovered waveform. Apart from any nonlinearity, lack of bandwidth and/or AM (amplitude modulation) rejection will produce a similar effect.

Envelope or group delay distortion. Group delay is defined as the time taken for a particular point on a modulated wave envelope to propagate between two points in a communications link. In the absence of delay distortion, this will be the same for all frequencies in the bandwidth. Thus delay distortion generates envelope distortion.

Gain compression. Even a linear amplifier system produces nonlinearity as its saturation point is reached. This gain compression feature is shown in Fig. 18.3, where the degree of acceptable distortion is set by the input signal level that produces 1 dB of gain compression. In this particular example, the actual gain is 1 dB less than it would be if the amplifier had been linear and this point occurs at an input of 6 dBm. The higher the level of this value, then the lower will be the degree of distortion.

Harmonic distortion. If the transfer characteristic of a system is nonlinear, equal increments of input signal do not produce equal increments at the output. Because such distortion can be shown to produce

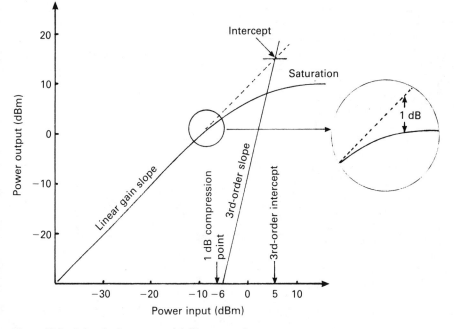

Figure 18.3 3rd-order intercept and 1 dB compression.

harmonically related frequencies this is described as amplitude, nonlinear or *harmonic* distortion. The distortion factor D is found by measuring the RMS voltage of the distorted waveform (M_1), filtering off the fundamental and measuring the RMS voltage of the residue signal (M_2). D is then equal to M_2/M_1 and quoted either as a percentage or in decibels. Assuming that the distortion follows a parabolic law, then for a sinusoidal input signal, the output will be of the form: $B_0 + B_1 \cos \omega t + B_2 \cos 2\omega t + ...B_i \cos \omega t$, where B_i, are the Fourier coefficients of the expansion of the series. The level of the ith harmonic distortion D_i is given by, $D_i = |B_i| / |B_1|$, while the total harmonic distortion (THD) is given by $D = \sqrt{D_2{}^2 + D_3{}^2 + ...D_i{}^2}$.

Intermodulation. When two sinusoidal signals are present simultaneously in a nonlinear network, beat notes in the form of sum and difference frequencies are generated. Although the waveshapes become distorted, the frequency components are not directly harmonically related. Some of the products of intermodulations are depicted in Fig. 18.4. Normally the second-order terms, $f_2 \pm f_1$, $2f_1$ and $2f_2$ fall outside the passband of interest and only the third-order terms, $2f_1 - f_2$ and $2f_2 - f_1$ cause problems. Intermodulation, which can be measured in a similar way to harmonic distortion, can in some cases be the more significant. Figure 18.3 shows the characteristic of an amplifier with a nominal gain of 10 dB. With increasing input the amplifier starts to distort and will eventually saturate. Over the linear part

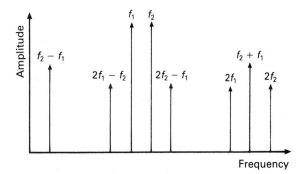

Figure 18.4 Intermodulation products.

of the characteristic, the gain slope is 1 dB/dB. Assuming that the amplifier is initially saturated due to over-driving, the output signal spectrum will contain harmonic components. If the input signal level is now reduced, the level of the harmonic components will reduce faster than will the fundamental. In fact, the third-order term will reduce with a slope of 3 dB/dB as shown. The slopes of the linear and third-order characteristics coincide at a point known as the *third-order intercept*, in this case at the 5 dBm input level. The higher this value, then the lower will be the third-order intermodulation component. If this point is known for any given fundamental output power, the level of the third-order term can be evaluated.

Incidental modulation (see also Modulation and demodulation, p. 186). When a carrier wave has been intentionally modulated in one of the standard forms (amplitude (AM), frequency (FM) or phase (PM)), later nonlinear processing will introduce an unwanted modulation component. For example, an FM wave can acquire an AM component. This can be measured by modulating the carrier to a standard depth of modulation or deviation and then, after processing, switching between two alternative demodulators to measure the levels of the wanted and unwanted components. The level of incidental modulation can then be calculated in the same way as for other forms of distortion.

Slew rate limiting. The *slew rate* for an amplifier is defined as the rate of change of the output voltage under large signal conditions. If this rate is exceeded by too large an input signal, then the output becomes distorted. Because this is related to the amplifier rise time, the slew rate is also related to the amplifier bandwidth. The relationship being given by slew rate $= 2\pi f_h V_{out(max)}$ where f_h is the upper 3 dB cut-off frequency. Testing is carried out using a high frequency square wave as input and measuring the slopes of the output signal relative to time, for both leading and trailing edges. Slew rate is then quoted for the slowest value.

Triple beat. When a complex communications system operates with a frequency multiplexed band of carriers, triple beat intermodulation can occur due to system nonlinearity. As the carriers are spaced in frequency by constant values, beats between sum and difference components of the adjacent carriers produce interference. The difference between the sum of the two adjacent carrier frequencies and the central carrier produces an interfering component at the central frequency.

1-dB compression (see Gain compression, p. 154)

Third-order intercept (see Intermodulation distortion, p. 155)

18.1.3 Interference distortion

Interference can be introduced into a system in a variety of ways, but in general only becomes obvious after demodulation. The extent of this nuisance can be measured by the system's *rejection ratio* to a particular form of interference which is given in dB by:

$$20 \log \frac{\text{Demodulator output voltage due to interfering signal}}{\text{Demodulator output voltage due to wanted signal only}}$$

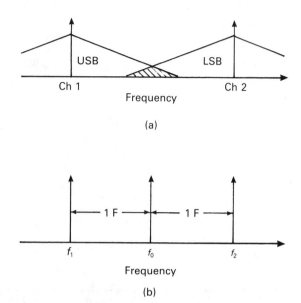

Figure 18.5 (a) Adjacent channel interference; (b) image channel interference.

Adjacent channel interference. If channel spacing is less than twice the base bandwidth of the signals being included in the frequency multiplex, then the upper and lower sidebands will overlap as shown in Fig. 18.5(a). The demodulator outputs of both receivers will then include unwanted components from the adjacent channel modulating signals which will introduce a form of distortion.

Cross-modulation. If two modulated signals are processed simultaneously through a nonlinear medium, the distorting effects will cause some of the modulating signal energy from each channel to be transferred to the other to produce interference.

Image channel interference. The superheterodyne (superhet) receiver deliberately beats the wanted signal together with one generated from its own internal local oscillator. The difference signal, or *intermediate frequency*, is then processed instead of the wanted signal. However, as shown in Fig. 18.5(b), there are two frequencies f_1 and f_2 that give rise to this effect: one is the wanted channel and the other a mirror image of it about the local oscillator frequency. This image is also known as a *second channel*, the wanted channel being the first.

Intermediate frequency (IF) breakthrough. If a strong signal at the IF is present at the input to a superhet receiver, there is a possibility that it can break through the front-end tuning system directly into the IF section.

As this is an unwanted signal it will result in interference.

Kendall effect (sideband folding). If the carrier frequency is so low that its lower sideband overlaps the upper end of the modulating signal baseband, it generates a form of interference that cannot be filtered out by a demodulator. This is very similar to *aliasing* in the digital sense. This condition also generates sideband folding distortion as shown in Fig. 18.6. When a carrier frequency f_c is modulated by a signal f_m, side frequencies of $f_c \pm f_m$ are produced. If the carrier frequency is too

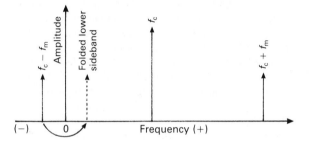

Figure 18.6 Sideband folding.

low, $f_c - f_m$ will be negative and reflect about zero, to produce a sideband component that appears to have been generated by a non-existent baseband component.

18.1.4 Noise measurements (see also Noise, p. 219)

In general, system performance can be characterised by three closely linked parameters: overall gain, bandwidth and noise figure. System gain is practically unlimited, but the usable level of gain is set by the bandwidth and the level of noise that can be accepted before it swamps the modulation or generates too many digital errors. Thus these three parameters should always be specified in a related manner. The *noise factor* or *figure* (F) describes how the noise generated within the system, due to amplification and processing, adds to the thermally generated noise already present at the input and further degrades the signal to noise power ratio (S/N). The noise factor F is thus the ratio,

$$\frac{S/N \text{ at input}}{S/N \text{ at output}}$$

and quoted either as a ratio or in dB.

The total noise output (N_0) of a system has two components, one due to the thermal noise at the input multiplied by the system gain (G), plus the internally generated noise (N_e).

Thus $F = \dfrac{N_0}{kTBG}$

and

$$N_0 = F(kTBG)$$

where k = Boltzmann's constant (1.37×10^{-23} J/K)
T = temperature in Kelvin
B = bandwidth in Hz
G = gain as a ratio.

The *excess noise* (N_e) added by the system is thus,

$$N_e = F(kTBG) - (kTBG) = (F - 1)(kTBG)$$

The excess noise ratio (ENR) or S/N ratio thus centres around the power measurement of N_0.

For a complex system consisting of several stages in cascade, the noise factor is given by:

$$F = F_1 + (F_2 - 1)/G_1 + (F_3 - 1)/G_1 G_2 \ldots$$

Carrier to noise ratio (C/N). In order to distinguish between the noise ratio of a receiver system before and after demodulation, it is usual to refer to the pre-detector ratio in terms of carrier to noise ratio. Then in this case, the S/N ratio refers to the post-detector signal quality.

Excess noise ratio. There are two basic methods of measuring this parameter.

(1) Connect a calibrated noise generator to the system input and a power meter at the output. Two measurements are taken, N_1 with the excess noise source cold and N_2 with the source at its normal working temperature. The noise factor F is then given by

$$\frac{T_1 - T_0}{T_0} \times \frac{1}{(N_2/N_1) - 1} \tag{18.1}$$

where T_0 = absolute zero or $-273°C$
T_1 = temperature of noise source.

(2) The method is similar to that above, but the input current of the noise generator is adjusted until the reading N_2 is just twice N_1. The second term in equation (18.1) thus becomes unity and F is given by $(T_1 - T_0)/T_0$, or alternatively by $F_{dB} = 10 \log 20IR$, where I = the generator current necessary to double the noise power, and R = input impedance of the system.

Noise temperature. The noise temperature (T_A) of an antenna can be measured using the Y-factor method. The receiver is first connected to a standard liquid-cooled noise source of temperature T_s so that the noise output N_s can be measured. The receiver input is then switched to the antenna and the noise output N_A is noted. Y can then be calculated from $Y = N_s/N_A = (I_s + T_R)/(T_A + T_R)$ and

$$T_A = \frac{T_s + T_R(1 - Y)}{Y} \qquad (18.2)$$

T_R the receiver noise temperature can be found in a similar way by connecting its input to a matched load at room temperature T_H, so that N_A can be measured: $Y' = (T_H + T_R)/(T_s + T_R)$ and

$$T_R = \frac{T_H - Y'T_s}{(Y' - 1)} \qquad (18.3)$$

By substituting for T_R in equation (18.2), T_A can be evaluated.

Psophometer. This is an instrument that can measure not only the noise in a communications link, but also weight its results to express the nuisance value. The weighting circuit is frequency selective so that the parameters can be adjusted to meet different situations.

Psophometric emf. This is an emf that is twice the psophometric voltage which is measured across a $600\,\Omega$ non-inductive resistor shunting the circuit at the point of measurement. The transmitting end is assumed to be terminated in the image impedance.

Psophometric power. This is the power absorbed by the $600\,\Omega$ non-inductive resistor from a psophometric emf source and defined as (psophometric voltage)^2/600 or

$$\frac{(\text{psophometric emf})^2}{4 \times 600}$$

Psophometric voltage. This is a subjective measure of the interference caused on a telephone circuit by noise. It is expressed as the equivalent level of an 800 Hz tone that would have the same nuisance effect.

SINAD ratio. This takes into consideration the total disturbing effects of both noise and distortion by using the ratio

$$\frac{\text{signal} + \text{noise} + \text{distortion}}{\text{noise} + \text{distortion}}$$

to obtain a more meaningful parameter.

18.1.5 Power measurements

The measurement of power in telecommunications systems produces a number of significant problems. The levels range from a few picowatts to several tens of watts, at frequencies ranging from dc to a few hundred gigahertz, with a wide range of load impedances and an almost infinite range of waveshapes. The problems are resolved by using a range of instruments to handle specific conditions. In general, since the power dissipated is proportional to the heat generated, power developed by different waveshapes can easily be measured indirectly. To resolve the problem of accuracy over a wide frequency range and variation of load impedance, measurements in communications systems are commonly taken at points having a standard impedance of $600\,\Omega$. A level standard is then set at 1 dBm or 1 mW into $600\,\Omega$. A variety of different sensor heads are available to handle the wide ranges of power levels and frequencies that might be encountered.

Bolometer. A small resistive element enclosed in an evacuated glass envelope which is capable of being heated from an electromagnetic power source, the increase in temperature which can be measured being proportional to the power absorbed. Useful when used in conjunction with bridge circuits to measure low levels of radio frequency power.

Dicke's radiometer. A small resistive element that is useful for measuring microwave noise power by comparison with the noise from a standard source.

Seebeck effect. When two wires of dissimilar metals are joined to form a loop and the two junctions are held at different temperatures, a current will flow round the circuit to produce an emf, the magnitude of which is proportional to the temperature difference.

Thermistors. These resistor-like devices made from manganese or cobalt dioxide have a significant exponential increase in conductivity when exposed to a power source, a feature useful for power measurement. Sensitivity to overload and consequent burn-out restrict their use to relatively low power measurements, and generally at low frequencies.

Thermocouples. The device described above under Seebeck effect forms a thermocouple which can be used to measure temperature and, indirectly, power dissipation. A range of metallic pairs are used for power measurement under different conditions. Semiconductor devices such as Schottky barrier diodes are also used in this respect. However, modern *planar doped barrier diodes* have a much lower unit-to-unit variation and better frequency response. These are usually restricted to power levels below about $10\,\mu\text{W}$, but have a good overload recovery capability.

Thin film thermoelectric (TFT) sensors. These devices are formed using semiconductor thin-film techniques and consist of an array of diodes which form a junction. These deliver a dc voltage output that is linearly proportional to the RF power dissipated. They combine the sensitivity and speed of the bolometer, together with the accuracy and stability of a calorimeter. They have a wide dynamic range, freedom from errors due to noise and drift and a negligible sensitivity to parameter variation with temperature.

18.2 Digital systems

The measurement of digital system parameters falls into two distinct categories. When a binary signal has to be transmitted for any significant distance, it has to be modulated on to a radio frequency carrier. The complex signal then takes on analogue characteristics which must be measured using analogue techniques. After such a signal has been demodulated, it becomes necessary to test and measure in the digital domain and these techniques are quite different. Although the binary signal is a simple two-state condition, the ideal digital signal does not exist in practice. Processing a square wave through any device is equivalent to passing the signal through a low-pass filter. Pulses then become distorted as they lose their high frequency components and are further corrupted by the addition of noise. In digital signalling, the signal to noise ratio (S/N) of the analogue systems is replaced by the concept of *energy per bit* (E_b), *per watt of noise power, per unit bandwidth* (N_0). If R and B are the system transmission bit rate and bandwidth, respectively, this ratio can be extended further to give:

$$\frac{E_b \times R}{N_0 \times B} = \frac{\text{Carrier power}}{\text{Noise power}} \quad \text{or } C/N \text{ ratio}$$

Obviously the energy per bit can be increased by increasing the signal amplitude and/or by increasing the bit period. The latter, however, reduces the signalling speed. This relationship is expressed in Shannon's rule for channel capacity, $C = B \log_2(1 + S/N)$ bit/s, where C is the channel capacity and B its bandwidth. As in an analogue system, noise destroys information, but in the digital case, it leads to bit errors, expressed as the *bit error rate* (BER), usually shown in the form of BER $= e \times 10^x$ bits per second. *Out of service* testing is commonly used and this involves measuring the C/N ratio that produces a specified BER. Basically the

technique requires a pseudo-random binary sequence (PRBS) generator to provide the input baseband signal, together with a noise generator and a variable attenuator to degrade the C/N in a controlled manner and to simulate the channel conditions. The output bit stream is compared at the receiver with the same uncorrupted PRBS using an exclusive OR gate. As

Table 18.1

A	B	F
0	0	0
0	1	1
1	0	1
1	1	0

shown in Table 18.1, the gate output F only goes to logic 1 when the received bit stream (A compared with B) is in error. A simple timed count then gives the BER for the given C/N ratio.

In service testing which involves performance analysis while the channel is carrying normal traffic is economically more efficient (see Distortion, interference and noise, pp. 154, 156, 157).

18.2.1 Cathode ray oscilloscope (CRO) type measurements

Digital storage oscilloscope (DSO). These devices are designed to capture, store and analyse repetitive signals or single events. Sampling rates as high as 5 GHz allow for the resolution of single events of duration as small as 200 ps. The typical technique involves storing the input signal in a serial *charge coupled device* (CCD) (see Semiconductor devices and technology, p. 292) shift register and then clocking the samples out via an analogue to digital convertor (ADC) at a lower speed to digitise the signal. The lower speed gives a high resolution at the ADC without sacrificing the high speed of data capture. The data from multiple input channels can be stored in separate memories, one of which is commonly used as a reference for comparison purposes. The triggering technique allows for the precise definition of the section of waveform to be captured. Pre-triggering allows for the events leading up to the particular trigger moment to be captured and analysed, while post-triggering allows small details of complex waveforms to be analysed. *Noisy* signals can be captured by using a running average algorithm that minimises the effect of the uncorrelated noise component on the signal. By using computer control, this instrument can form part of a continuously running automatic test equipment (ATE) system.

Logic analyser. These devices have the capacity to be used for the analysis of the simultaneous logic state and timing of more than 100 parallel channels. The state listings can be shown directly as a set of time related waveforms, or as disassembled machine code instructions, usually in binary or hexadecimal, to simplify the tracing of a program flow. These are displayed as a set of data states/values plotted against machine/clock cycles. High speed events such as *glitches* (spurious noise pulses) of less than 5 ns duration (even lower for fewer input channels) can be displayed. Data acquisition is via low capacitance, high impedance probes and for a serial input data stream, an interface is provided to perform serial to parallel conversion. Data sampling can be performed synchronously or asynchronously for state or timing analysis, respectively. Extensive random access memory (RAM) or means of interfacing to a computer is usually provided for data processing and comparison. Triggering is by a number of digital words that can be selected to precisely define the start and stop points of the data which is to be acquired. With such instruments it is important to distinguish between their sampling and decision rates. If the sampling rate is say 50 MHz then the device must be able to evaluate and store an event within a period of 1/50 MHz or 20 ns, otherwise information will be lost.

Signature analysis. When a prototype system has been constructed and proved to function correctly, it can be made to execute a sequence of instructions in a repetitive fashion. By monitoring each node of the system, data can be acquired about the correct logical activity. A *signature analyser* converts each node data into a unique hexadecimal code that represents the 'signature' at a correctly functioning node. System documentation is then prepared, which includes a circuit diagram with all the correct signatures appended. In servicing, a faulty component can then be identified as a device that produces an error output from correct input signatures.

Vector modulation analysis. These software and micro-computer-controlled instruments are used for the analysis of multi-level, high bit rate, digitally modulated signals. The modulated waveforms are sampled to provide displays that are based on phase and magnitude (vector) and are thus able to process the in-phase (I) and quadrature (Q) components of such signals. Because of the computer control, these devices are capable of performing statistical analysis of channels carrying either live or simulated traffic. The major advantage of such instruments is the ability to differentiate between the effects of the various sources of bit errors: noise, inter-symbol-interference, timing jitter or incidental modulation.

18. 2. 2 Distortion, interference and noise

Shannon's law for channel capacity, $C = B \times \log_2 (1 + S/N)$ bits per second, refers to an ideal channel. If data at a bit rate higher than C is attempted, the bit error rate (BER) rises rapidly. It is thus important to be able to identify the sources of signal corruption in a communications channel.

Analysis of binary signals by eye displays. If the demodulated binary data signal is applied to the Y input of an oscilloscope whose X timebase is locked in a particular way to the data rate, the resulting trace becomes a series of superimposed signal transitions as represented by Fig. 18.7, which also shows how *eye* patterns are formed. The height H and width W of the eye are a function of the slopes of the data transitions, which are in turn dependent upon the degree of noise and distortion to which the signal has been exposed. The *eye height* is defined as the ratio: difference in levels between worst-case 0s and 1s, to the difference in levels between a long sequence of 0s and 1s. From Fig. 18.7 the eye height $H = h/b \times 100\%$. The eye width W and *jitter* J are also significant in the interests of a low error rate. The greater the width of the eye, the lower will be the BER after resampling. W is defined as the ratio of eye width to the data bit period, or $W = d/T$. Jitter, which is due to inter-symbol-interference and timing errors, also affects the BER and is evaluated as $J = T(1 - W)$. The principle of an 'eye' pattern display unit can be explained with the aid of Fig. 18.8. A clock signal is derived from the data stream and its frequency divided by four. This is then low-pass filtered to obtain a sinusoid at $\frac{1}{4}$ of the bit rate. The sinusoid and the original data stream are applied to the oscilloscope X and Y inputs as shown. By adjusting the phasing of the sinusoid, the displayed Lissajous figure becomes an eye. The major disadvantage of eye height measuring is that the pattern is a summation of the effects of all the noise and distortion and cannot identify the particular source of bit errors.

Constellation analysis of multi-level signals. The constellation analyser is an instrument that is capable of distinguishing between the various causes of bit errors in multi-level (QAM) signalling systems. The display is a plot on the real and imaginary axes, representing the in-phase and quadrature signal components, respectively. The coincident points in the signal vector space produce a bright-up effect on the CRT face. An N-level signal produces an N-point square of bright regions, the constellation, where each point represents a particular phase/magnitude relationship from the complex modulated signal. Repeated display of the data stream integrates the effect to produce patterns similar to

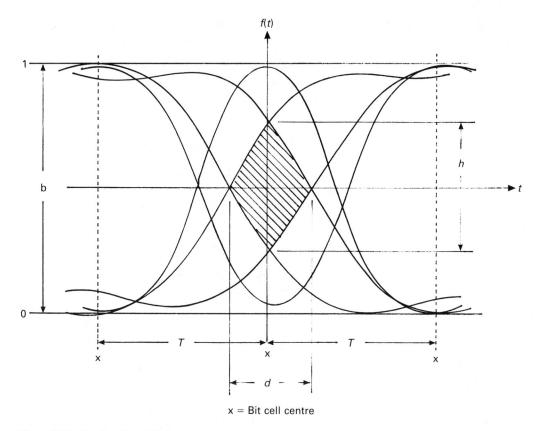

Figure 18.7 Details of 'eye' display.

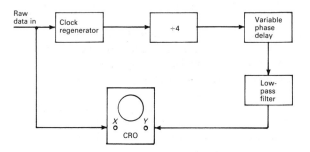

Figure 18.8 Principle of 'eye' display unit.

those shown in Fig. 18.9. Due to timing and other system errors, the bright regions spread out around some average position. The overall summation of noise and distortion produces an effect similar to that shown in Fig. 18.9(a), where the ratio $x/d \times 100\%$ is equivalent to the eye height. The angular rotation from the ideal position as shown by the cluster at (b) represents the effect of phase angle errors between the data stream and the decoder clock recovery circuit. This is described as a *lock-angle error*. An error between the I and Q phases can arise at either the transmitter or in the receiver decoder section and this is described as a *quadrature*

error. The total angular error can be evaluated by doubling the angle measured in (c). In a practical situation, a constellation would display the summation of all the errors, but these can be separated by careful analysis.

Noise power ratio (NPR). Specifically applies to frequency division multiplex (FDM) systems. The measurement of noise interference in wideband telecommunications systems is based on the fact that random noise evenly distributed over the system baseband closely simulates the wideband multiplex signal encountered under operational conditions. A noise generator which simulates the traffic at a specified level of loading is then applied to the system input. A quiet channel is produced by introducing suitable filters into the generator circuit and the channel output now represents the noise level due to the residual thermal noise and intermodulation in the equivalent operating channel. The filter is taken out of circuit and the noise output level again measured. The NPR is the decibel ratio of the two measured values. Switching off all the channels in the system allows the *intrinsic* noise in the wanted channel to be measured.

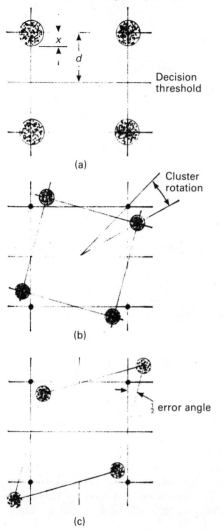

Figure 18.9 Constellation diagrams: (a) equivalent to eye closure; (b) lock angle error; (c) quadrature angle error.

18.2.3 ISDN testing (see also ISDN, p. 86)

Testing of this highly complex digital system would be very difficult if it were not modular in concept. Each subscriber's terminal is connected to line via the S-bus or interface and each line is connected to the exchange via the U-bus or interface. These points provide suitable nodes for overall testing and monitoring to ensure conformance with international operational standards. The line between the U and S buses can be tested for bandwidth and frequency response using a sweep oscillator and level test meters at opposite ends. The network between the U interface and the exchange can be checked for bit error rate using a pseudo-random bit pattern generator and error detector, while the section containing the subscriber's network terminator (NT) can be tested by connecting a network simulator to its

interface. In addition to hardware testing, it is important to exercise the system control using a protocol tester. This must analyse the complex data formats, control the ISDN protocols, emulate and test the software implementation of the control signals such as common channel signalling No. 7 (CCS-7).

18.3 Miscellaneous terms

Attenuator. A calibrated attenuator can be used to measure system power gain by using a backing-off mode. The attenuator is placed between system output and a power meter and the attenuation adjusted until the meter reads zero. The system gain is then numerically equal to the attenuation setting.

Averaging or ensemble averaging. This is a technique used to maximise a signal level in a noisy communications channel. An output is produced by successively averaging corresponding sample points from an input data set. In the *linear* mode, each data point has equal weighting, but in the *exponential* mode the weighting of each point depends upon the time lapse since its occurrence.

Convolution (see also Signals, p. 304). A sliding weighted averaging technique used in the time domain for smoothing non-periodic waveforms. This is equivalent to filtering in the frequency domain.

Correlation (see also Signals, p. 305). A sliding weighted averaging technique that is used to detect the time delay and similarity between input and output pseudo-random binary signals.

Decibels (dB). The unit that is used to compare the levels of two powers and is equal to 0.1 Bel. The ratio is expressed as, $n\,\mathrm{dB} = 10\,\log_{10} P_2/P_1$. Decibels may also be applied to other units that form an *intensity ratio*. Provided that the two powers are developed across the same values of resistance, the ratios can be extended to voltages and currents as follows:

$$\mathrm{dB} = 20\,\log_{10} V_2/V_1 \text{ or } 20\,\log_{10} I_2/I_1$$

Many variants of dB are used for special applications and include:

dBA	for sound pressure levels weighted according to the A scale
dBc	a decibel ratio relative to the carrier (c) level
dBi	the power level relative to an *isotropic* source
dBK	a figure of merit relative to K
dBm	power ratio relative to a level of 1 mW

dBm0	power ratio relative to 1 mW at a point of zero relative level
dBm0p	as dBm0, but psophometrically weighted
dBm0ps	as dBm0p, but weighted for sound transmission
dBp	a power ratio that is psophometrically weighted
dBq	absolute voltage level of noise in an audio channel relative to a reference voltage defined by international standard and measured with a quasi-peak noise meter without a weighting network
dBr	dB ratio relative to some specified reference level
dBv	dB ratio relative to a level of 1 volt
dBW	dB ratio relative to a power level of 1 watt
dBpW	dB ratio relative to a power level of 1 picowatt
dBμ	dB ratio relative to a level of 1 μV

Glitch. Jargon term used to describe usually a single unwanted spike of spurious signal. Often generated by an interference source.

Insertion gain or loss. When a network is connected between a signal source and a load, there will be a gain or loss of level at the load, depending upon the nature of the network, this being described as the *insertion gain* or *loss* of the network. Provided that the network maintains the impedance match that existed before its insertion, then the technique can be used to monitor or test the system performance.

Neper. The propagation constant for a transmission system is complex, consisting of a real attenuation constant and an imaginary phase constant. If logs to the base e are used to express the attenuation as $\log_e I_1/I_2$, where I_1 and I_2 are the input and output current, respectively, the loss is expressed in nepers. Strictly, this unit is reserved for dealing with transmission lines and 1 neper = 8.686 dB. As this is a relatively large unit, decinepers (dN) are commonly used.

Power spectral density. A concept used for the evaluation of the power distribution of random signals such as noise. It is usually expressed as a two-dimensional plot of the mean square noise voltage per hertz against frequency.

Predistortion. If a system's distortion is precisely known, it can be countered by predistorting an input signal in an inverse manner before processing (see also Pre-emphasis/de-emphasis, p. 362 and Television signal processing and systems, companding, p. 361).

Protection ratio. When planning a radio communications link, it is common practice to define the minimum level of service that is acceptable, expressed as the ratio of wanted carrier to interference (C/I) power ratio. Usually the level of service will be described in terms of BER for a digital system or a subjective level of just acceptable interference in the case of an analogue system, for a given C/I ratio. In a practical service, this level will be exceeded by a value known as the *protection margin.*

Quantisation distortion units (QDUs). This is a subjective measure of the sampling error that is introduced when an analogue signal is converted into a digital format. To relate this to dB, for an end to end audio link that uses digital processing, an analogue S/N ratio of 30 dB is approximately equal to 14 QDUs. Each A/D conversion and A law to μ law conversion adds about 0.5 QDU to the degradation.

Scalar measurements. A one-dimensional measurement such as voltage, either peak–peak or RMS, etc., frequency, time, etc.

Spurious free dynamic range (SFDR). The amplitude range between a fundamental frequency and the first spurious noise component. Measured in decibels below the fundamental level (−dBc).

Tram lines. Used as an alternative to eye height measurements. If the raw data stream is applied to the Y input and the timebase set to run at a much lower frequency, the data transitions will be displayed in a manner that gives rise to this descriptive title. Overshoot and undershoot associated with the 1s and 0s level will produce bright parallel bands. The eye height is then evaluated from the ratio of separation between the bands to the average height between the 1s and 0s level.

Weighting. In many situations, the measured signal to noise ratio does not adequately equate with the perceived degree of interference. Weighting curves are then applied to convert the objective measurement into a subjective value, that more nearly specifies the nuisance value.

18.4 Useful references

Bennett G.H. (1976) *Pulse Code Modulation and Digital Transmission*, Marconi Instruments, Chelmsford.

Compston I.K. *Relating Constellation Parameters to Digital Radio Performance*, Hewlett Packard, Wokingham.

Czech J. (1965) *Oscilloscope Measuring Techniques*, Philips Technical Library.

Hewlett Packard A Signature Analysis Case Study, Application Note 222-10, Hewlett Packard, Wokingham.

Tant M.J. (1974) *The White Noise Book*, Marconi Instruments, Chelmsford.

19 Memories

A memory element is a multi-state device that has the ability to retain its present state until forced to change by an input of energy. In communications applications, a memory consists of a group of cells, each of which is capable of taking up one of two states. A memory is thus a binary device. In general, memories fall into one of three categories, depending upon the nature of the stored energy. Magnetic stores are used chiefly as secondary or backing stores, while optical memories, which are a comparatively new concept, are used mostly for archiving purposes. The major devices used for communications purposes are based on semiconductor technology. A memory that loses its stored data when the power is removed is described as being volatile. A non-volatile memory is thus one that retains its data under power-down conditions. (See also Computers in communications, p. 66 and Semiconductor devices and technology, p.284.)

19.1 Magnetic memories

Small ferrite beads or cores which are capable of being magnetised in alternate directions along one axis, to form a non-volatile memory, have been much used in the past. These are threaded in a matrix fashion, on to a series of cross-wires, with a core mounted at each intersection. The cross-wires not only form a unique address of each core in the X, Y plane, but also provide means of magnetising the core with the desired polarity to signify a 1 or a 0. Several core planes are assembled into a cubic form to allow the storage of eight bits at the same corresponding address. Because of the large physical size of these memories, they are now used only in extreme temperature and radiation environments.

Bubble memory. These non-volatile devices are constructed from a substrate of non-magnetic gadolinium–gallium–garnet (G^3) on one side of which is grown a very thin layer of pure magnetic garnet. On to this, a number of very thin chevron shaped nickel–iron or Permalloy magnets are deposited. Two very strong permanent magnets are then placed on either side of the substrate with the flux lines perpendicular to the chip. These cause many of the very small magnetic domains in the garnet to line up with the same flux polarity as that produced by the magnets. The remaining domains take up the opposite polarity and, because of the strong field, become shrunk into minute *bubbles* which line up below the chevron patterns. A rotating magnetic field is generated from two orthogonal coils (at right angles) which are driven by sine wave currents. This causes the bubbles to be moved around within the garnet. For data storage, the presence of a bubble signifies a 1, while no bubble indicates a 0, bubbles being generated or destroyed with a small electromagnet. To read out the data, the magnetic bubbles are circulated across detector electrodes to generate output pulses that represent the logic 1s. These devices have the advantage of large storage capacity without the complexity of motor drives found in some other magnetic memories. By comparison with the devices, the read and write operations are relatively slow, the control circuitry is complex and consequently bubble memories are expensive.

Floppy disk memories. The basic element of this range of devices, is a thin layer disk that is coated with a magnetic material and permanently enclosed in a low friction envelope. Disk casings of 3 in, 3.5 in, 5.25 in and 8 in width are available. When placed in its drive, the disk is clamped to a spindle and rotated typically at 360 rpm, as indicated in Fig. 19.1. A read/write head permanently contacts the disk surface through a slot cut in the casing and the magnetic surface is divided into a number of circular tracks. These are further sub-divided into a number of sectors, the numbers depending upon the recording standard in use. An index hole in the disk, which is sensed by an opto-electronic detector, is used to indicate the start of each track. Sectoring is controlled in one of two ways. On the larger disks hard sectoring is common, with index holes used to indicate the start of each sector. The smaller disks use soft sectoring where only the first sector is indicated, the others being determined by division within the disk handling software. The data is recorded on to the disk surface, typically using frequency shift keying (see also Modulation and demodulation, p. 189) or one of the self

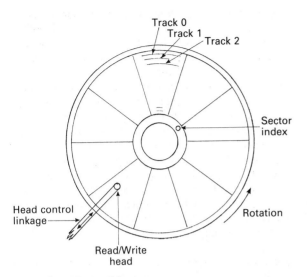

Figure 19.1 Floppy disk.

clocking (synchronising) codes (see also Codes and coding formats, p. 57).

Hard disk memories (Winchester). Typically, up to ten, 12 in aluminium disks are clamped to a hub in a permanent stack and enclosed in a sealed housing to form a dust free environment (see Fig 19.2). Each disk or platter surface carries its own read/write head and the whole assembly is interchangeable as a pack. Each disk is nickel plated and highly polished to minimise surface imperfections. Each surface is then coated with a fine film of magnetic ferric oxide or cobalt to provide the memory element. Each set of corresponding tracks on each surface is referred to as a *cylinder*. Thus with a multi-head configuration, a particular track is selected by a cylinder and a head number. Usually one surface, known as the *servo surface*, is reserved for speed control of the drive motor and position control of the heads. The heads do not make physical contact with the disk surface, but *fly* a few microns above it. In use, the disk assembly rotates at a typical speed of 3600 rpm to give

an average access time to the data of less than 20 ms. A very large data capacity of up to 500 Mbytes is available, but because of the high degree of mechanical precision necessary, such disk drives and packs are very expensive. Mini-Winchester drives of 3.5 in and 1.75 in diameter are available at lower cost and a typical disk pack can provide up to 10 Mbytes of memory capacity.

Magnetic tape memory. This tape is manufactured in the same way as floppy disks. A flexible mylar backing, about 10 to 12 microns thick, carries a thin ferric oxide or cobalt magnetic layer of about 3 microns thickness. The bulk material is then slit into appropriate widths. There are two basic versions using this concept. One uses narrow, 0.25 in wide tape with the data recorded in a bit serial fashion across the full width of the tape. The technique is very similar to audio recording and in fact, audio type cassettes are often used for small or portable data systems. Large mainframe systems use tape up to 2 in wide, which can carry several 8-bit parallel tracks across the tape width. The read/write heads make intimate contact with the tape and, at the simplest level, the polarity of the record current controls the polarity of the recorded signal which is used to signify a binary digit. The modulation coding format varies between systems and ranges from *double frequency* or F2F, where two pulses per bit period represent a 1 while one pulse per period signifies a 0, to *Kansas City* modulation, where eight cycles at 2400 Hz or four cycles at 1200 Hz represent a 1 or 0, respectively. Longitudinal recording is most common, in which each bit of the data forms a linear magnetisation along the tape surface. In order to increase the packing density of the data, a recently developed technique records each bit at right angles to the surface, so that the magnetic pattern is vertical to the surface. Chief advantages of magnetic tape memory are non-volatile storage of a large amount of data at low cost. The major disadvantage is that as the data is recorded in serial form, access is relatively very slow. For mainframe type systems, the tape handling mechanism is very complex. It has to maintain constant tape tension with constant tape speed and ensure smooth start up and stop operation without breaking the tape.

Winnie. Jargon term applied to Winchester disk systems.

19.2 Optical memories

This most recent and developing technology is characterised by a very large data packing density and these memories are used in one of two forms. The permanent *write once*, *read many* or WORM device that

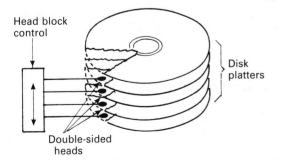

Figure 19.2 Hard disk (Winchester).

is effectively a read only memory, and the more recent erasable types. Disks consist of a glass or plastic platter that carries a very thin sensitive vacuum deposited layer on one side, to act as the memory. This layer is then coated with a transparent layer for protection and two such disks can be glued back to back to form a double-sided memory. The data is recorded by a laser beam in a continual spiral track. To ensure synchronism, some disk systems include a pre-recorded sync track. The general principle is shown in Fig. 19.3. About 10 mW of laser light energy is focused on to the sensitive surface to write the data. The exact effect depends upon the technology used. Reading is accomplished by using the same laser but at a much reduced power level of typically 0.5 mW. The reflected light energy is split from the main laser beam and focused on to a photo-detector to regenerate the data. To ensure correct focus, a final lens is mounted in a moving coil assembly that is driven by part of the servo system, the drive signals for this and other positional control being obtained from the recovered data signal path. Disks are available in several sizes, from 5 in to 12 in diameter, these being capable of storing upwards of 650 Mbytes and 1 Gbytes, respectively. Such a large capacity produces a longer average access time than other systems, but the typical half second is not too troublesome, as these memories tend to be used for archive purposes.

Dye polymer technology. Certain organic dyes, including alloys of zinc and silver, respond to light energy in a very convenient way. When a very thin layer is exposed to laser light, local heating causes microscopic pits or craters to be formed. The reflectance from the surface now varies according to the pit pattern. The reflected light energy has a different and varying phase from the incident beam and so can be extracted as a data signal. Other organic dyes change colour as the laser beam causes local spots to change between the crystalline and amorphous states. The colour changes again vary the reflectance. This latter technique produces erasable and reusable disks. A *digital paper* tape has been developed from this technology. It uses a thin flexible polyester base as a substrate that carries a very thin layer of dye polymer. The laser beam produces permanent holes in the polymer according to the data bit pattern to provide a cheap read only memory.

Heat sensitive technology. In this case, the sensitive layer is an alloy of tellurium and selenium with a light doping of arsenic to give more accurate control of the melting point. During the *write* mode, the laser beam burns microscopic holes in the sensitive layer and these are detected by the lower power beam during a *read* operation. Erasure is achieved by using just enough laser energy to melt the layer without forming holes. The molten regions then cool very quickly to solidify into a stable amorphous state. Re-writing ensures final erasure by transforming the amorphous domains back into the crystalline state.

Magneto-optical technology (M-O). This technique is alternatively referred to as *optically assisted magnetic recording*. A thin layer of amorphous gadolinium–iron–cobalt is used for the memory and the whole surface is initially magnetically polarised in the same plane but at right angles to the surface. When heat from a laser beam is applied simultaneously with a magnetic field from the other side of the disk, this small region reverses polarity and becomes frozen in this state. When plane polarised light is reflected from a highly polished electromagnetic surface, the light becomes elliptically polarised. This is the Kerr effect. A read operation is therefore performed using polarised laser light to detect the field changes. For erasure, the surface is again laser heated but with the applied magnetic field polarised as in the original sense.

19.3 Semiconductor memories

These devices are extensively used because they are capable of economically storing large amounts of data in a small space, with very short access times. An array of memory cells can be arranged either in the linear fashion of the shift register, or as is more common for the larger

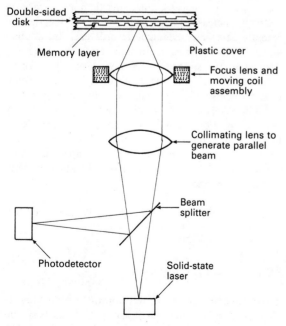

Figure 19.3 General principle of laser scanned memory disk system.

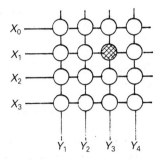

Shaded cell address $X_1 Y_3$ (0010, 0011)
(row 2, column 3)

Figure 19.4 Memory addressing.

memory units, in a square matrix. In the latter case, each cell can be uniquely addressed by its column and row line numbers as indicated in Fig 19.4. Once addressed, the binary state of a cell can be changed as appropriate, by the application of electrical energy. Semiconductor memories are made using various technologies (see also Semiconductor devices and technology, p. 294, according to the required characteristics. TTL technology is not much used because each cell occupies too much chip space (described in the jargon as *real estate*) which reduces the memory capacity and the relatively high power and heat dissipation. CMOS devices are used extensively for their high density and the low power requirement, while ECL technology devices are used for their high speed, even though the higher power dissipation reduces the memory capacity. Generally, these devices fall into one of two categories, either volatile or non-volatile memories.

Charge coupled device (CCD) memory (see also Semiconductor devices and technology, p. 292). Sometimes described in the jargon as *bucket brigade devices*, these memories are constructed in a linear fashion with a wide range of capacity, typically from 128 up to 65 536 bits. CCD memories find many applications in image or television signal processing, where they act as image sensors, delay lines and filter elements. Each cell is constructed in the form of a potential well beneath a gate electrode and capable of storing either an electric charge or no charge to represent a 1 or a 0. Charges can be shifted through the memory in the serial fashion of a shift register. As the charge on a cell can leak away, the memory contents need to be *refreshed* periodically and this is done by recirculating the data. An important parameter for such serial memories is the *latency*. This is described as the time taken to locate the first bit in a stored block of data and depends upon the memory length. Once this bit has been found, the access time is then very short, in the order of 200 ns.

Programmable logic device (PLD) (see also Semiconductor devices and technology, p. 290). This tends to be the generic term used to describe a device that is variously known as a programmable logic array (PLA), an electrically reconfigurable array (ERA), or integrated fuse logic (IFL). It is a user-programmable device that is capable of carrying out complex logic and control functions. It is based on a read only memory concept that uses the address and data lines to provide the inputs and outputs, respectively. Each array consists of groups of AND, OR, EX-OR and inverter gates, plus programmable interconnection links. Early devices were one time programmable (OTP), but the technology has now given way to devices that can be erased and reprogrammed many times. Two variations of these are available and they can either be erased by exposure to ultraviolet (UV) light for up to 20 minutes, or electrically erased and reprogrammed *in situ*, in a matter of seconds. This latter type has no need for the quartz window in its package that is required for UV erasure and so is the cheaper of the two concepts.

Random access memory (RAM). These are so described because any memory location can be written to or read from in any order. Two types of RAM are in use and both are generally volatile. However, this problem can be overcome by the addition of a lithium or similar back-up battery contained within the chip package.

Dynamic random access memory (DRAM) (see Fig. 19.5(a)). This version uses a capacitor as its memory cell with binary 1s and 0s being stored as a charge or no charge, respectively. Typically only one transistor is required per cell for charging, plus two others for read and write control. Each DRAM cell is thus physically very small and this provides a high data packing density with very small power consumption. The major disadvantage of the DRAM is the leakage of charge from the small capacitance. This has to be countered by *refreshing* the charge at least every 2 ms.

Static random access memory (SRAM) (see Fig. 19.5(b)). This RAM uses a flip-flop circuit as its basic memory cell and this can be set or reset to define a binary 1 or 0, respectively, and consequently does not require a refresh operation. As this device requires four transistors per cell, it occupies a greater area of silicon than the DRAM and therefore provides a smaller memory capacity.

Video random access memory (VRAM). This is a RAM variation designed specifically for image or video signal processing. For output, the data is transferred from a row of cells into a shift register during the line blanking interval. Then while this serial bit pattern is

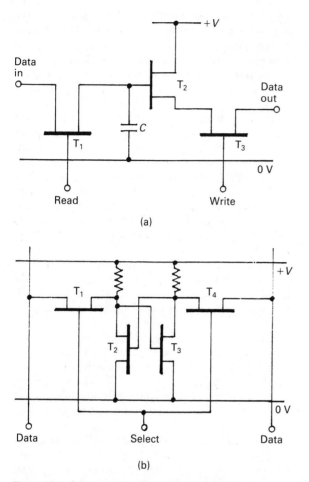

Figure 19.5 Memory cells: (a) DRAM; (b) SRAM.

output to provide the horizontal scan signal, the normal I/O (input/output) port is available for read or write operations. Thus the memory remains available for the whole time for both normal read and write and for the video display as well.

2/4 port static random access memory. The two versions of this concept allow either two or four addresses to be read from or written to independently, to link asynchronous systems running at different speeds. Simultaneous access is allowed to any memory location from any port to provide a significant increase in processing speed.

Read only memory (ROM). These devices are designed to store permanent data in a non-volatile manner so that for a particular input address, the output bit pattern will always be the same at each power-up. The initial technology provided a one-time programmable (OTP) device which had to be programmed during manufacture. However, this has given way to a range of variants,

each with some particular advantage. Sophisticated encryption functions can be incorporated on the chip to improve the security of the stored data and prevent illegal copying of certain ROMs. Certain user-programmable ROMs also carry an embedded code known as the *silicon signature*. This identifies the manufacturer and the particular device type.

Electrically alterable read only memory (EAROM). Sometimes known as *read mostly* memory, this device can have its data changed by the application of suitable control signals. Unlike some other types, these memories can be erased and reprogrammed *in situ*, in selected locations or blocks, instead of having to erase the complete memory.

Electrically erasable programmable read only memory (EEPROM). Devices are usually based on a two-transistor cell which can be totally or partially erased on a byte-by-byte basis *in situ*. They are particularly useful for operations where occasional program updating is required. A typical device has a data capacity up to 64 Kbits, access time less than 200 ns, data retention period up to 10 years, and can withstand 10 000 erase/rewrite cycles. Devices are more expensive than EPROMs and have a lower data capacity.

Erasable programmable read only memory (EPROM). These are based on MOS technology (see also Semiconductor devices and technology, p. 294) which uses a floating gate, beneath which can be stored an electrical charge to represent the binary state of each cell. Reprogramming can only be carried out after complete erasure by exposure to UV light for up to 20 minutes, thus preventing *in situ* reprogramming.

Flash erasable programmable read only memory (flash EPROM). These also use MOS floating gate technology with the buried charge being used to represent the binary state of each single transistor cell. This EPROM can be completely erased in less than 1 second (in a *flash*), by applying a strong electric field to the floating gates. The devices are characterised by large data capacity (1 Mbit), 10 years data retention, less than 150 ns access time, 100 000 erase/rewrite cycles and relatively low cost.

Programmable read only memory (PROM). Also known as field programmable ROM. The earliest type of user-programmable device which once programmed cannot be erased.

ROM programming

Avalanche injection programming. A user-programming system used for MOS devices where an electric charge is induced beneath the gate electrodes of selected cells, by an avalanche injection technique. As such a charge can easily be nulled, the technique can be reused.

Fusible link programming. A virgin memory consists of a full set of interconnections between input and output lines, the interconnections being provided by fusible links. Programming is achieved simply by blowing the unwanted links.

Mask programming. The required program pattern is imprinted on the memory silicon using a photographic mask, during final stages of manufacture. The technique is very expensive and so is only suitable for large-scale productions.

Programming technique. As supplied, all cells of a memory are in the 1 state and so it is only necessary to set those cells which have to contain a zero. With normal power supplied to the memory, each cell is adddressed in turn and a programming voltage applied as a short pulse. The cell is then verified and, if correct, programming passes to the next cell. If a cell has not changed state, a further pulse is applied and this is repeated until verified correct. By using this technique of pulsing and verifying, the programming is carried out in the shortest possible time.

Silicon disk technology. Also known as *solid state disk* or *virtual disk* systems. A number of semiconductor random access memory chips are organised into an array that mimics the behaviour of a magnetic disk system. As this is completely electronic, there is no motor to create vibration or read/write heads to wear as found in conventional disk drives. Memory capacity up to 40 Mbyte at high data transfer rates is commonly available, with access speeds many times that of any disk system.

19.4 Miscellaneous terms

Cache memory. A relatively small but fast access memory, located between the processor and the main memory. Typically about 8 Kbytes, located on the microprocessor chip and designed to duplicate the contents of the most used locations from the main memory, thus increasing the processing rate. The cache stores a copy of the data and also its address which is referred to as a *tag*.

Chequerboarding (checkerboarding). Certain schemes that are designed to maximise the use of memory, are organised by allocating memory sections to a particular process. When such processes are completed, the sections can be reallocated. In practice, not all the space is used and so as time progresses, spare sections of the memory will be unallocated, but are not contiguous. These smaller sections cannot be used and the memory map takes on the appearance of a chequerboard.

Content addressable memory (CAM). A conventional look-up table uses an address to locate a data item. The CAM operates in the reverse way. A data search word is loaded into a latch, the address count is cycled and a comparator is used to locate a *match* between memory locations and the latch contents. When a match is found, the corresponding address is loaded into a *snapshot* register. If multiple matches are found, the lowest address is indicated.

Demand paging. A technique designed to resolve some of the problems associated with *segmentation*. The memory is divided up into a large number of relatively small segments each of the same size, called the *pages* or *page frames*. An address translation table is held in RAM, so that the *memory management unit* (MMU) can quickly load the required pages from secondary memory, on demand.

Electron beam accessed memory. The device consists of a cathode ray tube (CRT) type structure, but with the screen made from an array of MOS circuit elements. These store data as charge packets written by an electron beam. Each element of the surface has a unique address and the beam can be scanned using horizontal and vertical deflections in the manner of a CRT. Read-out is achieved by re-scanning each element with the beam. If an element holds a charge, the reading beam current will be lower than it would have been had the initial charge been zero. Thus read-out, which is destructive, occurs in a serial form. This type of memory has to be periodically refreshed and is volatile. While this form of memory is very bulky and requires several power supplies, it is capable of holding upwards of 40 Mbits with an access time of a few microseconds.

Errors. These can occur between memory write and read operations for a variety of reasons. *Hard* errors are failures that can be duplicated by repeating the conditions that gave rise to the error in the first place. These are most likely to be due to mechanical defects. *Soft* errors are due to random effects such as noise and cannot be duplicated.

Ferroelectric memory. A semiconductor random access memory with fabrication characteristics similar to that of the dynamic RAM. The capacitance dielectric is replaced by a compound that has a memory which makes the device non-volatile. The material that is typically used is a lead zirconate titanate derivative that can be ionised in one of two directions, under the influence of an electric field. The ferroelectric cell has a capacitance that is several orders of magnitude greater than that of the conventional DRAM cell. These memories may therefore have a smaller cell size and a greater packing density for a given chip area. The memory is read by pulsing each cell and sensing the magnitude of any change of state.

FIFO (first in–first out) memory or register. These devices are organised as a stack of word-wide registers that operate on a *push down*, *pop up* basis and in such a way that the first word written in, becomes the first word to be read out.

Fly-by memory (check only configuration). An error detection/correction technique that may be applied to words read from a memory. If an error is detected, it is flagged so that a correction routine can locate the errored bit and complement it before the word is placed on the data bus.

Fly-through (correct always configuration). An alternative error checking routine that assumes all words read from memory will be errored. Words read from memory thus pass through an error detecting/correcting circuit before being placed on the data bus.

Grow back. This form of error affects certain types of ROM. The very small fuse links that are blown during programming can reform by metal migration, under the influence of stray electric fields that may develop across the gap.

Josephson junction memory. A Josephson junction is formed from two superconducting metals separated by an extremely thin insulating layer. This form of memory is thus a *cryogenic* device. If an array of such junctions are held near to a critical temperature, the cells are superconducting. However, the application of a magnetic field causes a cell to switch state, conduction ceases and the voltage across it changes. Thus the data is stored as a variation of magnetic field and read out by detecting the voltage across the appropriate cell. Since a Josephson junction can switch state in a few picoseconds, these may well form suitable memory elements for computers used in space.

LIFO (last in–first out) memory or register. Similar to a FIFO but organised on the basis that the last data item input becomes the first item to be read out.

Memory management unit (MMU). A useful element of multi-program operations that provides a hardware method of allocating memory to the user. The MMU reallocates programs to separate areas of memory so that they do not interfere with each other. In a multi-user system, a coprocessor is often employed to manage this task. The MMU can also protect various memory spaces from unauthorised users and aid the protection from deliberate corruption.

Paged memory. This technique divides a process which might require a memory space greater than the main memory into a set of equal sized areas or *pages*. When the process is being executed and an address is encountered that is not currently in the main memory, an interrupt signal is generated that causes the operating system to be entered. The memory management section then determines which page holds this address so that the appropriate program segment can be transferred into the main memory. Because the effective memory space is now that of the secondary store, there is a vast increase in memory capability (see Virtual memory, p. 171). An operating system based on this concept is known as a *virtual memory* operating system.

Persistent memory. Data items may be stored in various memory elements, located not by address, but by an object identifier. This is sometimes referred to as a *tag byte* and forms part of the total data word.

RAID (redundant array of inexpensive disks). This is a technique that has been developed to process data in high speed data acquisition or imaging systems using relatively low cost devices. Each RAID consists of five disk drives operating in parallel and in a synchronised manner. Four drives are used to handle 32-bit words from the host computer, which is split across four drives, one byte to each. The fifth drive is used to generate and store the parity bits for improved data integrity. This is in addition to any cyclic redundancy check (CRC) that is already provided within the original data stream. The drive system is coupled to the host computer usually via a small computer systems interface (SCSI) system, which allows for data rates as high 10 Mbytes/second. Five levels of integration are in use to cover a range of data rate transfers, data block sizes and data security, etc.

Refresh. The charge on the capacitance of a DRAM cell can leak, particularly during a read operation. To preserve the integrity of the data, it is therefore necessary to read and rewrite the state of each cell, typically every 2 ms. This is achieved by comparing the voltage level of each cell in a cyclic manner with a voltage that is half way between the 0 and 1 levels. If a cell voltage is greater than this, it is recharged, otherwise it is set to 0 volts. Some systems are capable of refreshing several rows of cells at the same time, without the involvement of the CPU. This is then referred to as a *transparent* or *hidden* refresh cycle.

Segmentation. This involves dividing the addressable memory space into a number of segments which may be of different sizes. These are exchanged between the RAM and disk as and when needed. If large segments are exchanged for smaller ones, the usable memory becomes fragmented (see Chequerboarding, p. 169).

Tag RAM memory. A device designed to support a cache memory system. An on-chip comparator compares the contents of the addressed RAM location with the current input data. A match/miss signal is then generated to signify the success or otherwise of the comparison.

Testing of memories. Apart from the test requirements of the normal voltage/current, access times and synchronism parameters, it is necessary to establish the integrity of the memory cells. Since errors between read and write can be pattern sensitive, even a 1 K RAM would have to be tested with 2^{1024} different test patterns. Since such testing would take an intolerably long time, an alternative procedure must be used. First a fixed pattern of all 1s, all 0s or alternating 1s and 0s is used to detect major problems. Then a *galloping* 1 or 0 pattern (a *galpat*) is applied. In this test a 1 is written into the first cell and all the others tested for corruption. This pattern is then applied to every cell in turn. Finally, the galpat is changed to 0.

Virtual address space. This is defined as the size of memory that can be accessed by a user program and which is restricted to an allowable range of addresses.

Virtual memory. Large areas of addressable RAM space is costly: disk-based storage is very much cheaper. Therefore by using a disk system to support the memory, the processor can be made to *cheat*. The large part of the addressable memory on the disk is now known as a *virtual memory*. This is achieved because the processor has the ability to detect accesses to memory pages not currently being held in RAM. When virtual memory detects such an access, it then, independently of the program, automatically transfers the required new pages into RAM, storing the previous RAM contents as necessary in the process. To the user, all memory appears to be in RAM, but in fact only that part of the program currently being executed is in RAM, the rest is in the disk store.

19.5 Useful references

Bartee T.C. (1981) *Digital Computer Fundamentals*, 6th edition (ISI), McGraw-Hill, New York.

Millman J. (1979) *Microelectronics: Digital and Analog Circuits and Systems*, McGraw-Hill, New York.

Vears R.E. (1988) *Microelectronic Systems Checkbook*, Butterworth-Heinemann, Oxford.

Zaks R. (1981) *From Chips to Systems*, SYBEX, Paris.

20 Microwave devices

The microwave wavelengths range from 1 metre down to 1 centimetre, corresponding to a range of frequencies from 300 MHz to 30 GHz. Above 30 GHz the phenomenon is referred to as the millimetre wave range which at present, apart from optical techniques, is little used for communications purposes. For components and devices, the range between about 300 MHz and 1 GHz forms something of a transition region. The shorter wavelengths, together with lower periodic and transit times, plus component self-capacitance and lead inductance, all act to increase the losses in components of the conventional construction that are used for the bulk of communications at lower frequencies. For this reason, devices that are intended for use above 1 GHz tend to have a completely different appearance to those used below 300 MHz. The frequency ranges used for microwave communications are commonly known by the American Radar Engineering Standards. In Europe however, another classification may be used and this can give rise to confusion (see Frequency bands in use, p. 137).

20.1 Active devices

20.1.1 Amplifiers

Solid-state semiconductor devices are almost universally used for receiver applications up to 30 GHz and for transmission up to around 500 MHz. For high power transmission systems above 1 GHz, the thermionic vacuum valve or tube device predominates. The conventional amplifier power efficiency parameter, particularly for high power applications, is often replaced by a *figure of merit* (FOM), given by the ratio:

$$\frac{\text{Peak ac power output}}{\text{dc power input}} \times 100\%$$

a value that can exceed 100%.

High electron mobility transistor (HEMT). Also referred to as low-dimensional structures, these devices are constructed using very thin heterojunctions of gallium arsenide and gallium aluminium arsenide. The epitaxi-ally grown layer is only a few atoms thick, so that the interface is almost a two-dimensional structure. Electrons which collect in this region due to the electric field behave very differently, because they only have two degrees of freedom of movement. This results in an electron mobility that is a factor of three better than that for silicon. HEMT devices can therefore act as amplifiers up to a frequency in excess of 60 GHz, with a noise factor of less than 1 dB and with a gain of more than 10 dB.

Inductive output tube (IOT) amplifier (see also Klystron, p. 173). Although originally designed for use in UHF television broadcast transmitter power output stages, these devices now have many other applications. The IOT devices have a FOM as high as 130%, which means a considerable saving in operating power costs. Like the klystron, the IOT is a beam tube but with a number of operational differences. An electron gun is used in conjunction with a robust grid which produces negligible primary or secondary emission. Because the grid is positioned very close to the gun, it has a high degree of sensitivity. The RF drive signal is applied via the biased grid electrode so that it acts directly on the beam, which is also electromagnetically focused. This results in electron density modulation or bunching. The bunching electron beam is accelerated directly into the output cavity under the influence of the final anode voltage. (The IOT requires no intermediate drift tubes or cavities.) The overall effect is to significantly reduce the amount of wasted beam energy, which would otherwise have to be dissipated at the collector electrode.

IOT amplifiers are available to handle peak output powers of more than 60 kW and with a power gain exceeding 20 dB. This can be achieved for a single device, over the frequency range of about 450 MHz to 1 GHz with a linear bandwidth of 16 MHz. Combined cooling systems of forced air and pumped water are used to dissipate the excess heat, air being used for the gun and cavities, with water for the body and collector.

Klystrode (see also Klystron, p. 173). This device combines the technology of the thermionic tetrode with that of the klystron. Figure 20.1 shows the construction of the klystron with which this device can be compared. Basically it is a two-cavity amplifier with a very short beam length and hence a good high frequency response. Tuning can be achieved by varying the volume of the main cavity. The output signal is extracted klystron fashion, using a probe, from a second cavity that is coupled to the first via an *iris* aperture. A grid electrode is positioned between the cathode and anode to provide an input. This radio frequency signal density modulates the electron stream that is maintained in a dense beam by the focusing effect of the external magnetic field. The electrons which *bunch* due to the modulating effect, are accelerated along the drift tube and through the resonant cavity, where the amplified signal is extracted. The shorter structure relative to the klystron results in a smaller device that also requires less magnetic power. This provides an amplifier capable of producing in excess of 60 kW of output power, with an efficiency figure of merit in excess of 120%, together with a wide bandwidth power gain which can exceed 23 dB.

Klystron. The klystron uses the principle of velocity modulation of the electron beam to overcome the problem of transit time losses. The basic construction is shown in Fig. 20.1. A high velocity beam of electrons is emitted from a heated cathode and accelerated through the annular anode where it passes along a drift tube to the collecting electrode, a magnetic field being required to constrain the beam density. Amplification is obtained when the beam flows through the annular shaped resonant cavities (typically four or five in number), which can be tuned to give a bandpass characteristic by varying their volume. The radio frequency input signal is injected into the first cavity and the alternating positive and negative voltages cause the electron velocities to increase and decrease, producing bunching. This becomes more defined as the beam passes through successive cavities. Two basic types are available: low power devices with cavities contained within the vacuum system and high power devices with external cavities. Klystron power efficiency is relatively low, typically 45% maximum for an external cavity device, but power outputs up to 75 kW can be achieved using water or vapour cooling of the collector electrode. Two developments have improved klystron efficiency. Reducing the collector voltage (*depressed collector*) reduces the wasted power at the collector but, at the same time, increases the reflection of secondary electrons that cause a reverse flow through the drift tube to reduce the tube gain. Segmenting the collector and using up to four different EHT voltages, ranging from typically 4 kV to 25 kV for each section, can restore the gain. This results in the lower energy electrons being

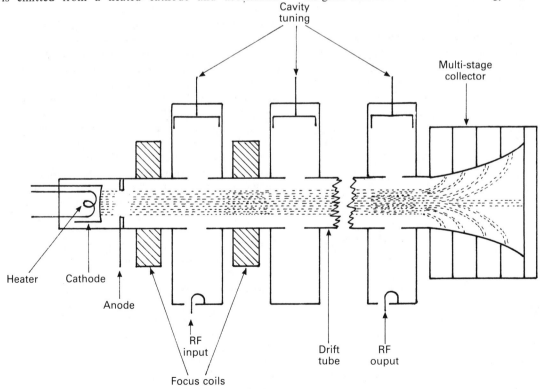

Figure 20.1 Klystron construction.

collected by the lower potential electrodes, producing less heat in the process. These devices are known as *multi-stage, depressed collector* (MSDC) klystrons.

Magnetron. This is basically an annular diode structure that produces microwave power due to oscillations. The anode block contains a number of resonant cavities, the cathode is heated to produce electrons and these two electrodes are contained within a vacuum. To this basic structure, which is shown in Fig. 20.2, is added a transverse magnetic field produced by a powerful permanent magnet. Electrons pass from cathode to anode under the influence of the applied voltage (which may be dc or pulsed) and the magnetic field, following a cycloidal path as shown. If the electron phasing is correct, then nearly all the dc energy will be converted into RF energy as the electrons pass the cavities. Since the latter have a very high Q factor, considerable output power can be produced by a probe placed within one of the cavities. Due to reverse bombardment of the cathode, the heaters may be switched off once operation has started. This avoids overheating and leads to an improvement in power efficiency, which may be as high as 60%. RF power output increases with both an increase in magnetic flux density and anode voltage. Although these devices are a useful source of microwave power, they tend to be too bulky for many modern applications.

Parametric amplifiers (see also Amplifiers, p. 20). Whereas the conventional amplifier uses dc as its energy source, the parametric amplifier is powered by ac energy from a source known as the *pump circuit*. This is used to modulate the voltage-dependent capacitance of a varactor diode. Signal, pump and idler circuits are connected in parallel across the diode, with the pump oscillator working at a frequency higher than the signal, while the idler circuit is tuned to the difference frequency. Signal input and output circuits are then transformer coupled to the signal circuit. The pump oscillator adds charge to the diode capacitance while the voltage across it is high and extracts it when the voltage is low. The signal circuit adds charge when this voltage is low and extracts it when high. Energy is thus transferred from the pump circuit to the signal output circuit. The amplifier gain, which may be in the order of 20 dB, is given approximately by the ratio of pump to signal frequencies.

Pulsed klystron. For television transmission applications, the klystron has been traditionally operated with a constant beam current. It has been discovered that the beam power could be reduced during the vision-only period by pulsing the anode, without serious loss of output power. This is achieved by introducing a pulsing electrode that is capable of switching the beam with a relatively low voltage. Beam current is then only maximum during the very short synchronising pulse period. This technique has provided means of increasing the klystron power efficiency up to 65–75%.

Reflex klystron. This single cavity version of the klystron amplifer can also be used as a microwave oscillator. Instead of a collector, the end of the drift tube carries a *repeller* electrode which is held at a potential negative with respect to the cathode. This reflects the electron beam back along the drift tube and through the cavity a second time. Bunching occurs at each pass to give up energy to the cavity which provides the output signal. The input signal is applied to a grid electrode positioned close to the cathode. The device can be tuned over a relatively narrow frequency range by varying the volume of the cavity. Oscillations are set up when the forward and reflected beam currents are in phase and this is a feature of the repeller to cathode voltage.

Solid state power amplifiers. Solid state devices have the considerable advantage of low voltage operation and when VMOS technology (see also Semiconductor devices and technology, p. 294) is used, such transistors can support a significant current. Compared with thermionic devices, however, the individual power output is very limited. To overcome this problem, many transmitters use a modular construction for the power output stages, using many single stages in parallel. Using this technique (up to 300 single stages have been used) power outputs in excess of 30 kW at frequencies to 1 GHz are possible. The parallel mode has

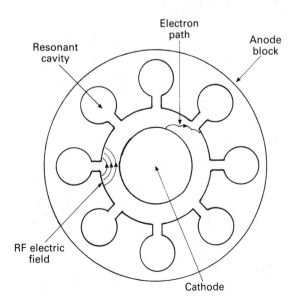

Figure 20.2   Magnetron construction.

a very significant advantage. If a single section fails, the loss of output power is practically negligible.

Travelling wave tube amplifier (TWT). The general construction of this amplifier is shown in Fig. 20.3. A heater and cathode assembly generates an electron beam which is focused and accelerated towards a collector electrode by suitable potentials, lhe beam passing through a slow wave helix structure. The radio frequency input signal is applied to the cathode end of the helix and interaction between the beam and the RF wave creates an amplified signal at the output end. The beam concentration is maintained by the use of an external magnet structure in the manner shown in the diagram. To improve power efficiency, the device can be equipped with a multi-stage depressed collector and the focus electrode can be used to blank the beam for pulsed operation. High power amplification over the frequency range 2–40 GHz with a gain as high as 70 dB are possible. The power output can be as high as 500 W, rising to 1 kW for pulsed operation.

20.1.2 Diodes

Diode applications for microwave communications generally fall into three categories: detectors, mixers and single-port oscillators. For detector circuits, there are two important parameters that need to be considered. The *open circuit voltage sensitivity* ($V_{S_{oc}}$) which describes the slope of the transfer function working into an open circuit. This describes the diode efficiency in terms of converting microwave input power into an output signal voltage. It is usually measured in units of mV/mW or mV/μW. The *tangential sensitivity* (P_{TS}) describes the noise performance

of the detector diode working into a matched load. It represents the lowest input signal level relative to 1 mW (0 dBm) required to double the detector noise output. This parameter is quoted for a video system as:

$$P_{TS} = \frac{BR_v TF}{V_{S_{oc}}}$$

where B = video amplifier bandwidth
R_v = video resistance
T = temperature in K
F = video amplifier noise figure
$V_{S_{oc}}$ = open circuit voltage sensitivity.

Mixer stages are notoriously noisy and so it is important to use nonlinear devices that are selected for a low noise factor. For some time, the Schottky barrier diode, with its low threshold of 0.3 V, has been popular. However, there have been. some recent diode developments that provide better parameters. These have been designed for surface mounting and thus have less self-capacitance and lead inductance, allowing them to operate efficiently up to frequencies in excess of 40 GHz.

Backward diode. A special form of tunnel diode in which the tunnelling reaches only to the early stages so that the negative resistance feature is practically non-existent. Below a threshold of about 500 mV, the forward current is very low. A relatively large reverse current flows for a very small applied voltage, so that the V/I characteristic is approximately the reverse of that for a conventional diode. This type can be used at frequencies in excess of 10 GHz. It has a high degree of sensitivity and is suitable for detector applications with very small input signals. It is also suitable for mixer

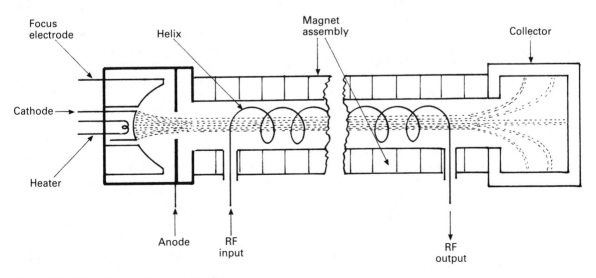

Figure 20.3 Travelling wave tube construction.

operations that require only low levels of local oscillator signal.

BARITT diode (barrier injection transit time diode). In contrast to other diodes, this device uses a pnp type structure and operates in the *punch-through* mode. It consists of two Schottky barrier contacts separated by a semiconductor layer. If the applied voltage is high enough, injected holes drift across the semiconductor region at a frequency dependent upon the transit time. The V/I characteristic has a negative resistance region which allows the device to be used as a source of microwave energy. A capacitive delay occurs due to the forward biased junction and this tends to make it less efficient than the IMPATT diode.

Esaki diode. See Tunnel diode, p.177.

Gold bonded diode. A point contact diode formed between a gold whisker wire and n type germanium. When fused with a current pulse, a pn junction of very low forward resistance is produced.

Gunn device. This device is not a diode because the phenomenon takes place within the bulk of the gallium arsenide semiconductor material. If an electric field in excess of about 35 V/cm is applied across a very thin sample, low energy electrons from the conduction band transfer to a satellite band with a higher energy state but with lower mobility. This leads to negative resistance and the generation of a travelling electric field domain that flows between cathode and anode. The pulse repetition frequency depends upon the transit time and the thickness of the sample. The equivalent circuit of the Gunn device is a negative resistance of about -5 ohms in series with a capacitance of about 0.2 pF. To make the device oscillate, the load connected to it should have:

(a) a series reactance that tunes the device to resonance;
(b) a series load resistance less than the modulus of the negative resistance;
(c) at frequencies other than resonance, a load resistance greater than the modulus of the negative resistance.

Provided that points (b) and (c) can be met, a variable reactance can be used to tune the circuit over an octave frequency range. Oscillators that work at frequencies up to 45 GHz and producing an output power of about 500 mW can be generated from an operating potential as low as 12 V.

Hot carrier diode. A *hot* carrier is a conduction electron that possesses an energy level greater than that due to the temperature of the crystal lattice in which it resides.

A hot carrier diode is a rectifying junction formed between a metal and a semiconductor (a Schottky diode). When forward biased, majority carriers in the n type region (electrons) that possess energy greater than the Schottky barrier, the hot carriers, can cross the barrier and provide a current flow. The number of hot carriers increase with the forward bias and so the current increases rapidly. The diode has a negligibly small reverse current and its high sensitivity makes it an ideal detector device.

IMPATT diode (impact ionisation avalanche transit time device). Also known as a READ diode. This device is constructed from silicon or gallium arsenide using two heavily doped p and n type regions. These are separated either by a lightly doped n type or an i type semiconductor. A dc voltage is used to bias the pn junction close to avalanche breakdown so that an applied microwave signal voltage drives the diode in and out of the avalanche mode. This generates an electric field that traverses the intrinsic region to provide a source of microwave energy. The combination of transit and avalanche times result in an ac output current that is 180° out of phase with the applied signal (negative resistance), at an appropriate resonant frequency. Output power in excess of 1 W can be produced at a frequency of greater than 20 GHz, but the diode requires an operating potential in the order of 120 V.

Mott diode. A development of the Schottky barrier diode concept. The diode is formed by a co-planar construction in gallium arsenide (GaAs) and designed for mixer operation up to 100 GHz, with a noise figure of about 7 dB. A thin, lightly doped GaAs layer is deposited over a thicker more heavily doped, low resistivity base layer. An abrupt interlayer interface ensures minimum self-capacitance and series resistance: typical figures are in the order of 0.05 pF and 5 ohms, respectively, and these are maintained over a wide frequency and voltage range. Diodes are made for direct mounting on a substrate (flip-chip) with gold bonding pads, so that lead inductance is virtually zero.

PIN diode. Usually constructed in silicon, this device is formed from two heavily doped p and n regions, separated with an intrinsic (i type) or a lightly doped p or n region. When forward biased beyond a threshold of about 8 V, the diode develops very low forward resistance. As the reverse leakage current is very low, the device performs as an efficient switch that is useful for such functions as band switching, from about 1.5 MHz up to 40 GHz. The minimum operating frequency and switching speeds are interrelated. This is because the diode cannot distinguish between a signal and a switching waveform that are changing at about the same rate. Fast switching can be achieved, but at the expense of reduced power handling. Below the forward threshold voltage, the diode exhibits a dc bias current dependent resistance, a feature that allows it to be used as a voltage-controlled attenuator. The small self-capacitance is largely independent of the dc bias up to about 10 GHz. Beyond this frequency, the diode begins to respond to an RF signal, more like a varactor diode.

Read diode. See IMPATT diode, p. 176.

Schottky diode. These diodes are formed from a rectifying contact between a metal and a semiconductor. When the metal, usually aluminium, is alloyed to an n type silicon sample, the metal acts as a p type impurity to form a pn junction with a forward biased threshold of 0.3 V. With forward bias, electrons flow from the semiconductor into the metal where a plentiful supply of electrons already exist. Because of this, the charge carrier storage time is negligible. This feature allows specially constructed whisker contact diodes to operate as mixer devices up to at least 100 GHz.

TRAPATT diode (trapped plasma avalanche triggered transit-time). The operation of this diode is based on an avalanche effect and the finite transit time of charge carriers. During part of the oscillatory cycle, a hole–electron plasma (collection of free electrically neutral carriers) is generated. An interaction takes place that results in a low field region being trapped between two high field regions. The diode is most useful as either an amplifier or an oscillator below about 10 GHz, where the longer drift region of the IMPATT diode generates heat dissipation problems. Oscillator power outputs in excess of 80 W at 5 GHz have been reported with amplification gain in excess of 5 dB.

Tunnel diode. These diodes are constructed from germanium or gallium arsenide, with highly doped p and n regions which results in a very narrow depletion layer. According to classical physics, an electron must have an energy at least equal to the barrier potential before it can cross a junction. Due to the very narrow depletion region, however, electrons with a lower energy level can cross, giving the impression that they have tunnelled through the barrier. This results in a negative resistance region in the V/I characteristic which allows the diode to be used either as a microwave oscillator or an amplifier. Because tunnelling takes place at close to the speed of light, the device is also used as a switch with a fast changeover time of about 50 ps.

20.2 Passive devices (see also Transmission lines and waveguides, p. 376)

Cavity resonator. At lower frequencies, a parallel combination of inductance (L) and capacitance (C) can be used to provide a resonant circuit, which will develop maximum voltage at a particular frequency. The progressive reduction in the values of both L and C will produce an increase in the resonant frequency. Figure 20.4 shows the limiting condition where C has been reduced to a pair of parallel plates, while L has been reduced by connecting the plates together with parallel straps that behave as inductors. The ultimate limit is reached when the two plates are connected together with the four sides of a box. This structure is a resonant cavity, and if energised at the appropriate frequency, will develop maximum voltage across the points shown. As the electromagnetic energy is reflected off the internal sides of the cavity, the resonant wavelength is determined by the length of the diagonal. In a practical case, the cavity can be tuned by including a piston that can be moved to vary the enclosed volume. The unloaded Q factor, which is in the order of 15 000, is given by:

$$Q = 2\pi(\text{total energy stored in cavity/energy lost per cycle due to resistivity of walls})$$

From this it can be deduced that Q is proportional to the ratio of cavity volume to interior surface area.

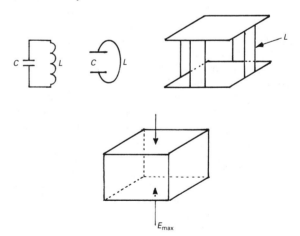

Figure 20.4 Derivation of the cavity resonator.

Circulator. The most commonly used circulator, shown schematically in Fig. 20.5(a), has three ports each spaced by 120°. When each port is correctly terminated, a signal input at one port appears as an output at the next, the third port being isolated from the signal. The behaviour

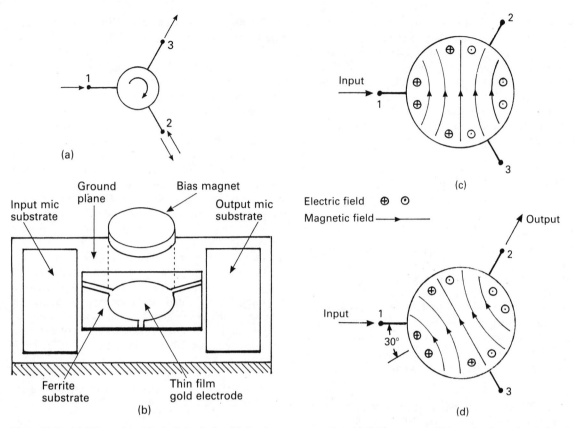

Figure 20.5 (a) Schematic symbol of circulator; (b) circulator construction; (c) field patterns within a circulator due to input signal alone; (d) due to input signal plus magnetic bias.

at each port changes in a cyclic manner as shown in Table 20.1.

The device is suitable for separating incident and reflected signals. The insertion loss in the transmissive direction is typically less than 0.5 dB, while the attenuation at the isolated port is better than 20 dB. For example, port 1 might be driven from a transmitter, while port 2 provides the antenna feed. In the receive mode, port 2 becomes the input, with the circulator passing the signal to port 3, to provide isolation between transmitter and receiver of better than 20 dB. For this high power application, the circulator would be positioned within the antenna waveguide feed. The basic construction of a low power device is shown in Fig. 20.5(b). A small thin slab of ferrite acts as a substrate for

Table 20.1   *Circulator port* $(^1/_0)$

	Ports	
Input	*Output*	*Isolated*
1	2	3
2	3	1
3	1	2

a thin-film deposited gold electrode. A permanently magnetised magnet sits on top of this to generate *Faraday rotation* within the structure. Signal inputs and outputs are provided via *microwave integrated circuits* (MIC). The behaviour depends upon the combined effects of the magnetic fields due to the signal power and the permanent magnet. Figure 20.5(c) shows the lines of the electric and magnetic fields due to the signal power and without the bias magnet. Ports 2 and 3 are at the same potential and are effectively isolated from port 1. By the application of a correctly chosen magnetic bias level, shown in Fig. 20.5(d), the field components become rotated by 30°. Port 3 is now at a null in the electric field, while ports 1 and 2 are at equal but opposite potentials. Conduction can now occur between ports 1 and 2 but with port 3 isolated from the signal. The signal circulation can be reversed by reversing the bias field.

Couplers (usually constructed from waveguide, microstrip, or dielectric guide)

Branch line or 90° hybrid. When input power is applied to port 1 of the coupler shown in Fig. 20.6(a), the signal

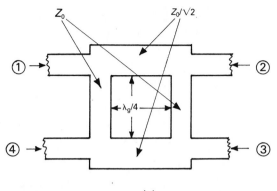

(a)

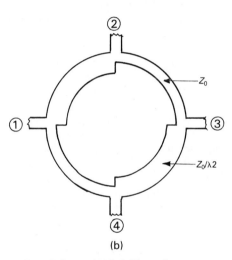

(b)

Figure 20.6 Branch line or 90° hybrid couplers.

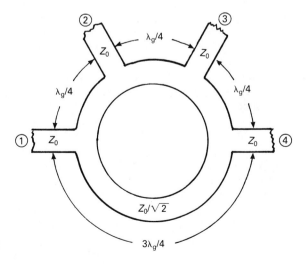

Figure 20.7 Hybrid ring or 'rat-race'.

can arrive at port 4 by two paths. One is $\lambda_g/4$ long and the other $3\lambda_g/4$ long. Because the signals at port 4 are in anti-phase, they are self-cancelling and port 4 is isolated. The output power divides equally between ports 2 and 3, and since these are separated by a strip $\lambda_g/4$ long there is 90° of phase difference between the signals. The alternative construction shown in Fig. 20.6(b) behaves in the same way because the four ports are spaced by $\lambda_g/4$. The structures shown have a bandwidth typically about 10% and are used extensively in balanced mixer circuits where a pair of quadrature signal feeds are needed. The attenuation depends upon the ratio of the impedance of the series and shunt arms. That shown in the diagrams provides a coupler with an attenuation of 3 dB.

Hybrid ring or rat-race. This device consists of an annular ring $1.5\lambda_g$ long, with four ports spaced $\lambda_g/4$ apart as shown in Fig. 20.7. Signal power applied to port 1 can circulate either clockwise or anticlockwise to reach ports 2, 3, and 4. Both paths to port 4 are $3\lambda_g/4$ long so that port 4 provides an in-phase output. The paths to port 2 are either $\lambda_g/4$ or $5\lambda_g/4$ and are again in-phase so that port 2 is also an output. At port 3 the path lengths differ by $\lambda_g/2$ to produce an anti-phase condition, so that port 3 is isolated. In a similar way, an input to port 3 will not couple with port 1. As well as the ability to combine two signals, this device can also divide a signal into two equal parts. When two unequal signals are added, the sum appears at the output port, while the difference is generated at the isolated port.

'Lange' or 90° hybrid coupler. A transformer type of coupling can be achieved in the manner shown in Fig. 20.8. When a 3 dB coupling is needed, the 90° hybrid or *Lange* coupler shown in Fig. 20.8(b) can be used. This consists of at least four inter-digitated fingers with *air bridge* connections to provide a tighter coupling. This structure typically has an octave bandwidth. Power coupled to the signal port is equally divided and transmitted to the coupled and direct ports, but with a phase difference of 90°.

Proximity coupler. The structure shown in Fig. 20.8(a) also provides a transformer coupling with 90° of phase difference but with lower efficiency than the Lange coupler.

Rotamode directional coupler. This device combines the properties of the directional coupler with those of the cavity resonator. Typical applications include the combining of two signals for transmission over a common transmission line and as a bandpass filter. The coupler consists of a number of cylindrical cavity resonators coupled via iris apertures in the common

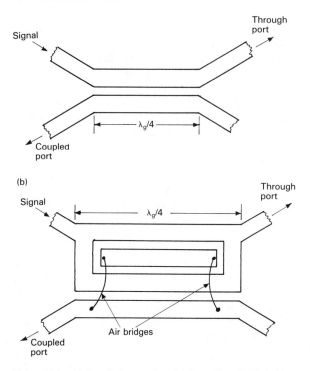

Figure 20.8 (a) Proximity coupler; (b) 'Lange' or 90° hybrid coupler.

walls as shown in Fig. 20.9. When a signal is applied to the input stripline coupler, the potential difference between it and the cavity wall excites the cavity to induce electric and magnetic fields that interact with the output stripline coupler to provide the output signal. When the feeds to the two couplers are impedance matched and resistively terminated, two input signals can be linearly combined, with the maximum of

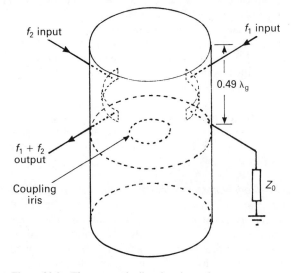

Figure 20.9 The rotamode directional coupler.

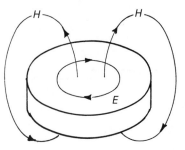

Figure 20.10 Field patterns in a dielectric resonator.

isolation between the two driver circuits. Different sized cavities are used to obtain bandpass filter characteristics.

Dielectric resonator. These are ceramic devices made as small discs of sintered barium titanate, that are used at microwave frequencies in much the same way as quartz crystals are used at lower frequencies, to control the frequency of an oscillator; directly and without frequency multiplication. They are much smaller than resonant cavities and have a similar degree of stability. The resonator is placed in the oscillator feedback path, with coupling being achieved via the magnetic field that is indicated in Fig. 20.10. Dielectric resonator oscillators (DRO) suffer less frequency pulling than an unstabilised oscillator, a characteristic that can be further improved by the use of an isolator. DROs can be tuned over a range of about 1% (100 MHz in 10 GHz) by perturbing the resonator's magnetic field by varying an air gap between the disc and the circuit's metallic enclosure.

Filters. Devices with different filter characteristics can be constructed using microstrip techniques and a representative sample are shown in Fig. 20.11. The high impedance strip of length $1 < \lambda_g/4$, connected between the two low impedance sections as shown in Fig. 20.11(a), behaves as an inductor. The two low impedance sections have significant shunt capacitance so that the combination acts as a low-pass filter. In Fig. 20.11(b), again $1 < \lambda_g/4$, but now the low impedance is connected between two high impedances which act as shunt inductors. The overall effect is thus that of a high-pass filter. An open circuit stub whose length is a half wavelength is a high impedance at each end. (This is also true for integer multiples of $\lambda/2$.) Thus the half wavelength stub shown in Fig. 20.11(c) behaves as a high impedance to signals that make it so. At frequencies on either side of this the stub impedance falls, so that less signal voltage will be developed across it, producing a bandpass characteristic. If the stub is reduced to $\lambda/4$ however, it develops a very low impedance and thus acts as a bandstop device. Figure 20.11(d) shows a series of half wavelength strips (or

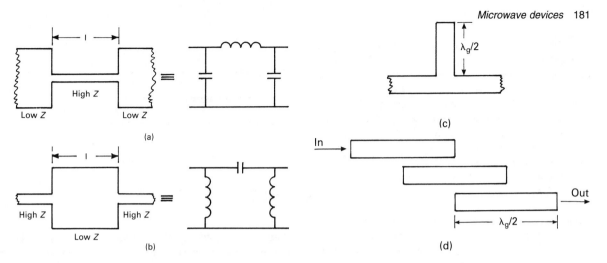

Figure 20.11 Filters: (a) low-pass; (b) high-pass; (c) bandpass filters: single open circuit stub; (d) coupled resonators.

resonators), edge or parallel coupled via the electro-magnetic field. The bandwidth of such a filter can be extended by varying the lengths of the resonators around $\lambda/2$.

Impedance transformer (quarter wave transformer). A strip of conductor that is just $\lambda/4$ long has a resistive impedance at both ends, one very high and the other very low. This impedance variation allows such a device to behave as an impedance matching transformer. In Fig. 20.12(a) a quarter wavelength strip of impedance Z_2 is used to provide such a match where $Z_2 = \sqrt{(Z_1 Z_3)}$. Z_2 is thus the geometric mean of Z_1 and Z_3.

Alternatively, a tapered section of line may be used as shown in Fig. 20.12(b) to effect an impedance match.

Isolator. An isolator is a two-port device that functions in a similar way to the circulator, with similar levels of insertion loss and attenuation. In fact, a circulator can be used for this function if its third port is terminated in a resistance equal to its characteristic impedance. Used in this way, the reflected energy is absorbed in the resistive load. Such a device is ideal to use in the

coupling path between a microwave oscillator and its load, to minimise frequency pulling.

Microwave integrated structures

These elements of microwave circuits resemble the printed circuit boards used for lower frequency operation. However, the behaviour is more closely related to that of transmission lines. A planar construction is used because it is compatible with solid-state components and provides low loss and stable circuits, which are very reproducible and hence cost-effective. The dielectric substrate carries the electro-deposited metallic conductors which are commonly gold on a thin film of chromium for better adhesion, or copper. The substrates include alumina, sapphire, quartz and rutile, plus proprietary materials such as Rexolite, Cuflon and RT Duroid. The relative permittivities for these range from around 2.2 to 11.5. The factors that affect the choice of a particular substrate include operating frequency, thermal conductivity, mechanical stiffness, cost, fabrication tolerances and reproducibility, and surface finish which is important to conductor adhesion. Alumina is popular for operation up to about 10 GHz. The microwave energy propagates in the dielectric and the surrounding air, rather than through the conductors, which simply act as waveguides. The combined structure thus has an *effective permittivity* (ϵ_{eff}) somewhere between that for air and the substrate. It is shown in the literature that

$$\epsilon_{eff} = (\lambda_o / \lambda_g)^2$$

where λ_o and λ_g are the wavelengths in free space and the guiding structure, respectively. Low dielectric constant substrates have the advantage of producing longer circuit elements, which eases the reproduction of dimensional tolerances. With rising frequency the propagation tends to concentrate more in the substrate

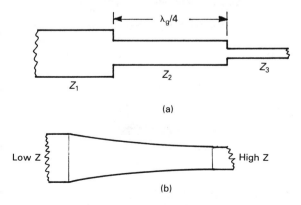

Figure 20.12 Impedance matching transformer.

and less in the air, so that λ_g tends to increase with frequency. The thickness of the conductor elements must be such that they are capable of supporting any dc supply current and dissipating the resulting heat. The elemental lengths are obviously wavelength dependent and their characteristic impedance is related to the relative permittivity (ϵ_r) and the thickness of the substrate and the width of the conducting elements. The relationship that is quoted in the literature is very complex and best approached using computer-aided design (CAD).

Coplanar waveguide. With this structure, the metallisation is formed on one side of the substrate only. Each conductor strip has two grounded parallel strips deposited on either side so that more energy flows in the central conductor and less in the substrate. At higher frequencies, the open nature results in radiation from the surface, but the structure is particularly suitable for lower frequency applications using components that are surface mounted across the intermetallic gaps.

Finline or E-plane structure. This structure, illustrated in Fig. 20.13, consists of a thin dielectric substrate, which for low cost applications is typically made of glass-fibre reinforced PTFE with copper metallisation. It is mounted so that it bridges the broad walls of a rectangular waveguide where it couples directly with the electric field. The conductor patterns which form the various circuit elements may be formed on one or both sides of the substrate. Fabrication of the structure is fairly simple and it is compatible with surface mounting and solid-state technology. Finline losses are very small so that the structure can operate at frequencies at least up to 100 GHz.

Imageline. This structure is formed by attaching a rectangular sectioned strip of dielectric to a metallic ground plane. Typical dielectrics include alumina, boron nitride and high resistivity silicon. At very high frequencies (a few hundred GHz), the dielectric to air interface, produces the features of a waveguide. Although the Q factor is very high, it is not very compatible with active devices.

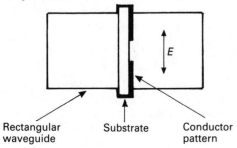

Rectangular waveguide Substrate Conductor pattern

Figure 20.13 Finline or *E*-plane structure.

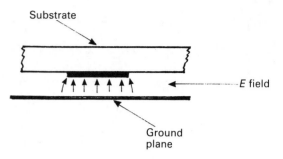

Substrate —— *E* field —— Ground plane

Figure 20.14 Inverted microstrip.

Inverted microstrip. In the version shown in Fig. 20.14, the substrate simply supports the guiding conductors in air, so that the effective permittivity for the structure is close to that of air ($\epsilon_{\text{eff}} = 1$). Thus λ_g is relatively longer than for other structures and this allows for operation at even higher frequencies. Compared with conventional microstrip, this also provides a wider conductor for a given value of characteristic impedance, a feature that not only aids heat dissipation, but also relaxes fabrication tolerances.

Microstrip. As shown in Fig. 20.15, this structure is formed on both sides of the substrate, the conductor pattern on one side and a continuous ground plane on the other. The electric field is set up almost entirely within the substrate. The characteristic impedance of each conductor strip is dependent upon the frequency, the width (w) of the strip, the height (h) and the relative permittivity (ϵ_r) of the substrate. Typical approximate formulae for this relationship are given in the literature as:

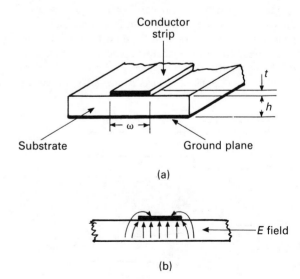

Conductor strip t h ω Substrate Ground plane

(a)

—— *E* field

(b)

Figure 20.15 Microstrip waveguide construction; (a) important dimensions; (b) *E* field diagram.

$$\epsilon_{\text{eff}} = \frac{\epsilon_r + 1}{2} + \frac{\epsilon_r - 1}{2}(1 + 10h/w)^{-0.555}$$

and

$$\epsilon_{\text{eff}} = \frac{\epsilon_r}{0.96 + \epsilon_r(0.109 - 0.004\epsilon_r)[\log(10 + Z_o) - 1]}$$

With suitable components and substrate, microstrip circuits can function satisfactorily at frequencies up to 60 GHz.

Slotline. The dielectric substrate of this transmission line type structure, is metallised on one side only and this carries an etched slot to provide the guiding mechanism. It is not a particularly favoured technique as characteristic impedances below about $50\,\Omega$ are difficult to fabricate and the Q factor is lower than many other types.

Suspended stripline. In this type of structure, the ground plane of the inverted stripline concept is made to totally enclose the substrate with its single-side metallisation. To ensure correct operation at the higher frequencies, the top and bottom sides have to be mounted close to the substrate and this makes for difficulties when mounting active devices on such a structure.

Trapped inverted microstrip. This is a further development of the inverted stripline concept. The ground plane is formed into a channel shape so that the conductor pattern is effectively screened on three sides. The technique has been used at frequencies as high as 100 GHz.

20.3 Miscellaneous terms

Magnetostatic surface wave filter (MSSWF). This is the microwave equivalent of the surface acoustic wave filter used for lower frequencies, but functions on a different principle. The device consists of a thin rectangular gadolinium gallium garnet (G^3) substrate upon which a thin film of yttrium iron garnet (YIG) has been grown.

The centre section is loaded with, typically, three very thin silver discs to act as magnetostatic wave resonators and the two ends carry narrow parallel strip transducers also made of silver. An external magnetic field is applied normal to the structure and an increase in field strength results in an increase in the filter centre frequency. The principle of operation is based on ferrimagnetic resonance and interaction between the input signal and the permanent magnetic bias. The filters can be used up to about 20 GHz, with very low insertion loss, wide tuneable frequency range and with good out of band rejection of spurious signals.

Microwave integrated circuit (MIC). The term usually used to describe the structure of metallic conductors formed into circuit elements such as inductors, transformers, filters, etc., that are constructed on an insulator substrate.

Monolithic microwave integrated circuit (MMIC). The term used to describe the complex circuits that can be fabricated using semiconductor planar technology for operation up to 30 GHz. The technique is usually based on gallium arsenide and the complex multi-layer circuits can include transistors, diodes, small value capacitors and even inductors. Circuits with a degree of complexity that allows the down convertor for a Ku band satellite receiver to be constructed on a single substrate have been produced.

20.4 Useful references

Bhartia P. and Bahl I.J. (1984) *Millimetre Wave Engineering and Applications*, Wiley, London.

Connor F.R. (1972) *Introductory Topics in Electronics and Telecommunication (Wave Transmission)*, Vol. 3, Edward Arnold, London.

Edwards T.C. (1981) *Foundations of Microstrip Design*, Wiley, London.

Thomas H.E. (1972) *Handbook of Microwave Techniques and Equipment*, Prentice Hall, Englewood Cliffs, NJ.

Tsutsumi M. (1987) A study on the magnetostatic wave resonators using the YIG film. *Trans. IEICE Japan*, **J70-C** (8), 1214–1216.

21 Mixer signal processing

Two distinctly different mixing concepts are used in communications and these are based either on linear or nonlinear techniques. Generally the linear technique is restricted to audio and video programme production, while the nonlinear concept is very largely used for frequency changing.

21.1 Linear mixing

Many input signals sources are linearly added together to produce a composite output signal that ideally contains no intermodulation products. For example, a multi-track recorder can be used to record separate instrumental tracks. The mixer unit can then combine these to produce an orchestral effect.

Analogue mixer
Each channel of the mixer will carry equalisation and gain networks to balance the relative frequency responses and amplitudes, together with a *fader circuit* that allows a smooth transition and blend between channels. For video production, there is the additional need to maintain inter-channel timing and synchronisation. Vision signal mixing is closely allied to *chroma-keying* (see also Image processing, p. 141). In the analogue domain, maintaining linearity is particularly demanding. When the case of stereo audio is considered, the problems of *cross-talk* and intermodulation become even more critical.

Digital mixer
While processing in the digital domain provides a greater degree of flexibility, with reduced distortion problems, this is only gained at the expense of circuit complexity. For uni-polar (varying dc) signals, as exemplified by video, the addition process can easily give rise to overflow or overloading. For bi-polar signals such as audio, an offset value can be added to each digital sample, but this offset binary concept can also easily lead to an overflow. It is therefore common to use the 2s complement technique to represent positive and

negative values, with the most significant bit (MSB), 0 or 1, being used to signal a positive or negative value, respectively.

21.2 Nonlinear mixing

This technique involves the introduction of two signals into a nonlinear circuit with square law characteristics, so that harmonics and intermodulation are deliberately generated. Often diodes are used for this purpose, but field effect devices (FETs) have a more nearly square law mutual characteristic. This then allows some particular product of mixing to be extracted using a filter network. The concept forms the basis of the *superheterodyne receiver* and the frequency translation process is shown in Fig. 21.1.

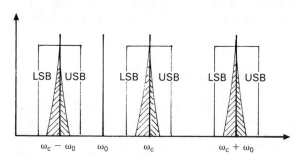

Figure 21.1 Frequency translation and reciprocal mixing.

21.2.1 Analogue techniques

Additive mixing. The principle of this method is shown diagrammatically and mathematically in Fig. 21.2, where the two input signals are $V_{\text{in1}} = A_{\text{c}} \sin \omega_{\text{c}} t$, the carrier frequency and $V_{\text{in2}} = B_{\text{o}} \sin \omega_{\text{o}} t$ provided by a local oscillator. After adding and squaring these inputs, the output of the circuit is filtered to remove unwanted products of mixing, to leave either the *sum* or *difference* frequency which is often described as the *intermediate*

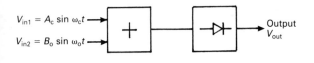

Square law device, $V_{out} = a + b V_n + c V_m^2 \ldots$
considering only the square term,
$V_{out} = c (A_c^2 \sin^2 \omega_c t + 2A_c B_o \sin \omega_c t \cdot \sin \omega_o t + B_o^2 \sin^2 \omega_o t)$
After filtering out the 2nd harmonic (square) terms.
$V_{out} = 2A_c B_o C \sin \omega_c t \cdot \sin \omega_o t$
$\quad = A_c B_o C [\cos (\omega_c - \omega_o)t - \cos (\omega_c + \omega_o)t]$

Figure 21.2 Additive mixing.

frequency. Thus the input carrier signal can be either up- or down-converted in frequency as required.

Multiplicative mixing. Figure 21.3 shows how the product of the two input signals V_{in1} and V_{in2} directly produces sum and difference signals. Because of this, it is easier to filter off the wanted mixing product and the circuit is less noisy than that of the additive mixer.

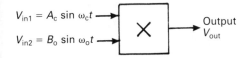

Multiplier output

$V_{out} = A_c B_o \sin \omega_c t \sin \omega_o t$

$\quad = \dfrac{A_c B_o}{2} [\cos (\omega_c - \omega_o)t - \cos (\omega_c + \omega_o)t]$

Figure 21.3 Multiplicative mixing.

Reciprocal mixing. Reference to Fig. 21.1 shows how the sidebands of the carrier signal ω_c are transferred to the sum or difference frequency by the heterodyne action with the oscillator signal ω_o. In a similar way, any noise component associated with the oscillator circuit that produces frequency or phase modulation of the output will cause the oscillator signal to contain its own sidebands, and these can be transferred to the sum and difference signal in the same way. Such *phase noise*, as this is usually described, can give rise to unwanted

adjacent channel interference being contained within the filtered output pass band.

Sub-harmonic mixing. Noise in frequency changer circuits arising from oscillator instability is a problem that worsens as the frequency increases. At microwave frequencies, the problem is often minimised by running the local oscillator at some sub-multiple or sub-harmonic frequency and using multiplication to achieve the required value.

21.3 Miscellaneous terms

Balanced mixers. The single nonlinear device circuit provides many intermodulation products thus increasing the complexity of the filters. By using two diodes or transistors in a full-wave configuration, the oscillator signal can be nulled from the output by balancing. A more effective circuit can be produced using a double balanced ring of four devices. In this case, both of the input signals are nulled from the output by the balance conditions. This configuration results in a lower conversion loss, with much less intermodulation.

Conversion loss. The ratio of the output signal level at intermediate frequency to the level of input signal at carrier frequency. Usually expressed in dB and stated for matched conditions; typically in the order of -5 to -7 dB.

Isolation. This parameter describes the isolation between the local oscillator circuit and the RF input and the IF outputs, respectively. It is therefore the ratios of the local oscillator signal input level to that measured at both carrier frequency input and intermediate frequency output under operating conditions. Typically this will be in the order of 35 to 40 dB.

21.4 Useful references

Manassevitsch V. (1980) *Frequency Synthesisers, Theory and Design*, 2nd edition, Wiley, New York.
Robins W.P. (1982) *Phase Noise in Signal Sources*, Peter Peregrinus, London.
Watkinson J. (1985–1987) *Digital Audio and Video*, Broadcast Systems Engineering, Link House Publications, London.

22 Modulation and demodulation

Information for communications purposes often falls within the low frequency sonic range and the transmission of such information using sound waves is virtually impossible. The low velocity of propagation through air (330 m/s) would produce long time delays and the signal attenuation would severely restrict the length of the communications path. Increasing power output would only add to the general level of interference between the channels sharing the relatively narrow bandwidth, and this in turn restricts the number of channels that are simultaneously available. These problems are overcome by superimposing the information on to a higher frequency signal that can be radiated as an electromagnetic wave. This process is known as modulation. Provided that this spectrum is suitably regulated, a wide frequency range, extending from about 15 kHz to 300 GHz, provides for very many communications channels. The attenuation is relatively very small and the propagation velocity about a million times higher $(3 \times 10^8$ m/s). Basically, the low frequency information signal is used to distort one of three features (amplitude, frequency or phase) of the carrier wave in a controlled manner, so that the complementary process of demodulation at the receiver can recover the original information.

22.1 Modulation

The study of this topic is complicated by the fact that there are many variations of the basic methods available that can be used to suit different conditions. While the analogue and digital baseband domains are very different, the modulation schemes have great similarity.

22.1.1 Analogue modulation (basic concepts)

Amplitude modulation (AM)
Figure 22.1(b) shows how the modulation signal modifies the carrier wave without affecting its frequency. The scheme has the advantages that it is a long

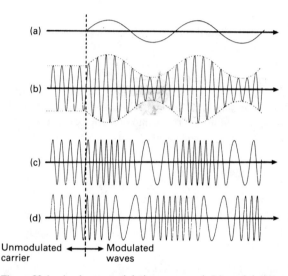

Figure 22.1 Analogue modulation compared: (a) modulating signal; (b) amplitude modulation; (c) frequency modulation; (d) phase modulation.

established, well understood scheme, which under equivalent operating conditions has the narrowest bandwidth. Offset against this are the disadvantages of susceptibility to noise and a poor energy efficiency. The amplitude of the carrier $A_c \sin \omega_c t$, is varied in sympathy with the modulating signal $A_m \sin \omega_m t$ so that the expression for the complex wave is given by $(A_c + A_m \sin \omega_m t) \sin \omega_c t$, which can be expressed as:

$$A_c \sin \omega_c t + \frac{mA_c}{2} \cos(\omega_c - \omega_m)t - \frac{mA_c}{2}\cos(\omega_c + \omega_m)t$$

where $m = A_m/A_c$ which is described as the *depth of modulation* or *modulation factor*. This expansion shows that the complex spectrum consists of *carrier* and *sum* and *difference* frequencies. The bandwidth is clearly the spectral range between $w_c - \omega_m/2\pi$ and $\omega_c + \omega_m/2\pi$. In general, the bandwidth of an AM wave is given as twice the highest frequency component in the modulating signal.

The power in the sidebands that contain the information depends upon the sum and difference components and the depth of modulation. When $m = 1$ or 100%, the sideband power is equal to half the carrier power, so that the transmission efficiency can be stated as the ratio, sideband power to total power or $m^2/(2 + m^2)$. If m exceeds unity, overmodulation produces distortion which generates harmonics to create interference. Thus under normal operating conditions m on average, varies between about 30% and 50%.

Frequency modulation (FM)

A study of Fig. 22.1(c) shows that unlike AM, the FM wave is of constant amplitude and with a frequency variation that depends upon the amplitude of the modulating signal. Furthermore, the power in the complex wave is constant and since the signal has a wider bandwidth, it has a better noise characteristic. Because even the carrier component contains information that is useful to the reciver, the method is also more power efficient than AM. These advantages have to offset against the fact that both transmitters and receivers for FM are more complex and expensive.

The basic expression for an FM wave is

$$A_c \cos[\omega_c \pm m \cos \omega_m t]t$$

showing that FM is a form of *angle modulation*. m is the *modulation index* which is the ratio of carrier frequency (f_c) change to modulating frequency (f_m) which produced the change. Typically, if $m > 1$ the signal spectra contain a large number of sideband pair components. This number increases with a reduction in the modulating frequency. The basic expression for the wave can be expanded using Bessel functions (see Signals, p. 302) to yield an approximate spectrum:

$$A_c m \cos \omega_c t \pm m[\cos(\omega_c + n\omega_m t) \pm \cos(\omega_c - n\omega_m)t]$$

where m is the modulation index and n is the order of the sideband pair. Compared to AM, the sideband components are phase shifted by 90°. Theoretically, the bandwidth of such a signal extends from zero to infinity. However, beyond some practical limit, the amplitudes of these components are less than about 2% of the carrier amplitude and can be considered insignificant. This is usually defined as the *Carson's bandwidth* which is given as $2[f_d + f_m]$, where f_d is the *frequency deviation*, which is the maximum change of frequency, and f_m is the highest modulating frequency permitted. Further, a *deviation ratio* is defined as $M = f_d/f_m$. It is shown in Fig. 22.2, that FM has a

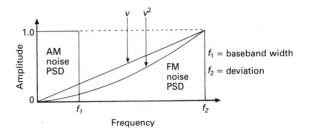

Figure 22.2 Signal to noise power ratio advantage of FM over AM ($= 3(f_2/f_1)^2$ or $3M^2$, where $M = f_2/f_1$ the deviation ratio).

signal to noise ratio advantage over AM and that this depends upon the deviation ratio M. For example, if $M = 5$, then the advantage is almost equal to 19 dB. An FM signal can be generated by modulating the voltage-dependent reactance of a varactor diode that forms part of an oscillator tuned circuit. The classic *Armstrong* method, however, recognises the 90° of phase difference between the sidebands in AM and FM waveform spectra. In this case, both carrier and modulating signals are fed to a balanced mixer to produce a double sideband suppressed carrier (DSBSC) signal. The original carrier is then phase shifted by 90° and then added to the two sidebands. After limiting to remove any AM component, the modulated signal is pure FM.

Phase modulation (PM)

Phase modulation is also a variant of angle modulation and is thus closely related to frequency modulation. However, the carrier signal phase is varied in sympathy with the modulating signal amplitude. Whereas a frequency modulator fed with a modulating signal of 1 V amplitude at different frequencies will produce the same carrier deviation with each signal, a phase modulator fed with the same signals will produce a frequency deviation that is proportional to the modulating signal frequencies. Thus maximum modulated carrier frequencies occur when the rate of change of modulating signal amplitude is greatest. This feature can be seen in Fig. 22.1(d). Like the FM wave, the complex PM waveform contains an infinite set of sideband pairs related in amplitude by Bessel functions (see Signals, p.302). It is thus a wide bandwidth system that has good noise characteristics. An approximate expansion of the waveform expression is given by:

$$A_c m \sin \omega_c t + m[\sin(\omega_c + n\omega_m)t - \sin(\omega_c - n\omega_m)t]$$

where n is the sideband pair order and m is the modulation index, given by the ratio of carrier frequency change to modulating frequency. For the practical reason of bandwidth restriction, m is fixed, usually at a value of 0.5 rad.

Analogue modulation (variants)

The majority of these variations of the basic modulation schemes have been devised to minimise the bandwidth required to transmit signals, chiefly audio, over standard telephone radio systems having a typical bandwidth of 300 to 3400 Hz.

AM/PM-VSB (amplitude/phase modulation–vestigial sideband). This complex form of modulation was devised for use with Group 2 facsimile machines in order to increase the transmission speed over standard telephone lines. A normal double sideband AM signal can be filtered to remove the high frequency components from one sideband, leaving a *vestige* of it to provide a vestigial sideband signal. This reduces the AM bandwidth and still leaves a signal that can be processed through a normal AM system. This typically saves about 30% of the bandwidth. A further saving can be made by introducing a phase modulation component. If peak white in the document represents a peak positive voltage and black is represented by zero volts, then a series of transitions between black and white represents a particular frequency. If alternate white peaks are inverted to produce a negative voltage, the fundamental frequency is halved. When this signal is used to modulate a sub-carrier, the negative peaks produce a carrier phase reversal, producing in turn, a white signal without any change of frequency or amplitude. Using this type of modulation, the transmission speed of the system can be doubled.

DSB (double sideband) modulation. Strictly, this term applies to any modulation system that produces mirror image sideband components in the complex spectrum. The term, however, is often used to describe the spectrum of the basic amplitude modulation signal.

DSBSC (double sideband suppressed carrier). The power efficiency of the double sideband AM signal can be improved by filtering or nulling out the carrier component from the modulation spectrum. This still leaves a radio frequency signal that is capable of electromagnetic propagation. Since all the transmitted power now represents information useful to a receiver, this can achieve up to 95% power saving under normal operating conditions. When zero modulation is present, the carrier signal is effectively switched off, showing a further power saving. The major disadvantage of this technique lies in the fact that the carrier signal has to be accurately reinserted at the receiver. This increases the complexity of the demodulation process. The *Cowan* modulator can be used to produce a DSBSC signal. This

takes the form of a bridge circuit of diodes that shunt the input audio signal. The diode action is switched by the carrier signal so that the output circuit contains samples of the audio input, with the carrier frequency nulled out, but with the wanted sum and difference frequency components. A more effective circuit provides balance to both the audio input as well as the carrier signals. This uses a balanced *bridge ring modulator* with the diodes in series with the audio and carrier inputs. The result is an output that contains only sum and difference frequency components.

FM^2 (frequency/frequency modulation). This technique involves frequency modulating a series of sub-carriers to form a frequency multiplex group. This group is then further used to frequency modulate a higher frequency carrier. The following is an example from television. Multiple sub-carrier frequencies spaced by 180 kHz are chosen from the range 6.3 to 7.94 MHz, and these are frequency modulated by a number of different audio channels using a low modulation index (typically 0.14 to 0.18) to minimise bandwidth. This group of signals are then frequency modulated on to a carrier frequency in the microwave band for transmission over a satellite link.

ISB (independent sideband). As the information to be transmitted is contained in each sideband of the normal double sideband AM signal, it is only necessary to transmit a single sideband. By using a pair of balanced modulators driven by the same carrier signal, two DSBSC signals can be generated. If the upper sideband is filtered from one and the lower sideband from the other, when the two modulation products are combined, each sideband contains independent information. This automatically doubles the channel capacity but at the expense of halving the signal power at the receiver. Thus the received signal to noise ratio will be 3 dB worse than for a double sideband transmission.

NBFM (narrow band frequency modulation). A study of a table of Bessel functions will reveal that for a modulation index m of about 0.1, there is only one significant sideband pair. The resulting bandwidth is thus the same as for AM and the signal can be demodulated with an AM receiver. Even for $m = 0.5$, there are only two pairs of significant sidebands. This then defines narrow frequency modulation.

QAM (quadrature amplitude modulation) (see also Digital QAM, p. 192 and Television signal processing and systems, p. 350). If two quadrature versions of the same carrier frequency are amplitude modulated, they can be added to produce a third carrier signal containing both modulations that can effectively occupy the same bandwidth as one modulated carrier. If the two carriers to be added have the form $A_c \sin \omega_c t$ and $B_c \cos \omega_c t$, then the third will have a relative amplitude and phase of $\sqrt{A_c^2 + B_c^2}$ and $\tan^{-1}(B/A)$ but at the same frequency.

SSBSC (single sideband suppressed carrier). Because the information transmitted in an AM signal is present in both sidebands, the removal of the carrier and one sideband would still leave an RF signal that carried the modulation. This represents a significant improvement in power efficiency and bandwidth conservation. If all the available power is concentrated into the one sideband, then a 9 dB improvement in signal to noise ratio relative to AM can be achieved, but at the expense of increased circuit complexity. The single sideband is first generated at a low frequency and then heterodyned by a mixer process into the desired transmission band. The three methods for generating SSBSC signals are as follows:

1st or filter method. This is the simple technique of generating a double sideband suppressed carrier signal first and then removing the unwanted sideband with a suitable filter. The difficulty lies in the design of the filter. This has to have a sharp cut-off to avoid removing part of the wanted sideband.

2nd or phasing method. This technique produces an SSBSC signal from two DSBSC signals in quadrature using two balanced mixers. One modulator is driven directly by the carrier and modulating signal, while the second is driven by the same two signals that have been phase shifted by 90°. When these two outputs are added or differenced, the resulting signal has the form $A_c \cos(\omega_c - \omega_m)t$ or $A_c \cos(\omega_c + \omega_m)t$, the lower or upper sideband, respectively. Because of the difficulty of accurately maintaining the phase shift over the modulating signal bandwidth, this technique tends to be restricted to the relatively narrow speech bandwidth of 300 to 3400 Hz.

3rd or Weaver method. This is a derivative of the phasing method developed for signals that contain a zero energy band close to dc. It is thus suitable for use over telephony speech circuits (300 to 3400 Hz). The system uses four balanced quadrature modulator devices operating in two stages. The first pair uses a sub-carrier in the centre of the audio band, typically 1.7 to 1.8 kHz, so that the lower sideband of modulation folds about dc, to occupy the zero energy range. A low-pass filter in each modulator output removes all signal components above the sub-carrier frequency, leaving the folded sideband to be input to the second quadrature modulator pair. This uses a higher carrier frequency that may fall in the transmission band. Addition or

subtraction of these two outputs provide, respectively, either an upper or lower sideband for transmission.

VSB (vestigial sideband). Spectrum conservation becomes particularly important when wide baseband signals, such as television, form the modulating signal. Single sideband techniques could be used, but the complex demodulators needed would add significantly to the cost of the receivers. By passing the double sideband AM signal through an asymmetric filter, the major part of one side band can be removed, leaving just a vestige of the original. Such a pseudo-double sideband signal can be processed with standard AM receiver circuitry but requires only about two-thirds of the original bandwidth.

22.1.2 Digital modulation

As with analogue transmissions, the main criterion for digital communications is the accuracy of the received information. Therefore the aim is again to maximise the effects of the system signal to noise ratio. It can be shown, that for any signalling pulse shape there is an inverse relationship between pulse width and pulse bandwidth. If the processing circuit bandwidth is less than the pulse bandwidth, distortion occurs and this leads to bit errors. To minimise this problem, the data pulses are usually prefiltered before modulation to produce a $(\sin x)/x$ or raised cosine shape. For analogue systems, the carrier to noise (C/N) or signal to noise (S/N) ratios provide a good indication of signal quality. For a digital system it is more convenient to consider the ratio of *energy per bit* (E_b), per *watt of noise power*, per *unit bandwidth* (N_o). The trade-off between bandwidth and bit rate is shown by the relationship between these two ratios as follows:

$$\frac{E_b}{N_o} = \frac{C}{N} \times \frac{B}{R}$$

where $N_o = N/B$, C/N is the carrier to noise power ratio in bandwidth B, and R is the serial bit rate.

The *bit error rate* (BER) is the ratio of the number of bits received in error to the total number of bits transmitted per second. It is often typically quoted as BER $= e \times 10^{-3}$ per second. Assuming that digital errors are due to Gaussian noise, the probability of a given error rate P_e, can be found from tables of *error functions* or complementary error functions. These are related as follows:

complementary error function$= 1 -$ error function
or erfc $\qquad\qquad = 1 -$ erf

Digital modulation (basic concepts)

Amplitude shift keying (ASK). In this case, the carrier amplitude is varied in discrete steps to represent the digital levels. The most common case involves binary signalling so that only two levels are permitted. A zero carrier level indicates a binary 0 and some amplitude a binary 1. Because of the on–off nature of the carrier signal, this is described as on–off keying (OOK). The modulated signal spectrum is similar to normal AM, with the bandwidth being twice the baseband bit rate. The error probability for the OOK concept in the presence of Gaussian noise is given by:

$$P_e(\text{OOK}) = 0.5\,\text{erfc}\,\sqrt{\frac{E_b}{4N_o}}$$

Frequency shift keying (FSK). This method involves switching the carrier between two predetermined values. One represents binary 0 and the other binary 1. The upper and lower frequencies each have a spread of energy due to the baseband signalling frequency, so that FSK obeys Carson's bandwidth rule of $2(f_d + f_m)$, where f_d is the difference between the two carrier frequencies and f_m is the baseband bit rate. The probability of error is given by

$$P_e(\text{FSK}) = 0.5\,\text{erfc}\,\sqrt{\frac{E_b}{2N_o}}$$

showing that FSK has a 3 dB noise advantage over OOK.

Phase shift keying (PSK) or phase reversal keying (PRK). This is essentially a single frequency method where the data bit stream causes a change of carrier phase. Typically for the binary concept, a 0 produces no change of phase, while a 1 produces a phase change of $180°$. The probability of error is given by:

$$P_e(\text{PSK}) = 0.5\,\text{erfc}\,\sqrt{\frac{E_b}{N_o}}$$

showing that PSK has a 3 dB noise advantage over FSK. This, together with the narrowest bandwidth, means that PSK is a popular technique.

Pulse modulation. A digital carrier signal consisting of a continuous stream of pulses can be modulated to convey information. The basic pulses can be varied in amplitude, duration or width, or relative position in time, to give rise to pulse amplitude modulation (PAM), pulse duration or width modulation (PDM, PWM), or pulse position or time modulation (PPM or PTM). PAM

is related to amplitude modulation and carries the same noise penalties, but PDM and PPM are derivatives of phase modulation and thus have a better noise performance. Of the two, PDM is the easier to use and so is the most popular.

Digital modulation (variants)

Most of these modulation variants have been devised to increase the data capacity of a transmission channel and at the same time minimise its bandwidth. The *spectral efficiency* of the channel is expressed in terms of *bits per second per Hz of bandwidth* (b/s/Hz). However, this is often obtained at the expense of increased circuit complexity and the employment of enhanced forward error control techniques (see also Information theory, p. 143 and Hartley/Shannon rule, p. 144).

ADPCM. See DPCM, below.

AFSK (audio frequency shift keying). A concept devised to allow data signals to be transmitted over standard voice-type telephone circuits, the binary signal frequency pair falling within the band 300 to 3400 Hz. Kansas City modulation is one example of this method. A binary 0 is represented by four cycles at 1200 Hz and a 1 by eight cycles at 2400 Hz. The bit periods are therefore of equal time duration.

CPM (continuous phase modulation). If frequency or phase shift keying is produced using two oscillators, the outputs of which are binary selectively switched, the output signal is described as phase discontinuous. However, it is possible to use the output from a single oscillator, in which case the signal phase varies continuously (at a high rate) between the two limiting values.

DPCM (delta or differential pulse code modulation) (see also Pulse code modulation, p. 91). For delta modulation or differential PCM systems, the modulating signal is coded by 1 bit, positive or negative, based upon whether a signal sample is greater or less than the previous value. Because only one bit per sample needs to be transmitted, the sampling rate can be increased significantly. This reduces quantisation noise, simplifies the receiver anti-aliasing filter and at the same time reduces the bandwidth required, relative to standard PCM. A 1-bit error simply produces a step of the wrong polarity in the output signal, which afterwards retains its correct general shape, as shown in Fig. 22.3. Such errors are thus likely to pass unnoticed. An overload effect can occur when the input signal amplitude changes by a greater than quantising step size during the sampling period. However, this only affects large amplitude, high frequency signal components which rarely occur in

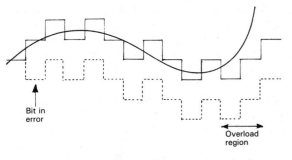

Figure 22.3 Delta modulation (bit error and overload).

audio signals. Adaptive modulation (ADPCM) systems can be used to overcome the overloading, shown in Fig. 22.3. A step-sizing circuit (a differentiator) continually evaluates the signal slope to select an appropriate step value. If the pattern of an analogue signal is known up to some point in time, then because of the correlation between successive values, it is possible to draw certain inferences about the future behaviour of the signal. This can be achieved by a *predictive* or extrapolation process. For predictive coding the signal is sampled at regular intervals, but as each sampling time approaches the probable value of the next sample is predicted. It is the difference between this prediction and the actual value that is coded as for PCM and transmitted. The receiver has to make the same prediction and add the same correction. The general principle of this technique is shown in Fig. 22.4(a) and unless the two predictors (transmitter and receiver) work on the same signal, errors can occur. The improved

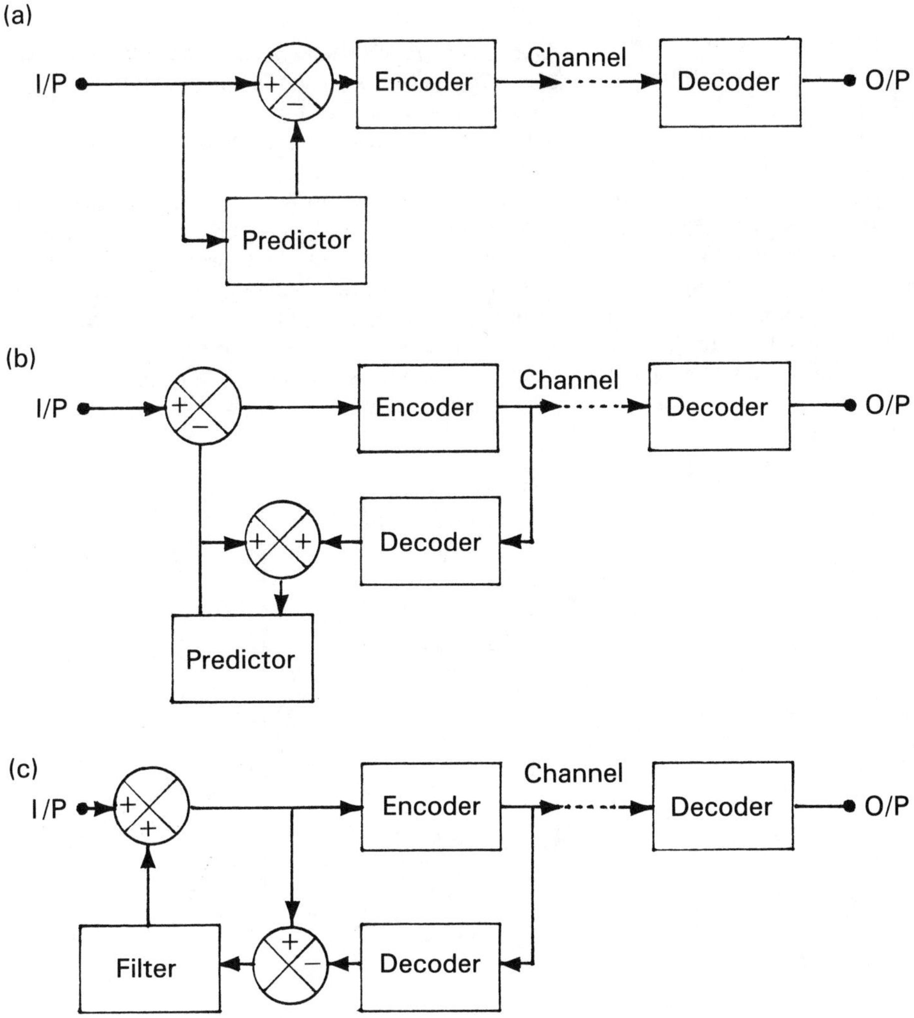

Figure 22.4 Predictive coding: (a) basic principle; (b) improved version; (c) integrator/filter version.

circuit shown in Fig. 22.4(b) allows both predictors to work on the decoded signal, to overcome the above problem. Now if no noise is added in the transmission channel, both will generate identical results. In the circuit shown in Fig. 22.4(c), the predictor is replaced by an integrator/low-pass filter, so making better use of the previous sample value because of the removal of sampling ripple. Although predictions made under any system must contain errors, the advantages claimed for DPCM can be explained as follows. In standard PCM, no prediction is made; or in other words, the next predicted value is zero. Since an approximate prediction must be better than no prediction at all, and because of the good correlation between successive samples, DPCM must lead to a better signal to noise ratio and hence fewer errors.

DPSK (differential phase shift keying). To detect binary PSK signals with minimum errors, the receiver must generate an accurate reference signal. DPSK systems ease this problem by using the phase of the previous bit as a reference for the current bit.

DSM (delta–sigma modulation). Delta modulation systems cannot handle signals with a dc component and have a signal to noise ratio that is sampling and modulation frequency dependent. Both of these disadvantages are overcome by using the delta–sigma technique, which employs an integrator circuit at the input to the modulator. The amplitude of the output pulses then conveys information about the signal amplitude and not just its variations.

FFSK (fast FSK) (see also CPM, p. 190). A frequency shift keying technique that uses two oscillators to provide the signalling pattern. The data stream then switches between the two frequencies in a discontinuous manner.

GMSK or GMFSK (Gaussian minimum (frequency) shift keying) (see also MSK, below). If the binary pulses are pre-filtered (before modulation) so that they have a Gaussian energy distribution, the resulting signal will have a greater spectral efficiency and be more resilient to co-channel interference.

MSK (MFSK) Minimum (frequency) shift keying. In order to minimise the spectral spread of the modulated signal about its limiting values, each binary pulse is pre-filtered before modulation to produce a sinusoidal shape.

Quadrature amplitude modulation (QAM) digital (see also Constellations, p. 160 and Analogue QAM, p. 188). The concept of quadrature phase shift keying (QPSK) can be extended as shown by Fig. 22.5(a). Eight separate phasors obtained from the same carrier frequency can be used to represent eight unique binary code patterns, each of three bits, thus theoretically doubling the information rate provided by a QPSK system. By allocating neighbouring vector codes according to a Gray coding scheme, any decoding errors can more easily be corrected. Although 3-bit sequences provide useful bandwidth compression, they are not particularly compatible with 8-bit bytes.

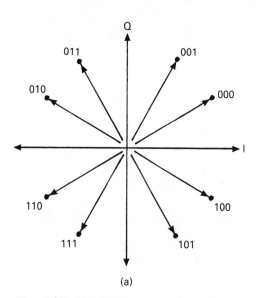

(a)

(b)

Figure 22.5   (a) 8 QPSK modulation signal space; (b) 16-QAM signal space.

Further bandwidth compression can be achieved by using 16-QAM, in which each of eight phasors separated by 45° can be amplitude modulated to one of two levels. This provides 16 phasors with each one representing a unique 4-bit binary code group. An alternative method uses four phasors, each with four permitted amplitude levels, to again provide 16 unique code patterns. In both cases, each transmitted symbol represents four bits.

Figure 22.5(b) shows a further version of 16-QAM, where each quadrant is divided into three vectors (with one permitted, two alternative amplitudes) to provide 16 unique 4-bit code groups as indicated. By scaling the x and y axes according to Gray coding, most of the neighbouring vectors, except X and Y, differ by only one bit. Using forward error correction, this arrangement provides about 2 dB improvement in the signal to noise ratio margin.

256-QAM systems giving eight bits per transmitted symbol are already in limited use and 1024-QAM is being researched. Figure 22.6 indicates the way in which 16-QAM signals are demodulated. The carrier recovery circuit ensures that the local carrier oscillator is locked in frequency to the incoming signal. In-phase and quadature versions of this signal are applied to the I and Q ring demodulator stages along with the input modulated signal. After bandpass filtering, the signals at point A have an analogue form but of discrete amplitudes. Synchronism for the data rate clock circuit is also obtained from point A. The two data slicer circuits recover the original digitally formatted signal and this is then progressively decoded (points A, B, C and D) to provide the original bit stream.

QPSK (quadrature phase shift keying). In this case, each quadrature phase shift of the carrier is used to convey two bits of data in the same bandwidth as one bit. Because channel noise can give rise to spurious phase changes in the signal, it is usual to arrange for sequential bit patterns to change in a Gray code pattern, where each 90° of phase swing represents a change in just one bit.

RBQPSK (reduced bandwidth QPSK). Conventional quadrature phase shift keying signal that has been heavily filtered after modulation, in order to significantly reduce the bandwidth required in the transmission channel.

RFSK (relative FSK). Initially the data signal generates normal FSK modulation so that $f_c + f_s$ represents a one and f_c a zero, where f_c and f_s represent a sub-carrier and shift frequencies, respectively. These two signals are then applied together with a carrier oscillator frequency f_o as inputs to a balanced modulator. The outputs thus become $f_o \pm (f_c + f_s)$ for a 1 and $f_o \pm f_c$ for 0. There are thus two pairs of side frequencies in the modulation spectrum, the inner pair representing logic 0 and the outer pair logic 1. The bandwidth is therefore determined by a 1 signal in the normal FSK manner. The use of two signals allows the signal and noise power to be separated at the receiver to produce a better signal to noise ratio and hence a lower error rate.

Spread spectrum modulation (see also Spread spectrum systems, p. 312). There are conditions when it is more important to transmit data accurately than conserve bandwidth. In these cases it is possible to use a particular binary pattern and its inverse to represent logic 1 and 0 respectively. Such signals can be resolved with little error even over very noisy channels.

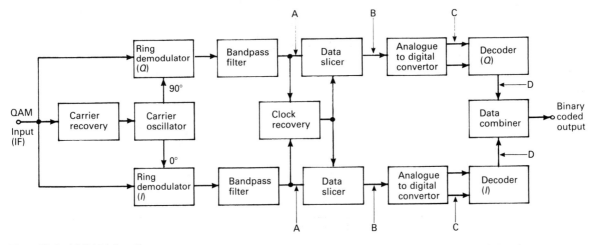

Figure 22.6 16-QAM decoding sequence.

22.2 Demodulation

Historically, the technique used to recover the information from a modulated waveform was described as *detection*. In fact, the actual demodulator stage of a superheterodyne (superhet) receiver is often described as the second detector, the first one being the frequency changer or mixer stage. In the interests of minimum signal distortion, analogue demodulators should have a linear characteristic. The different forms of modulation require different types of circuit, although there is often a similarity between those used for the digital version of a corresponding analogue modulation.

22.2.1 Analogue demodulation

Amplitude demodulation

The two common types of linear detector used are the *envelope* or *non-coherent* and the *synchronous* or *coherent* circuits. The former circuits are usually constructed from rectifier and filter stages and used for full carrier double sideband or vestigial sideband signals. The latter consist of multiplying stages and are used for suppressed carrier signals.

Envelope detector. Rectification of a complex AM waveform produces three signal components: the wanted modulation, a dc component proportional to the input signal amplitude and a carrier frequency component. If there is any nonlinearity in the detector characteristic, this will generate harmonics of both the wanted signal component and the carrier. It is a fairly simple matter to separate out the wanted from the unwanted signal components with resistance/capacity filter networks that have appropriate time constants.

Product detector. This method is very similar to multiplicative mixing where two signals are multiplied together in a nonlinear device. If one of the inputs is either a single or double sideband suppressed carrier signal and the other exactly the original carrier frequency, the output consists of sum and difference frequencies, of which the latter is the original modulating signal. The components can be separated out using simple filters. Any frequency error in the regenerated receiver carrier will produce distortion of the modulation component. In a similar way, a phase error will also create problems. If the phase error reaches 90°, the modulation signal will disappear completely. Because of this difficulty, some modulation schemes allow for a carrier signal, typically reduced by 26 dB from its original level, to be included in the complex waveform. This can then be used at the receiver, to synchronise the local oscillator source.

Vestigial sideband (VSB) demodulation. The VSB signal can be demodulated using simple envelope detection without undue distortion, provided that the signal spectrum has been pre-filtered to be *skew-symmetric* about the half amplitude level. This is shown in Fig. 22.7, where if area 'a' is equal to area 'b', there will be no distortion in the low frequency range of the modulating signal component.

Frequency demodulation

Changes of frequency due to the modulating signal have to be converted into changes of amplitude. This can be achieved simply by using a parallel resonant circuit tuned to one of the FM sidebands and then using envelope detection to recover the modulation. However,

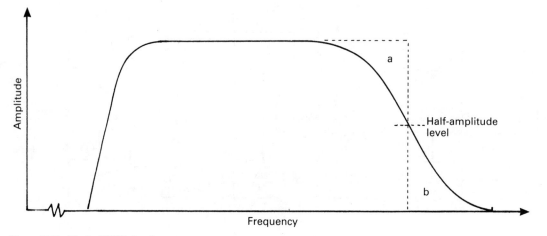

Figure 22.7 Typical VSB signal spectrum.

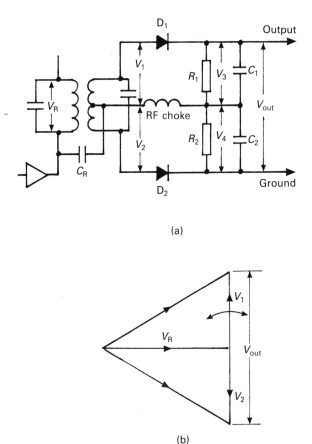

(a)

(b)

Figure 22.8   Foster–Seeley discriminator: (a) circuit diagram; (b) phasor diagram.

the sensitivity and linearity of *slope detection*, as this is commonly known, is usually unacceptable.

Foster–Seeley discriminator. A typical circuit diagram for this FM demodulator is shown in Fig. 22.8(a). This is fed from the intermediate transformer (IFT) primary winding, across which is developed a reference voltage V_R. This is coupled by capacitor C_R into the secondary circuit to develop an in-phase voltage across the RF choke. In addition, electromagnetic action results in two further voltages of equal magnitude, V_1 and V_2, being developed as shown. At resonance, V_1 and V_2 are 90° out of phase with V_R and are always anti-phase to each other due to the centre tap. The magnitude of the voltages applied to the output load circuit that consists of R_1C_1 and R_2C_2, via the two diodes, are due to the phasor sums $V_R + V_1$ and $V_R + V_2$. Thus at resonance, both diodes conduct equally to produce equal but opposite polarity voltages across the two halves of the load, so that the output voltage is zero. When the input frequency swings above resonance, the IFT primary circuit impedance becomes inductive so that the phase of

V_1 relative to V_R increases, while that of V_2 decreases. The effect is to cause the phasors V_1 and V_2 to swing clockwise about the end of phasor V_R as shown in Fig. 22.8(b). This results in the diode D_1 passing more current than D_2 so that voltage across the upper part of the load becomes greater than that across the lower part. The output signal therefore swings positively. When the FM input signal swings below resonance, the reverse action occurs as the primary impedance of the IFT becomes capacitive. This causes the output signal to swing negatively. If the centre point of the output load circuit is grounded, this discriminator will provide push–pull outputs. Provided that it is driven from a pre-detector limiter stage to remove any AM component, this circuit is capable of providing a useful level of output signal with good linearity.

Phase lock loop (PLL) demodulator (see also Threshold extension, p. 196). This circuit consists essentially of a phase detector, a voltage controlled oscillator and a feedback loop as is indicated by Fig. 22.11(c). When fed with an FM signal, the phase detector generates an output whose magnitude is proportional to the difference between the phases of this signal and that produced from the local oscillator. The feedback network thus ensures that the oscillator will be forced to change frequency so that it follows the frequency modulation. The error signal is therefore exactly the same as the original modulation signal which can be recovered from the feedback loop.

Ratio detector. The typical configuration of this circuit is shown in Fig. 22.9, where it will be seen that the reference voltage V_R from the primary of the input transformer is derived directly by transformer coupling of the *tertiary* or third winding. This is combined in the diode circuit with the two anti-phase secondary voltages V_1 and V_2 that are also obtained by transformer coupling. At resonance, the voltages $V_R + V_1$ and $V_R + V_2$ are equal and thus produce equal voltages V_3 and V_4 due to the series aiding effect of the two diodes. When the FM signal swings above resonance, the phase angle between V_1 and V_R increases, while that between V_R and V_2 decreases so that the phasor representing V_1 and V_2 swings clockwise around the tip of V_R as shown in Fig. 22.9(b). V_4 is thus less than V_3 so that the output signal V_{out} becomes negative. The reverse action occurs as the FM input swings below resonance. The magnitude of V_{out} thus depends upon the ratio of the voltages V_3 and V_4. Since the dc load part of this circuit has a long time constant due to the high value capacitor C, the voltage here cannot change very quickly. This has the effect of maintaining the voltage $V_3 + V_4$ essentially constant so that the level of $V_1 + V_2$ is automatically limited. Therefore no pre-detector limiter stage is

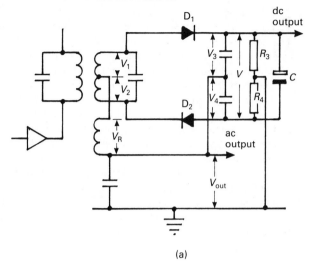

(a)

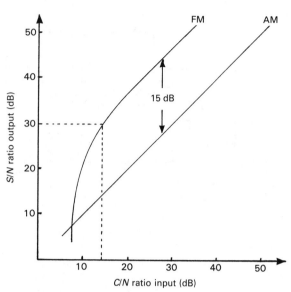

Figure 22.10   FM threshold effect.

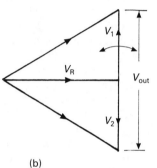

(b)

Figure 22.9   Ratio detector: (a) circuit diagram; (b) phasor diagram.

needed. Since the amplitude of this dc voltage is proportional to the amplitude of the FM input signal, it can be used to drive an automatic gain control system. The circuit provides a larger output than that of the Foster–Seeley circuit, but with a lower degree of linearity.

Threshold extension

Unlike AM detectors, FM demodulators can enhance the signal to noise ratio. The quality of the modulated signal is usually referred to by its carrier to noise (C/N) ratio, and after demodulation by its signal to noise (S/N) ratio. Ignoring any small signal degradation, the AM detector's S/N ratio at the output will be equal to its C/N ratio. Within certain limitation depending on the deviation ratio, an FM demodulator can give a 15 dB improvement over AM, provided that the C/N ratio is above the 14 dB threshold. These features are shown in Fig. 22.10 and the threshold is usually defined by a 1 dB compression effect. That is the value of input

C/N ratio at which the output S/N ratio is 1 dB less than it would be over the linear region. When the C/N ratio falls below this value, characteristic noise spikes becomes apparent, producing *clicks* on audio and *sparklies* on video channels. A number of techniques have been developed that allow this threshold to be reduced to around 5 dB. These function by reducing the detection bandwidth without reducing the signal energy. By combining analogue and digital processing techniques, video demodulators have been designed that have a theoretical threshold of 0 dB.

Dynamic tracking filter demodulator. The tracking filter of the system shown in Fig. 22.11(a) is an active device with a narrow bandwidth and with a centre frequency that is made to track the relatively slowly deviating FM carrier under feedback control. Since the detection noise bandwidth is just that of the tracking filter, the S/N ratio can be improved by about 6 dB.

Frequency modulated feedback (FMFB) demodulator. As indicated by Fig. 22.11(b), this demodulator is a type of frequency changer. Assuming that the mixer input is the carrier frequency f_c, plus the instantaneous deviation f_d, $(f_c + f_d)$, the output from the discriminator is a voltage v, proportional to f_d. When this is filtered and applied to the VCO with a nominal frequency of f_o, it will produce a change of frequency, say kf_d. The second mixer input is thus $f_o + kf_d$. The difference signal at the mixer output is therefore $f_c + f_d - f_o - kf_d$. Now $f_c - f_o$ is a new intermediate frequency f_i, so that the mixer output becomes $f_i + f_d - kf_d = f_i + f_d(1 - k)$. Since k must be less than unity, the deviation has been reduced to allow the use of a narrow bandwidth filter,

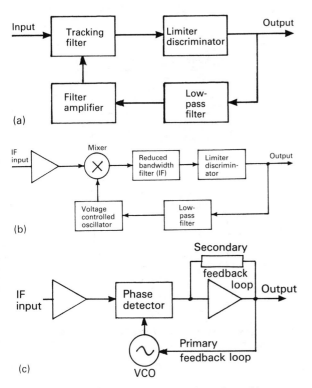

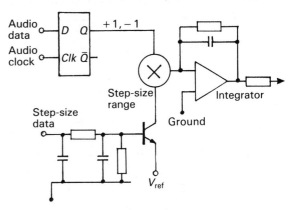

Figure 22.12 Adaptive DPCM demodulator.

Figure 22.11 Threshold extension: (a) dynamic tracking filter; (b) frequency modulated feedback; (c) phase lock loop.

thus reducing the detection bandwidth and hence the noise power.

Phase lock loop (PLL) demodulator. Figure 22.11(c) shows a basic PLL modified by the addition of a secondary feedback loop filter. Due to the feedback, the VCO is made to follow the deviations of the input FM signal. Its driving voltage is thus dependent upon the deviation, and so represents the modulation component. This is the basic action of the FM demodulator. The additional narrow bandwidth secondary feedback loop, in the form of an active filter, reduces the detection bandwidth to improve the S/N ratio.

Phase demodulation. Phase modulated signals can be demodulated using FM techniques. However, to produce an output proportional to the change of phase, the usual discriminator needs to be followed by an integrator circuit.

22.2.2 Digital demodulation

In general there are two methods of demodulation available for each modulation scheme, either coherent (synchronous) or non-coherent (asynchronous). The former is the preferred method because it typically

leads to a 3 dB improvement in signal to noise (S/N) ratio, with a corresponding improvement in bit error rate (BER).

ADPCM (adaptive differential pulse code modulation). After demodulation, the digital signal is separated into audio data bits and step-size control bits. The general principle of the technique is shown in Fig. 22.12. The audio bit stream is converted into bi-polar pulses and clocked into a multiplying stage, the step-size bits being used as the multiplying constant. The resulting signal is then converted into analogue format with a leaky integrator. This device is used to allow the output signal to follow the high frequency components of the original audio.

ASK (amplitude shift keying). Two non-coherent techniques are available to demodulate what is basically an amplitude modulated (AM) signal. If the input signal is large enough, then a simple diode/filter circuit can be used. However, for signals with an amplitude of less than about 0.7 V, a square law detector is more efficient. As shown in Fig. 22.13(a), the squaring action produces signal components at the data rate $f(t)$ and the second harmonic of the carrier. This latter is easily removed by a low-pass filter. The coherent method is basically the product detector, where the input signal is multiplied by its own carrier frequency. As shown in Fig. 22.13(b), multiplying again gives data signal $f(t)$ and the second harmonic of the carrier.

DPSK (DPCM) (differential phase shift keying). Only coherent techniques can be used with all forms of phase shift keying. For standard PSK demodulation, the decoder has to detect the input signal and compare it with a highly stable reference source. Any instability in this locally generated signal degrades the error rate.

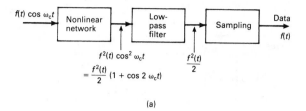

$$f^2(t) \cos^2 \omega_c t$$
$$= \frac{f^2(t)}{2} (1 + \cos 2\omega_c t)$$

(a)

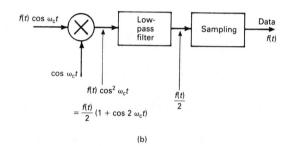

$$f(t) \cos^2 \omega_c t$$
$$= \frac{f(t)}{2} (1 + \cos 2\omega_c t)$$

(b)

Figure 22.13 Demodulation of ASK signals: (a) non-coherent; (b) coherent.

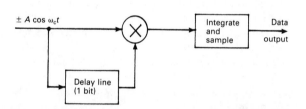

Figure 22.14 Delay line demodulation of DPSK (DPCM) signals.

With DPSK, this problem is avoided. Since phase differences between successive bits are used to carry the data, it is only necessary to compare the input data stream with itself, delayed by one bit period. The basic principle for this is shown in Fig. 22.14.

FSK (frequency shift keying). As indicated by Fig. 22.15(a), non-coherent demodulation involves producing two channels in the decoder, each processing the 0 and 1 frequency respectively. Envelope detection in each channel then produces a pulse to coincide with the transmission of each 0 or 1. The output bit stream is finally produced by combining the two channel outputs. The coherent technique uses two product detectors, one for each channel or bit carrier frequency as shown in Fig. 22.15(b). The decoded data stream is again obtained by combining the two channel outputs.

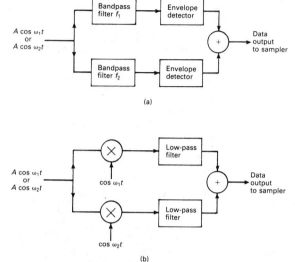

(a)

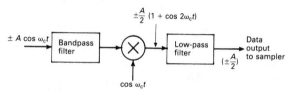

(b)

Figure 22.15 Demodulation of FSK signals: (a) non-coherent; (b) coherent.

Figure 22.16 Coherent demodulation of PSK signals.

PSK (phase shift keying). Only coherent decoding can recover the data from such a modulated signal. Figure 22.16 shows how such a signal is pre-filtered to narrow the detection bandwidth and minimise the noise, before being input to a product detector stage. Low-pass filtering is used to recover the data stream from the unwanted harmonic components of the carrier frequency.

QPSK (quadrature phase shift keying). Four-phase PSK doubles the bit rate of PSK systems without increasing the transmission bandwidth; however, the reduced spacing between signal phases increases the likelihood that noise can generate false decoding. Figure 22.17 shows the basic principle of a suitable decoder. The input signal is multiplied by sine and cosine versions of the carrier frequency in product detectors and the outputs are low-pass filtered. After comparison in threshold level detectors, the signals are combined to provide the data stream. The upper limb of the decoder decides whether the signal falls within the upper or lower

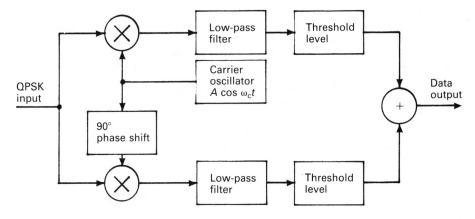

Figure 22.17 Demodulation of QPSK (4-phase) signals.

half of the phasor diagram, while the lower section decides whether it is in the left- or right-hand half.

RFSK (relative frequency shift keying). At the receiver, the two offset carriers that represent the data are down-converted to a suitable intermediate frequency (IF) and input to a product detector. This output produces sum and differences frequencies in the normal way. These are separated using suitable filters and the difference component is up-converted, filtered and input to a ratio detector to recover the data from the frequency changes. It is claimed that the sum channel contains only noise and this can be used to provide a gain control voltage to the IF stage to reduce the overall noise level. It is further claimed that this technique increases the signal level at the same time as it reduces the noise level, thus improving the S/N ratio and hence the BER from two directions simultaneously.

22.3 Modulation-efficiency systems

In broadcasting, the high cost of energy worldwide has concentrated attention on transmitter efficiency. With typical high power transmitters now exceeding power outputs of $500\,\mathrm{kW}$, even 10% saving contributes significantly to reduced station operating costs and a big environmental improvement. Greatest savings can be made with the poor efficiency AM systems, but worthwhile economies can also be made with high power FM transmissions. In addition to using efficiency modulation schemes to save power directly, there is also a desire to change to single sideband operation as soon as practicable. This mode is not only more power efficient, but also makes better use of the spectrum allocations. Power efficiency systems use some form of amplitude compression at the transmitter and comple-

mentary expansion in the receiver (see also Noise, companding, p. 223).

Digital modulation. The radio frequency (RF) power output stages of this type of transmitter consist of many relatively low power solid state amplifier (PA) stages coupled in parallel to produce the required total output power. These are operated in a Class D switching mode with the transitions occurring at zero current crossing points. During the negative excursion of the audio signal all PA stages are turned off, the negative half-cycle of this signal being replaced in the modulation by the natural flywheel action of the tuned tank circuit. The audio signal is converted into digital format using a high speed *flash* analogue to digital (A/D) convertor, typically using 12-bit words. A triangular dither signal is added to the audio before conversion in order to minimise the effects of quantisation noise. The 12 bits may be divided into two groups. The six least significant bits (LSB) retain their binary weighting, while the six most significant bits (MSB) may be alternatively weighted using a look-up table stored in a read only memory (ROM). This arrangement is used because it provides a smoother control over the output power. The digital audio signal is then used to switch on the appropriate number of output stages to provide the required instantaneous output power. Doubling the number of PA stages switched on, doubles the output voltage and quadruples the output power. Operated in this way, the system requires no high power modulated stage and RF power stages operate as the digital to analogue (D/A) convertor to produce the high power AM RF output signal. Apart from power efficiency improvements in the order of 90%, the parallel operation of many low power PA stages improves the system reliability. If one stage fails, the system continues to operate at only slightly reduced output power.

Doherty modulation. This concept, that has been considerably modified, was developed in the 1930s to provide a more efficient alternative to the AM techniques then in use. The system combines the outputs of two or more linear radio frequency (RF) power amplifier (PA) stages, each operating in the push–pull mode. Coupling is provided through an impedance inverting coupler that may consist of a quarter wavelength of transmission line or the equivalent L or T network. At low output levels and below some transitional threshold, the first PA stage is operating with the second cut-off. When this stage reaches saturation and maximum efficiency, the second PA stage starts to conduct and operates up to the peak output voltage level. In the classical Doherty system using two PA stages, the transition occurs at half the maximum output voltage. The power efficiency factor over standard AM techniques is improved by a factor of at least two for a two-stage system. Adding a third stage further improves the efficiency so that the average and peak efficiencies are practically equal at about 70%. In addition to use with full double sideband AM transmissions, the technique can be usefully applied to single sideband (SSB) transmissions.

Dynamic amplitude modulation (DAM) or volume dependent amplitude modulation (VDAM). With this concept, the carrier level is controlled by the depth of modulation. The carrier is operated at reduced level up to 40% modulation, the region where most of the transmitted power lies in the wasteful carrier. Beyond 40% the carrier level is allowed to increase linearly with the depth of modulation. This ensures that the average depth of modulation is always higher than it would be for normal AM, but producing a power saving at low modulation levels. The major disadvantage lies in the fact that the receiver signal to noise ratio is somewhat reduced at low modulation levels.

Dynamic carrier control (DCC). This technique, which produces carrier level compression at the transmitter, relies on the normal automatic gain control (AGC) function of the receiver to provide the necessary expansion. The full carrier level is transmitted during low levels of modulation, to ensure a good signal to noise (S/N) ratio. As the audio level increases, the carrier level is automatically reduced. The corresponding worsening of the S/N ratio at the receiver is masked by the higher level of output signal. The transmitter power consumption thus decreases with increasing depth of modulation. A significant power saving can be made during the transmission of modern music programmes that consistently produce higher levels of modulation. From the reception point of view, the system works better in areas of high signal levels and with medium

wave broadcasting. The carrier control signal has to have fast attack and slow release times in order to correctly control the level of compression.

Envelope elimination and restoration (EER). See Pulsam system (below).

Pulsam system. The overall power efficiency of a high frequency (HF) band transmitter can be improved if a switched mode high level modulator stage is used. A single sideband (SSB) signal generated by either of the three methods can be resolved into two components, an amplitude or envelope component and a constant amplitude phase component. This latter can be amplified in a high efficiency Class C stage, while the envelope can be separately amplified in a linear stage and used to high level modulate the amplified phase component. The resulting product output represents the reconstructed original SSB signal at a much higher power level. This is sometimes described as the envelope elimination and restoration technique. The major difficulty with this system lies in the matching of the characteristics of the two amplifier channels, this being necessary to avoid distortion and the generation of carrier harmonics. The concept represents the power saving advantage of SSB over full double sideband AM plus that gained by spectrum efficiency.

Pulse duration modulation (PDM). The voltage for the high power radio frequency (PA) amplifier is provided via a series switch whose duty cycle is dependent upon the level of the modulating signal. The switching action allows the PA stage to pass current only when it is in the low output resistance, high efficiency state. Thus the power efficiency increases with depth of modulation. The carrier level is set to deliver full power only at 100% modulation and the level falls back as the average depth of modulation falls. The system thus operates with depth of modulation always close to 100% so that overall power efficiencies in the order of 70% can be achieved. When the modulating signal falls to zero, it is arranged that the minimum carrier level falls to about −6 dB of its normal full power level to avoid complete loss of signal.

Pulse step modulation (PSM). A single sideband (SSB) signal generated by either one of the three methods is separated into its envelope and phase components in the EER manner. The amplitude component is then used to digitally switch on solid state power amplifier (PA) modules as the depth of modulation increases. The phase component is superimposed upon the radiating carrier frequency and the two components recombined at a higher power level in the final PA stage. The concept thus combines digital processing with the basic analogue principles to produce output power efficiencies of around 75%.

22.4 Miscellaneous terms

COFDM (coded orthogonal FDM) (see also OFDM, p. 202). This variant of OFDM uses fast Fourier transform (FFT) processing and convolution error coding at the modulator stage. When this is combined with complementary FFT processing and Viterbi decoding at the demodulator, the overall bit error rate is very low.

Since the COFDM spectrum has noise-like properties and the signal can be transmitted at relatively low power, it produces very little adjacent channel interference.

Controlled carrier modulation (floating carrier modulation) (also known as Hapug). A compound form of amplitude modulation (see Modulation-efficiency systems, p.199). The carrier signal is amplitude modulated in the normal manner and the complex wave is then further amplitude modulated by a signal that is the short-term average of the modulation. This results in an almost constant depth of modulation.

Deviation distortion. Distortion produced due to lack of bandwidth, poor linearity or AM rejection in an FM demodulator stage.

Deviation sensitivity. The characteristic of an FM demodulator measured in volts of output signal per hertz of deviation at the input. It is usually quoted as the smallest value of deviation that produces a specified output level.

Differential phase detection. In phase modulated signals the data is carried by the changes of phase of the carrier signal. This is demodulated by comparing each received bit symbol with the previous one that was transmitted.

Double modulation. A sub-carrier may be modulated by one signal and this modulated wave may then be modulated on to a higher carrier frequency (see Quadrature amplitude modulation, p.188).

Exalted carrier reception. A technique that can be used with transmissions where the carrier level has been suppressed in order to save power. This involves filtering off the residue of the carrier, amplifying and limiting it in a side chain to remove any trace of modulation or noise and then using the regenerated carrier in the demodulation process. The technique is useful in minimising the effects of selective fading in a channel or the distortion introduced by asymmetric sideband response.

Foldover distortion. Whenever a carrier is modulated, two sidebands are generated. If the carrier frequency is lower than twice the baseband width, then the high frequency components in the lower sideband become negative. These then fold about zero to produce a range of side frequencies that would have been generated by different baseband frequencies, to produce a form of distortion. Under normal demodulation conditions, these folded components cannot be separated from the wanted components that did not fold.

Incidental modulation. Any nonlinearity in a modulator stage gives rise to distortion and creates secondary modulation. For example, a nonlinear frequency or phase modulator will produce an amplitude modulated component in the complex wave.

Kansas City modulation (alternatively computer user tape system, CUTS). A variation of frequency shift keying (FSK) designed to allow low speed modems to operate over analogue telephone lines. Equal time periods are used but each period contains tone bursts of either four cycles at 1200 Hz or eight cycles at 2400 Hz to represent binary 0s and 1s.

Kendall effect. A kind of aliasing that affects analogue modulated signals. If the modulation baseband overlaps the carrier sidebands because the carrier frequency is too low, interference occurs and demodulation cannot then separate the original modulating signal from the sideband structure.

Linear detection. A demodulation process where the output signal is directly proportional to the input signal level over the operating range of circuit.

Modulation classification. The basic characteristics of radio frequency modulation systems are described by a three-symbol scheme. In general, the first symbol indicates the form of modulation, and the following two, the nature of the modulating signal and the type of information being transmitted.

Amplitude modulation (AM)

A1A On–off keyed telegraphy without the use of a modulating audio frequency.
A1B As A1A but using automatic keying.
A2A Telegraphy by on–off keying of an amplitude modulating audio frequency, or by on–off keying of the modulated signal.
A2B As A2A but using automatic keying.

A3C Facsimile transmission.

A3E Double sideband telephony.

A3F Slow scan television.

C3F High definition television.

H3E Single sideband (SSB) telephony with full carrier and amplitude modulation.

J3E As H3E but with suppressed carrier.

R3E As H3E but with reduced carrier.

Frequency modulation (FM)

F1A Telegraphy by frequency shift keying (FSK) without the use of a modulating audio frequency; one of two frequencies being transmitted at any time.

F1B Automatic telegraphy by FSK without the use of a modulating audio frequency.

F2A Telegraphy by on–off keying of an FM audio frequency; or by on–off keying of the FM transmission.

F2B As F2A but using automatic keying.

F3C Facsimile transmission.

F3E Telephony.

F3F Slow scan and high definition television.

Phase modulation (PM)

G1A Telegraphy by phase shift keying (PSK) without the use of an audio modulating frequency.

G1B As G1A but using automatic keying.

G2A Telegraphy by on–off keying of a PM audio frequency; or by on–off keying of the PM transmission.

G2B As G2A but using automatic keying.

G3C Facsimile transmission.

G3E Telephony.

G3F Slow scan or high definition television.

Pulse modulation (PM)

K1A Telegraphy by on–off keying of a pulsed carrier without the use of an audio modulating signal.

K2A Telegraphy by on–off keying of a modulating audio signal or by on–off keying of a modulated pulsed carrier, the audio frequency modulating the pulse amplitude.

K3E Telephony using AM pulses.

L2A Telegraphy by on–off keying of a modulating audio frequency or by on–off keying of a modulated pulsed carrier, the audio frequency producing pulse width or duration modulation of the pulses.

L3E Telephony with pulse width or duration modulation.

M2A Telegraphy by on–off keying of a modulating audio frequency or by on–off keying of a modulated pulsed carrier, the audio frequency providing position or phase modulation of the pulses.

Q2A As M2A but with audio signal producing angle modulation of the carrier during the pulses.

V2A As M2A and Q2A in combination or modulation produced by other means.

Modulation suppression. If a strong unwanted signal reaches an amplitude demodulator along with the wanted signal, the depth of modulation of the wanted signal is reduced by cross-modulation effects.

OFDM (orthogonal frequency division multiplex) (see also COFDM, p. 201). This concept consists of generating a large number of carrier frequencies (often as many as 500) with equal spacing. Each is digitally modulated with a sub-band of frequencies and then filtered to produce a $(\sin x)/x$ (sinc x) response as indicated in Fig. 22.18(a). The spectra of the individual neighbouring carriers thus overlap in the orthogonal manner shown. When these combine, the total spectrum becomes practically flat as shown in Fig. 22.18(b). The channel capacity approaches the Shannon limiting value and the spectrum behaves as a parallel transmission bus.

The allocated bandwidth is divided into N elementary frequencies and arranged to carry P program channels. There are therefore N/P interleaved elementary carriers which carry sub-band modulation in the manner shown. Either four- or eight-phase PSK or 64-QAM can be used for the digital modulation.

Pilot carrier (tone). In order to simplify the demodulation of single sideband (SSBSC) or double sideband suppressed (DSBSC) carrier transmissions, it is sometimes arranged for a low level of carrier to leak into the modulated signal. This can then be used at the receiver to regenerate and synchronise the locally generated demodulator carrier signal. (Also known as reconditioned carrier reception.)

Pulse interval modulation. Alternative term to describe pulse position modulation. The interval between pulses of a pulsed carrier transmission is varied in sympathy with the amplitude of the modulating signal.

Quiescent carrier modulation. An alternative term used to describe suppressed carrier modulation.

Reduced carrier transmission. See Pilot carrier, above.

Reference pilot. A single tone or frequency included in a frequency multiplex designed to facilitate test and maintenance. This signal is continually monitored and

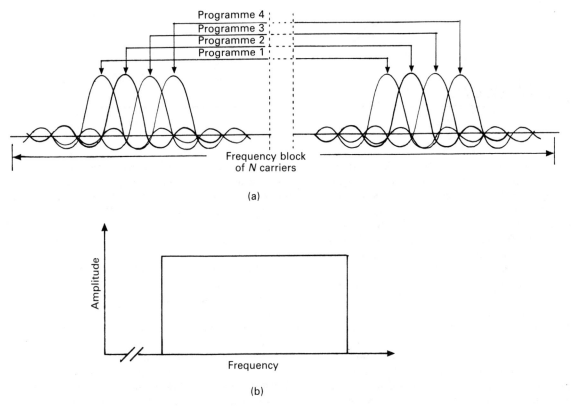

Figure 22.18 (a) Time domain of OFDM carriers; (b) OFDM frequency spectrum.

any change in its characteristics can be automatically verified and used to operate an alarm system.

Regulating pilot. A number of single frequency, sinusoidal signals may be distributed throughout a frequency multiplex structure. These are monitored at strategically placed points in the system and their levels used to automatically control the system gain/frequency characteristics.

Wobble modulation. A jargon term used to describe frequency modulation. Usually associated with the jitter/dither signal added to low frequency systems where the modulating signal has a dc component.

22.5 Useful references

Beech B. and Moor S. (1984) Threshold extension techniques. IBA Experimental and Development Report 130/84. Independent Broadcasting Authority (IBA), UK.

Betts J.A. (1978) *Signal Processing, Modulation and Noise*, Hodder and Stoughton, London.

Boksberger, H.U. (1986) SSB operation with the BB&C high power PSM transmitter. *Proc. IBC'86*, Brighton, UK, pp. 184–188.

Coates R.F.W. (1985) *Modern Communication Systems*, 2nd edition, Macmillan, London.

Connor F.R. (1977) *Introductory Topics in Electronics and Telecommunication, "Modulation"*, Edward Arnold, London.

Doherty W.H. (1936) A new high efficiency power amplifier for modulated waves. *Proc. IRE*, **24** (8), 1163–1182.

Lewis G.E. (1992) *Communications Services Via Satellite*, 2nd edition, Butterworth-Heinemann, Oxford, UK.

Pace S.D. and Brett J. E. (1986) Performance and implementation of SSB on a high power broadcast transmitter. *Proc. IBC'88*, Brighton, UK, pp. 133–136.

Raab F.H. (1987) Efficiency of Doherty RF power-amplifier systems. *IEEE Transactions on Broadcasting*, **BC-33** (3).

Schwartz M. (1980) *Information Transmission, Modulation and Noise*, 3rd edition, McGraw-Hill, New York.

Schwartz M., Bennett W. R. and Stein S. (1966) *Communication Systems and Techniques*, McGraw-Hill, New York.

Tyler V.J. (1958) A new high-efficiency high-power amplifier. *Marconi Review (UK)*, **12** (130), 96–109.

Weaver D.K. (1956) A 3rd method of generation and detection of SSB signals. *Proc. IRE*, **44** (12), 1703–1705.

Williams W.F. (1986) AM companding; a technique for dynamic carrier control of an AM transmitter. *Proc. IBC'86*, Brighton, UK, pp. 179–183.

23 Networks

Until the expansion of the use of computers in communications in the 1970s, typical systems were centred upon the telephony networks on the PTTs (the national post, telegraph and telephony organisations), public switched telephone networks (PSTN), and radio services. The PTT networks connect individual subscribers to a second specified or addressed subscriber, so that the service is often described as *narrow casting*. While the radio services might well be intended for a specified receiver, the very nature of the transmission precludes specific addressing, so that the term *broadcasting* applies. The introduction of computer to computer communications completely changes the concept of a network. The communication of a computer with its peripheral devices forms the basic or primeval network, messages being passed over the data bus under the direction and control of the address and control buses. Many bus-expansion systems have been developed, each for a specific application. At a higher level, all the computer-associated elements on a particular site may be interconnected to form a *local area network* (LAN). By using the carrier services of radio and the PTTs, this concept can be expanded to cover an area as large as 50 km diameter. This gives rise to the term, *metropolitan area network* or MAN. Further expansion makes use of the international radio and PTTs networks, to form a *wide area* (WAN) or global network. The actual description of a LAN, MAN and WAN is not easy to define as one definition tends merge into another. At the basic level, communications is often at baseband, thus forming a *narrow band* system. However, many popular networks are *broadband* and use radio frequency multiplexing to increase the throughput of data. Networking requires the use of a *modem* or *mo*dulator/*dem*odulator device to handle the data transfer. Different types of networks may be interconnected by using *gateways* placed at strategic points. The physical communications medium varies from twisted pair or coaxial copper cables, through radio links to optical fibres. The latter has the particular advantages of freedom from interference and electromagnetic effects and is capable of operating in very broadband networks. Networking makes better use of expensive peripheral devices, as these can be shared

between many users. The technique also increases the rate at which information can be disseminated throughout an organisation (see also Telephony and associated systems, p. 327, ISDN, p. 91, and Pulse code modulation, p. 91.)

23.1 Network management systems (NMS)

In practice, the many functions necessary to control and manage a communications network can be automatically carried out by a dedicated computer terminal attached to the network. The important features of such a scheme include the following:

Administration. This task is associated with the interface between users and the network. To ensure that the system is *user friendly*, it is important to include a menu of *help* instructions. For a public network, subscription and authorisation records need to be maintained in order to manage billing statistics, etc.

Control. An NMS scheme can carry out routine equipment testing to enable the network to be reconfigured to reduce the impact of fault conditions. If control is provided at a single node, the maintenance of a central record of the *quality of service* (QOS) means that the network service staff can be more effectively deployed.

Monitoring. In larger networks, it is important to be able to recognise areas of limitation. Therefore the system response time, bit error rates and noise conditions at various nodes, average and peak traffic flow, the loading of specific terminals and general network statistics need to be acquired. These can be of use when network changes or upgrading become necessary.

Network planning. During the operational lifetime of a network, it can become necessary to add additional nodes and modify the topology. Software packages that can be run on the central control computer terminal are available and these allow such development work to be planned for optimisation of resources. Provided that control and monitoring has been effective, the network statistics provide the basic input data for such a program.

23.2 Network topologies and protocols

Network topologies

Bus or highway systems. All communicating elements of the system are connected to a common cable in the manner shown in Fig. 23.1. Each node has a unique address and responds only to data that is sent to it. In network terminology, a node is defined as a connection, tap or branch point. As the cable acts as a transmission line, its two ends must be suitably loaded to prevent signal reflections, and because it carries no power, it is said to be inactive. The network is controlled by a common clock which must be positioned near to the middle of the cable. Its frequency then controls the rate of data transfers. If the network length is increased beyond some critical value, then the clock rate has to be reduced to maintain an acceptable bit error rate (BER). Each node is allowed direct access to the network, but in contention with all the other devices. Each transmitter device listens continually to the network. If a collision of data is detected, then transmission ceases and the control protocol takes over. It is usually easy to extend this type of network, or add or remove a terminal without interrupting the data flow.

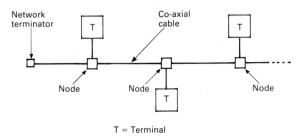

Figure 23.1 Bus or highway network.

Ring networks. As the name implies, all the communicating devices are connected to a *ring*, with each device or terminal being equipped with an access node that has a unique address in the manner shown in Fig. 23.2. Often no one node is responsible for network control and each has equal status. Transmission is normally uni-directional, although some bi-directional networks have been implemented. The ring usually carries power for the node interfaces and each node acts as a data regenerator, receiving the data on one side and

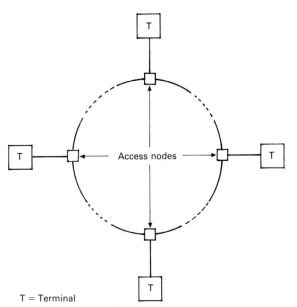

T = Terminal

Figure 23.2 Ring network.

retransmitting it to the next node in sequence. If the ring is broken either under fault conditions or to modify the terminals, then the data flow is interrupted. *Dual* and *braided* rings which have two parallel signal paths have been devised to overcome this problem.

Star networks. Unlike the bus and ring networks, all the message switching and control is carried out at the central *hub*, as indicated in Fig. 23.3. Since terminals can only communicate with each other via the hub, the configuration only finds favour where the data signals

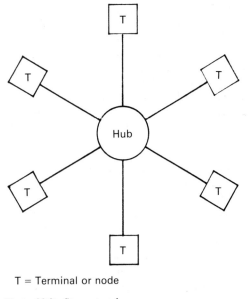

T = Terminal or node

Figure 23.3 Star network.

flow chiefly in one direction. Such an application would be the distribution of a television service.

Network protocols and access control

Modem protocols. A number of international standards have been derived, often from earlier manufacturers' *de facto* standards. The following is a general guide to these concepts. The data to be transmitted is organised into *packets* by a device described as a *packet assembler/ dissembler* (PAD). Each packet contains a *header*, a number of data bytes or octets, followed by a *footer*. The header can start with a series of sychronising bits, described as a preamble, followed by *destination and source addresses*. If the data block is variable, then the header may also contain information about the block length. The footer carries either parity bits or *cyclic redundancy check* bits. At the receiver, the header and footer are stripped off and the check sum recalculated. If this is in agreement with the transmitted check sum, then the data can be accepted as correct. Otherwise a *request for repeat transmission* can be made. *Smart* modems containing an embedded microprocessor are available and these can be used to maximise the data throughput under noisy conditions in an adaptive manner. The packet size may be varied according to the error rate. As the error rate rises, the packet length is reduced. Also such modems are capable of operating at a range of bit rates, starting at the highest rate, and automatically falling back to a lower rate under noisy conditions.

Hayes AT command set. This originated as a *de facto* standard devised by Hayes Microcomputer Products Inc. for use with smart modems and is now in the public domain. The term 'AT' is described as the *attention* prefix, and this precedes nearly all the commands in the language. This allows the user to execute a number of specified commands, either via software or through the computer keyboard. The modems are equipped with inbuilt *intelligence* and designed to operate with any personal computer. This obviates the need to adjust the computer interface by manually changing switches when it is required to communicate with a modem with different parameters.

Link access protocol-modem (LAP-M). This is a high level data link control (HDLC) procedure and forms a primary protocol related to LAP-B and LAP-D of the ISDN system. It provides error correcting procedures for modems using synchronous to asynchronous conversion. It is the primary protocol for modem error control adopted by the International Standards Organisation (ISO) and the Consultative Committee for International Telegraph and Telephones (CCITT) and supports the Hayes and MNP standard of V42

protocols. The standard ensures that future developments such as data compression and remote configuration are allowed for.

Microcom networking protocol (MNP). Originally a *de facto* standard developed by Microcom Inc., USA, that has been absorbed into the ISO/CCITT standards. Built onto a Hayes standard modem, this handles file transfers character by character rather than on a packet basis. By transferring the function of error control to modems with embedded intelligence, rather than leaving it to computer software, data throughput can be increased. The protocol is based on nine layers or levels. Levels 1 to 4 represent increasing data rates up to 1200 bit/s, with automatic fall-back to a lower level under noisy conditions. Level 5 is basic data compression. Level 6 is universal link negotiation statistical duplexing. Level 7 is enhanced data compression. Level 8 is piggyback acknowledgement – multiple selective negative acknowledgement. Level 9 is enhanced universal link negotiation.

Network protocols

Contention techniques. This allows each terminal exclusive use of a network, but in contention with all the other terminals under the control of a technique known as *carrier sense–multiple access* (CSMA). Carrier sense, because each terminal listens to the network to detect the absence or presence of a carrier. Multiple access, because all terminals share the same transmission medium. Thus a device will not transmit its own data until it detects a gap in the transmissions. It is possible that two terminals waiting to transmit, both detect the same gap and start to transmit together and cause a data collision. When this happens, both transmitters halt and *back-off* for a time before trying to retransmit. The control of this back-off period varies from system to system.

CSMA with collision avoidance (CSMA–CA). This is based on an empty time slot system. Each node on the network is given a priority listing from a controlling node. If it has data to transmit, it does so only at its allocated time slot. If it has no data available, then this time slot is reallocated to the node with the next highest priority. Extra time slots can be allocated to those nodes that have a great deal of data to transmit.

CSMA with collision detection (CSMA–CD). Data collisions can only occur during the short period after a transmission has started and before the signal has had time to reach all the other nodes on the network. This is the *collision window*. When a collision occurs, transmissions continue until all the nodes on the network recognise the burst of noise as a collision. The back-off period varies from a random interval, to a period calculated from the node address. Thus all terminals will back-off for different lengths of time.

Daisy-chaining. This is a term used to describe a technique where control of the network is passed from one node to the next. It is more formally known as *token passing*, rather in the manner of the *key* retained by the authorised train driver on a single track railway line.

Empty slot. With this protocol, several packets circulate around a ring network and these may be in-use or empty. When a node has data for transmission, it waits until it receives an empty packet. It then loads its data into the packet, with the destination address, its own address and sets a packet flag to *in-use*. This packet is then circulated until it reaches its destination, where it is read. The receiving node then sets a read flag in the packet and retransmits it back to the originator. This node then sets the packet in-use/empty flag to empty, before retransmitting it as an empty packet. A node is not normally allowed to reuse such a packet as this would tend to produce *network hogging*.

Polling. The network is controlled by a polling device that asks each node in turn if it has data awaiting transmission. If it has, then it is given sole use of the network. Various priorities can be allocated to ensure that each node has a fair share of network time. Since slow devices tend to *hog* the network, this concept is not popular.

Register insertion. With this technique, a node waiting to transmit first loads the data into a shift register. When it detects a gap, the register is switched in series with the network so that the packet is transmitted to the next node. Any data on the line then flows through this register until the packet returns to the originator for checking. The register is then switched out of circuit. Since the number of registers on the ring increases with the amount of traffic, propagation delays increase in proportion. This can be countered either by reducing the packet length or by introducing transmit and receive registers. In this latter case, when a gap is detected, the *transmit* register is emptied and any incoming data switched to the *receive* register. When the node finishes transmission, the receive register is connected into the line.

Token passing. A special packet which carries a specific bit pattern, circulates around the ring to act as a *token*. A node waits until it receives this token, which it retains and replaces with its own data for transmission to the next node. The data circulates until it reaches its destination, where it is removed. A *flag* bit in the packet is then set to indicate reception and the packet returned to the sender, where reference to the flag will signify receipt. A monitor station has to be included in the network to check that corrupted data or token does not continually circulate to clog the network.

Alternatively the token may be retained on the network, but with a modified pattern to indicate that it is in use by the node identified by the address in the following packet.

23.3 Network modulators/demodulators (modems)

A modem is the device that is positioned between a data terminal and the transmission medium (cable, fibre, radio). The modem accepts data input in the serial format and then processes this to generate a form suitable for transmission.

In the reverse direction, the modem performs complementary functions. For effective operation, the modem not only has to manage the communications facility, but also needs to have an inbuilt control feature. At the basic level, this involves a form of hand-shaking between the calling and called terminals. This produces a problem with radio links that have to provide a reverse channel. The round trip propagation delay, particularly via satellite links, can be troublesome.

Each modem has to:

- match the channel bandwidth;
- provide synchronism for the data stream at both ends;
- handle the modulation and demodulation processes;
- maintain the high quality or integrity of the data signal.

A wide range of modem types have been manufactured to CCITT recommendation standards. These range from early V21 devices running at 200 bauds, through to V42 at 19.2 Kbit/s and the developing V.fast running at a rate of 28 Kbit/s. By using data compression, V.fast can have an overall effective data rate of about 100 Kbit/s.

Various modulation techniques are employed, ranging from two-tone frequency shift keying (FSK), through variants of phase shift keying (PSK), differential PSK (DPSK) to multi-level quadrature amplitude modulation (QAM).

In general, all modems tend to be backward compatible with earlier devices to ensure an economical upgrade for the end user.

Some of the developing *smart* techniques include:

- *Channel probing*. This is used during the initial handshake period to probe the quality of the channel signal in order to select the optimum data rate.
- *Trellis coding*. This type of forward error control (FEC) can be used with soft decision decoding at the receiver to enhance the bit error rate (BER).
- *Pre-emphasis*. The transmitted signal can be conditioned adaptively according to the existing line signal quality.
- *Nonlinear mapping*. With multi-level QAM, the signal space at the edge of constellations tends to be more affected by nonlinear distortion. Therefore mapping these points to a wider spacing across the constellation improves the system performance.

Recommendation V32.terbo provides an interim technique that is positioned between V32 and V.Fast. This extends the basic V32 convolutional encoding and Viterbi decoding, by using 128QAM (seven bits) to achieve a data rate of 16.8 Kbit/s. This automatically falls back to the V32 rates of 9.6 Kbit/s or 4.8 Kbit/s under noisy conditions.

Recommendation V42 modems use 64QAM (six bits) together with data compression to achieve a data rate of 19.2 Kbit/s over standard dial-up telephone lines. The error control protocol includes a selective *repeat–request* technique. If a number of frames have been detected as being received in error, a single message can request the repeat of a given number of frames.

Modem data compression techniques. The bulk of the information handled by network modems is contained in text files. This means that there is a significant *a priori* information that most of the symbols used will be contained within the standard 128-element ASCII (American Standard Code for Information Interchange) code set. It is due to the probability of occurrence that this knowledge provides, that a number of data compression algorithms have been developed. The Lempel/Ziv algorithm recognises that many of the symbols will have a high probability of occurrence, for example, the vowels a, e, i, o, or u. Each of these can be represented by a short code with a maximum of three bits, while the character z would be allocated a full 8-bit code. This particular algorithm has been refined by British Telecom (BT). The BTLZ algorithm searches for long strings of characters such as *the, and, there*, etc., and then allocates short codes to these occurrences. The most commonly occurring words are held in an adaptively updated memory known as a dictionary. The high data rate is achieved by starting the search for string similarities that are longer than expected. The larger the dictionary, then the greater will be the throughput of data. While the LZ algorithm operates on a character comparison based technique, the BTLZ algorithm utilises character string comparisons.

The MNP7 (Microcom Networking Protocol) algorithm is similarly based on a dictionary code book. This calculates the probability of a particular string occurrence and then selects the equivalent code word from a look-up table. This technique is somewhat more memory intensive than the BTZL algorithm.

23.4 Network security (see also Computers in communication, p. 66)

The most serious affliction for any networked system is the intrusion by *hackers*. At best, this represents an illegal access to secret files and sensitive databases, but more importantly, it provides an opportunity for the theft of corporate information and money. In addition, the vindictive and malicious hacker may also introduce a virus that can create havoc and confusion throughout a database. It is therefore important that the network of any computerised database should be adequately protected to prevent any unauthorised access.

The *call-back* system provides a simple form of control. Each user has to make a call to the access control and provide a password. This is checked against a library listing to verify the password and ascertain the permitted level of access. If the caller fails to meet the required criteria, control terminates the call. If the caller is recognised as legitimate, he will be asked to terminate the call, whereupon, control will call-back to a pre-arranged address number to establish the communications link.

Physical keys and passwords alone form a poor defence because both may be easily bypassed by a determined attacker. In the most secure system, the data will be stored in an encrypted form (see also Encryption and decryption, p. 100).

For telephone line based systems, leased lines are more secure than dial-up lines because these simply connect two specified locations. In a similar way, private packet based systems are more secure than public PTT networks. Local area networks (LANs) are the most vulnerable to attack and should be protected by the use of encryption, passwords and physical keys.

23.5 Network systems

Bus extension systems

The central processor unit within a computer utilises a number of parallel path networks for its own communication and control. These paths which cater for *data*, *addresses* and *control* are referred as the *bus system*.

Extending these buses into the outside world expands the range of influence of the processor. In general, this is achieved by constructing the computer as a series of plug in cards mounted in a *back-plane*. Bus extension concepts developed from a need to integrate various modular electronic devices into a computer controlled environment, modular automated test equipment (MATE) being a particular influence. The rate of data transfers is governed by the bus length so that increasing the area of influence reduces the data throughput. Variations on this basic concept has caused many *de facto* standards to be developed. Some of these that have achieved the status of international standards have been developed either by a specific manufacturer or by groups of manufacturers with a common interest. Most but extension schemes are covered by IEEE (Institute of Electrical and Electronic Engineers) standards. Because of the wide variations and continually changing technology, it is impossible to provide an exhaustive discourse on this topic.

AT/XT bus (PC bus). This represents the main communications link within IBM-compatible personal computers. The original XT version was developed for 8-bit data systems, with 20 address lines to handle 1 Mbyte of addressable space. It has a maximum data transfer rate of 800 Kbyte/s. The AT version (also known as Industry Standard Architecture (ISA)) is fully compatible with the XT bus and was provided for 16-bit systems with 24 address lines to cater for 16 Mbytes of addressable memory. This raised the maximum data transfer rate to 1.6 Mbyte/s.

Bitbus. This is related to the later development, Fieldbus. It is based on a master/slave hierarchy and employs a self-clocked synchronous system with two self-generated data rates of 62.5 Kbit/s and 375 Kbit/s. The maximum path lengths at these rates are 1200 m and 300 m, respectively. By introducing suitably spaced repeaters into the network and using the lower data rate, the path length can be increased to 12 km.

By using an external phase locked clock coupled via a twisted pair of cables, the data rates can be increased to 500 Kbit/s and 2.4 Mbit/s. However under these conditions, the maximum path length is restricted to about 30 m.

The system uses SDLC (synchronous data link control) protocol which allows for 250 nodes, but by using the RS 485 interface this restricts the system to 128 nodes per cable. When using repeaters, the full complement of nodes can be handled at 62.5 Kbit/s, reducing to 56 nodes at the 375 Kbit/s rate.

EISA (extended industry standard architecture) bus. This is a 32-bit version of the AT bus, with a maximum data transfer rate of 33 Mbyte/s. It functions as a shared resource in a PC-based network with its architecture divided into memory and I/O (input/output) buses.

Fieldbus. A network system designed for the management and control of a manufacturing or processing plant. It is designed around the three lower layers of the OSI model (data link, application layer and physical layer) and is compatible with manufacturing automation protocol (MAP). The use of two way digital communications between the controlling computer and the system transducers that have embedded intelligence, allows for their continual re-calibration and drift compensation. On-line diagnostics can be applied, with the computer running self-test programs. In large area plants, this central monitoring allows remote sensing and rapid location of any fault conditions. The network even allows for the system to be reconfigured from a central control. The communications protocol is described as HART (highway addressable remote transducer). This allows several sensor devices to be multiplexed on to a single network cable. Two alternative data rates are specified: a low rate of 31.25 Kbit/s and a high rate of 1 Mbit/s.

Futurebus. This represents an architecture designed for use with sub-systems that contain embedded processors and is considered to be the ultimate level of back-plane extension. The concept supports the growing needs of artificial intelligence (AI), computer-aided design (CAD), plus computer graphics and animation. The standard provides for 32-, 64- and 128-bit processors with data transfer rates as high as 100 Mbit/s. In order to minimise interference, the board edge connectors have an earthed ground plane running between the parallel contacts.

General purpose interface bus (GPIB). Also known as the IEEE-488 or Hewlett Packard Interface Bus (HPIB) after the originators. The bus was designed specifically for instrumentation purposes and consists of eight parallel, bi-directional data lines, three control lines and five lines for general interface control. The maximum data transfer rate is 1 Mbyte/s. Up to 15 devices can be attached to the bus at any one time, but the total transmission line length must not exceed 20 m. Three types of devices may be connected and these are described as either *talkers, listeners or controllers*. For example, a voltmeter may be described as a talker, a printer as a listener, and a computer as a talker, listener or a controller. Although the network may contain more than one controller, one is designated as master and allocates the bus to individual devices in turn.

Hewlett Packard interface loop (HPIL). A standard developed to handle medium speed, portable instrumentation requirements. The communications link consists of a two-wire serial loop connecting each device to a controller. Messages are transmitted in blocks of 11 bits: eight bits of data, plus three control bits. These latter define the way in which a device must operate before it can send data over the network.

I^2C bus (inter-IC). A small standardised two-wire bus system configured as a ring network and devised for the inter-communication between integrated circuits. It can be adapted for use in the control of analogue or digital systems, ranging from a television receiver to industrial applications. Because only two lines are dedicated to control, the cost in terms of additional pin-outs on each IC is small. The two lines provide for bi-directional serial data transfers up to a rate of 100 Kbit/s plus a clock signal. The bus operates on a master/slave basis but allows for several devices to act in turn as master, each one providing its own clock signal. An arbitration procedure prevents bus-conflict or contention between devices, thus allowing only one master to gain control at any one time. Each IC on the bus has a unique 7-bit address which is contained in first byte of each data transfer. Each device on the bus compares this byte with its own address and only if a match is found will it respond to the master that transmitted the address. The number of data bytes that follow is unrestricted and the slave acknowledges the received data on a byte-by-byte basis. Due to its self-capacitance the bus is limited to a length of about 3.5 m. Provision is also made to allow the use of devices of different semiconductor technologies, e.g. C-MOS, bipolar, etc.

Micro channel architecture (MCA). A development of the IBM Corporation to expand the applications of personal computers into the realms of multi-tasking. The standard can be used both with 16- and 32-bit computers to increase the data transfer rate through a network up to a maximum of 40 Mbyte/s.

Multibus system architecture (MSA). The multibus concept was developed by Intel Corporation in the late 1970s and became a *de facto* standard for multiple processor operations. Currently, this specification and its successor Multibus II are covered by an IEEE standard. It was designed for real-time processing on an open system and is thus not microprocessor dependent. Multibus I allows for 16-bit data plus 24-bit address bus space, while Multibus II has been extended for use with 32-bit computing devices. A maximum data transfer rate of 40 Mbyte/s can be achieved.

NuBus. A flexible 8-, 16-, or 32-bit bus structure with similar characteristics to EISA and MCA. Because there is no need to run configuration software or make switched or jumpered connections, the system is very simple to install. The original data transfer rate of 37.5 Mbyte/s has been upgraded to 80 Mbyte/s. When additional hardware is installed the bus permits direct memory access (DMA).

Profibus. This bus system, which is covered by the DIN 19245 standard, has many similarities with Bitbus. It was designed to operate in the industrial control environment and provide communications between computers, programmable logic controllers and sensors. Profibus will run on a range of computers, including those using the OS-9 operating system. It is a multi-master/multi-slave, real-time, asynchronous system using a token bus protocol and an RS485 interface (although the latter could change). The data transfer rate ranges from 9.6 Kbit/s to 500 Kbit/s. The system software is object orientated using an object dictionary in the form of a look-up table. The objects include such obvious names as *oil pressure*, *air temperature*, *valve 1*, etc., which are listed in the dictionary together with the appropriate linked address. The look-up table can also provide barred communications between certain nodes. An extended software protocol can be used to allow synchronous sensors to be added to the system.

Small computer systems interface (SCSI). This network facility, pronounced *scuzzy* in the jargon, is a standard interface bus that allows computers to communicate with any peripheral device that carries embedded intelligence. The standard is covered by the American National Standards Institute (ANSI) and has developed into SCSI-2. Different types of device can be connected in a daisy chain fashion via a common 50-way cable, both ends of which must be correctly terminated. All signals on the cable are thus common to all devices. To avoid bus contention, each device connected to the bus is given a unique 'SCSI' address, with each address being allocated a degree of priority. The bus carries 8-bit data, 1-bit parity and 9-bit control lines, to provide a maximum synchronous data transfer rate of 5 Mbyte/s. The maximum bus length is dependent upon the type of bus driver/receiver used. For single-ended devices, the length is restricted to 6 m. This can be extended to 25 m through the use of differential driver/receivers. SCSI-2 has been extended to increase the data rate to 10 Mbyte/s with either 16- or 32-bit processors. The maximum data rate is thus nearer to 40 Mbyte/s. The protocol has also been expanded to include *tagged commands*. This allows the execution of queued control commands according to a prescribed sequence.

SBus. This bus was developed by Sun Micro Systems Inc. for operation with RISC (reduced instruction set computer)-based processors and workstations. It provides a 32- or 64-bit data bus with a maximum data transfer rate of 80 Mbyte/s.

STEbus (Standard Eurocard). This inherently simple bus is covered by an IEEE standard and is basically devised as an 8-bit system, with 20 address lines. However, 16- or 32-bit processors can be catered for by transferring data in two or four, 8-bit blocks. The maximum synchronous data transfer rate is in the order of 5 Mbyte/s. The bus is practically processor independent and can operate in a multi-processor environment of industrial control systems. All devices on the bus are effectively connected in parallel and non-daisychained. The fact that this bus is compatible with the VMEbus and can easily be interfaced to it, shows the degree of versatility incorporated in the system.

Superbus. This concept is based on a so-called *scalable coherent interface* (SCI) which is hard-wired, to break away from the back-plane technique. It is thus anticipated that this will be the successor to Futurebus and provide data transfer rates as high as 1 Gbyte/s in multi-processor applications.

Turbochannel. This bus system was devised by the Digital Electronic Corporation (DEC) to operate with 32-bit RISC-based processors and DEC workstations. It provides a maximum data transfer rate of 80 Mbyte/s.

VERSAbus. This concept was designed in the late 1970s by Motorola Corporation to integrate modular electronics sub-systems. In the early 1980s, Mostek and Signetics Corporations, in cooperation with Motorola, extended the specification to produce the VMEbus.

VMEbus (virtual machine Eurocard). This IEEE standard bus system provides a good example of the way in which the technology has developed over the space of little more than a decade. VME which was derived from VERSAbus now provides the basis for Futurebus. It is designed around a back-plane concept to provide a data transfer rate of up to 40 Mbyte/s for 16- and 32-bit computers with 32 bits of address space. It provides fast real-time processing in the multi-processor environment but lacks many of the attributes needed for instrumentation applications. In addition to the three signal groups, data, address and control, three further groups are defined. These include, arbitration, interrupt and general utility signals. A VME extension for instrumentation (VXI) has been developed to overcome the instrumentation problems. This can cope with the most demanding test and measurement applications and still provide data transfer rates approaching 40 Mbyte/s. A multiple extension interface (MXI) has been defined to extend the VXIbus across multiple mainframes and to couple individual personal computers to the system, using multiplexed address and data lines. However, for such applications, the data transfer rate is reduced to around 20 Mbyte/s.

Cambridge ring. This network, constructed from two pairs of twisted telephone pair cables, functions on the empty slot technique. Typically four slots circulate unidirectionally around the network, being passed between repeaters placed at each node. As these are actively connected to the ring for the whole of the operational time, the necessary power is supplied over the ring. One of the nodes is allocated to a fixed monitor station which creates the empty slots at start up, clears corrupted data packets and manages the error reporting function. The success of this type of ring depends upon the cable characteristics. Any discrepancy in the lengths of the four wires can introduce a signal phase difference; the self-capacitance and attenuation can cause distortion, all leading to an increase in bit errors. To counter these problems, the repeaters are usually spaced at intervals of less than 100 m. Provision is also made to introduce sections of optical fibre into the ring where it is necessary to traverse regions of high electrical interference. The raw data rate around the ring is 10 Mbit/s, but due to the addition of house-keeping bits, etc., the actual rate is nearer to 1 Mbit/s. The basic data coding depends upon the voltage transitions on the cable pairs, which nominally carry a dc potential of 28 V. A logic 1 is signified by a change of state on both pairs of wires, while a logic 0 produces a change of state on one pair only. For a succession of zeros, the state changes alternate between pairs. The data is organised into *minipackets* each of 40 bits (originally 38 bits). These contain two data bytes, source and destination addresses, response and control bits. Each circulating slot carries one minipacket. The address space for each node occupies eight bits allowing for 256 possible network stations. The all-zero pattern is reserved for the monitor and the all-ones indicates a minipacket that is broadcast to all stations. Communications around the ring is further organised into packets each containing up to 1024 data minipackets (2048 data bytes). Each packet is made up as follows, one header minipacket, one routeing minipacket, up to 1024 data minipackets, plus a check sum minipacket. The check sum is a 16-bit pattern obtained from all the minipackets, computed modulo-2^{16}-1, with end around carry.

Cambridge fast ring. In this development of the basic Cambridge ring, which also operates on the empty slot protocol, the raw data rate has been increased up to

100 Mbit/s, a feature that has been achieved chiefly by improvements to the data handling at each node. Each node device is a serial–parallel–serial convertor to match the serial network data requirement to the parallel needs of each terminal. Each node can support a small cluster, the devices of which are connected in a byte-wide ring fashion. The basic structure of the network still provides either for twisted copper pairs or optical fibre, but the nodes are now powered locally. The network modulation scheme also has the same characteristics in that a transition occurs at each clock period. This allows synchronisation to be achieved from the data stream. Each network, which has gateways or bridges to other networks, also contains a monitor station to maintain the slot structure, handle test and maintenance procedures and generally manage the network quality of service. The size of each data packet has been increased to 32 bytes, a feature that helps to increase the data throughput. A packet is made up as follows: a start of packet bit always set to 1; a full/empty bit which signifies if the slot is in use or not; a monitor bit used to delete lost packets by the monitor station; and a start bit that is always set to 0. This is followed by the destination and source addresses each of two bytes. Following the 32 bytes of data, each packet ends with 12 bits for CRC (cyclic redundancy check) and four response bits, the latter being used to signify to the source, destination and monitor, that the data has been received correctly or otherwise.

Ethernet system. This is a half-duplex send–receive system based on a highway or bus. This is constructed from a length of 50 ohms co-axial cable which must be terminated at each end in a suitable load. The typical maximum cable length is 2.5 km, but this can be extended by interfacing other similar networks via suitable gateways. The maximum number of nodes or stations permitted on each sector is 1024. The raw data rate is 10 Mbit/s and each node on the network is driven by a separate 20 MHz clock. The Manchester code format is used to ensure that the clocks are synchronised to the data stream. The bit cell is divided into two, the second part containing the true bit value and the first its complement. Network access control uses the *carrier sense, mutliple access/collision detect* technique. All nodes listen continually to the network to detect a suitable time to transmit data. If two stations detect the same clear period, each will transmit and then detect a collision. Both will continue to transmit for a further 80 ns so that all nodes can recognise that a data collision has occurred. The transmitting nodes then back-off for a random period before trying to retransmit. If a further collision occurs both back-off for a longer period. After 16 attempts, the node concerned logs a system failure. The transmission frame format consists of a maximum

of 1526 bytes made up as follows. A clock synchronising preamble of alternate 1s and 0s consisting of seven bytes, followed by a similar eighth byte but ending with two 1s, the last 1 representing a start bit. Destination and source address fields each of six bytes. The first bit in each is described as a *multicast* bit. When this is set to 1 in the destination address, all terminals within a group or block must respond to the message. In the source address, this bit is always set 0. Although only 1024 stations are permitted per network and these could be addressed by just ten bits, the 47 bits available could address more than 1.4×10^{14} unique nodes. This has been arranged so that stations in different networks have different addresses and that owners of the patent rights can exert some control over the development of Ethernet. Twenty-three bits of the address field are thus allocated by Ethernet, leaving the remaining 24 bits for user allocation. The following two bytes are used to identify the data type and length. The data field is variable between the limits of 46 and 1500 bytes. Each frame ends with four bytes of cyclical redundancy check (CRC). At the receiver, the bits following the start bit are stripped off and the CRC recalculated and compared with that transmitted, any disagreement being reported to the local microprocessor.

The Thin Ethernet or Cheapernet system operates at the same speed and in the same modes but uses the thinner 75 ohms coaxial cable. This restricts operations to a shorter network.

Fibre distributed data interface (FDDI) network. This network, which operates as a token ring, uses optical fibre as the transmission carrier. It has been designed so that it can be coupled through suitable gateways into Ethernet or other standard ring networks. The carrier wavelength is typically 1300 nm and this can easily accommodate the bit rate of 100 Mbit/s. Rings of greater than 100 km circumference can be constructed, and with up to 500 nodes per ring. The network consists of two separate fibres which connect to each node. This allows the network to be reconfigured under fault conditions at any node. A 4B/5B coding system is used to aid synchronisation. Each 4-bit symbol is coded into five bits. Of the 32 bit patterns available, 10 are discarded as having too few data transitions, six are used for control purposes and the remaining 16 represent the original symbol. Expanding the data from four bits to five, produces a baud rate of 125 Mbit/s. Although this coding is less rich in transitions than the Manchester coding which would require a bandwidth of 200 MHz for the same data rate, the slightly worsened synchronisation is a good trade for the lower bandwidth. Unlike normal token rings, the FDDI ring node releases the token immediately it has transmitted its data. This utilises the ring bandwidth and time more effectively so that many messages can be circulating simultaneously. This loses some of the error checking capability but this is recovered by making each node continually monitor the token rotation time. This effectively defines a time slot for a node to transmit.

FDDI-2 system has been designed to expand the flexibility of the concept and cater for both packet and circuit switched communications. This supports the 64 Kbit/s circuit switched voice channels of an ISDN system. This is achieved by using a transport mechanism described as a *cycle* of 125 μs length. Of the 100 Mbit/s rate, 98.304 Mbit/s is available to share between packet and circuit switched applications. The remaining 1.696 Mbit/s is used for synchronisation, header and cycle delimiting (house keeping).

FDVDI (fibre distributed video/voice data interface) is under development to integrate the data, video and voice services. It is expected to have a data rate of 2.48 Gbit/s.

Synchronous optical network (SONET) or synchronous digital hierarchy (SDH) (an example of a global network). This is a developing world-wide standard to allow interconnection of the broadband ISDN (B-ISDN) services. This has become necessary because of the near saturation of the 2 Mbit/s PCM hierarchies, and the difficulty of inserting or dropping a particular channel from each multiplex. The common interface is a flexible structure capable of handling bit rates from 50 Mbit/s up to 2.4 Gbit/s currently and 12.8 Gbit/s in the future. Because of the optical link, different types of equipment can be connected at either end. The system is defined as a *transport vehicle* designed to carry a pay load with the basic transport mechanism being described as *virtual container*, each container being capable of carrying the European 2, 8, 34 and 140 Mbit/s or the American 1.5, 6 and 45 Mbit/s traffic. A *network node interface* (NNI) is described as the interface between two national or international networks. The *user network interface* (UNI) must be capable of carrying 150 Mbit/s to handle high definition television (HDTV) signals. To allow efficient use of this synchronous system, the basic rate is defined at 150 Mbit/s and a *synchronous transmission mode* (STM) as the basic rate extension as STM-*n*, for $1 < n < 16$ so that the data rate extends from STM-1 at 300 Mbit/s to STM-16 at 2.4 Gbit/s.

23.6 Recommended standard (RS) interfaces

In order to simplify the problem of interconnecting a wide range of devices, a number of standard interconnections have been defined for various applications.

Centronics interface. This is probably the most commonly used parallel interconnection, being originally designed by Centronics Corp. for use with their printers. It is based on a 36-pin connector and cable in the manner shown in Table 23.1.

Table 23.1 *Centronics interface*

Signal pin	Return pin	Signal
1	19	Data strobe
2	20	Data bit 1
3	21	Data bit 2
4	22	Data bit 3
5	23	Data bit 4
6	24	Data bit 5
7	25	Data bit 6
8	26	Data bit 7
9	27	Data bit 8
10	28	Acknowledge
11	29	Busy
12	30	Paper end
13		Select
	31	Input prime
14		Supply ground
	32	Fault
15		Osctx
16		Logic ground
17		Chassis ground
18		+5 volts dc

Note. Pins 12, 13, 14, 15, 18, 31, 32, 34, 35 and 36 vary in function. They are commonly used for printer auxiliary control and error reporting. The cable should consist of twisted pairs for the signal lines 1 to 12 and the corresponding ground returns.

Current loop operations. Many serial data driven, low input impedance devices require to be driven from a current source. Two common values are in use, either 20 or 60 mA, with a logic 1 or 0 being represented by the presence or absence of the current.

GPIB. Designed for instrumentation purposes and uses a standard 24-pin connector and cable as shown in Table 23.2.

RS-232C. A well-established standard for communication path lengths of less than about 15 m and data rates below 20 Kbit/s. The interface is based on a 25-pin connector and cable in the manner shown in Table 23.3. In many applications, not all of the facilities are used. Although still commonly referred to as RS-232C, this standard has, since 1986, been covered by RS-232D.

Table 23.2 *GPIB interface*

Pin	Signal	Pin	Signal
1	Data I/O 1	13	Data I/O 5
2	Data I/O 2	14	Data I/O 6
3	Data I/O 3	15	Data I/O 7
4	Data I/O 4	16	Data I/O 8
5	EOI	17	REN
6	DAV	18	DAV ground
7	NRFD	19	NRFD ground
8	NDAC	20	NDAC ground
9	IFC	21	IFC ground
10	SRQ	22	SRQ ground
11	ATN	23	ATN ground
12	Shield	24	Logic ground

Key:	ATN	Attention
	DAV	Data valid
	EOI	End or identify
	IFC	Interface clear
	NDAC	Not data accepted
	NFRD	Not ready for data
	REN	Remote enable
	SRQ	Service request

RS-442 and 443. These were introduced to overcome some of the problems associated with RS-232C, the maximum bit rates being in the order of 10 Mbit/s and 100 Kbit/s, respectively. The mechanical standard is generally the same as for RS-485 and RS-449 which is shown in Table 23.4. Again, not all the facilities need be used in any particular application.

RS-485. This standard was devised for the interconnection of multiple digital transmitters and receivers to a common cable. As indicated by Table 23.5, it is intended for use with a balanced cable network and for signals that might have a dc component.

23.7 Standard interconnects and internetworking

MAP (manufacturing automation protocol) (see also OSI, p. 214). General Motors Corporation of America was one of the first manufacturers to recognise the need to integrate many disparate production systems into a controlled environment. Such systems include production line computers, numerically controlled machine tools, robotic assembly devices and automated stores control. This latter concept leads to *just in time* (JIT) delivery of components on the assembly line from outside suppliers, reducing the costs for stock in hand and the consequent warehousing requirement. In addition, the worldwide manufacturing of this organisa-

Table 23.3 *RS-232C interface*

Pin	Signal	Source	Key
1	–	–	Protective ground
2	TXD	DTE	Transmitted data
3	RXD	DCE	Received data
4	RTS	DTE	Request to send
5	CTS	DCE	Clear to send
6	DSR	DCE	Data set ready
7	–	–	Signal ground
8	DCD	DCE	Data carrier detect
9	–	–	Reserved for data set test
10	–	–	Reserved for data set test
11	–	–	Unassigned
12	–	DCE	Secondary received signal detector
13	–	DCE	Secondary clear to send
14	–	DTE	Secondary transmitted data
15	–	DCE	Transmission signal element timing
16	–	DCE	Secondary received data
17	–	DCE	Receiver signal element timing
18	–	–	Unassigned
19	–	DTE	Secondary request to send
20	DTR	DTE	Data terminal ready
21	–	DCE	Signal quality detector
22	–	DCE	Ring indicator
23	–	DTE/DCE	Data signal rate selector
24	–	DTE	Transmit signal element timing
25	–	–	Unassigned

tion required international communications links through satellites and PABXs (private automatic branch exchanges) connected to the national PTT (Post, Telephone and Telegraphy) systems. MAP, a developing world standard that provides for future expansion, is a sub-set of the OSI (open systems interconnect) concept and conforms to the OSI model at all levels. The protocol allows for the standard *file transfer, access and management* (FTAM) and the *manufacturing messaging specification* (MMS).

OSI (open systems interconnect)

This concept has been developed to enable dissimilar and normally incompatible devices, to exchange data by means of an agreed set of protocols. It is based on a defined system of interconnections and interactions between the seven functional layers or levels of the model shown in Fig. 23.4. The layers below level 3 are related to system hardware and electrical interfaces, while the layers above are technology independent and

Table 23.4 *RS-449 interface*

Pin	Signal	Pin	Signal
1	Shield	20	Receive common
2	Signalling rate indicator	21	Unassigned
3	Unassigned	22	Send data
4	Send data	23	Send timing
5	Send timing	24	Receive timing
6	Receive data	25	Request to send
7	Request to send	26	Receive timing
8	Receive timing	27	Clear to send
9	Clear to send	28	Terminal in service
10	Local loopback	29	Data mode
11	Data mode	30	Terminal ready
12	Terminal ready	31	Receiver ready
13	Receiver ready	32	Select standby
14	Remote loopback	33	Signal quality
15	Incoming call	34	New signal
16	Select frequency	35	Terminal timing
17	Terminal timing	36	Standby indicator
18	Test mode	37	Send common
19	Signal ground		

Note. In general, this table is also applicable to RS-422, RS-423 and RS-485 standards.

Table 23.5 *RS-485 standard interface*

Pin	Signal
1	Common ground
2	RTS-A
3	RTS-B
4	Tx(Rx)-B
5	Tx(Rx)-A
6	CTS-B
7	CTS-A
8	Rx-B
9	Rx-A

Figure 23.4 OSI 7-layer model.

related mainly to software. Interaction between terminals occurs at the same level, i.e. level 3 to level 3.

Level 1. This layer relates to the movement of data bits from the computer terminal to the network, by defining the mechanical, electrical and functional characteristics. For example, the RS-232C interface.

Level 2. The data link control layer defines the way in which a terminal can gain access to the network. The protocol arranges data bits into suitable and acceptable sized blocks, organises error control and, in general, provides reliable and accurate data transfers.

Level 3. The network layer is primarily responsible for setting up and maintaining connections and moving data. In addition it must add the necessary destination and routeing information to provide for the links across other inter-connected networks. For example, circuit or packet switching.

Level 4. The transport layer is basically responsible for network security from end to end and the quality of service. It also defines how the various nodes are to be addressed and the way in which connections can be made or broken.

Level 5. This layer effectively operates as an interface between the transport and applications layers. It also allows control software to be written in a network-independent manner. Together, levels 3, 4, and 5 are often described as the network *subnet* level.

Level 6. This layer handles the interchange of data and information by translating the data formats, codes and syntax applicable to the session and application layers.

Level 7. This layer defines the services available to the user, such as file transfer, access and management (FTAM), distributed databases, worldwide messaging and electronic mail, and the provision of a user-friendly interface so that the system is transparent to the user.

SNA (systems network architecture). This originally represented the IBM Corporation's *de facto* standard for networking the hardware of IBM manufacture. But since this has been expanded to allow integration with ISDN (Integrated Services Digital Networks), which is OSI compatible, SNA is fast becoming an open system, so that larger SNA networks have been developed. The SNA model is very similar to that of OSI, but based on three blocks. The functions of the lower two equate approximately with levels 1 to 6 of OSI, while the upper block covers applications in a manner similar to OSI. A further expansion of this standard is described as *systems applications architecture* and this provides an even greater degree of flexibility.

Inter-networking

As the use of LANs expands, the need for inter-networking increases. Since different types of LAN operate with different electrical parameters and operating protocols, this can create significant problems. However, these may be resolved by using hardware devices that carry varing degrees of intelligence. These devices are known variously as bridges, brouters, routers and gateways.

Bridges. These function at layers 1 and 2 of the OSI model and so effectively couple LANs of the same type together and so are protocol independent. Bridges read all passing data packets but only interpret the source and destination addresses. The effect of *bridging* two LANs is thus simply to extend the original.

Brouters. As the name suggests, these devices are capable of performing the functions of both bridge and router.

Routers. These provide some bridging functions but operate at layer 3 (network level) of the OSI model. They contain a higher degree of intelligence and so

provide a better traffic control and routeing than a bridge. Again, they are normally protocol independent and equipped with look-up tables to deal with routeing in a more efficient manner. Routers also have a store and forward capability, a feature particularly useful during periods of congestion.

Gateways. This is the most complex of inter-networking elements and with the highest degree of inbuilt intelligence. Operating at the higher levels of the OSI model, these thus allow the interchange of information between completely different environments.

23.8 Miscellaneous terms

Abstract syntax notation.1 (ASN.1). This concept operates at the presentation layer of the OSI model. It defines the exact form of a language syntax between communicating processes. Part of this is a *transfer syntax* which converts variables, commands and requests into a form that is hardware independent. This overcomes such problems as converting between ASCII and EBCDIC code formats.

Cluster. Term used to describe a terminal configuration where several devices are connected to the same drop cable.

Concentrator. A device used to divide a high rate data channel into several lower rate channels. The division may be on a dynamic basis depending upon the traffic level. Also known as an *intelligent time-division multiplexer.*

Data compression. A technique used to increase the data throughput by removing redundant or repetitive data according to a suitable algorithm, the missing information being reconstituted at the receiver.

Deterministic network. Each user has a periodic guaranteed time slot for transmission.

Discrete access. An access method that can be used with star networks whereby a terminal has direct access to the switching capability.

Distributed queue dual bus (DQDB). This network configuration, which was devised for metropolitan area networks (MAN), consists of two uni-directional buses connected as an *open ring* (i.e. each bus is not directly interconnected). It provides for full duplex operation between any two nodes and the bandwidth available provides for signalling speeds in excess of 155 Mbit/s. Each transmission cell or packet which is compatible with B-ISDN (broadband ISDN), consists of a 5-bytes header plus 48 bytes of message. The header bytes include bits for control, priority and an error check sequence.

Drop cable. The cable providing the final link between a terminal device and the network proper.

Dynamic routeing strategy. An adaptive method used for directing messages through a network. If a route is busy or disabled, then an alternative routeing can be provided.

EARN. European Academic Research Network.

EDI (electronic document interchange). The use of standardised documents to speed up international economic and commercial data interchange.

EDIFACT. EDI for Administration, Commerce and Transport.

Euronet. The data communications network provided for the European Economic Community (EEC) by its member telecommunications authorities. It is based on packet switching and linked to the public switched telephone network (PSTN).

Explicit access. Each terminal is allowed access to the network but only at specified times.

Inter-network router. Used for communications between sub-networks. Such routers operate at the network layer of OSI and only pass messages to the correct sub-network.

Jabber. The jargon term used to describe the continuous sending of random data from a terminal under fault conditions.

JANET. The Joint Academic Network designed to link all academic and research centres.

Loopback. A diagnostic technique used whereby a test message is sent to a specific terminal. This is immediately retransmitted back to the originator for checking.

Multi-drop. A technique for connecting several terminals to a computer using a master and several slave modems. Some form of polling is necessary to ensure unique addressing.

Network time delays. The time taken for a message to be delivered to its unique address depends upon the following factors:

Access delay. A data packet at the head of a queue has to wait for a convenient time gap.
Propagation delay. As electrical signals traverse the network at a velocity of about 2×10^8 m/s, this will normally be the smallest element of the total delay.
Queueing delay. Once a packet is ready for transmission, it has to take its place in the interface device, behind the other packets already waiting.
Transmission time. This value depends upon the packet length and the network bit rate.

Neural networks. A conventional computer requires that its problems are completely and accurately described before it can produce a solution. A neural-based machine is able to define the problem for itself by exploring the connections with the problem before reinforcing those links which would tend towards a useful result. The human brain is referred to as a neural network which solves problems by parallel processing. The basic element of the brain is the neuron and these communicate across synapses or synaptic gaps. The collective functioning is determined by the way in which neural units are physically connected. A brain may have 10^9 neurons each with as many as 1000 synaptic links to other neurons. Artificial neuron networks represent an electrical system of interconnected elements for the parallel processing of data. This developing concept will find applications in speech pattern recognition, artificial intelligence and other devices that need to learn by experience.

Non-deterministic network. A network where access to the transmission medium within some specified period, cannot be guaranteed.

Non-persistent. A term used to describe a station under CSMA control that does not attempt to re-transmit immediately following a data collision.

Persistent. A term used to describe a CSMA controlled station that immediately tries to re-transmit after it has detected a data collision.

Queued packet synchronous exchange (QPSX) (also known as switched multi-megabit data service (SMDS) and distributed queued dual bus (DQDB) system). This developing concept is covered by an IEEE standard and is designed to deliver up to 150 Mbit/s to customers' premises using the public network and forms the basis of a metropolitan area network (MAN). The plan allows for such networks to be interconnected on a national basis to provide voice, data and video services. The design supports the asynchronous transfer mode (ATM) which makes it compatible with broadband ISDN.

RACE (Research on Advanced Communications in Europe). An EEC funded research programme devised to develop and implement broadband communications services and plan a pan-European network using optical fibres. The programme provides for research into optical fibre switching and network management for high quality video, telephony, data and other television services such as video-conferencing.

SCPI (standard commands for programmable instruments). A concept that allows instruments that are not primarily GPIB compatible to use the GPIB bus system. This conforms to the IEEE-488.2 standard.

TCP/IP. Transmission control protocol/internet protocol. A communications protocol that functions at level 3 and 4 of the OSI model. The protocol is used for internetwork routeing and quality of service.

TRANSPAC. The French PTTs public data network that provides such services as facsimile and teletex.

Wireless LANs. The conventional local area network (LAN) utilises wires or optical fibres as a common carrier medium; however, other possibilities exist. Low microwave frequencies (less than about 10 GHz) can provide for data rates as high 10 Mbit/s, millimetric waves at around 60 GHz could support several 10 Mbit/s channels, while infra-red beams could support even greater data throughputs. The area covered by such a scheme would be restricted by the low allowable power radiation. The data rates of such systems tend to be restricted by inter-symbol-interference and multipath propagation problems that arise due to reflections within the building. Because of the wide bandwidth available, channelling can easily be provided by using spread spectrum methods and code division multiple access, a technique that significantly improves the system security.

Wireless in-building network (WIN). This service is provided by a low power (less than 25 mW) radio communications link operating at around 18 GHz. This frequency is capable of supporting data rates as high as 15 Mbit/s. Because the microwave signal propagation is somewhat constrained by the building materials used, the energy is mostly restricted to within the structure.

The concept can be used to support any of the hard-wired types of network but introduces a security problem. This can be overcome by using encryption, but at the expense of a lower data rate.

The use of low power within a building is necessary to meet the World Health Organisation (WHO) standards on radiation.

10BaseT. The IEEE standard (802.3) which is used to specify the connection of a terminal unit to an Ethernet via a medium attachment unit (MAU) over unshielded twisted pair wiring is known as 10BaseT. This allows data to be transmitted over existing telephone lines and still provide for the basic bit rate of 10 Mbit/s. The line impedance over the maximum link length of 100 m should lie between 85 and 110 ohms, a value acceptably close to that of 100 and 120 ohms cable commonly available in the USA and Europe, respectively. For maximum data rates, the link length should be kept to a minimum.

Since all unshielded pair cables act to some extent as antennas, interference may be a problem. In these cases, the standard permits two variants: 10Base5 (IEEE 802.3) using thick, and 10Base2 (IEEE 802.4) using thin, coaxial cables.

23.9 Useful references

Heath S. (1989) *VMEbus Users Handbook*, Butterworth-Heinemann, Oxford.

Hopper A., Temple S. and Williamson R. (1986) *Local Area Network Design*, Addison Wesley, Reading, MA.

Milne A.C. (1986) *Interfacing Standards for Computers*, IEEE Monograph, Institution of Electrical and Electronic Incorporated Engineers, London.

Rutkowski A. (1985) *Integrated Services Digital Networks*, Artech House, London.

Sluman C., Maynard J. and Dand A. (eds) (1988) *Networks*, Blackwell Scientific, Oxford.

Stallings W. (1987) *Handbook of Computer Communications Standards*, Vol. 2, Macmillan, New York.

STEbus Manufacturers and Users Group, c/o ARCOM Ltd, Clifton Road, Cambridge, UK.

Wesley W.C. (1986) *Distributed Processing Systems*, Vols. 1 and 2, Artech House, London.

24 _Noise_

In the communications sense, noise is best described as a phenomenon that destroys information. In the electronic sense, it represents signals of any form that are unwanted, including signals from random variations of electrical charge from all sources, interference from signals that are required by other users and even distortion to the wanted signal. The presence of such noise is therefore often the limiting factor in signal detection and this gives rise to the term signal to noise power ratio (S/N ratio) which defines a system sensitivity to noise. Noise and interference are different in nature. Noise appears as randomly varying voltages that are not related in either phase or frequency. By comparison, interference tends to be periodic and regular. When viewed on an oscilloscope, noise appears to be very _peaky_ and, on average, these peaks have a very short duration. In addition, the _crest factor_ or peak to RMS ratio is high, in the order of 5:1. (See also Information theory, p. 143, Electromagnetic compatibility, p. 97 and Meausurement of system parameters, p. 152)

24.1 Classification of noise

For technological reasons, noise is often divided into two classes.

Artificial or man-made noise. Such interference tends to be impulsive and is mainly due to electrical machines and arcing or sparking contacts. The interference is regular in nature and can usually be suppressed at source using appropriate filters.

Natural noise. This form of noise is chiefly due to cosmic electromagnetic radiation that invades a system through either an intentional or accidental aerial. In addition, all circuit components generate thermal noise that has the same characteristics. Such noise is completely random and its behaviour is usually assessed by statistical processes. This form of noise covers the whole of the electromagnetic spectrum and cannot be eliminated. Only by clever design can its effects be minimised.

24.2 Hartley–Shannon channel capacity

The Hartley-Shannon law shows that the channel capacity (C bits per second), the channel bandwidth (B) and the signal to noise ratio (S/N) are related by the equation:

$$C = B \log_2(1 + S/N) \, \text{bit/s}$$

The ability to trade channel capacity, bandwidth and S/N ratio is particularly important in system design, where the channel transmitter is power rather than bandwidth limited. Although this specifically relates to a digital channel, a similar argument can be applied in the analogue case. Whereas the analogue signal quality degrades gracefully with an increase in noise level, the digital channel is normally unaffected until some threshold level is reached, when the communication suddenly _crashes_.

24.3 Noise parameters

Average noise factor. The effective noise factor F over the bandwidth under consideration.

Correlation (of noise sources). Due to the random nature, noise voltages from different systems will be completely independent in amplitude, phase and frequency. Such voltages are described as being uncorrelated. If, however, two signals each affected by different noise sources are passed through a common circuit, then this common element will destroy some of the independence to produce a degree of correlation. This is best expressed as follows:

$$\overline{v_t^2} = \overline{v_1^2} + \overline{v_2^2} + 2\gamma v_1 v_2$$

where $\overline{v_t^2}$, $\overline{v_1^2}$ and $\overline{v_2^2}$ are the total and individual mean square noise voltages and γ is the correlation coefficient.

γ may take values between -1 and $+1$. When γ is zero there is no correlation, when γ is negative there is a phase cancellation effect and when γ is positive the noise effects are directly additive.

Equivalent noise resistance (see also Thermal noise, p. 222). Because the noise power produced by a circuit element depends upon its ohmic resistance value, the power can be reduced to an equivalent resistance that produces the same magnitude of noise power at the same temperature and bandwidth.

Excess noise (see also Thermal noise, p.222). The noise power generated by a source or system that exceeds that noise level generated from a body at the standard temperature of 290 K. Often expressed simply as $F-1$, where F is the absolute noise figure or factor.

Maximum available noise power (see also Thermal noise, p.222). When a noise source impedance is correctly matched to a load, there will be a maximum transfer of noise power given by noise power $= kTB$ watts, where $k =$ Boltzmann's constant $(1.38 \times 10^{-23}$ J/K), T is the absolute temperature (K) and B is the bandwidth (Hz). If any impedance mismatch occurs, then the noise power transferred will be less than this maximum.

Noise factor of figure (F) (see also Noise temperature, p. 220). This is a measure of the degradation in S/N ratio introduced by a device or system (either amplifying or attenuating) during signal processing. It is thus the ratio between the input and output S/N ratios. Or alternatively expressed as:

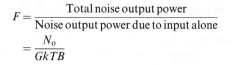

$$F = \frac{\text{Total noise output power}}{\text{Noise output power due to input alone}}$$
$$= \frac{N_o}{GkTB}$$

where N_o is the noise output power, G is the gain (or attenuation) and kTB is the maximum available noise power.

A further alternative equation derived from these is that F is the ratio of the actual noise output power from a practical system to the noise output power from a similar ideal noise-free system. As F is simply a power ratio, it may also be expressed in decibels.

Noise temperature. The maximum noise power (P_n) available from a resistor over a bandwidth B Hz is kTB watts, where T is the absolute temperature of the device. This equation can be rearranged to give $T_n = P_n/kB$, where T_n is the effective temperature of

the resistor that produces the same level of noise. This concept can be extended to other noisy sources not necessarily associated with a physical temperature. For example, an antenna collects noise as random electro-magnetic radiation and this may be associated with an *equivalent noise temperature*. If a noise power P_n watts is received by the antenna over a bandwidth B Hz, then the antenna has an equivalent noise temperature, $T_a = P_n/kB$. Thus the noise factor F can be further expressed as:

$$F = (T_s + T_r)/T_s = 1 + T_r/T_s$$

or

$$F - 1 = T_r/T_s$$

to give the *excess noise temperature* (T_s and T_r being the source and receiver noise temperatures, respectively). When evaluating the performance of systems with noise factors of less than about 5 dB, the comparison scale becomes cramped. The graph shown in Fig. 24.1 is a plot of the above relationship based on the standard noise temperature of 290 K. This shows how the use of the equivalent noise temperature provides an expanded scaling.

Spot noise factor. The noise factor F considered for a single frequency only.

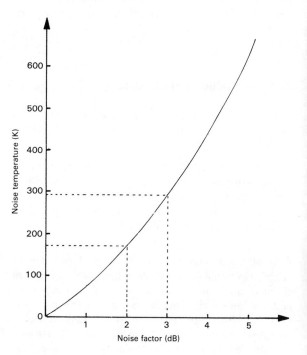

Figure 24.1   Noise factor/temperature relationship.

Standard noise temperature. This has been defined by international agreement to be equal to 290 K or 17°C, the typical ambient room temperature.

24.4 Noise spectral density

Gaussian or white noise. A random noise where the power per unit bandwidth is constant. That is, noise signals of equal amplitude are probable over the whole electromagnetic spectrum. Such noise has an energy distribution described statistically by the Gaussian distribution curve, Fig. 24.2:

$$y = 1/\sqrt{2\pi}\exp(-x^2/2)$$

for which the following points are of interest:

(1) When $x = 0$, $y = 1/\sqrt{2\pi}$ or approximately 0.4.
(2) The curve is symmetrical about the y axis.
(3) The area under the curve is unity.
(4) y approaches zero as x tends towards infinity.
(5) For this normal distribution, the area under the curve represents the probability that $x(P(x))$ falls within a range of values. For example:

$$P(x) - \infty < x < +\infty = 1$$

$$P(x) \text{ is negative} = P(x) \text{ is positive} = 0.5$$

In general, the probability that x lies between a and b is given by the area bounded by a, b and the curve.

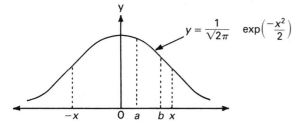

Figure 24.2 Gaussian distribution curve.

Pink noise. Such noise has a power density distribution that falls with increasing frequency at the rate of 3 dB/octave or 10 dB/decade.

Red noise. Such noise has a power density distribution that falls with increasing frequency at the rate of 6 dB/octave or 20 dB/decade.

Rayleigh distributed noise. When Gaussian noise is band limited, the distribution becomes a one-sided amplitude/frequency distribution governed by the curve

$$P(a) = a/N\exp(-a^2/2N)$$

where a is an amplitude and N the total noise power.

24.5 Specific types of noise

Flicker or 1/f noise. A form of noise that affects low frequency amplifiers. The spectral amplitude is proportional to K/f, where K is a constant that is to some extent current dependent and f is the frequency. It is therefore important that very low frequency, low noise amplifiers are operated at very small current levels. The noise arises in transistors from generation/recombination effects on the surface of the semiconductor materials and from temperature fluctuations in other devices.

Microphony. The mechanical vibrations of the electrode structure within a thermionic device can give rise to a noise component. In a similar way, the mechanical vibration of a length of co-axial cable produces a variation of capacitance that also generates a noise effect.

Phase noise. When the output of an oscillator is viewed on a spectrum analyser, the display should be a single spectral line at the appropriate frequency. If this signal is phase or frequency modulated (PM or FM), then a comb-like sideband structure would be displayed. Noise signals can reach the oscillator amplifier stage via several paths and so will produce a form of modulation. However, the sidebands will be very close together so that a triangular noise spectrum will be displayed on either side of the oscillator signal frequency.

Partition noise. When an electron stream separates as it does in bi-polar transistors or thermionic valves, the current divides in a random manner giving rise to a noise component. For this reason, field effect transistors and triode amplifiers are preferred at very high frequencies.

Popcorn noise. A jargon term that describes a noise feature that affects certain transistor amplifiers. It is most noticeable below about 100 Hz and occurs in bursts in a random fashion. The *pops* may be absent for several minutes and then appear several times per second. It is thought to be aggravated by surface contamination of the semiconductor material.

Recombination noise. The recombination of holes and electrons in bi-polar transistors occurs in a random manner to produce a noise component. This is more noticeable at higher frequencies and then tends to fall in proportion to the frequency squared.

Shot noise (also known as Schottky noise). This form of noise is due to the random fluctuations in the emission of electrons from the cathode surface of thermionic devices and the random diffusions of charge carriers in transistors. The magnitude of the mean square noise current is therefore given by

$$\overline{i_n^2} = 2eI_{dc}B,$$

where e $= 1.6 \times 10^{-19}$ coulombs
I_{dc} $=$ dc current in amps
B $=$ bandwidth in Hz.

Sky noise. This term is used to describe the combined effects of *galactic* and *atmospheric* noise temperatures. As indicated by Fig. 24.3(a), the noise temperature seen by an antenna from earth depends upon the direction in which it is pointing. Looking at an elevation of 90° the antenna will see the background or residual noise from space of about 4 K. Parallel to the earth surface the noise temperature will be the standard 290 K. As shown in Fig. 24.3(b) galactic noise falls quite rapidly and above about 1 GHz it is not too significant. Between about 1 and 12 GHz, a low noise window exists that is extensively used for space communications. Over this region the noise temperature is typically in the order of 4 K, but again depending upon the angle of elevation of the antenna. At about 22 and 60 GHz, there are resonant peaks due to scattered radiation caused by water vapour and absorption by oxygen molecules, respectively.

Thermal noise (also known as white or Johnson noise). This is the most dominant noise of all and extends across the complete electromagnetic spectrum. It is caused by the thermal oscillations of charge carriers in a manner similar to Brownian motion. This noise form was evaluated experimentally by J.B. Johnson and theoretically by H. Nyquist and the results published in 1928. Based on rigorous thermodynamics, the mean square noise voltage is given by:

$$\overline{v_n^2} = (4RBhf)/(\exp(hf/kT) - 1)$$

where $R =$ resistance in ohms
$B =$ bandwidth in Hz,
$f =$ frequency in Hz
$T =$ absolute temperature (K)
$h =$ Planck's constant 6.62559×10^{-34} J s
$k =$ Boltzmann's constant 1.38×10^{-23} J/K

For all except very low temperatures and very high frequencies, this equation can be simplified to

$$\overline{v_n^2} = 4kTBR$$

which can be rearranged to give the mean square noise current $\overline{i_n^2} = 4kTB/R$.

24.6 System noise performance (Friis' equation)

For a system consisting of several stages of signal processing, the overall noise factor F_o is affected not

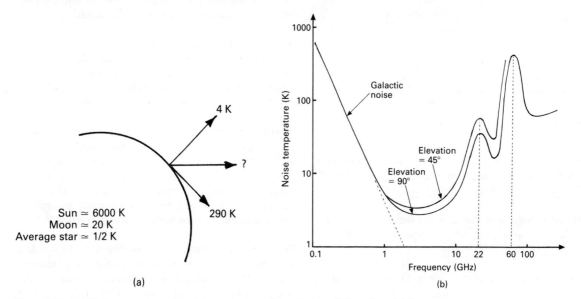

Figure 24.3   (a) Noise temperature 'seen' by an antenna; (b) variation of sky noise with frequency.

only by the signal to noise ratio at the input, but also by the noise contributions from the individual stages and their gain factors. That is, the noise generated by any particular stage will be amplified by the succeeding stages to be added to the total noise. Friis' equation states this relationship as follows:

$$F_o = F_1 + (F_2 - 1)/G_1 + (F_3 - 1)/G_1G_2 + \ldots$$

where F_1, F_2, etc., are the noise factors of the individual stages and G_1, G_2, etc., are the corresponding gain values. If any stage introduces a loss, i.e. behaves as an attenuator, then the noise factor for this stage is given by $F = L$, where L is the numerical loss ratio.

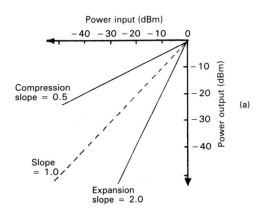

24.7 Systems designed to combat noise

Companding. Companding is the compound term used to describe the *compression* and *expansion* of the dynamic range of a signal amplitude. This is achieved by using nonlinear amplitude-dependent amplifiers with characteristics similar to those shown in Fig. 24.4(a), before and after transmission. The compressing amplifier in this example has a slope of 0.5. The receiver expanding amplifier therefore must have a slope of 2, to give overall unity response. Figure 24.4(b) shows the overall response of the system and can be used to explain the noise reduction properties of the concept. Signals above the level 0 dBm (1 mW) are assumed to be either non-existent or unaffected by companding. During the compression stage, the dynamic range is halved by lifting the amplitudes of the lower level components. Assume that there is −30 dBm (1 μW) of noise present in the transmission channel. This would have swamped the original low level signals, but now is only just comparable with them. The expansion process at the receiver effectively depresses the low level signal components and the noise to restore the dynamic range to its original value, at the same time improving the signal to noise ratio.

Lincompex. This is a transmission method that is used to combat noise and signal fading over long distance HF band radio telephone channels. Compression that varies at the syllabic rate is used at the transmitter to obtain an almost constant amplitude audio signal. The compressor stage also produces a pilot signal that represents the varying degree of compression and this is used to frequency modulate (FM) a 2.9 kHz sub-carrier. This tone and the compressed audio signal is then transmitted over a standard 3.4 kHz radio channel. The receiver decoder separates the two components and demodulates the FM signal to provide the expander circuit control

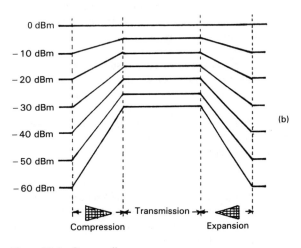

Figure 24.4 Companding.

signal, which restores the audio to its original dynamic range.

Pre-emphasis/de-emphasis. An unfortunate feature of frequency modulation (FM) systems is that noise is most troublesome at the higher baseband frequencies. This is due to the differentiating action of the FM demodulator that reduces the noise power spectral density at the lower end. If a low-pass filter is added following the demodulator, this will cut the high frequency noise component and also the high frequencies in the modulation baseband. The attenuation of the HF signals can be compensated for by boosting these in a complementary manner before transmission. The action of this system which is shown in Fig. 24.5 can lead to as much as 6 dB improvement in FM signal to noise ratio.

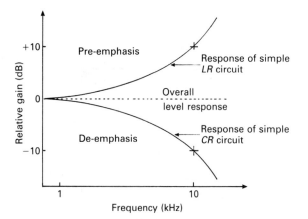

Figure 24.5   Pre-emphasis/de-emphasis characteristic.

24.8 Miscellaneous terms

Harmful interference. Defined by international agreement as that level of interference that seriously degrades a communications service or endangers one of the navigation or safety services.

Masking. In certain cases, the measured value of noise is greater than its aurally perceived value. This is because the wanted signals swamp or mask the noise effect.

Permissible interference. The level of interference below which a service will always meet some specified performance criteria.

Protection ratio. A figure of merit for a transmission service based on the carrier to interference (C/I) ratio that produces just perceptible interference. The protection margin then specifies the value by which the actual C/I ratio exceeds this level.

Quantisation noise. Whenever an analogue signal is converted into a digital code, the approximation process always leads to amplitude errors. These form the noise component of the quantisation process.

Quasi-impulsive noise. White noise with a superimposed impulsive component.

Residual noise. According to quantum theory, when the electromagnetic spectrum is analysed in a dark, screened vacuum chamber, there will be small random irregularities. In most practical cases this is of no account.

Signal+noise to noise ratio. This is often used to describe the sensitivity of a receiver system. It is expressed as the minimum input signal level necessary to produce a given output level for a certain value of signal to noise ratio at a specified degree of selectivity.

SINAD noise ratio. In all systems, nonlinearity of signal processing creates distortion and this is an unwanted form of noise. The SINAD ratio therefore expresses in dB the *signal + noise + distortion to noise + distortion* obtained at the output of a system that is driven by a modulated carrier.

Weighting. In many cases, the actual annoyance factor of a noise component will not be consistent with its measured value. To allow for this, many noise measurements are therefore weighted using a non-linear scale factor.

Zipper noise. A jargon term used to describe the noise generated by signals passing through digital filters while the coefficients or other parameters are being changed in an adaptive manner. This tends to be most troublesome with recursive (IIR) filters, probably due to the multiple feedback connections.

24.9 Useful references

Betts J.A. (1978) *Signal Processing, Modulation and Noise.* Hodder and Stoughton, London.

Connor F.R. (1976) *Introductory Topics in Telecommunications: Noise,* Edward Arnold, London.

Johnson J.B. (1928) Thermal agitation of electricity in conductors. *Phys. Rev.,* **32**, 97.

Lewis G.E. (1992) *Communication Services Via Satellite,* 2nd edition, Butterworth-Heinemann, Oxford.

Mosteller F. et al. (1970) *Probability with Statistical Applications,* 2nd edition, Addison-Wesley, London.

Nyquist H. (1928) Thermal agitation of electric charges in conductors. *Phys. Rev.,* **32**, 110.

Schwartz M. (1980) *Information Transmission, Modulation and Noise,* 3rd edition, McGraw-Hill, New York.

25 *Optical communications, devices and systems*

As worldwide communications develop, the radio section of the electromagnetic spectrum is quickly becoming saturated. Even the use of metallic conductors is being overloaded. Wide band analogue and high speed digital signals are extremely bandwidth hungry. The conventional electronic communications services are therefore becoming very expensive to support under such pressures. Historically, communications have been driven to higher and higher frequencies in order to meet this growing demand. Light beams that can be simply switched on and off, or even deflected, have long been used for low speed signalling purposes, but now the concept of guided light energy using optical glass or plastic is set to provide for a massive communications expansion. Because of the very high frequencies involved (typically about 10^{13} to 10^{16} Hz) it is common to consider optical communications in terms of wavelength instead. Typically the range of light wavelengths from about 350 nm up to 750 nm is available within the visible range. But due to certain difficulties, the range currently being used is in the order of 1350 nm to 1550 nm in the infra-red region. (Shorter wavelengths down to about 550 nm are being explored.) This represents a bandwidth in the order of 28×10^{12} Hz, a range greater than that currently available to radio communications by a factor of about 100. By utilising fibres as a guiding medium, each light wavelength becomes available for reuse over and over again. As the need for wideband systems increases and the optical fibre production expands, the cost of the fibre falls and when its many advantages are taken into consideration, the small cost advantage of copper cables begins to dissappear. Optical fibre forms an almost ideal transmission line with the following important advantages:

- The transmission losses are very small.
- The bandwidth is greater than any other guiding medium.
- Because it is an electrical insulator, fibre does not radiate and is immune to electromagnetic interference.

- The concept has a very high degree of immunity to cross-talk between adjacent fibres.
- It can operate in hostile or hazardous environments.
- Its raw material, silica (sand) is the second most abundant element on earth.

The insulating properties of the glass fibre produce the one major disadvantage, and means that metallic conductors need to be included to power the repeater amplifier needed over long cable runs. However, these can form the strength members that are necessary to aid the laying of fibre cables. For shorter cable runs of about 50 to 100 m, cheaper plastic polymer fibres with a greater degree of attenuation are available. Typical fibre applications range from use in local area network sections, where there is a high degree of electrical intereference, to trans-oceanic telecommunications cables.

Probably the most significant development in optical fibre networks is the project represented by TAT 12 and 13 (Trans-Atlantic Telephone cable). Each consists of two fibre pairs designed to provide a complete ring network between the USA, Canada and Europe. Each fibre of a pair will be capable of providing a bit rate of at least 5 Gbit/s for voice, video, digital images and mainframe to mainframe computer communications, each with an individual capacity of 155 Mbit/s. The operating wavelength will be at about 1500 nm, with repeaters spaced about every 50 km. These will operate as optical amplifiers to avoid the need for double conversion between light and electronic signals. The system is designed to allow for only three failures in 25 years of operation and will use the synchronous digital hierarchy (SDH) concept (see also Telephony and associated systems, p. 336).

The basic concept involves generating a high intensity light source from a laser or light emitting diode, modulating either the intensity, frequency, phase or polarisation and launching the energy into a guiding fibre. A light sensitive detector at the far end then recovers the modulated signal. The precise type of modulation depends upon the system circumstances, but intensity modulation is very common for both analogue and digital services. In general, the analogue system is limited by signal to noise ratio which leads to a narrower bandwidth and shorter cable run than could be expected with a corresponding digital system.

25.1 Driver/source devices

The chief function of the driver/source device is to act as an efficient transducer to convert the signal electrical energy into light energy. An ideal source should have a linear light output/drive current characteristic, the emission at a suitable wavelength should be capable of wideband modulation, the emitted light should have a narrow band/linewidth in order to minimise the losses within the fibre, and have parameters with good temperature stability. In addition, the device should be low cost, reliable and compatible with modern circuit components. Of the three classes of device that might be used for this function, only the laser and light emitting diodes are suitable, the incandescent lamp being unsuitable because it produces a wideband, almost continuous spectrum. The light output from a laser, which is produced by stimulated emission, is monochromatic and coherent. The output from a light emitting diode (LED), which is spontaneous due to the application of a voltage source, is monochromatic but incoherent. Because of the random phasing of the emitted photons, the spectral band/linewidth is about 100 times wider than that of the laser. Gas lasers, because of their bulk and complex power supply requirement, are completely unsuitable for these applications.

25.1.1 Lasers

NOTE. All laser equipment should carry a permanent legible label warning of the dangers to unprotected eyes. In certain cases, the radiation is invisible. For safety reasons, protective safety spectacles that have an optical density of at least 6 for the appropriate wavelength should always be worn.

The term laser is an acronym derived from light amplification by the stimulated emission of radiation. The device is basically an optically transparent cavity with a reflecting surface at one end and a partial reflector at the other. The stimulated emission is produced by the recombination of the injected current carriers within the cavity. This sets up oscillatory standing waves between the two ends and some of the energy emerges through a partial reflecting surface. Common semiconductor devices are diodes fabricated

from *heterojunctions*. These provide a confinement layer for the charge carriers which reduces the volume where radiative recombination can occur and so increases the output power. When the forward current through the diode exceeds some threshold level, often as low as 20 mA for about 2.5 V of forward bias, the lasing action occurs. Increasing the current produces a linear increase in light output up to some saturation level. If the current is allowed to fall below the threshold level, then lasing ceases. Such a device is described as a continuous wave (CW) laser. Pumped or pulsed lasers, where the cavity has to be excited from a flash lamp source to produce emission, are not suitable for these applications.

Buried heterostructure (BH) laser. A very narrow mesa stripe is buried by etching and deposition within the high resistivity region of a double heterojunction structure. This then forms the active region. The very small dimensions ensure single mode operation of this laser, with a very low threshold current level and good linearity. Output power of around 1 mW can support a modulation bandwidth in excess of 2 GHz.

Channelled substrate lasers. This concept includes devices that are described as *constricted double heterojunction* and *plano-convex waveguide* types. In each case, the narrow active region is buried by etching and deposition, within the double heterojunctions, during fabrication. The structures result in single mode lasers that can produce power outputs around 20 mW for a threshold current of about 30 mA.

Distributed feedback (DFB) laser. The complex heterostructure of this device is shown in Fig. 25.1. An optical grating forms part of the guiding structure, the corrugations of which produce a period variation in the refractive index seen by the propagating wave. This introduces feedback and removes the necessity for a reflective end. When the whole of the active length is formed in this way, the feedback is said to be *distributed*. The emission wavelength is controlled by the periodicity of the optical grating and the addition of a *mesa stripe*, as shown, improves the radiated power output.

Large optical cavity lasers. To obtain continuous wave (CW) power outputs in excess of about 25 mW, it is necessary to use a very thin active layer, or to provide a large optical cavity. This type of laser utilises additional guide layers that form reverse biased pn junctions to improve the current carrier containment.

Multimode lasers. The output spectra of a laser with a broad area active layer consists of a series of spectral lines each separated by a few tenths of a nanometre. The device is thus said to be a *multimode* laser.

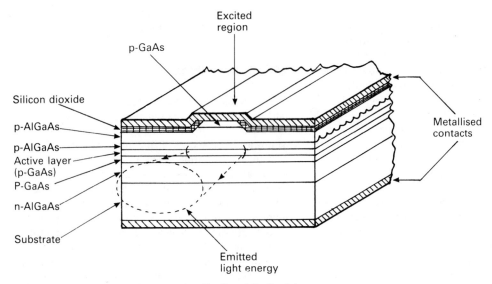

Figure 25.1   Section through mesa stripe distributed feedback laser.

Single mode lasers. Single spectral line, single mode devices are necessary for long haul, wideband systems. These can be fabricated by reducing the cavity width so that only one transverse mode can be supported. In addition, the length must be reduced so that the separation frequency between adjacent multimodes is larger than the transition linewidth.

Slab laser. This represents a developing area of laser technology and may provide a bridge between gas and semiconductor laser systems. The device is formed around a spectral geometry lasing crystal that is capable of providing several kilowatts of output power. The essential features of the simple construction are a crystal, a pumping cavity and two resonator mirrors. The slab laser is compatible with optical fibre couplings and, apart from the water cooling pump, contains no moving parts. The output wavelength is typically around 1060 nm.

Stripe geometry lasers. The basic double heterojunction laser produces light output from the whole width of the active region. This makes it necessary to roughen the side faces to minimise unwanted side emission. The broad emission area also reduces the power density and makes it difficult to couple to a cylindrical optical fibre. These problems are solved by reducing the effective width of the active layer. The stripe geometry is implemented in the manner depicted in Fig. 25.2, by introducing a high resistance area on either side, either by proton bombardment, or by forming an oxide isolation region. At the same time, this fabrication technique reduces the threshold level for lasing.

Transverse junction stripe lasers. A device that is formed on a semi-insulating i-type substrate to minimise the temperature-dependent leakage current. It is fabricated from a combination of homo- and heterojunctions that confines the carriers to a very small active region. This results in a device that produces a relatively high output power for a low threshold current.

25.1.2 Light emitting diodes (LEDs)

Energy applied to semiconductor materials causes the generation of hole/electron pairs. In a complementary manner, energy is released when recombination occurs. In germanium and silicon, this results in vibration of the molecular structure and heating within the crystal. However, in materials such as gallium arsenide, the spontaneous energy released when an electron falls from the conduction band back into the valence band, appears as radiated light. The energy gap between the two bands controls the wavelength (colour) of the emission as indicated in Table 25.1.

Table 25.1

Material	Band gap (eV)	Emission wavelength (nm)
Gallium arsenide	1.43	910
Gallium arsenide phosphide	1.91	650
Gallium phosphide	2.24	560
Silicon carbide	2.5	490
Gallium nitride	3.1	400

In general, the efficiency increases with an increase in forward current and falls with a rise in temperature. Above a threshold of about 1.6 V of forward bias, the light output is proportional to the forward current. Relative to lasers, LEDs are low optical power devices with a small modulation bandwidth (typically less than 100 MHz). Offset against this are the considerable advantages of a simple device of low cost, reliable operation with low temperature dependence, and requiring only simple circuits. Of the four variants available, the dome and planar types are not commonly used for communications purposes.

Burrus diodes or surface emitters. These devices are fabricated from a double heterojunction structure with a well etched deep into the substrate. These two features combine to constrict the radiance to a small angle through the substrate, by a combination of electrical and optical confinement. The fibre can then be efficiently coupled into the well.

Edge emitters. These devices are constructed in a manner similar to lasers, as indicated in Fig. 25.2. The heterojunctions provide a good guiding structure and the stripe geometry helps to reduce the lateral radiation, both helping to increase the radiation intensity. Relative to the surface emitter device, these diodes produce about three times the power output, making them suitable for medium haul, medium bandwidth systems. A typical diode is capable of providing about 1 mW of optical power for a forward current of about 500 mA, with a modulation bandwidth of several hundred megahertz.

25.2 Optical fibres

Although glass is optically transparent over the visible part of the spectrum, its signal losses are relatively high. For fibres made from silica glass, the attenuation can be as low as 0.15 dB/km over the infra-red range of 1.35 to 1.55 μm. For fluoride-based glasses the losses are even lower, at wavelengths around 2.5 μm. Such characteristics allow communications systems to be tuned to minimise the installation costs of fibre against the number of repeaters necessary for long distances. Due to the very small diameter of these fibres, multi-fibre cables can readily be constructed to maximise the transmission bandwidth to installation costs ratio. Such cables need to include metallic strength members to avoid cable laying stresses that could lead to fibre fracture. In addition, the cables have to be sealed against water ingress which would cause an early deterioration of the fibre characteristics.

25.2.1 Fibre characteristics

Attenuation
The system power budget depends upon the efficiency and sensitivity of the transmitter and receiver devices

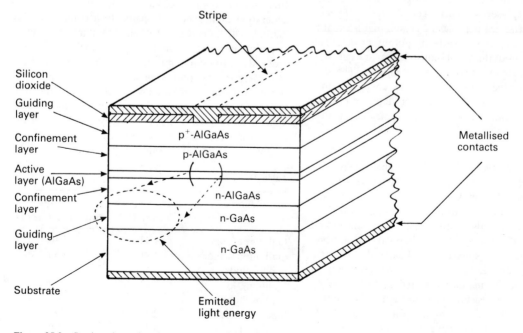

Figure 25.2   Section through stripe geometry, edge-emitting LED.

and the losses in the fibre, this latter element being quite complex.

Absorption. This loss mechanism is due to the molecular structure of the medium and results in some of the transmitted optical power producing heat within the glass. This can be considered in two ways. Intrinsic absorption occurs in pure glass due to the molecular resonances within the glass. Extrinsic absorption is caused by resonances within impurities, such as chromium, copper or hydroxyl (OH) ions trapped within the glass during manufacture. OH absorption is practically negligible over the range 1.3 to 1.6 μm.

Bend losses. If the fibre path is curved beyond some limiting value, then its guidance nature deteriorates and some of the energy will be radiated into the space surrounding the fibre.

Dispersion. A pulse of light contains a wide range of spectral components and these propagate through the fibre with differing velocities. Therefore the pulse energy spreads out in time to give rise to inter-symbol interference, loss of signal level and degradation of signal to noise ratio.

Scattering. These losses are caused by energy from one propagating mode being transferred to an alternative leaky mode, nearly always due to imperfections and impurities in the glass. Scattering losses have been characterised as either a *linear* or *nonlinear* phenomena. Included in the former is the dominant loss due to random variations of refractive indices over distances that are small compared with a light wavelength and known as *Rayleigh* scattering. Variations of the reflective interface and any non-cylindrical geometry of the fibre, over distances comparable with the light wavelength, give rise to *Mie* scattering. Nonlinear scattering occurs more frequently when using high levels of power, giving rise to the generation of spurious frequencies so that energy is lost from the original carrier. *Brillouin* scattering is the modulation of the light carrier by molecular vibrations due to heating. *Raman* scattering is similar but usually has a higher power threshold level and creates a wider modulation range (see Raman effect, p. 230).

Modal propagation

A typical optical fibre is constructed in the bi-cylindrical manner indicated in Fig. 25.3 and Fig. 25.4, the glass used for the inner core having a higher refractive index than for the outer cladding. The behaviour of a ray of light directed into the core can be predicted from Snell's law which states that $n_1 \sin \theta_1 = n_2 \sin \theta_2$, where n_1 and n_2 are the refractive indices of two media forming a reflective layer and θ_1 and θ_2 are the angles of incidence

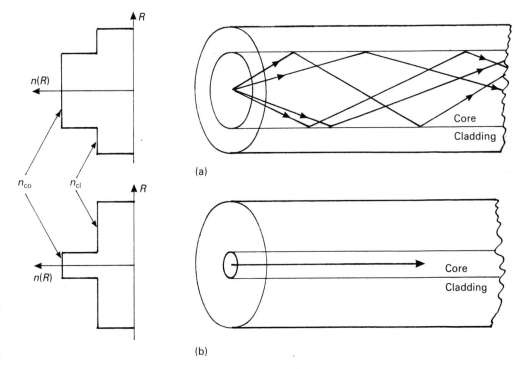

(a)

(b)

Figure 25.3 Step index fibres: (a) multimode; (b) monomode.

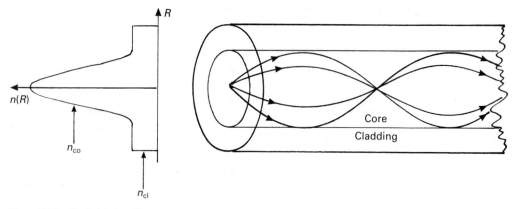

Figure 25.4   Graded index fibre.

and refraction to the normal, respectively. For two practical media, a critical angle θ_c will be found when total reflection occurs at the interface and this is given by $\sin\theta_c = n_{cl}/n_{co}$, where n_{cl} and n_{co} are the refractive indices of the cladding and core, respectively. For example, if the cladding and core refractive indices are 1.4 and 1.5, respectively, then the critical angle is approximately 69° to the normal. Any ray directed into the fibre at an angle greater than this will therefore be trapped and transported to the far end.

Monomode step-index fibre. Typically the inner core is made less than 10 μm diameter so that only one mode of wave can propagate in the manner shown in Fig. 25.3(b). The outer cladding diameter must be more than 10 times this value to minimise losses. These fibres are suitable for wide bandwidth, long haul systems.

Multimode step-index fibre. De facto standards of 50 μm core and 125 μm cladding diameters have been established for telecommunications operations. As several modes can propagate simultaneously in the manner shown in Fig. 25.3(a), the scattering and absorption losses are higher, making this type suitable for short haul, bandwidth restricted, low cost applications.

Multimode graded-index fibre. In this type the core glass has a radially graded refractive index. As indicated in Fig. 25.4, the rays tend to take a sinusoidal path through the core with the fastest components taking the longest path. As these rays can be considered to represent the different modes travelling through the fibre, they all have the same transit time. Thus the losses are relatively very small. Telecommunications de facto standard diameters have been agreed for core and cladding diameters of 50 μm and 125 μm, respectively. The best grades of this type are suitable for long haul, wide bandwidth systems.

Raman effect. When monochromatic light is passed through a medium such as glass, some of the light energy scatters. If the output spectrum is examined, it will be found to contain not only the original wavelength, but also other lines at a lower energy level that are spaced by a constant amount. The extra spectral components are described as Raman lines. This spectrum spreading is due to the loss or gain of energy to the photons, as a result of their interaction with the vibrating molecules of the medium through which they are passing.

25.2.2 Fibre technology

Jointing

Various jointing techniques are available and they all rely upon forming smooth and square end faces by a cleaving process. Simple butt joints are not suitable for anything but short haul systems, but clamps that interpose a spherical lens between the two ends provide a significant improvement. A further simple approach uses a vee groove in which the two prepared ends are clamped before being sealed with an ultraviolet (UV) light curing epoxy adhesive. As this technique relies upon alignment of the outer cladding, any eccentricity of the core will exagerate the jointing loss. However, joint losses of less than about 0.5 dB can be regularly achieved with this technique. The preferred alternative is fusion splicing, the technique shown in Fig. 25.5. Light is injected into one fibre end using the bend technique and then extracted from the other in a similar way. The two ends are then manipulated under microprocessor control, until maximum light level is detected at the output. At this point, an electric arc is struck, the glass fuses and a permanent joint is effected with a typical loss of less than 0.05 dB. Core/cladding eccentricity produces no errors with this technique. Glasses with different refractive indices have different molten surface tensions, so that the cores tends to align automatically during the fusion state.

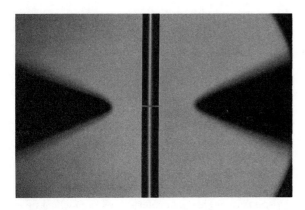

(a)

(b)

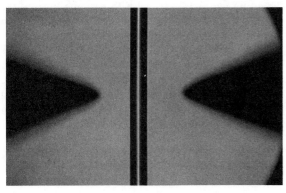

(c)

Figure 25.5 Fusion spliced optical fibres: (a) fibre ends aligned; (b) arc struck; (c) splice completed (Courtesy of BICC Research & Engineering Ltd).

Manufacturing

Glass for optical fibres has to be produced from ultra pure chemical compounds and in an ultra clean processing environment to avoid the impurities that degrade system performance. There are two basic methods of production, each with a number of variants. The lower melting point glasses are generally produced in the liquid or molten phase, while the high temperature glasses tend to be fabricated using the vapour phase technique.

Liquid or molten phase production. Oxides and carbonates of suitable elements in powder form are mixed in the correct proportion to give the required refractive index. These are then heated in a silica-lined furnace and the molten glass used to produce rod- and tube-shaped preforms. The rod forming the core material is inserted into the cladding tube and the latter is collapsed to form a cylindrical glass component. From this, the glass fibre is pulled after heating in a second furnace. An alternative process uses a double crucible technique in which the core glass is pulled through the molten cladding glass, thus coating the core with the cladding. These techniques are used for step index fibres with a relatively large core. However, the latter technique can be modified to produce a degree of graded index fibre where the refractive index can be modified by ion exchange when the glass is in the molten phase.

Vapour phase production. There are four important variants of this technique and all are used to produce glass of a very high degree of transparency. Glass material is laid down using an oxidation process on a substrate while the constituent elements are in the vapour phase state. In this way, glass can be built up in a series of layers with very carefully controlled refractive indices by the addition of dopant gases. The process is applicable to the production of both step and graded index fibres and provides fibres with low loss and wide bandwidth characteristics.

Modified chemical vapour deposition (MCVD). Glass layers are deposited on the inside of a silica tube due to the through flow of a suitable gaseous mixture and under the influence of an external traversing heating source. Glass is deposited layer by layer with refractive indices that are controlled by the nature of the gaseous mixture during that particular pass. The cladding and core are built up in successive stages and from this preform, a graded index fibre can be produced.

Outside vapour phase deposition or oxidation process (OVPD or OVPO). Deposition takes place on the outside of a rotating mandrel which is contained within a suitable gas flow. The actual deposition is produced by hydrolising the gas with an oxy-hydrogen flame. This leaves a fine sooty deposit of silica on the cooler mandrel. Glasses of various refractive indices are produced by changing the gaseous mixture. After several hundred passes, the mandrel is removed from the preform, which is then collapsed and sintered, before being drawn into a fibre. Highly accurate graded index fibres can be produced with this technique.

Plasma-activated chemical vapour deposition (PCVD). This is a variation on the MCVD technique but using various plasmas either heated through a 2.45 GHz microwave cavity or from inductively coupled RF heating sources to provide the necessary energy. Due to the use of this form of energy, very accurate and concentric layers can be laid down without rotating the silica tube. Fast traversing of the heating system produces very thin layers of various refractive index glass (depending on the constituent gases) to give excellent grading.

Vapour axial deposition (VAD). This technique forms the basis of an almost continuous process. Vaporised gases are directed from burners on to the end of a rotating fused silica target which is continually pulled vertically. Flame hydrolising produces a fine soot that provides a boule shaped porous preform. As the preform grows, the upper region can be sintered and, from this, a fibre can be drawn in a continuous process. Variation of refractive index is achieved by varying the gas flow conditions as in the other methods.

25.3 Photodetector/receiver devices

These devices, whose function is to transform energy in the form of light into electric signals, are probably the most important element in the system. Their sensitivity, noise performance and efficiency set the limits of the system power budget. Photons of light energy incident upon such a device, excite electrons in the valence band to generate current carriers. The important features of these devices include, high sensitivity, linearity, short response time, low noise level, low bias voltage, high reliability, low cost and small physical size. While photomultipliers and vacuum photodiodes meet some of these requirements, their bulk and high operating voltages make them unsuitable. The devices used for optical fibre communications systems are therefore all semiconductor types.

25.3.1 Detector/receiver characteristics

Dark current. Ideally a detector that is contained in darkness should produce zero photocurrent. However, since there will always be some thermally generated current carriers, this dark current will have a finite value. In general, this value is higher for the semiconductor materials with the lower bandgap.

Long wavelength cut-off. The absorption of the photons that generate the photocurrent is strongly dependent upon the light wavelength and the semiconductor bandgap. A threshold of detection termed the long wavelength cut-off is often used to define this relationship as $\lambda_c = hc/E_g$, where $h = 6.626 \times 10^{-34}$ J s, Planck's constant, $c = 3 \times 10^8$ m/s, and $E_g = $ bandgap in ev.

Quantum efficiency. This is simply the ratio of the number of photoelectrons collected to the number of incident photons per unit time.

Responsivity. This parameter is a more practical figure of merit for any photo-sensitive device. It is the ratio of the output photocurrent to the incident optical power that produced it and measured in amps/watt. At any particular wavelength, the responsivity is directly proportional to the quantum efficiency.

25.3.2 Detector/receiver devices

Avalanche photodiode (APD). The quaternary wafer depicted in Fig. 25.6 results in a very high internal electric field when the diode is reversed biased by about 100 V. Hole–electron pairs generated by light energy excite new pairs by impact ionisation – the avalanche effect. This results in internal current carrier gains in the order of 1000 times relative to a simple pn junction. The profile shown is adopted to act as a guard ring in order to minimise edge leakage currents that would reduce efficiency. APDs are available to cover the range of about 0.8 μm to 1.6 μm and with response times in the order of 200 ps. Some avalanche noise is produced because the ionisation process has a random nature.

PIN diodes. These diodes have a layer of intrinsic (I type) material interposed between the p and n sections. This increases the width of the depletion region and increases the photon absorption efficiency (quantum efficiency). Diodes are available to operate at wave-

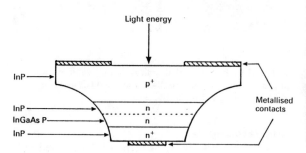

Figure 25.6 Section through heterojunction avalanche photodiode.

lengths between about $1.1\,\mu m$ and $1.8\,\mu m$, while some based on gallium arsenide (GaAs) technology have dark currents as low $0.2\,nA$ and with response times in the order of $100\,ps$.

PN junction diodes and transistors. Photo-sensitive devices have a very thin diffused layer. The photons of light energy penetrate this to generate the current carriers under reverse bias conditions. The magnitude of the current depends not only on the level of light energy, but also on the level of impurity doping and the reverse bias voltage. The diode does not have the internal gain of the transistor. Neither silicon nor germanium devices will operate efficiently at wavelengths greater than about $0.9\,\mu m$. Diodes have a peak current of around $600\,\mu A$ at about $40\,V$ of reverse bias. By comparison, transistor peak currents are in the order of about $15\,mA$ with $20\,V$ dc supply. Both devices are reasonably linear to changes of light levels.

Reach-through avalanche photodiodes (RAPD). The electric field at avalanche breakdown should be as low as possible to minimise the random noise component of the process. In the reach-through diode, a very lightly doped, almost intrinsic layer, is interposed, and this has the effect of removing some of the excess field strength. These devices have less internal gain than the standard APD, but with a somewhat lower noise level. Response times of less than $500\,ps$ at wavelengths of between 0.8 and $1.0\,\mu m$ can be achieved.

25.4 Miscellaneous terms

Bandwidth–length product. The information-carrying capacity of a fibre system can be described in terms of this parameter. For a given quality of service, the transmission length and signal bandwidth can be exchanged on a simple trade-off in the manner of the Hartley–Shannon law.

Bend loss. When the path of a fibre is curved, the angle of incidence within the guide changes. If this angle falls below the critical value, total reflection will not occur and some energy will escape through the cladding.

Coherent detection. This technique, which is based on the superhet radio receiver, can be beneficial to long haul, wideband monomode transmission systems. It requires that the received optical signal is mixed with a second locally generated, unmodulated optical signal. Then a difference frequency can be extracted from the mixing products to form either an intermediate frequency (IF) for further processing, or if the local and transmitted signals have the same wavelength, then

this homodyne mixing yields the baseband modulation component directly. The process is suitable for all forms of direct light modulation, both analogue and digital. The major advantage of the concept lies in the ability to increase the local oscillator power up to a level that gives an effective improvement to the sensitivity of the following detector stage. Offset against this is the requirement that both the source carrier generator and the local oscillator should produce very narrow linewidth signals. Used on multimode systems, spurious mixing products are likely to degrade the received signal to noise ratio.

Coupler/multiplexer (see also T-coupler, p. 235) Optical waveguides can be formed within a glass substrate so that optical signal paths can be split or combined in a variety of ways. Figure 25.7 shows a simple 1- to 2-way coupler or splitter. The waveguide pattern is formed on a special glass substrate using photolithographic and masking techniques, similar to those used in semiconductor fabrication. The final waveguide is formed by an ion exchange process that takes place in a molten salt bath when thallium ions are exchanged for those of sodium and potassium ions contained within the glass. Fibre pigtails are then attached to the ports using an index matched adhesive. At the typical operating wavelength of 800 to 1550 nm, the through loss is only in the order of 0.2 dB and the 3 dB split of one signal into two paths can be achieved within ± 0.1 dB.

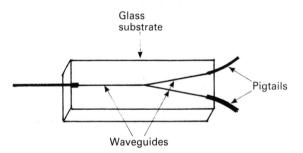

Figure 25.7 Optical multiplexer/splitter/coupler.

Erbium-doped fibre amplifier (Raman amplifier). This device uses the Raman effect to amplify a light signal directly. A loop of about 20 m of silica fibre is doped with trivalent erbium and linked into the optical fibre signal path in the manner shown in Fig. 25.8. The low level signal to be amplified is combined with a high level pump signal using a coupler. The combined signal then passes through the length of doped fibre. The pump source commonly uses a wavelength of 980 nm or 1480 nm and requires a few tens of milliwatts to provide about 50 mA of drive current. As the two signals pass through the doped section, the weak signal extracts

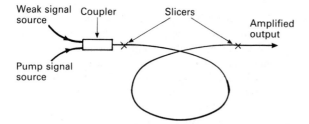

Figure 25.8 Erbium-doped fibre amplifier.

energy from the pump signal through stimulated emission and ionic excitation of the molecular structure. The wanted signal thus becomes coherently amplified as it passes through the doped fibre. These amplifiers have a high degree of linearity and can easily handle data rates as high as 100 Gbit/s.

Heterodyne/homodyne detection. See Coherent detection, p. 233.

Neodymium-doped yttrium–aluminium–garnet (Nd YAG) laser. Although not commonly used for fibre optical communications purposes, this non-semiconductor device is included to show the versatility of the YAG material. A small rod of this compound carrying dielectric mirror end surfaces can be optically pumped at one end and made to emit laser light at the other, typically at wavelengths between 1 and 1.35 μm and with a very narrow linewidth.

Numerical aperture. This is a parameter that defines the acceptance angle or cone for an optical system. In the case of a fibre it is described as the sine of the maximum possible coupling angle. In mathematical terms it is thus given by $N_a = \sqrt{n_{cl}^2 - n_{co}^2}$.

Optical delay lines. When processing optical signals it is sometimes necessary to introduce delays into the signal path (for example, the equivalent of equalisation in an electronic circuit). By using silicon semiconductor technology, suitable coiled lengths of fibre can be fabricated. Such guiding structures can be formed using a silicon substrate with layers of undoped and arsenic doped silica. At the typical operating wavelength of 1550 nm, delays of about 100 ps can be formed on a 2 cm square area.

Optical space switch and wavelength division multiplexing (WDM) (see also T-coupler, p. 235). An optical guiding structure formed into a Y shape with a very narrow angle and fabricated from a compound such as lithium niobate (LiNbO$_3$), can, if coupled into a fibre system, be used to split a light signal into two parts. The structure can also be used in reverse to combine two modulated light signals for transmission over a single

fibre. If the two signals are of different wavelengths, then the resulting signal spectrum on the common fibre is that of a wavelength division multiplex. If this structure is equipped with parallel conductor strips, then an applied electric field will cause the refractive index of one path to increase while that of the other falls. The biasing voltage therefore causes the light energy to be propagated along the path with the higher refractive index, thus switching the common input signal on to either of the output fibres.

Optical time domain reflectometry (OTDR). An optical fibre system produces signal reflections at a point of refractive index change, in the same way as a conventional transmission line does at points of impedance mismatch. By intermittently pulsing a transmission fibre, reflections will be produced by backscatter at couplings, joints and breaks. These can be displayed either on an oscilloscope or chart recorder to provide a record of the system status. Then from a knowledge of the propagation velocity, the positions of any undue reflections can be interpreted as a distance along the fibre. This technique, which is suitable for both field and laboratory, provides a simple way of testing a complete system from either end. Any faults that occur within a *dead zone* of about 5 m from the testing end are difficult to locate. This problem is however simply resolved by inserting about 5 m of known good fibre between the system and the instrument.

Raman fibre amplifier. A high intensity light beam at a frequency of ω_p, the pump frequency, is combined with a second beam at frequency ω_s known as the Stokes beam, in a transparent medium. The frequencies are chosen so that the difference frequency ω_d coincides with a resonant frequency for molecular vibration within the medium. This results in energy being transferred from the pump signal to the Stokes beam to provide amplification. Optical power outputs approaching 100 mW can be achieved in this way.

Refractive index (n). The ratio of the phase velocity of light waves in free space to that obtained in a given medium.

Semiconductor laser amplifier. In intensity modulated coherent systems where the signal dispersion is small and the attenuation is chiefly due to fibre losses, the optical cavity of a laser can be used to provide amplification. This avoids regeneration of the original electronic signal, reduces repeater costs and allows for predetector amplification to improve sensitivity. Typical optical power outputs in the order of about 3 mW can be obtained in this way.

Soliton. A wave or pulse that does not disperse or lose its shape as it propagates through a medium. By comparison, a practical light pulse contains more than one wavelength, so that as the pulse travels through the fibre, these components travel at differing rates creating dispersion. Soliton pulses that have a duration of less than 1 ps can propagate through 10 000 km of fibre and still retain their shape. By using pulse position modulation, such pulses should be capable of providing very high speed digital services over great distances and without repeaters.

T-coupler or tapping element. An optical component designed to combine the light energy of two fibres. It may also be used to split a single light source into two parts.

Wavelength division multiplex (WDM). The optical transmission equivalent of frequency division multiplex whereby the bandwidth is shared between carriers of different wavelengths.

Wavelength and time division multiplex (WTDM). The very wide bandwidth available on an optical fibre is readily available through the use of a hybrid multiplex hierarchy. A number of baseband signals, such as television, may be digitally modulated into time division multiplex (TDM) time slots in the conventional PCM manner. A group of similar structures can then be wavelength division multiplexed (WDM) on to a single optical fibre. By using laser sources in the 1300/1500 nm wavelengths, it is possible to generate carriers with a separation as small as 4 nm. Sixteen such carriers are capable of supporting a gross bit rate greater than 40 Gbit/s.

25.5 Useful references

Allan W.B. (1980) *Fibre Optics*, Oxford University Press, Oxford.

Kao C.K. (1988) *Optical Fibres*, Peter Peregrinus, IEE, London.

Midwinter J.E. (1979) *Optical Fibres for Transmission*, Wiley, London.

Senior J. (1985) *Optical Fiber Communications, Principles and Practice*, Prentice Hall International, London.

Due to the rapid expansion in the use of computers in communications and control, and the adoption of many portable devices, the security of power supplies against data corruption and loss of control has become particularly important. The range of such protection varies from small on-board back-up batteries to support volatile memories, to the provision of back-up supplies to protect mains power driven equipment. Such *uninterruptible* power supplies can be designed to give protection for periods varying from a few minutes up to several hours. The use of high speed digital computer devices that operate from low voltage, high current dc supplies, which may have to support in excess of five times the normal operating current at switch-on, helps to increase the radiation of electromagnetic interference and pollute the mains supplies. Even battery-driven mobile equipment is not immune to these problems. (See also Electromagnetic compatibility/interference, p. 97.)

26.1 Batteries and portable supplies

There are now many chemical compound pairs available to battery manufacturers and these are selected to provide the required battery parameters. Batteries are categorised as either being *primary* or *secondary*. Basically the former are *throw-away* devices when discharged, while the latter are rechargeable. No attempts should be made to incinerate or recharge primary cells, as this can lead to an explosion. Recent technological developments make it economical to recycle these batteries to recover the heavy metals such as mercury, lead and cadmium used in their production. Due to the toxicity of certain chemicals, the rechargeable batteries are considered to be more environmentally friendly.

Battery capacity is specified by the product of discharge time and a constant current. This quantity is thus expressed in *ampere hours* (Ah), the discharge time being limited by the *end-point* or *end of discharge voltage* (EODV). The rated capacity is commonly quoted for 5, 10 or 20 h discharge time. Thus C_{10} is the rated capacity for a constant current discharge to the end-point voltage in 10 h. The rated current I_{10} would be 1/10 of the rated

capacity in Ah. For rechargeable cells, discharge beyond the end-point voltage should not be exceeded too often as this abuse tends to reduce the useful lifetime. The recharge rate typically ranges between 150% and 300% of the Ah capacity.

Alkaline cell

A non-rechargeable primary cell developed from the zinc carbon cell but uses an alkaline electrolyte of potassium hydroxide, a granulated zinc anode and mercuric oxide as a depolariser. The chemicals are contained within a sealed steel casing which does not take any part in the chemical process so making the device leak-proof. The cell has a rated voltage of 1.5 V and an end-point of 0.9 V. Cells with a capacity of up to 8 Ah are available and all have a long shelf-life and the ability to support relatively high currents over a wide range of operating temperatures.

Large value capacitor

Although a capacitor is not a cell in the accepted sense, the device is capable of holding an electric charge and therefore behaves in the manner of a rechargeable cell. Capacitors with a value of 1 to 5 F are now available, in sizes of less than 1 cubic inch. If charged slowly through a relatively high resistance, then the very low self-discharge rate can provide back-up for such as a C-MOS memory for several weeks.

Lead acid battery

Large rechargeable batteries suitable for use with *float charge* uninterruptible power supplies. Generally available as 6 or 12 V units in multiples of 2 V cells. These have a high charge capacity and an ability to withstand abuse. The electrodes when charged consist of plates of lead dioxide (+ve) and spongy lead (−ve), with an electrolyte of sulphuric acid or a thixotropic acidic gel compressed within the plate structure. During discharge both plates tend towards lead sulphate, water is produced and the specific gravity of the acid falls from about 1.2–1.28 to 1.15, a value that governs the discharged state and an end-point voltage about 1.7 V. The recharging process which reverses the chemical reaction must be carried out at constant voltage. This generates gases, chiefly hydrogen, which need to be adequately vented. The typical float charge voltage is 2.25 V per cell. If this figure is exceeded, then gassing becomes excessive.

Lithium based cell

Lithium is a highly reactive alkaline metal which degrades rapidly into lithium hydride with the release of hydrogen gas even in humid air. However, if suitably sealed, these cells have the highly desirable feature of producing about 3 V per cell. In the commonly available types, pure lithium forms the anode and either manganese dioxide or sulphur dioxide is used for the cathode, the electrolytes being lithium perchlorate dissolved in an organic solvent or methyl cyanide with a small amount of lithium bromide added to improve conductivity respectively. Each cell is contained within a stainless steel casing to resist corrosion. The rated voltage is typically 3 V per cell but some types can produce up to 3.3 V on open circuit. The lithium polycarbon monofluoride and the lithium thionyl chloride are further variants that are available and provide open circuit voltages of 3 V and 3.7 V, respectively. All these devices provide a high energy density with capacities up to 11 Ah, together with a long working life over a wide range of operating temperatures.

Lithium ion cells

This rechargeable cell development avoids the direct use of lithium because it oxidises too readily. The carbon based anode is formed from a mixture of compounds at about 1100°C and then electrochemically doped with a lithium compound. The cathode is formed from a mixture of the compounds of lithium, cobalt, nickel and manganese. A non-aqueous mixture of the carbonates of propylene and diethyl is used for the electrolyte. The anode becomes doped with lithium during the initial charge and lithium ions migrate through the electrolyte to be absorbed by the cathode during discharge. This cell is nominally rated at 3.6 V and has a long self-discharge period, typically falling by 30% after 6 months. It has a storage capacity of about three times that of the NiCd cell and without the memory effect. The recharge period is typically about 3 h and the cell can withstand at least 1200 charge–discharge cycles.

Manganese silver chloride cells

These non-rechargeable cells are completely inert until immersed in water, at which point they reach full operating voltage within a fraction of second. They provide a high energy density, with an indefinite storage life, without performance degradation, over a wide range of operating temperatures. These properties make them particularly suitable for marine search and rescue equipment.

Mercuric oxide cell

A cell that is capable of providing a rated 1.35 V and was available in both large and button cell forms. Gives a good voltage stability over the temperature range of −30°C to +70°C. Because of the high energy density and small size, these devices contributed a great deal to the miniaturisation of electronic equipment. This type is now largely obsolescent.

Nickel cadmium (Ni–Cad) cell

These cells, which are the workhorses of portable rechargeable power sources, consist basically of a positive pole constructed from a steel tubular mesh, coated with solid nickel hydroxide. This is surrounded by a second perforated steel tube into which powdered cadmium is sintered. The electrolyte, potassium hydroxide, is in a gel form. An alternative cell that is smaller for a given capacity, the mass-plate type, is constructed without the steel framework. During discharge, the hydroxide ions oxidise the cadmium to produce the energy, this reaction being reversed during the recharging process. The steel casing is equipped with an automatic resealing vent to allow gases to escape during charging. It is recommended that these cells be charged from a constant current source. When this is not possible, then maximum charge rate should not be exceeded at any time. As supplied, these devices have only a residual charge and therefore require charging before entering service. Recharge cycles up to 10 000 times have been recorded but this depends on the working conditions. Repeated overcharging, high operating temperatures and discharge abuse considerably reduce the lifetime. The fully charged voltage of 1.25–1.3 V provides a rated voltage of 1.2 V per cell, the typical end-point voltage being 0.9–1.0 V. These cells tend to have a relatively high self-discharge and the shelf-life, without charging, is therefore short. In use, a Ni–Cad is likely to provide only about 25% of the operational time as a similar capacity alkaline cell, before it requires to be recharged for about 14 h. Operation below about 0°C is poor and recharging should not be attempted below this temperature.

Nickel cobalt (Ni–Co) cell

This rechargeable cell has been integrated into a photovoltaic system for powering mobile equipment indirectly from sunlight. The chemical reaction takes place between cobalt, nickel hydroxide and water to produce a cell with a rated terminal voltage of 1.28 V. During discharge, the cobalt becomes oxidised to produce the energy. This reaction is reversed during the recharging process which, if carried out from a 1.5 V per cell source, does not produce any gassing. The chief advantages over the Ni–Cad cell is that it produces about the same terminal voltage but with twice the power density and at about the same production costs.

Nickel–hydride cells

This group of rechargeable cells include the nickel–hydrogen and the nickel–metal hydride (typically caesium hydride). These provide a significantly higher energy density (typically up to 2.5 Ah), a greater discharge depth and are more robust against overcharging than Ni–Cad cells.

Solar cell

When a pn junction is irradiated with light an electric field is developed that is responsible for a photo-emf. For example, a silicon pn junction of about $4\,cm^2$ can produce an open circuit voltage of 0.6 V and a short circuit current of around 100 mA in bright light. Arranging many such cells in series parallel produces a solar panel of the type commonly used to power space craft. Power outputs in excess of 7.5 kW can be produced in space from the solar constant of about $1.4\,kW/m^2$ with a conversion efficiency in the order of 15%.

More efficient solar cells can be fabricated from compounds such as gallium arsenide (GaAs) and gallium antimonide (GaSb), that have a greater bandgap than silicon. These materials produce a better response to infra-red radiation and the addition of individual lenses to each cell gives rise to a further improvement.

Zinc carbon cell

This 1.5 V cell that is now practically obsolescent has a carbon positive pole, a zinc case that forms the negative pole and a gel of ammonium chloride as an electrolyte. Since the case zinc takes part in the chemical reaction, it ultimately disintegrates and leaks. This source of corrosion is not acceptable in present types of electronic equipment.

26.2 Stabilised/switched mode power supplies

The conventional transformer, rectifier and filter type of power unit is not compatible with the requirements of modern semiconductor equipment. Generally these operate from low voltage, high current supplies and this makes considerable demands upon the stability of output voltage and tends to increase the amplitude of the ripple component. This latter problem can be minimised by the use of a series regulator device in which a monitor circuit compares the output voltage

(v_{out}) with the voltage from a reference source to generate a difference or error signal. This is then used to control the level of v_{out} via a regulator circuit. However, this only adds to the heat dissipation problems because the regulator stage is operating in the linear mode. Although such a stabilised power supply can be adapted to provide either constant load current or voltage, the concept still requires the use of relatively large and heat dissipative components. To minimise heat dissipation the series device should be switched between saturation and cut-off. In both of these states, either the volts drop across the device or the current through it is practically zero, hence it dissipates very little power. The conductive period is used to charge a reservoir capacitor that in turn continues to provide the output voltage during the cut-off period. Increasing the switching frequency to well above that of the mains supply, typically above 10 kHz, significantly increases the ripple frequency so that even a smaller filter can provide better ripple suppression. This *switched mode* power supply (SMPS) is operated in a pulse width modulated (PWM) manner under the control of the monitoring circuit. As the load current increases, the load voltage tends to fall and this action causes the regulating circuit to increase the *duty cycle* or on period, so that the output voltage can rise to meet the demand of the load. The mean value of the output voltage v_{out} is thus proportional to the product of v_{in} and the duty cycle ($t_{on}/(t_{on} + t_{off})$). SMPS systems are smaller, dissipate less heat, are lower cost and more compatible with modern semiconductor circuits than any previous concept. The control sections of SMPS can operate with either one or two control loops. The single loop concept is described as the *voltage mode* where the action is as described above. However, sudden changes in load current (transients) can cause inductor saturation. Since the series inductor current is proportional to the input voltage, the introduction of second loop to monitor changes of input voltages can terminate the conduction period before saturation sets in. This technique is referred to as the *current/voltage mode*. There are four popular methods of implementing control of SMPS and these are described as follows.

Flyback or ringing choke convertor. This circuit, shown in Fig. 26.1, is fed from a rectified and unfiltered ac source (raw dc) via a switching transistor. During the conductive period, current flows through the inductor L_1 into the reservoir capacitor C_1 to provide the output voltage v_{out}. During this period, a magnetic field builds up around the winding of L_1. Immediately the switch turns off, this field collapses and the energy released produces an emf that keeps the current flowing, but now through the single fast diode D_1 to continue to supply a falling v_{out}. The change in load voltage is sensed by a

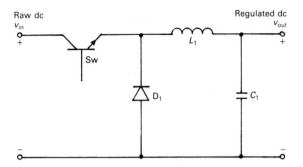

Figure 26.1 Basic flyback switched mode supply.

control loop which forces a change in the mark to space ratio of the switch. This then stabilises v_{out} with the minimum of ripple. To minimise the radiation of radio frequency interference (RFI), this circuit needs to be driven from a mains supply via a fairly large transformer having a gapped core to prevent saturation. The circuit is not particularly suitable for driving loads where high peak currents are encountered.

Forward convertor. In this circuit, shown in Fig. 26.2, the switching stage is isolated from the diode circuit by the transformer, an arrangement which places a lower voltage stress on the switching transistor. During the conductive period, the reservoir capacitor C_1 is charged via D_1 and the choke L_1. The magnetic field developed around L_1 collapses when the switch turns off and the discharge energy continues to flow but through the diode D_2, in the manner of the flyback convertor. The output voltage v_{out} is maintained with a small ripple by varying the duty cycle of the switch, under the influence of the feedback control system. This circuit is usually modified by the addition of a third transformer winding and series diode to shunt the switching transistor and its winding. This helps to recover energy that would be wasted from the magnetic field surrounding the transformer. The technique is more efficient than the

flyback convertor circuit and is capable of handling loads with high peaky current demands.

Resonant mode convertor. With this technique, the switching duty cycle or pulse width is held constant. This allows switching speeds as high as 1 MHz to be used, with the attendant easing of the ripple filtering problem. The circuit is operated near to its self-resonant frequency and feedback control is used to frequency modulate the switching frequency. Increasing the switching frequency increases v_{out} and vice versa. The large increase in operating frequency allows for the use of very much smaller magnetic and capacitive components, to show a considerable reduction in size, weight and cost.

Cúk convertor. This SMPS unit consists of two sections similar in form to that shown in Fig. 26.2, but are capacitively coupled. The energy stored in the first inductor due to the dc input, is transferred by the switched coupling capacitor into the output section inductor to provide the dc output. The circuit is effectively a dc-to-dc transformer whose transfer ratio (step up or step down) is governed by the duty cycle of the switch. Since the power unit is designed to operate in a programmable system, the switching frequency can be synchronised to the system clock which may be as high as 12 MHz. The first stage acts as a pre-regulator to provide a fixed voltage across the second series regulator. The system which can be constructed from small components, provides an overall efficiency that can be greater than 90% and generates very little radio frequency interference.

26.3 Uninterruptible power supplies (UPS)

Line conditioner. A device that consists essentially of a tuned transformer that compensates for slow variations of mains input voltage. By the addition of suitable filtering stages, the effects of voltage surges and mains-borne interference are removed to provide a clean sine wave output voltage. The device is relatively large and consumes power even when off-load. As a UPS, the device only provides for power reductions of durations that are fractions of an input cycle.

Off-line UPS. These are battery back-up systems that have to be switched when a power loss is detected. The changeover time, which is the summation of sense and switching periods, may be 5 to 10 ms. This may not be acceptable under some circumstances.

On-line UPS. Mains power is used to continually float charge batteries that are coupled in parallel with the

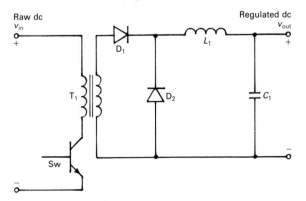

Figure 26.2 Basic forward convertor switched mode supply.

mains derived dc output. When a power failure occurs, the system can continue to function automatically for several hours. When the mains supply is restored, the system automatically reverts to the charging mode.

Voting type UPS. See Off-line UPS, p. 239.

26.4 Miscellaneous terms

Battery memory. See Voltage depression phenomena, p. 240.

Black out. A total unpredictable and complete loss of power.

Brown out. Reduction in mains power supply that may be sufficient to result in the complete failure of some systems while others may still be able to run at reduced output. Such situations may be planned.

Charge/diode pump circuit. The circuit shown in Fig. 26.3 forms the basis of a voltage doubler (multiplier). On the first positive going half-cycle of the input, C_1 charges to peak input. When the supply goes negative, this charge is transferred to C_2 by the action of the diode switches. Thus the output voltage v_{out} is pumped by the action of C_1 into the reservoir C_2. The circuit is particularly adaptable as it can be driven by a digital signal to provide a low level, dc control signal. This circuit has much in common with the multi-stage Cockcroft–Walton multiplier used to obtain very high voltages.

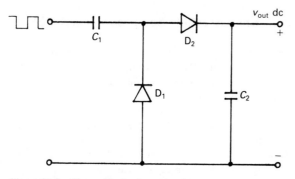

Figure 26.3   Charge/diode pump circuit.

Depolariser. The chemical reaction within primary cells results in hydrogen collecting at the positive pole. This can polarise the charge within the cell to neutralise the output voltage. This is prevented by surrounding this pole with a substance that oxidises the hydrogen, a depolariser.

Electronically resettable fuses (multifuses). These solid-state devices have a positive temperature coefficient and their resistance can jump from a few milliohms to several ohms when the current flowing exceeds some threshold value. This tripping current is typically 50% above the normal rated current. These devices thus behave as fuses that have the ability to reset when the high level of current is reduced.

Fuel cell. This is a device that produces electrical energy by the oxidation of a fuel. The cell is rechargeable only by the replenishment of the fuel. The simplest cell consists of oxygen and hydrogen brought together over catalytic electrodes (see Zinc air fuel cell, p. 241).

Peak in-rush current. At switch-on when the charge on the reservoir capacitor is zero, the peak current that flows is the charging current, equal to $V_p \omega C$. Therefore diodes and fuses must be rated to accept this. As the value of C is reduced to reduce this current, the magnitude of the output ripple will increase. The value of the reservoir capacitor is thus a compromise.

Power factor correction. The power factor is defined as the ratio of the true power consumed from an ac mains supply to the apparent power delivered, the difference between the two representing wasted power. The true power is equal to the product of apparent power and the power factor. Only in the case of a true resistive load is this ratio equal to unity. Most practical loads tend to be inductive and this produces a phase angle between the two components. The power factor is then given by the cosine of this phase angle. Power factor correction, aimed at recovering the wasted power, involves coupling a suitable valued capacitor in parallel with the load to reduce the phase angle to zero.

Snubber. Small low value capacitors wired in parallel with rectifiers to by-pass radio frequency interference currents to earth.

Voltage depression phenomenon or battery memory. This shows itself as an apparent loss of capacity in a rechargeable cell. After a fully charged cell has been in service for a short time, the terminal voltage will suddenly drop by about 0.1 V. It will then continue to operate practically normally up to the end of discharge voltage. A fully charged cell that is only partially discharged before being recharged and then cycled in this way a number of times will often produce this phenomenon at the same voltage level. Hence the term *memory*. The overall effect is to produce a cell that has to be recharged more frequently.

Voltage reference sources

Band gap reference. This is derived from the highly predictable base to emitter voltage for a transistor, which is based on the energy band gap voltage for crystalline semiconductor material. At 0 K this has values of 0.785 eV and 1.21 eV for germanium and silicon, respectively. As an example of the low temperature dependence of this, at normal room temperature the value only falls to 1.204 eV for silicon. By the addition of suitable gain amplifiers, voltage reference devices can be made with levels of 2.5, 5 and 10 V with a very high degree of stability.

Buried diode. Integrated circuit regulator devices can contain a diode/heater pair buried within the substrate. In operation, the typically 600 ohm heater resistor maintains the diode at a constant temperature and hence provides a constant reference voltage. Because of the heater power drain, this concept is not popular with battery operated systems.

Zener diode. These diodes are designed to operate beyond the reverse breakdown point. Then a variation in the reverse current produces only a very small change in the voltage drop across the diode. For many purposes this is acceptable as a constant voltage. However, the property can be improved if the diode is driven from a constant current source such as a transistor.

Zinc air fuel cell. The plates are effectively made of zinc and air, the latter being obtained by passing air through a porous nickel electrode and using potassium hydroxide as an electrolyte. The energy is obtained by converting the zinc into zinc oxide. These cells provide about five times the energy density of lead acid cells.

26.5 Useful references

Finnegan T.S. (1991) Cúk: the best SMPS. *Electronic World and Wireless World*, January, 69–72.
Linear Circuits (1989). *Data Book*, Vol. 3, Texas Instruments Ltd, Manton Lane, Bedford, UK.
Patchett G.N. (1970) *Electronic Power Supplies*, Pitman, London.
Voltage Regulator Handbook (1982), National Semiconductor Corp, California.

27 Propagation

The propagation of electromagnetic waves through the space surrounding the earth depends upon many phenomena, some quantifiable according to a definite theory and some unpredictable except in a statistical manner. The nature of the earth's surface in the immediate vicinity, the frequencies in use (see also Frequency bands in use, p. 137), the prevailing meteorological conditions, and many extra-terrestrial features all aid to make this line of research still very rewarding. Much knowledge of such propagation has thus been acquired by observation in a pragmatic way. The use of computers with modelling software now produces very acceptable sets of predictions from a given set of circumstances. Like antenna design, predictions about the behaviour of electromagnetic waves is based on radiation in free space. This provides a useful reference against which practical measurements can be compared.

Every point on the wavefront radiated from an antenna can be considered as the source of a secondary wave and these combine to form a new wavefront. This is expressed by Huygen's principle which states that the resultant displacement at any point, due to the super-position of individual waves of a system, is equal to the sum of the displacements of the separate waves at that point. At a distance far from the radiating source, the wavefront over a relatively small area can be considered to be that of a plane wave. The literature shows that under such circumstances, the electric field strength E for a radiated power P, at distance d, is given by:

$$E = \frac{(30P)^{1/2}}{d} \text{ volts per metre}$$

If the power is assumed to be 1 kW (a common standard), then this simplifies to, $E = 173.2/d$ V/m, showing clearly that the field strength is inversely proportional to path length.

Electromagnetic waves consist of energy in the form of electric (E) and electromagnetic (H) fields that are interdependent. The waves propagate through space due to source energy, with the two fields acting at right angles to the direction of propagation. Such waves

occupy three-dimensional space and are said to be *orthogonal*. By convention, the *plane of polarisation* of these waves is associated with the E field. Vertical and horizontal polarisations are used extensively in terrestrial communication frequency reuse systems. Here the E field is either vertical or horizontal with respect to the earth's surface.

Figure 27.1 shows the *Poincaré sphere*, a type of spherical quadrant, with centre O and the edges representing the x, y, z axes. Consider O as the source of energy, generating an electric field propagating in the arbitrary direction OP. At the surface of the sphere, the wavefront is perpendicular to the direction OP. Due to the angles θ and ϕ there will be components of the E field in both horizontal (E_x) and vertical (E_y) directions, the total field being given by $E = (E_x^2 + E_y^2)^{1/2}$. These

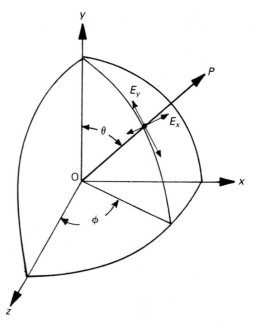

Figure 27.1 The Poincaré sphere and electric field vectors.

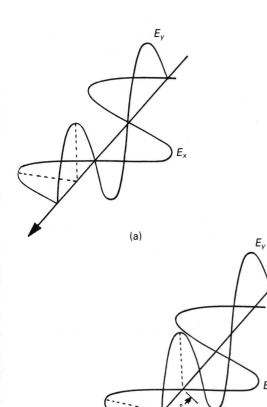

Figure 27.2 Wave polarisation: (a) linear; (b) circular.

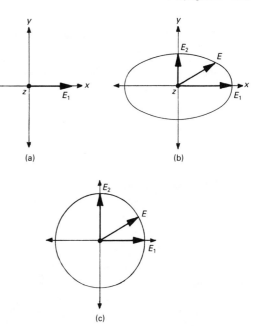

Figure 27.3 Polarisation vectors: (a) linear; (b) elliptical; (c) circular.

$E_1 = E_2$, linear polarised at $45°$ in the x, y direction.

$\delta = \pm 90°$
$E_1 \neq E_2$, elliptical polarised (Fig. 27.3(b))
$E_1 = E_2$, circular polarised (Fig. 27.3(c))

using the following convention:

(a) δ positive, left-hand circular (LHC) or elliptical polarised,
(b) δ negative, right-hand circular (RHC) or elliptical polarised.

two components, shown as being sinusoidal in Fig. 27.2, can be represented by:

$$E_x = E_1 \sin(\omega t)$$

and

$$E_y = E_2 \sin(\omega t + \delta)$$

where E_1 and E_2 are the amplitudes of the waves polarised in the x and y directions, respectively, and δ is the time-phase angle by which E_y leads E_x. In Fig. 27.2(b) this is shown as $-90°$.

The polarisation of the wave depends on the relative magnitudes of E_1, E_2 and δ as tabulated:

$\delta = 0$
$E_1 = 0$, linear polarised in the y direction
$E_2 = 0$, linear polarised in the x direction (Fig. 27.3(a))

It should be pointed out that the convention for the direction of circular and elliptical polarisation is that adopted by the Institute of Electrical and Electronic Engineers (IEEE). A clockwise rotating wavefront approaching a receiver is defined as being left-hand circular polarised. This is the opposite of the definition taken from classical optics.

The *Poynting vector* describes the power P in the wavefront per unit area. Thus $P = E \times H \sin \theta$ watts/metre (W/m), where E and H are the peak instantaneous magnitudes of the two fields and θ is the phase angle between them.

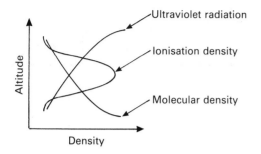

Figure 27.4   Ionisation formation.

27.1 Ionosphere

Ionisation
Ultraviolet radiation from the sun which propagates as electromagnetic waves of very short wavelength, causes gas molecules in the earth's atmosphere to ionise, different gases being ionised by different wavelengths. As indicated by Fig. 27.4, the radiation loses energy as it penetrates deeper into the atmosphere. It is also shown that the density of gas molecules available for ionisation decreases with an increase in altitude above the earth's surface. At some altitude, these opposing mechanisms will achieve a pseudo-equilibrium state where the ionised molecular density will be a maximum. However, as the ionisation is wavelength dependent, this maximum will not be clearly defined. In addition, cosmic radiation and other phenomena produce further ionisation.

Ionospheric model
Due to the variation in response of the ionsophere to different frequency signals, it is convenient to consider it as a series of layers of varying density ionisation. The layers are not clearly defined as suggested by Fig. 27.5, as the altitudes vary considerably around the earth's surface. The ionosphere is generally considered to extend from an altitude of about 40 km up to about 400 km. Below the ionosphere there are two layers known as the *troposphere* and *stratosphere*. These form the region known as the *atmosphere* with the division at about 10 km altitude. The troposphere is a region in which there is considerable variation in both air pressure and temperature, the latter falling fairly linearly by about 100 K in about 10 km. The stratosphere region is relatively calm and the temperature remains fairly constant up to about 40 km when it begins to increase with height. Over both of these regions the pressure falls in a regular manner. The three important layers of ionisation are referred to as the D, E and F layers. The E and F layers are sometimes referred to as the Kennelly-

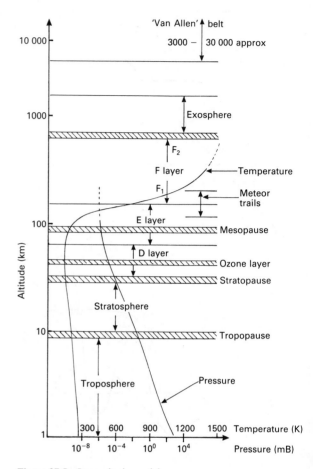

Figure 27.5   Ionospheric model.

Heaviside and Appleton layers, respectively, after the scientists that carried out most of the early investigative research in this subject. The D layer is the least ionised of the three, extending from about 40–90 km altitude and largely existing only during daylight. It reflects signal frequencies below about 50 kHz and severely attenuates those at higher frequencies when passing through it. The E layer extends from about 100–150 km and has a maximum density during daylight. It remains only weakly ionised during night time. The degree of ionisation falls during winter so that the E layer practically disappears during the longer dark nights. However, while present, the average height remains fairly constant all year round. The F layer that extends from about 150–400 km altitude is the most strongly ionised and experiences considerable variation. During night time this exists as a single layer at about 300–400 km, but during daylight splits into two parts. The F1 layer extends most strongly from about 180–220 km for both summer and winter days, while the F2 layer which extends from about 300–400 km during night time in both summer and winter, falls to about 250–350 km on a winter day and peaks at about 350–450 km or even

500 km during summer daylight. Other phenomena that affect ionisation and consequently propagation are meteor trails at altitudes between about 90 and 110 km and aurora effects between 50 and 1000 km.

Wave propagation

The behaviour of electromagnetic waves which travel in straight lines when not influenced by ionisation, is both frequency and refractive index dependent. At very low altitudes (less than about 10 km), the ionisation is practically zero and the refractive index is close to unity. At higher altitudes, the refractive index falls with increasing ionisation density. At frequencies below about 100 kHz, the change of ionisation density between zero and maximum occurs within a distance that is small compared with the wavelength of the signal. To such a signal, this layer looks like a perfect reflecting surface. Thus signals with frequencies up to about 500 kHz travel as a *surface wave* between the lower ionosphere and the earth's surface. In particular, the lower frequencies (16–30 kHz) can, in this mode, provide worldwide communication links, propagation always being stronger over water than dry land. At higher frequencies, the signal wavelength becomes sufficiently short compared with the changes in ionisation density so that the medium behaves more like a dielectric, with a continually changing refractive index.

Signal frequencies around 500 kHz begin to penetrate deeper into the ionosphere and the changing refractive index introduces refraction. This gives the waves a continually curved path until they become reflected back towards earth, thus producing the phenomenon known as a *sky wave*.

Generally, but with certain limitations, the higher the frequency in use, then the greater will be the altitude at which reflection occurs. Frequencies up to about 1.5 MHz have both a significant surface or ground wave that can be used for short distance communications, and a sky wave component. This latter, due to the reflection back to earth effect, covers a greater distance. The distance between the end of the ground wave and the first point of reflection is known as the *skip distance*, or alternatively as the *dead space*. If the wave energy is high enough, then frequencies higher than about 500 kHz can produce successive reflections between the ionosphere and earth to give worldwide coverage in a series of *hops*. For frequencies from about 3–30 MHz, the propagation is almost entirely by sky wave.

Above about 25–30 MHz, the waves no longer experience reflection in the ionosphere and so escape into space. Frequencies above this range are therefore used extensively in terrestrial communications links, with path lengths restricted to line-of-sight distances. Such waves are often referred to as *space waves*.

It is often convenient to consider wave propagation in terms of rays rather in the manner of light beams. The angle at which the wave is launched into space is often referred to as the *angle of incidence* which is the complement of the angle of elevation. As the angle of incidence is increased, then for a given frequency, a *critical angle* may be reached at which the ray is no longer reflected and escapes into space. This is termed the *escape ray*. The critical angle is obviously frequency dependent and the frequency at which the critical angle reaches zero is known as the *critical frequency*.

27.2 Propagation anomalies

These features represent much that is problematical to propagation in communications engineering. Some of the phenomena are helpful, some cause interference, but all tend to be only weakly connected to any one physical property of the atmosphere/ionosphere.

Advection (see also Inversion, p. 246). Basically the term relates to the transfer of energy from point to point, due to the drift of a horizontal gas stream, in this case air. Advection inversions occur mostly over the sea and low-lying coastal areas. These are created by the circulation of a warm dry air mass over the cooler and moist air near to the water. With the absence of wind there will be no mixing of the two air masses and the situation can then exist for several days. Advection inversions give rise to extended propagation in terrestrial space wave communications up into the SHF range. The effect then gives rise to serious co-channel interference for such services as television, but can enhance the communications capabilities of other services. This can sometimes be predicted a few hours in advance, from meteorological data about temperature changes and humidity.

Aurora. Sporadic intense radiation from the sun can interact with the earth's magnetic field, particularly close to the polar regions, to give rise to a visible effect. This is most noticeable following the observance of a large sun spot. It gives rise to considerable interference and distortion of radio signals right up into the UHF range. Signals beamed into such a region tend to be reflected in a north–south direction and can extend well beyond the normal range for any given frequency. However, the distortion produced only allows for narrow band data or code signals to be received with any degree of reliability.

Chordal hop (see also Sporadic E propagation, p. 246). A jargon term used to describe signals that become trapped and conveyed within an ionised layer or duct, eventually to be reflected back to earth. The effect, which is largely unpredictable, produces greater than normal propagation distances for signals in the VHF range.

Inversion (see also Advection, p. 245). Within the atmosphere, temperature normally decreases with altitude. When this reverses, it affects the distribution of the water vapour. The temperature/humidity feature then becomes horizontally layered to produce sudden changes of refractive index. This causes signals in the VHF/UHF range to be carried well beyond the normal horizon, giving rise to co-channel interference to the fixed services. It is possible for several inversion layers to form at different altitudes simultaneously.

Meteor scatter. When meteors burn up within the ionosphere, typically in the E layer, they leave behind short lived, highly ionised trails which are capable of supporting signal reflections and scattering. Typical maximum frequency range varies from about 40–400 MHz and propagation can extend over about 2000 km. Due to the short lifetime, communications must occur in bursts and this restricts the mode to slow speed code or data transmissions. Some meteor activity is random but others occur in regular annual showers, as shown in Table 27.1. The dates refer to 1994, but these change only marginally on a year-by-year basis.

Table 27.1 *Regular meteor showers*

Meteor shower	Period
Lyrids	April 19–25
May Aquarids	April 24–May 20
Delta Aquarids	July 15–Aug 20
Perseids	July 23–Aug 20
Orionids	Oct 16–27
Taurids	Oct 20–Nov 30
Leonids	Nov 15–20
Geminids	Dec 7–16
Ursids	Dec 17–25

Radiation cooling. When a calm cloudless night follows a warm day, the land mass cools fairly quickly. This forms ducts near to the earth's surface that cause enhanced propagation and hence co-channel interference to some services. The phenomenon is often short lived and is always less troublesome when an undulating earth surface exists between transmitter and receiver.

Sporadic E propagation. Unpredictable and particulate radiation from the sun produces abnormal ionisation within the E layer. The most probable cause of such propagation is wind shear in the upper atmosphere. This generates clouds of ions moving in opposite directions which give rise to an ionically dense reflecting layer in the form of a duct. This in turn provides VHF and UHF signals with an unusually long propagation path length. The effect occurs mostly during summer months and only affects relatively small patches of the ionosphere.

Subsidence inversion. Air in an anticyclonic weather system (winds rotating outwards from an area of high pressure) can descend slowly, gaining heat by compression; a subsidence. When the air pressure below such a mass produces equilibrium, the warm air spreads out to form a subsidence inversion layer. This generally occurs at an altitude of about 2 km. If such an inversion descends any lower, it can form an intense layer that can last for several days. The lower such an inversion descends, then the stronger becomes the trans-horizon propagation with increased co-channel interference problems. Such an inversion can form over land or sea and can generally be predicted a few days in advance, from radiosonde data from the upper atmosphere.

Trans-equatorial skip. The F2 layer produces the greatest propagation distances for signals in the HF band. This mode can support multi-hop propagation which can lead to multi-path reception problems at the greater distances. Towards dusk, the F1 and F2 layers start to merge and the maximum usable frequency (MUF) can rise by as much as 50% on its daytime value. This gives rise to occasional exceptional propagation distances between the northern and southern hemispheres. The south to north path tends to peak about mid-afternoon in winter, while the reverse path peaks 3 to 4 hours later.

Tropospheric scatter. The troposphere is in a continual state of turbulence which produces local variations in refractive index. Waves passing through such a region are then scattered in all directions. When the wavelength of the signals are small compared with the irregular eddies, scattering takes place in a forward directed cone. Enhanced over-the-horizon propagation can cover distances of up to 2000 km at frequencies as high as 3 GHz. Tropo-scatter links operate at around 2.5 GHz to provide reliable communications. The system requires the use of two high gain directive antennas for transmitter and receiver, both of which are beamed on a point at about 1–2 km altitude and approximately half-way between the two stations. Typically for every watt of power transmitted about one picowatt is collected at the receiver. Apart from this power inefficiency, the scattered energy simply adds to the general ionospheric background interference.

Weather fronts. A weather front often contains a region of super-refractivity and this produces enhanced trans-horizon propagation at VHF/UHF. It is usually associated with a region within a weakening anti-cyclonic system and the co-channel interference that it produces is usually short lived.

27.3 Miscellaneous terms

Adaptive and real time channel evaluation system. This concept involves continually monitoring the characteristics of several frequencies, usually within the HF band, with synchronised frequency agile transmitters and receivers. This improves the use of the spectrum, by selecting the highest suitable frequency. Adaptive signal processing allows the use of a micro-processor-controlled sub-system to select filter band-width, demodulator type, data rate and choice of frequency, to maximise the throughput of information.

Auroral oval region. Term used to describe the profile of the region of intense ionospheric disturbance over the North Polar region. The shape is thought to be the result of the distribution of the land mass.

Dellinger fade out (named after J.H. Dellinger, USA). Random and unpredictable large eruptions on the solar surface produce a large increase in ionisation extending down to the E layer. This results in the complete disappearance of sky waves in the HF frequency range 1–30 MHz for a few hours. The effect is more intense near to equatorial latitudes and only affects the daylight hemisphere.

Diffraction. When electromagnetic waves pass the edge of an object they bend or diffract. This causes the range of radio signals to extend beyond the optical horizon, and gives rise to the term *radio horizon.*

Earth radius factor. Due to the effects of refraction, the earth's radius appears to be larger than its physical value by a factor of 4/3. The ratio of radio horizon to optical horizon distances is related to this factor.

Exosphere. This region lies beyond an altitude of about 400 km. The density is such that an air molecule moving directly outwards has an even chance of colliding with another molecule or escaping into space.

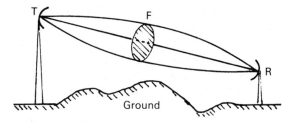

Figure 27.6 Fresnel zone.

Fresnel zone (referring to the line-of-sight path between two microwave antennas). An elliptical zone between the two antennas as shown in Fig. 27.6, where the total path distance nowhere varies by more than half of the operating wavelength. The concept is extended to describe the distance by which the direct wave clears any intervening obstacle such as a mountain peak. If the total path distance between transmitter, peak and receiver is 1 wavelength greater than the direct distance, then the clearance is said to be two Fresnel zones.

Gyro or gyromagnetic frequency. The frequency that has a period equal to that of the revolution of an electron in orbit and under the influence of the earth's magnetic field. Due to variations in field strength, this value varies between about 1.25 and 1.4 MHz. Wave propagation within this range is greatly attenuated.

Ionospheric absorption. Attenuation of the energy in a radio wave due to the interaction between it and gas molecules. Deviative absorption describes the appreciable bending that occurs in an ionospheric layer at close to critical frequency. Non-deviative absorption describes the condition where little or no bending occurs as the wave passes through an ionised layer. The D layer produces the highest level of absorption and this tends to decrease with an increase in operating frequency.

Ionospheric cross modulation. Nonlinearities within the medium can produce nonlinear absorption. This can lead to the modulation on a strong signal being transferred to a weaker carrier. Sometimes described as the *Luxembourg effect.*

Ionospheric focusing. A variation in the curvature of the ionospheric layers can give rise to a focusing/defocusing effect at a receiving antenna. This may produce either an enhancement or attenuation in the received field strength due to signal phase variations.

Ionospheric sounding. The properties of the ionosphere can be evaluated by the near vertical transmission of pulsed radio frequency signals and then measuring the time taken for the reflected pulse to return to earth. The *virtual heights* for different frequencies can be obtained by changing the carrier frequency.

Lowest useful high frequency.
The lowest frequency in the HF band that can be successfully used for point-to-point communications services: depends upon the current state of the ionosphere, the absorption, the gains of antennas in use, the type of service (analogue or digital) and the noise level. The final choice is therefore a compromise.

Magneto-ionic double refraction.
The combined effects of the earth's magnetic field and atmospheric ionisation causes linear polarised waves crossing the interface to split into *ordinary* and *extraordinary* components. These are generally elliptically polarised and in the opposite sense. The degree of polarisation depends upon the relative directions of the wave propagation and the magnetic field. In general, the ordinary wave vector which is least affected rotates anti-clockwise if the angle between the earth's field and direction of propagation is acute, and clockwise if the angle is obtuse. The opposite effect applies to the extraordinary wave vector. As these waves travel different paths with different attenuations and phase velocities, they arrive at a receiving antenna with phase relationships that are varying and this leads to a multi-path interference effect.

Maximum usable frequency (MUF).
This term generally applies to the HF band and point-to-point communications services. It represents the highest frequency that can be reliably used. The MUF can be calculated from the product of critical frequency and an MUF factor which is a function of the transmission path length.

$F_{\text{muf}} = F_0 \sec i$, where F_0 is the critical frequency and i is the angle of incidence, $\sec i$ being described as the MUF factor. (This is sometimes described as the secant law.) For path distances greater than about 1000 km, a further correction is included, so that $F_{\text{muf}} = F_0(k \sec i)$. The MUF is predictable and tabulated, based on monthly averages and sun-spot activity.

Mesosphere.
(1) The region of the earth's atmosphere between the stratosphere and the thermosphere (40–80 km). (2) The region of space between the ionosphere (400 km) and the exosphere (1000 km). Alternatively sometimes considered to be part of the exosphere.

Modified refractive index.
A correction factor that can be applied to take into consideration the curvature of the earth. It is the sum of the refractive index of air at height h above sea level and the fraction h/r, where r is the mean earth radius.

Multipath propagation.
Patches of the ionosphere with variable ionisation and physical structures, such as mountains or tall buildings, can all give rise to signal reflections that affect the HF, VHF and UHF bands. Such reflected signals arrive at a receiving antenna over different path lengths with different phase relationships that give rise to distortion, interference and ghosting.

Near vertical incidence propagation.
HF band signals radiated from an antenna in an almost vertical direction can provide reliable communications over medium distances. In addition, this concept is very useful where *mountain hopping* is required.

Optimum working frequency (OWF).
To minimise the effect of the unpredictable variable component in the maximum usable frequency (MUF), and to ensure reliable communications, the OWF is calculated as 85% of the MUF.

Refractive modulus.
The difference between the modified refractive index and unity expressed in parts per million, the parameter being measured in M-units.

Round the world echo.
Low frequency signals can propagate around the world to arrive at a receiving antenna some time after the direct signal. If the two components arrive from opposite directions, the long path signal is referred to as a backward echo. Alternatively it is known as a forward echo when the two components arrive from the same direction.

Selective fading.
Rapid changes in ionisation density can give rise to fading that varies with frequency. Wide band frequency modulated signals, in particular, suffer distortion from this phenomenon.

Silent or skip zone.
The region of the area surrounding a transmitter that is not reached by the ground wave. See also Dead space, p. 245

Solar cycles.
Although the production of flares and eruptions from the sun's surface occurs in a random manner in the short term, they repeat fairly predictably over an 11-year cycle. Cycle 22 started in September 1986 and the activity peaked during 1990. Such solar activity chiefly affects propagation in the HF bands. A close relationship has been noted between *abnormally quiet days* (AQD) (days when solar activity is minimal) and the magnitude of the following peak. During such periods, the MUF can rise intermittently as high as 50 MHz, and the 21 and 26 MHz bands can sustain reliable communications for considerable periods. This correlation, which holds good as far back as 1885, the

earliest year for which data is available, allows future peaks to be predicted for several days ahead.

Solar winds. Streams of highly energetic particles that flow out from the sun during solar flare activity. The interaction of these with earth's polar magnetic field gives rise to the aurora effect.

Standard M-gradient. The refractive modulus increases linearly with altitude above the earth's surface and the slope of this is taken as a standard for comparison purposes. The normal value of this parameter is 0.12 M-units/metre.

Standard radio atmosphere. An atmosphere that has a standard M-gradient.

Standard radio horizon The radio horizon that would exist due to propagation through a standard radio atmosphere.

Standard refraction. The value of the refraction in a standard radio atmosphere. In a practical situation, super- and sub-refraction values are obtained.

Sun spots. The average energy output that reaches earth from the surface of the sun is represented by the *solar constant* of about 1.35 kW/m². Eruptions on the surface of the sun produce solar flares, giving rise to considerable variability in this value in the short term. Sun spots are therefore areas that produce this effect, giving rise to the intense magnetic activity that has such a profound effect on the propagation of radio waves.

Thermosphere. The region of the upper atmosphere in which the temperature increases with altitude.

Van Allen belts. Two layers of charged particles emitted from the sun that are trapped within the earth's magnetic influence. These are named after the discoverer, J. Van Allen. The inner layer exists from about 2400 to 5600 km altitude and consists of secondary charged particles. The outer layer lies between about 13000 and 19000 km and is thought to consist of the original particles released from the sun's surface.

Virtual height. For a given frequency, the apparent height of a reflecting layer is calculated from the period of time between the transmission of a signal and the reception of its ionospheric reflection. The signal is assumed to travel with the velocity of light for the whole time. However, as the ionisation increases with height, the velocity will fall somewhat so that the virtual height is always greater than the true height.

Wave tilt. As a surface wave propagates it becomes diffracted because it sees the earth's surface as an edge. Furthermore, the imperfect conductivity of the earth absorbs energy from the wave. Together these act to cause the wave to tilt in a forward direction.

Whistler. The descriptive jargon term given to a radio effect produced by the electromagnetic radiation from a lightning strike. This radiation tends to follow the lines of the earth's magnetic field and gives rise to a whistle of descending frequency when heard on an extra low frequency radio receiver.

27.4 Useful references

Bunney R.W. (1986) *Television DXers Handbook*, Babani, London.

Glazier E.V.D. and Lamont H.R.L. (eds) (1958) *The Services Textbook of Radio*. Vol. 5, *Transmission and Propagation*, H.M.S.O, London.

I.B.A. (1982) I.B.A. Technical Review No 17, Independent Broadcasting Authority (IBAUK), London.

Radio Society of Great Britain, (1988) *Radio Communication Handbook*, RSGB, London.

Schanker J.Z. (1990) *Meteor Burst Communications*, Artech House, London.

Sources of astronomical and meteorological data

Central Administration of Swedish Telecommunications, Development Department, S-12386 Farsta, Sweden.

Environmental Data Services, National Oceanic and Atmospheric Administration (NOAA), Climate Center, Ashville, North Carolina 28801, USA.

FTZ, Forschungsgruppe Ionosphäre, Postfach 5000, D-6100 Darmstadt, Germany.

Ionospheric Prediction Service, PO Box 702, Darlinghurst, 2010, Australia.

Marconi Propagation Services, Baddow Research Laboratory, Great Baddow, Essex, CM2 8HN.

Meteor Section, British Astronomical Association, Burlington House, Piccadilly, London W1V 0NL.

Radio Research Laboratory, Ministry of PTT, 2-1 Nukui-Kita-Machi, 4-chome, Koganeishi, Tokyo 184, Japan.

Service des Ursigrams, CNET, Observatoire de Paris, F-92190 Meudon, France.

Sunspot Index Data Centre, Observatoire Royal, 3, Avenue Circulaire, B-1180 Brussels, Belgium.

28 *Quality and reliability/ quality assurance (QA)*

Quality is often defined as the measure of how well a product or system conforms to its claimed specification. Reliability is then the measure of how long the product or system continues to meet this specification. Quality and reliability are of great interest to the end user because this directly affects his confidence that the product or system will meet his expectations. These broad parameters can be evaluated from statistical evidence obtained by the manufacturer's quality assurance programme. Just as an analogue signal can be digitised for processing by sampling and a highly accurate version of the original regenerated afterwards, a statistical sampling scheme, supported by well tried and tested statistical tables, can give an accurate assessment of the state of a population. Statistical data is therefore acquired through inspection and testing, using industry standard sampling plans supported by such national and international bodies as International Standards Organisation (ISO), British Standards Institution (BSI) and American National Standards Institute (ANSI). In addition, life testing of randomly selected batches can expand the information in this data base of product statistics. This can then be applied to *failure analysis*. A further important feature of user confidence is the concept of traceability. That is, component parts including the raw *materials** of a system have all been produced and tested under a recognised quality assurance scheme so that lifetime failures of all elements can easily be traced back to the original source scheme. Such information feedback can thus be used to improve the overall reliability of a product or system. Since original system design must take such statistics into consideration, a worst-case design methodology must be employed. Because product reliability can be assured, the concept of *just in time* (JIT) stocking of assembly lines becomes viable. This allows an assembly process to flow smoothly with the minimum number of component inputs, thus significantly reducing the cost of *work in progress*. Basically a QA programme involves testing and inspection of a specified number of components from each batch of a product and rejecting the whole batch if the number of defectives exceeds some level. Since the samples are selected at random, the technique leads to either a *consumer* or *producer risk* as each batch has a risk of being wrongly accepted or rejected. Confidence achieved by a QA programme can minimise the cost to both. The number of samples per batch needs to be large enough to give confidence and small enough to be cost effective. All of these features are assured if an industry standard scheme is employed.

28.1 Miscellaneous terms

Accelerated life testing. Normal life testing involves the manufacturer operating components/systems as near as possible to the end user's applications and conditions. However, if the accepted lifetime is very long, then data gathered by this technique is too time consuming so that the test process has to be accelerated. For example, devices designed for intermittent operation would be run continuously, thus compressing the life span. Electronic components that are intended for continuous operation may be tested with an increased load or at an elevated temperature. Further environmental tests which simulate extreme conditions of temperature, humidity, pressure, shock, vibration, and abnormally dirty conditions in a time cycled way, might be used.

Arrhenius model (acceleration factor). This is often used, particularly with electronic components, to correlate the failure rates of accelerated testing with those that would be obtained in a normal life span. The acceleration factor (*AF*) is given by:

$$AF = \exp(E/K(1/T_1 - 1/T_2))$$

*In this context, this word is often spelt 'materiel' due to US military connotations.

where E = the activation energy for semiconductor material and typically 1 eV

K = Boltzmann's constant $(8.61 \times 10^{-5}\,\text{eV/K})$

T_1 = lower temperature K

T_2 = upper temperature K.

Thus testing for 1 h at 125°C would be equivalent to testing for about 26 h at 85°C.

Acceptable quality level (AQL). The concept of AQL has been commonly adopted as a parameter of batch sampling and is obtained from a plot of actual percentage defects per batch against probability of acceptance. A commonly set value is 0.95 as shown in Fig. 28.1, but some other value close to this may be set. The diagram shows that the probability of this particular batch being accepted with exactly the AQL percentage of 0.7% defects is 0.95. The producer's risk that such a batch will be wrongly rejected is therefore 0.05 or 5%.

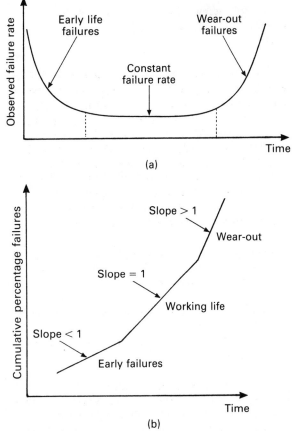

Figure 28.2   Device failure rate/time curves: (a) bath tub diagram; (b) Weibull cumulative failure plot.

and this is described as the *wear out region*. The characteristics of the bath tub diagram may also be described by the cumulative plot of percentage failures against time, the Weibull model. This is shown in an idealised way in Fig. 28.2(b), where a slope of unity represents the constant failure rate range of the bath tub. Slopes less than, or greater than, unity represent the decreasing and increasing failure rates of the infantile mortality and wear out regions, respectively. By comparison, this model has the more simple mathematical equation given by:

$$R(t) = \exp(-((t - \gamma)/n)^{\beta})$$

where $R(t)$ is the fraction of components surviving at time t, i.e. the reliability; β = slope of curve; n = the time at which 63.2% of components have failed, and γ is described as the location parameter and usually equal to zero.

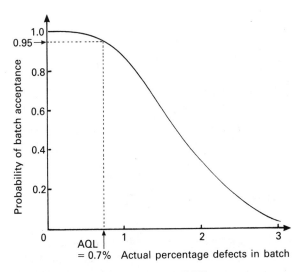

Figure 28.1   Acceptable quality level (AQL).

Bath tub diagram. This is a jargon term used to describe the time/failure rate curve of Fig. 28.2(a), which shows the typical lifetime failures for components or systems. The high levels of early failures, often described as *infant mortality*, are usually due to manufacturing faults, deficiencies in quality control scheme, design deficiencies, and misuse by or the inexperience of the end user. These failures are fairly quickly eradicated and the failure rate settles down to a constant, often zero level over the normal working lifetime. Towards the end of the useful service life, the failure starts to rise once more

Burn-in. The jargon term used to describe the application of thermal and other forms of stress during the early stages of life testing. This removes many of the infantile failures from the test procedures and enhances the reliability.

Confidence limits and intervals (see also Normal distributions below). For a normal distribution of samples, the standard deviation is used to set *confidence limits*: 68.26% of all samples will fall within the interval of ±1 standard deviation. Increasing the interval to ±2 or ±3 standard deviations increases this figure to 95.44% and 99.72% of the samples, respectively. The interval ±2 standard deviations is usually described as the 95% confidence limit, because there is slightly more than a 95% confidence level that any device will lie in this range and is therefore of acceptable quality.

Double sampling. A sampling plan that provides for two sample sizes and two levels of allowable defectives. This might be represented as follows:

first sample = 100, accept number = 1, reject number = 4
second sample = 100, accept number = 4, reject number = 5

The implication of this is that a batch of 100 devices will be inspected and if the number of defectives is less than 2, or more than 3, the batch will be accepted or rejected as indicated. If the number of defective devices is more than 1 but less than 4, a second sample of 100 will be inspected. If the total number of failures in the two batches now does not exceed 4, then both batches will be accepted.

Down-time (see also MTTR, below). Jargon term used to describe the length of time that a system is out of action between failure and repair.

Failure in time (FIT). For high reliability devices the usual parameter for failure rate is the FIT. This represents the number of failures in 10^9 device hours. 1 FIT = 0.0001% failures/1000 hours.

Failure rate. There are two parameters of interest:

(1) The average failure rate given by the ratio of the number of device failures per device hour during test.
(2) The instantaneous failure rate given by the ratio of number of devices failing per hour and the number still surviving at the instant *t*.

The fraction of the devices that have failed at time *t* is given by $F(t) = 1 - R(t)$, where $R(t)$ is the reliability.

Lot tolerance percentage defectives (LTPD). This term describes the percentage of defects, at and above which it is almost certain that the batch will be rejected. This point on the characteristic, shown in Fig. 28.3 at 3%, is also known as the *rejectable quality level* (RQL). As with all random sampling, there is a small probability that a batch will be wrongly accepted due to the selection of untypical devices. This is described as the consumer's risk and is often set at 0.1, so that the probability of rejection is 0.9 (1–0.1). The characteristic also shows the percentage of defectives that the user is prepared to accept. This level is often set by the process average, which in this case is 1%. This point also defines the *producer's risk* that a good batch will be wrongly rejected.

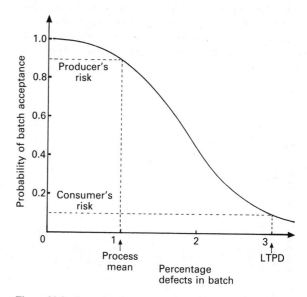

Figure 28.3 Lot tolerance percentage defects.

MTBF or MTTF (mean time between failures or mean time to failure). For a system with a constant failure rate of 0.001 per hour, it would be expected to fail and have to be serviced every 1000 hours. Thus the average or mean time between failures would be 1000 hours. MTBF is therefore the reciprocal of the constant failure rate. When it is not practicable to replace a failed component, it is more usual to refer to the mean time to failure.

MTTR (mean time to repair). See Down-time, *above.*

Normal distributions (Gaussian or Laplacean distributions). The distribution of random numbers or variables is very accurately described by the normal distribution curve. For simplicity, only the positive quadrant of this function is shown in Fig. 28.4. The equation to this curve:

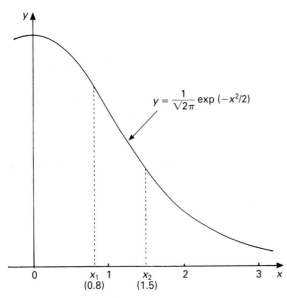

$$y = \frac{1}{\sqrt{2\pi}}\, \exp\left(-x^2/2\right)$$

Figure 28.4 The normal distribution curve for *x* positive.

$$y = (1/\sqrt{2\pi})\exp(-x^2/2)$$

gives the function the following properties:

(1) symmetry about the *y* axis;
(2) total area between curve and *x* axis is unity;
(3) *y* tends towards 0 as *x* tends towards $\pm\infty$;
(4) when $x = 0$, $y = 1/\sqrt{2\pi}$ or approximately 0.4;
(5) area to left of *y* axis $= 0.5$;
(6) area to right of *y* axis $= 0.5$.

The probability of any random number *x* occurring in any particular range is given by the area bounded by the curve, the *x* axis, and the two limiting values. Hence the probability that *x* lies between $-\infty$ and $+\infty$ is 1, or a certainty. Values for such areas are listed in normal distribution tables, an extract of which is shown in Table 28.1.

Table 28.1 *Part of a normal distribution table*

x	*Area A(x)*
0.0	0.0000
0.2	0.0793
0.4	0.1554
0.6	0.2257
0.8	0.2881
1.0	0.3413
1.2	0.3849
1.4	0.4192
1.5	0.4332

From Table 28.1 we can derive probabilities as follows:

The probability that *x* lies between 0 and 0.8 $= 0.2881$
The probability that *x* lies between -0.8 and $+0.8 = 2 \times 0.2881 = 0.5762$
The probability that *x* is greater than $0.8 = 0.5 - 0.2881 = 0.2119$
The probability that *x* lies in the range 0.8 to $1.5 = 0.4332 - 0.2881 = 0.1451$

Random variables, described as above, have the following important statistical properties: the mean $\mu = 0$, the variance about the mean $\sigma^2 = 1$, and the standard deviation $(\sqrt{\sigma^2}) = 1$. In a practical sampling case, the set of values would have a non-zero mean and non-unity variance and standard deviation. Such a range of values can be transformed as follows so that the basic normal distribution table can be used. A set of variables Z will have a mean of μ_z and standard deviation of σ_z, and these are transformed into a set of X values using the shift and squeeze relationship, $X = (Z - \mu_z)/\sigma_z$. The sampled values can be used to find the mean and standard deviation as follows. The mean is simply the arithmetic average of the series of measured values, while the standard deviation is given by:

$$\sigma_z = \sqrt{\frac{\sum_{i=1}^{n}(z_i - \bar{z})^2}{(n-1)}}$$

where $z_i - \bar{z}$ represents the differences between each value and the mean $\bar{z}$, squaring removes any negative signs, and finally summation, division by $n - 1$ and square root gives the variance.

The tables also show that the ranges:

$\pm 1\sigma$ contains 68.26% of a population
$\pm 2\sigma$ contains 95.44% of a population
$\pm 3\sigma$ contains 99.72% of a population
$\pm 3.09\sigma$ contains 99.90% of a population

Pareto curve or chart. This is named after the Italian economist who used it to study national wealth, but is now often used as a tool of failure analysis. Any device off a production line can suffer from a range of defects, some of which will be more repetitive than others. The chart represents a plot of cumulative percentage defectives against specific defect categories. For a given period of time, the total number of defectives under each category are organised into a ranked list with the smallest number at the bottom. Each sub-total is then converted into a percentage of the total defectives. These are then summed cumulatively from top to bottom to give cumulative percentage defectives. When these are plotted against category of defectives in the same order, the result commonly has an exponential curve and shows graphically the important production areas that are generating most failures. The chart is often described as an 80/20 one, because 80% of the failures tend to come from 20% of the categories.

Pressure cooker. Jargon term that describes the test that exposes devices to a high temperature, moisture-laden atmosphere, at a pressure twice that of normal. The test exposes any susceptibility to galvanic corrosion due to chemical reactions within component encapsulation.

Process control chart. A control chart with similar limits to that shown in Fig. 28.5 provides a useful way of monitoring the quality of a production process. It plots successive samples against predetermined limits or warning lines. If two successive samples fall within either inner and outer warning bands, or one sample falls outside of the outer line, then the process will be halted and readjusted to maintain most of the samples within the two inner warning lines. Using the limits shown, these two cases have approximately equal chances of occurring $(4.56\%)^2 = 0.208\%$ and 0.28%, respectively.

Rejectable quality level (RQL). See Lot tolerance percentage defectives (LTPD), p. 252.

Reliability (see also Failure rate, p. 252). Numerically reliability $R(t)$ is given the ratio of the number of components surviving at time t, to the total number of components at the beginning of life testing. Hence $R(t) = 1 - F(t)$, where $F(t)$ is the failure rate.

Sequential sampling (see also Double sampling, p. 252). As the concept of double sampling is extended into *multiple sampling*, the sample size becomes progressively smaller. Ultimately this reduces to a sample of one. Each sequential sampling plan gives two numbers, one called the *handicap H* and the other the *penalty P*. After inspecting each item, a score is calculated given by:
$$score = (H + \text{number of acceptable items found})$$
$$- (P \times \text{number of defectives found})$$
The batch is accepted if the score reaches $2H$, and rejected if the score falls to zero or is negative.

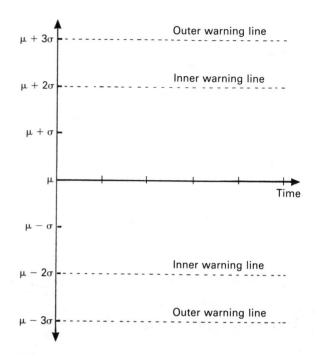

Figure 28.5 Process control chart.

Temperature cycling. A test that exposes components or devices to wide changes of temperature in a cyclic manner. Typical upper and lower temperatures are 150°C and −65°C, respectively.

Thermal shock. A temperature cycling technique whereby the test temperatures change rapidly, often in less than 10 s.

Weibull distribution. See Bath tub diagram, p. 251.

28.2 Useful references

Caplen R.H. (1988) *A Practical Approach to Quality Control*, Business Books, London.

Haber A. and Runyon R.P. (1973) *General Statistics*, 2nd edition, Addison-Wesley, London.

IEEIE (1990) Monograph, *A Statistics Cookbook*, Institution of Electronic and Electrical Incorporated Engineers, London.

Juran J.M., Gryna F.M. and Bingham R.S. (1974) *Quality Control Handbook*, McGraw-Hill, New York.

Mosteller F.M., Rourke R.E.K. and Thomas G.B. (1973) *Probability with Statistical Applications*, 2nd edition, Addison-Wesley, London.

29 Radar and navigation systems

RAdio Direction And Ranging (RADAR) and direction finding (DF) systems function in a complementary manner. A radar system is used by a base station to locate a traveller's relative position, while a DF system is used by the traveller to determine its position relative to the location of the base station (see also Satellite systems, p. 275.)

A radar system relies on the reception of a reflected signal from a given target, while a DF system operates by receiving a signal direct from a transmitter. Due to the dual path length of the radar signal, only a small fraction of the transmitted power arrives back at the receiver. Thus radar systems require the use of considerably greater transmission power levels. However, in the interests of obtaining accurate directional information and economy of signal power, both radar and DF systems rely on the use of highly directional antennas.

29.1 Navigation systems

29.1.1 Antenna polar diagram DF systems

Azimuth information about the arrival of an electromagnetic wave at a receiving antenna can be obtained by both amplitude and phase measurements. Direction finding therefore involves either evaluating the voltage developed on a very directional rotatable antenna or by taking time/phase difference measurements. Although both methods have much in common, most modern systems adopt the latter technique. Operation in the VHF or UHF bands is often adopted to ensure that DF systems can operate with small but highly directional antennas.

Figure 29.1 shows the antenna polar responses of commonly used systems. Figure 29.1(a) represents *maximum* signal DF, where the antenna is turned until

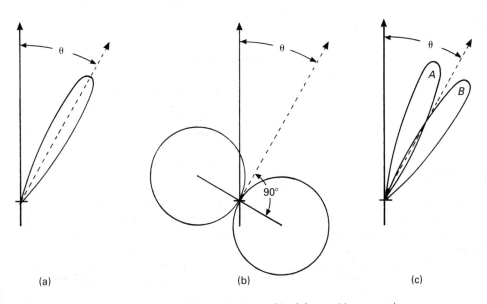

(a) (b) (c)

Figure 29.1 Polar diagram response DF: (a) maximum; (b) minimum; (c) comparative.

the received signal level reaches a maximum. The beam is then pointing towards the beacon transmitter. The accuracy of this system is relatively low because small angles of movement close to maximum produce very little change in signal level.

The system represented in Fig. 29.1(b) uses an antenna with a figure-of-eight polar response whose null is much more sensitive to small variations of azimuth angle. However, sensitivity can be affected by noise at the zero signal level and the polar diagram introduces an ambiguity of ±180°.

The problems introduced by both of the above systems can be countered by using two antennas with overlapping responses as shown in Fig. 29.1(c). The two antennas are turned together and their output voltages fed via suitable amplifiers to a comparator circuit. If the two outputs are of equal amplitude, then the azimuth angle lies along that shown in the diagram. This concept is very much more accurate because the response increases and decreases rapidly as the pair of antennas are turned in either direction from maximum.

The ambiguity of the azimuth angle found with the minimum response system can be removed by combining the uni-polar response of an auxiliary rod antenna with that of the figure-of-eight as indicated in Fig. 29.2. Although the resulting cardioid response has poor directional properties, the high front-to-back ratio when swung through ±90° will easily resolve the true azimuth angle.

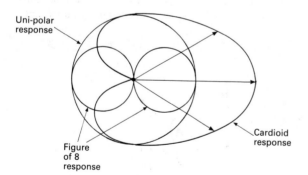

Figure 29.2 DF sense response curves.

29.1.2 Decca Navigator System

This is an example of a phase measuring DF system. The base stations operate in groups of four transmitters, with one acting as a master. Each transmitter continuously radiates on one of four frequencies, either 70, 84, 112, or 126 kHz (all multiples of 14 kHz). This creates a pattern of electromagnetic wave energy around each transmitter, with peaks and troughs separated by a few kilometres, the mapping of which is referred to as *lanes*. Thus each

pair of transmitter antennas becomes the foci of a set of hyperbolic lanes. A position within three lanes can then be automatically calculated by a mobile receiver by phase comparison. Maximum accuracy of a fix (about 300 m) can be obtained within about 150 km of the controlling transmitters, but the system can provide useful information at distances up to about 500 km.

29.1.3 LORAN-C system (LOng RAnge Navigation)

This system was designed as a long range (about 1500 km) maritime navigational aid. It is based on three widely spaced high power transmitters, radiating accurately timed 10 μs pulses within a bandwidth of ±10 kHz centred on a carrier frequency of 100 kHz. Again, pairs of transmitters produce hyperbolic electromagnetic energy patterns. A LORAN receiver measures the time of arrival of these pulses to evaluate the distances from each transmitter and, using triangulation, it automatically calculates the position. The 10 μs, 100 kHz pulses can be processed directly and without demodulation using the processing power of a transputer. Using an averaging technique, such a system can resolve the necessary trigonometrical calculations to provide a fix accurate to within about 300 m.

29.1.4 Single station locators (SSL)

This technique utilises the HF bands and reflections from the F2 layer to provide both azimuth and range information at a single DF site. If the height of the reflecting layer is known, then the elevation angle of the received signal can be used to evaluate the range of the distant transmitter, while its bearing is given by the azimuth angle.

29.2 Radar systems

These systems locate distant objects by making use of the reflected radio frequency (RF) energy emanating from them resulting from the radiation of a high power microwave transmitter equipped with a highly directive antenna.

Figure 29.3 shows how this basic principle can be applied in practice. A pulse of RF energy is radiated from a rotating directional antenna in the search for a distant object. When this energy impinges on an object it is scattered, some being reflected back towards the source, where it is received a short time (t_d) later. The total dual path delay is 6.66 μs per km or 12.36 μs per nautical mile. Knowing the propagation velocity, it is possible to evaluate this time delay in terms of distance or range, the bearing of the object being obtained directly from the azimuth angle of the antenna.

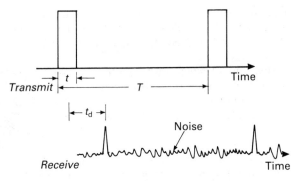

Figure 29.3 Radar transmission and reception waveforms.

The minimum and maximum ranges of these systems depends upon the pulse width and its repetition frequency. Referring to Fig. 29.3, if the time delay t_d is less than the pulse width t, then the reflected pulse will be masked by the transmitted pulse and lost. If the time delay is greater than the pulse repetition period, an ambiguity arises because the reflection may have been generated by one or more transmission pulses. However, by continually monitoring an *out-of-range* object and making use of knowledge about its reflecting properties, the use of high speed computer processing makes it possible to remove some of this uncertainty.

High peak power for radar systems is achieved by using a suitable pulse repetition frequency and pulse duty cycle. Again referring to Fig. 29.3, the duty cycle is given by:

$$\frac{\text{Pulse width}}{\text{Pulse period}} = \frac{t}{T}$$

and

$$\frac{\text{Average power}}{\text{Duty cycle}} = \text{Peak power}$$

Thus for a pulse repetition frequency of 1 kHz, a duration of 2 μs and average power of 200 W, the peak pulse power is 100 kW.

29.2.1 Doppler shift equation

To evaluate the frequency shift that is observed when an electromagnetic wave is emitted from a moving object, consider the following:

A stationary transmitter emits a wave with a frequency f towards a stationary target, for a time t, over a range r. The number of waves emitted is then ft. The reflection from the target arrives back at the source after covering a distance of $2r$. Thus $2r = ft$, and $r = ft/2$, and the number of waves emitted $= ft\lambda/2$. If at the same time the target is moving towards the

transmitter with a velocity v, in time t the waves will be compressed into a shorter distance r' with a shorter wavelength λ', giving rise to an increase in frequency. Thus $r' = ft\lambda'/2$. Therefore $r = r' + vt$ and

$$\frac{ft\lambda}{2} = \frac{ft\lambda'}{2} + vt$$

$$f\lambda = f\lambda' + 2v$$

$$\frac{\lambda}{\lambda'} = 1 + \frac{2v}{f\lambda'} = \frac{f'}{f}$$

where f' is the apparent frequency.

$$\frac{f'}{f} - 1 = \frac{2v}{f\lambda'}$$

$$\frac{f' - f}{f} = \frac{2v}{f\lambda'}$$

where $f' - f$ is the change in frequency. Therefore the Doppler shift is given by:

$$\frac{2v}{\lambda'} = \frac{2vf'}{c}$$

Now since $f' - f$ is very small compared with f, the Doppler shift is approximately equal to $2vf/c$.

If the target approaches at an angle of θ, then the approach velocity is given by $2vf \cos\theta/c$.

29.2.2 Range equation

Referring to Fig. 29.4, a transmitter radiates with a power P_T from an antenna with a gain of G_T. An object at range r thus receives a power flux density

$$PFD_T = \frac{P_T G_T}{4\pi r^2} \qquad \text{W/m}^2$$

If the object has a reflective area of α square metres then it reflects a power equal to αPFD_T watts, to produce a

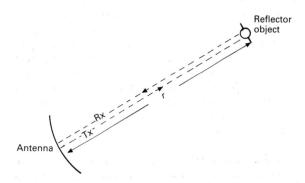

Figure 29.4 Radar ranging principle.

second PFD (PFD_R) at a co-sited receiving antenna. Thus

$$PFD_R = \frac{P_T G_T \alpha}{(4\pi r^2)^2} \qquad \text{W/m}^2$$

If the receive antenna has an effective area of A square metres and a gain of G_R, then since $G_R = 4\pi A/\lambda^2$, the received power is given by:

$$P_R = \frac{P_T G_T \alpha}{(4\pi r^2)^2} \times \frac{G_R \lambda^2}{4\pi}$$

and

$$\frac{P_R}{P_T} = \frac{G_T G_R \alpha \lambda^2}{(4\pi)^3 r^4}$$

showing that the range depends upon the fourth root of the transmitted power.

Due to the use of high peak powers, large antennas and careful choice of operating frequencies, the free space attenuation is generally insignificant and in the order of 1 to 2 dB.

If the minimum usable signal to noise ratio is S/N, then $P_R = S$ and $N = kTB$, where k is Boltzmann's constant $(1.38 \times 10^{-23}\,\text{J/K})$, T is absolute temperature and B is the system bandwidth. Then the range r is given by:

$$r = \sqrt[4]{\frac{P_T G_T G_R \alpha \lambda^2}{(4\pi)^3 kTB(S/N)}}$$

29.3 Miscellaneous terms

Adaptive radar. Generally the pulse rate, spectral shape and width for any given system are fixed, and define its performance. However, in the interests of obtaining more accurate information, certain parameters may be modified under the control of a system computer. The extent of the changes that can be made are dependent upon the number of targets present and the parameters include the data gathering rate, power output and receiver sensitivity. Additionally, the antenna may be directed onto targets in a specific order and its dwell time on each target varied.

ADSEL (ADdress SELected system). An airborne control system in which a ground station has the facility to transmit interrogating codes to selected

aircraft. The addressed aircraft then transmits high speed data which includes confirmation of its navigational course and speed, its altitude and fuel status.

Airport surveillance radar (ASR). A system that continuously scans the surrounding area to provide information for air traffic control. It provides distance and bearing information about all aircraft within its vicinity.

AM/CW radar. This is a variation on the continuous wave (CW) systems. Energy is transmitted on two frequencies f_1 and f_2 so that a single target will return $f_1 \pm f_D$ and $f_2 \pm f_D$, where f_D is the Doppler shift that allows the system to distinguish between stationary and moving targets. The range, which has a linear relationship with the phase difference between the two returned frequencies, can be calculated unambiguously from:

$$r = \frac{c\phi}{4\pi(f_1 - f_2)}$$

provided that ϕ is less than 2 radians, where r is range in metres, c is velocity of propagation and ϕ is the phase difference in radians.

Complications arise where many targets are present simultaneously, but these can be resolved by making use of the Doppler components.

Azimuth angle. The angular deviation from the North or South Pole bearing measured in the horizontal plane (often referenced to the magnetic poles).

Backscatter. The reflection of electromagnetic energy from a target back towards the transmitter antenna.

Bistatic radar. A system in which the transmit and receive antennas may be separated, often by many kilometres. A small transmitter antenna is usually supported by a larger receive antenna to maintain the same overall system gain. Because the transmitter radiates less energy it is more difficult to detect. Although the receive antenna is larger, this part forms a passive element that is difficult to detect by electronic means.

Cell. A radar cell is defined by the area in which targets can be detected and is determined partially by the antenna beamwidth. In a pulse radar system it is additionally set by the pulse length, and for an FM system, by the length of time that the transmission frequency remains constant.

Chirp radar. A system using pulses that are usually linearly frequency modulated. The returned pulses are time compressed so that there is an apparent system power gain proportional to the pulse compression ratio or dispersion factor.

Clutter (grass). A jargon term that is used to describe the interference due to noise and spurious reflections from the ground or sea surface. This is visible on a radar display device and tends to mask the genuine returns.

CW radar. Continuous transmission of energy on a single frequency produces a continuous reflection from a target, whose velocity can be determined from the Doppler shift. The target range can only be determined if two or more radars are used in cooperation.

DABS (Direct Addressed Beacon System). An alternative title for ADSEL systems.

DATATRAK (automatic vehicle location system (AVLS)). This system is designed to track moving vehicles to within 25 m. Each vehicle is fitted with a vehicle locator unit (VLU) consisting of low frequency (LF) receiver, VHF or UHF transmitter, one small antenna and a microprocessor, all driven from the vehicle low voltage power supply.

Time division synchronised position locating signals are radiated from a network of LF transmitters separated by about 150 km, each vehicle being identified by its uniquely defined time slot.

The VLU measures the phase difference between the signals from pairs of transmitters and uses the information to generate hyperbolic navigation lanes. From these, the microprocessor calculates the vehicle location, speed, direction and system status. This information is then broadcast through the VHF or UHF transmitter to a base station. Here the information is collated and transferred to a regional control centre (RCC) via private (leased) telephone lines.

At an RCC, the data may be displayed either as Ordnance Survey grid references or overlaid on a digitised map. Data logging is carried out using either hard disk store or printer.

An accurate vehicle location can be obtained from three transmissions, but if more are available, this improves the accuracy and gives an added degree of system redundancy. Alarm time slots are also provided within the transmission sequence to provide increased monitoring in emergency situations.

Doppler radar. Any radar system designed to determine target velocity using Doppler shift.

DORAN (DOppler RANge). A Doppler radar system devised for tracking missiles.

Early warning radar. Any system designed to provide warning of approaching aircraft or missiles at the greatest possible distance.

Elevation angle. The vertical deviation angle of the target bearing relative to the horizontal plane.

FM radar. The transmission carrier is continuously frequency modulated within limits, so that when any reflected signal is received and compared with the instantaneous transmission frequency, a beat note is generated that is directly proportional to the range. Like all continuous transmission systems, these have a power limitation problem, but the technique has been used successfully for radio altimeters.

Fruit. A jargon term used to describe any spurious responses created by secondary interrogation. In a monopulse SSR system, asynchronous replies may be received as a result of interrogation by another ground station.

H radar. An aircraft navigation system in which the user interrogates two ground stations in order to determine its position and distances.

IFF (identification friend or foe). A secondary surveillance radar (SSR) system that allows a ground station to transmit regular pulse pairs described as P_1 and P_3, on a carrier of 1030 MHz, and via a rotating antenna with narrow azimuth and wide elevation beams. Each pulse pair is uniquely coded according to international agreement. An aircraft carries a transponder that continually decodes these interrogation signals and only makes a reply when it recognises the code. It then transmits its own unique code on a frequency of 1090 MHz to allow the ground station to identify that aircraft.

Intersection. The technique used by two or more DF receiving stations to locate the position of a single transmitter, as in the case of a distress call.

LADAR or LOPPLER. A Doppler radar system that uses coherent laser light instead of a radio beam.

Minimum discernible signal. This is defined as the smallest signal that produces a discernible response. It is related directly to the transmitted power level and the sensitivity of the receiver.

Minimum radial distance. A parameter that measures the ability of a system to determine the separation distance between two targets on the same bearing.

Monopulse radar. In theory only a single pulse is necessary to determine the radar range of a target. In practice, a series of such pulses are averaged to obtain a more accurate assessment of bearing and range.

Monostatic radar. A system that uses a single antenna for both transmit and receive. The transmitter and receiver circuits are linked to the antenna by a T/R cell. This acts as an isolating switch to give either complete monopoly of the antenna but at different times.

Moving target indicator (MTI). Fixed targets produce return signals with the same RF characteristics on each pulse. Moving targets, however, produce returns with phase relationships that vary from pulse to pulse. By using phase sensitive detector circuits and suitable signal processing, it is possible to distinguish between stationary and moving targets.

Multistatic radar. A secondary surveillance radar system that uses four ground stations located in precisely known positions. One acts as a master to synchronise the system and the others as slaves. The master transmits to interrogate an incoming aircraft and activate its transponder. The difference in time of arrival (DTOA) of the replies are measured at each receiving slave station and these are then used to determine the aircraft position to an accuracy of about 10 m.

Omega system. A now obsolescent navigation system that utilised a quasi-continuous carrier frequency of nominally 10 kHz. Because of the ground wave propagation that this frequency provided, the system gave worldwide coverage, but with an accuracy limited to about 1 mile.

Over the horizon radar (OTHR). Conventional radar systems operating at frequencies of 1 GHz upwards are restricted to line-of-sight applications which at low azimuth angles provide only relatively small coverage. By using frequencies in the range of about 5 to 28 MHz, reflections from the ionosphere can be used to extend coverage to almost 3000 km. These systems therefore use forward scatter from the ionosphere and back scatter from a distant target. At such relatively low frequencies, bistatic systems have to be employed with very long antennas to achieve the necessary gain and beamwidth. Because of the vagaries of propagation, it is necessary to be able to vary the operating frequency seasonally and with time of day. This makes it very difficult to avoid creating interference with the normal HF communications services.

Passive radar. These systems use the microwave or infra-red radiation emitted from a target to detect its position and are therefore easily concealed.

Performance figure. A system parameter that is very much related to range. It is based on the ratio of peak transmission power to minimum discernible signal at the receiver.

Precision approach radar. A system used for the guidance of ships or aircraft in fog. It gives very accurate guidance at close quarters.

Primary radar. Systems whereby the power in the received return signal has been generated by reflection from a target of a small part of the primary transmitted energy.

Pulse modulated Doppler radar. Conventional pulse modulated radar detects both fixed and moving objects. The constant reflections from the fixed objects can be eliminated from the display by using a Doppler system with an antenna operating with short dwell times in azimuth.

RAdar beaCON (RACON). A radar transceiver that transmits coded information when interrogated by a radar signal. An operator can obtain bearings from the replies.

Radar beacon system (RBS). An alternative title for IFF.

Radio beacon. A radio transmitter located in a precisely known position radiates a steady directional beam along a fixed path. This provides directional guidance for either ships or aircraft. Alternatively, the transmitter, whose location is precisely known, radiates from an omni-directional antenna so that DF bearings may be taken.

Radar parameters for typical applications.
Long range air traffic control:
10 cm/23 cm. 500 kW to 2 mW peak power. 2 to 4 μs pulse duration.
Low power portable:
3 cm/8 cm. 1.5 kW. FM/CW. Dual pulse, 0.5 μs at 1000 Hz for accuracy to 30 km; 0.08 μs at 2500 Hz for accuracy to 2 km.
Meteorology, storm avoidance, weather detection:
3 cm/10 cm. 40 kW to 500 kW. Pulse duration 0.5 to 1 μs.
Precision approach/airfield control:
3 cm. 20 kW to 400 kW. Pulse duration 0.2 μs.

Radar range. This system parameter is usually defined as the maximum distance at which a target can be defined by at least 50% of the transmitted pulses.

Resection. A DF technique that is complementary to intersection. The traveller takes bearings from DF transmitters to obtain navigational information.

Secondary radar. Systems whereby the received return is transmitted by the target's transponder after inter-rogation by the primary transmitter.

Secondary surveillance radar (SSR). The generic term used to describe such as the IFF, RBS, and similar systems.

SONAR (SOund NAvigation and Ranging). A radar type system that uses ultrasonic frequencies to detect underwater objects and for making meteorological measurements at very high altitudes.

Synthetic aperture radar (SAR). A system devised for use with satellite-borne earth monitoring radars, the term being derived from the way in which the antenna functions. An antenna just a few metres long at an altitude of about 800 km can provide an earth resolution in the order of 5 m or less. The radar systems operate typically at 5.3 GHz and exploit the relative movement of the space craft. The satellite emits a stream of radar pulses but by the time a particular return echo is produced at the receiver, the antenna has moved several kilometres along its track. The effect therefore mimics an antenna several kilometres wide, with high gain and very narrow beamwidth. Highly accurate maps of the earth's surface can be produced after digital processing of the returned signals.

Volumetric radar. Three-dimensional positional plots of several targets can be produced by using two radar systems simultaneously, each one separately providing the azimuth and elevation information.

Woodpecker. The jargon term given to a Russian OTHR system because of the staccato chirp that it produced in HF receivers. The system operated at peak powers of about 20 MW, radiating 100μs pulses at 10 HZ. The operating frequency varied between 5 and 20 MHz and created severe interference to the normal HF services over a bandwidth of about 100 kHz.

29.4 Useful references

Lynn P.A. (1987) *Radar Systems*, Macmillan New Electronic Series, Macmillan Education, London.

Mattos P. (1992/3) G.P.S. *Electronics World and Wireless World*. December 1992, 982; January 1993, 29; February 1993, 147; March 1993, 210; April 1993, 296; May 1993.

Mazda F.F. (1989) *Electronic Engineers' Reference Book*, 6th edition, Butterworth-Heinemann, Oxford, UK.

Skolnik M.I. (1989) *Radar Handbook*, McGraw-Hill, New York.

Stevens M.C. (1988) *Secondary Surveillance Radar*, Artech House, Boston.

30 *Radio frequency receivers*

The major aim of any communications system must be to ensure the highest level of fidelity of the signals received at a distance, over any transmission medium. The receiver is thus the final and most important element in the chain, to ensure the accurate and near error free transmission of information. (See also Noise, p. 219.) The major requirements of any receiver can therefore be summarised as follows:

(1) to select the wanted signal from the many others present in any given waveband;
(2) to recover the information from the modulated wave;
(3) to present the information in a suitable manner, e.g. audio, video, data, etc.;
(4) To carry out these functions without degrading the received signal to noise (S/N) ratio or bit error rate (BER) by more than is minimally necessary.

Since noise power is proportional to bandwidth, it is important that the receiver bandwidth should not be greater than that required to just accommodate the wanted signal. Apart from the bandwidth requirement and the method of demodulation, it is of little import whether the baseband signal is analogue or digital, the modulated wave carrying the information being of an analogue nature in either case.

A receiver needs to be responsive to a wide dynamic range of signal amplitudes. The lowest level is set by the thermal noise generated by the receiver input circuits and the simultaneous noise present within the transmission medium due to atmospherics and man-made interference. In today's congested communications bands, these requirements are probably best met by adopting a receiver that operates using the *supersonic-heterodyne* (superhet) principle.

30.1 Basic superhet receiver

Figure 30.1 shows the basic organisation of this type of receiver. The input radio frequency (RF) signals are amplified and preselected with bandpass filters to provide one input to the mixer or frequency changer stage. The RF stage also helps to reduce interfering radiation from the local oscillator, which provides the second input. This circuit is tuneable so that the nonlinearity of the mixing process can produce sum and difference frequencies at the output. By synchronising the RF and the local oscillator tuning, these two products of mixing represent a constant frequency that carries the wanted modulation signal. Either of these two frequencies, known as the intermediate frequencies

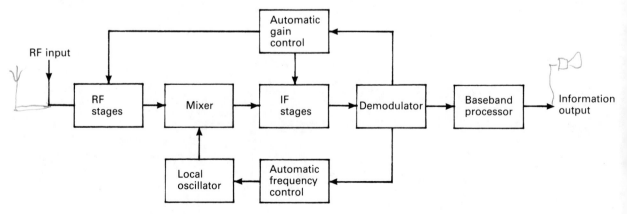

Figure 30.1 Typical superhet receiver.

(IF), can be selected by the IF stage filters for amplification, the selected frequency being the first channel, while the alternative is referred to as the *second* or *image channel*. The frequency spacing between the two is twice the IF value. The intermediate frequency is chosen to be as low as possible in order to obtain high signal gain with good stability. Further, since the mixer stage is the noisiest in the receiver (due chiefly to the conversion loss equal to the ratio of RF to IF signal levels and its nonlinearity), pre-mixer RF amplification helps to maintain a good S/N ratio. Following selection and amplification in the IF stage, the signal is demodulated and processed at baseband to provide the information output. In order to minimise the mis-tuning effects of drift in the local oscillator, it is common to provide a feedback loop that automatically controls the frequency (AFC) to maintain the wanted signal at the demodulator input on-frequency. To combat the effects of fading in the propagation path, the receiver is equipped with an automatic gain control (AGC) circuit to adjust the IF stage gain according to the average level of the signal being processed. Further, this loop is often extended to include the RF amplifiers to protect the receiver from overload by strong local transmissions.

30.2 Double/triple frequency conversion receivers (see Figure of merit, *below* and Interference rejection ratios, *below*).

In order to maximise the selectivity and S/N ratio, and minimise the effects of interference, the superhet concept is often extended to two or three conversion stages. The first stage of frequency conversion normally produces a fixed value of IF, while the second and subsequent conversion stages may be tuneable over the band of interest. At the first stage of conversion two possibilities exist, either the sum or difference frequency of the mixing products can be selected to provide *up-converting* or *down-converting* systems and both are in use. Up-conversion improves the rejection of the first image frequency but produces a higher value first IF. High gain in this stage, therefore has to be sacrificed in the interests of amplifier stability. The effects of frequency drift, caused by a change of temperature, can be largely overcome by combining both up- and down-conversion in the same receiver, if the oscillator frequencies are all controlled from the same basic crystal oscillator. Where frequency synthesis is employed for the local oscillator, the use of a high local oscillator frequency improves the system design by reducing the octave range of this circuit. This then allows the up-conversion receiver to provide a better dynamic range.

30.3 Figure of merit

All receivers are prone to effects of interference and some forms arise from using the superhet concept, while others would occur no matter which type of receiver was used. Shannon's rule which can be expressed as:

$$C = B \log_2(1 + S/N) \quad \text{bit/s}$$

where C is the channel capacity in bit/s, B is the bandwidth under consideration and S/N is the signal to noise ratio, shows how the receiver design is capable of maximising the trade-off between these three parameters in the interests of signal fidelity.

Interference rejection ratios are defined by the formula:

$$20 \log \frac{\text{Wanted signal voltage}}{\text{Interference signal voltage}} \quad \text{dB}$$

interference being caused by the adjacent or image channels, or by a strong signal present in the waveband of interest at the intermediate frequency. This latter form is referred to as IF breakthrough. Adjacent channel rejection is chiefly a function of the IF stage selectivity, while image channel rejection is a function of the RF selectivity, which itself is a function of the choice of intermediate frequency. (Image frequency equals twice the IF from the wanted channel.) This is shown by the relationship:

$$\text{Image rejection ratio} = 1/\sqrt{1 + (Qy)^2}$$

where Q is the Q factor of the pre-mixer tuned circuits and

$$y = f_i/f_o - f_o/f_i$$

where f_i and f_o are image and resonant frequencies, respectively.

IF breakthrough is avoided by selecting a little used frequency and this is aided by international agreement. Table 30.1 lists some of the common values in use.

Noise related factors are associated with the thermal noise generated largely in the first stages of a receiver and that present on the antenna system. The receiver's S/N ratio or noise factor (F) is a measure of how much it degrades the input S/N ratio. The perfect receiver would have an F value of 1 or 0 dB. The system G/T (gain/temperature) ratio is very much controlled by the antenna. The factors that affect this ratio are the antenna gain, its equivalent noise contribution and the equivalent noise temperature of the system. The importance of this figure of merit lies in the choice that it presents to the designer. A particular G/T ratio

Table 30.1   *IF frequencies in common use*

Frequencies	Applications
85 kHz	2nd/3rd IF in communications receivers
455–470 kHz	Domestic AM receivers
1.4–1.6 MHz	1st/2nd IF in communications receivers
10.7 MHz	Domestic FM receivers
33.4/38.9 MHz	Sound/vision IF PAL TVsystem G
33.5/39.5 MHz	Sound/vision IF PAL TV system I
41.25/45.75 MHz	Sound/vision IF NTSC TV
134–135 MHz	2nd IF Satellite TV convertors
950–1750 MHz	1st IF Satellite TV convertors

can be obtained with a high gain antenna, or a smaller, cheaper device combined with a lower noise specified system.

30.4 Frequency synthesis and digital control

Figure 30.2 shows the basic principles of this sytem. The local oscillator signal is provided by the phase locked loop (PLL) consisting of voltage controlled oscillator, dividers, filter and comparator. The *prescaler* (n) is used to extend the range of operation. The basic reference frequency is provided from a crystal oscillator whose output frequency is divided down to provide an input to the phase comparator. An example from television gives a useful explanation of the principle. The reference frequency is provided from a 4 MHz crystal oscillator, the output of which is divided by 1024. The receiver local oscillator thus locks to multiples of 3.90625 kHz. When a channel is selected, control logic converts this into a frequency and the phase comparator causes the

local oscillator to be adjusted until the output from the variable divider (*m*) is equal in phase and frequency to 3.90625 kHz. If the prescaler has divided the local oscillator frequency by 16, the actual tuning steps become $3.9062 \times 16 = 62.5$ kHz. Thus by choosing suitable reference frequencies and division ratios, any tuning step size can be generated. Often for a communications receiver, the step size is as small as 10 Hz and this gives the operator the impression of almost continuous tuning. The output from the local oscillator can also be fed to a frequency counter that has been offset by the IF value. The counter output can then be used to provide a digital read-out of the tuned frequency. Because the division ratios are scaled in binary, it is quite easy to adapt this concept to digitally controlled tuning either from a keypad or a shaft encoder. A further digital extension provides for the *scanning receiver*. This type of receiver can be made to scan tune a particular waveband automatically and stop when a signal with predetermined characteristics is found. The tuning of this type of receiver is very stable due to the fact that all injection frequencies are obtained from the same high stability crystal oscillator. Problems that can arise include excessive noise in the form of jitter from the oscillator and spurious beat notes that can be generated from the high speed synthesiser divider circuits.

30.5 Miscellaneous terms

Direct conversion receiver (homodyne or zero IF). If the local oscillator frequency is made equal to the wanted RF signal frequency, then the output of a mixer stage will be the modulation from the wanted carrier signal. The mixer thus acts directly as a demodulator. This avoids any image channel or IF break-through interference and simplifies the tracking of local oscillator

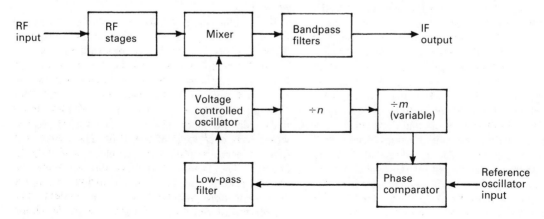

Figure 30.2   Basic frequency synthesis tuner unit.

and signal frequency tuned circuits. Adjacent channel sideband interference is easily removed by simple low-pass filters. When used in the higher frequency bands the oscillator stability can become a problem. This, however, can be resolved by using harmonic mixing generated from an overtone oscillator stage. Two strong signals with sum or difference frequencies that fall in the RF band may cause interference. In general, such intermodulation is generated by signals with frequencies of $mf_1 \pm nf_2$ where m and n are integers, typically less than 4. A further significant advantage of the zero IF principle is that it allows the use of extensive digital signal processing, a feature particularly valuable in portable equipment.

Diversity reception. Two possibilities exist. Space diversity, where two receivers are separated by a few wavelengths, or frequency diversity, where the same modulation is applied simultaneously to two adjacent carriers. Since both signals are unlikely to experience fading at the same rate and time, the stronger of the two outputs can safely be accepted as the wanted signal.

Frequency agile receiver. In certain secure transmission systems it is required to change channels quickly and in a random manner under the control of a pseudo-random binary sequence. Such signals can be received and the information recovered by using a receiver with a similarly tuned, digitally controlled local oscillator.

Frequency hopping. The jargon term that describes the behaviour of a frequency agile communications system.

Homodyne. See Direct conversion receiver, p. 264.

Optimum intermediate frequency (IF). The design of any superhet receiver has to take into consideration the problems of image channel and IF interference. In multiple conversion superhets this problem is particularly difficult, but the use of the optimum IF relationship

$$(IF_1)^2 = IF_2 \times f_{RF}$$

ensures the best compromise.

RAKE receiver. Multipath propagation effects appear as noise superimposed on a wanted signal and will seriously impair the quality of analogue transmissions. In digital applications, this problem leads to inter-symbol interference (ISI) and corrupted data. The RAKE receiver concept has been developed to turn this problem to advantage with code division multiple access (CDMA) systems. This receiver uses multiple demodulators that simultaneously operate on several versions of the same signal which is time-offset by a range of values. The outputs of the demodulators are then combined and used to determine the correct code pattern, but with a considerably improved S/N ratio.

Sensitivity/selectivity characteristic. This may be expressed as the minimum input signal voltage necessary to produce a given output level for a certain signal to noise ratio and at a specified degree of selectivity. This is also known as the *signal + noise to noise* ratio.

Spurious free dynamic range (SFDR). The amplitude range between a fundamental frequency and the first spurious noise component expressed in $-$dBc.

Sub-harmonic mixing. It is most important that the local oscillator stage of a microwave receiver should be highly frequency stable. If this signal is generated at some lower sub-multiple of the required value, doubling or even tripling can be achieved in the mixer circuit to provide the correct conversion. Since the technique produces a higher conversion loss, the cost of the improved frequency stability is compromised by a higher level of mixer noise.

Tuned radio frequency receiver (TRF). If a communications channel occupies only a single or a narrow band of frequencies, then the RF selectivity can be produced by fixed tuned circuits. This simplifies tuning and the required RF amplification can be maximised for the particular application.

Wadley drift free loop. An analogue synthesiser technique that uses triple mixing and a variable oscillator for tuning purposes. The first two mixer oscillator signals are derived from the same highly stable crystal oscillator circuit based on a 1 MHz fundamental frequency. Up- and down-conversion in these two stages provide frequency drift cancellation. The front end tuning varies in increments of 1 MHz bands so that the third mixer stage simply tunes across this range in a bandspread manner.

Zero IF receiver. See Direct conversion receiver, p. 264.

30.6 Useful references

Kennedy G. (1985) *Electronic Communication Systems*, 3rd edition, McGraw-Hill, New York.

Lewis G.E. (1992) *Communication Services via Satellite*, 2nd edition, Butterworth-Heinemann, Oxford.

Markus J. (1982) *Communications Circuits Ready-Reference*, McGraw-Hill, New York.

Rohde U. and Bucher T.T. (1988) *Communications Receivers: Principles and Design*, McGraw-Hill, New York.

Schwartz M. (1980) *Information Transmission, Modulation and Noise*, 3rd edition, McGraw-Hill, New York.

31 Satellite systems

Satellite communications systems are based on the relay systems that have developed for terrestrial communications, except that the receiver/transmitter, known in this case as the *transponder*, is located in space.

In spite of the high costs involved, the flexibility and advantages gained allow satellites to make a valuable contribution to worldwide communications.

All the propagation parameters are well defined and can be accurately modelled mathematically. A single satellite can provide communications coverage for almost one-third of the earth's surface, using much less radio frequency power than would be required for an equivalent terrestrial system. Even though the signal attenuation due to the long path lengths is high, it is fairly constant. A *signal fade margin* allowance of only about 3 to 5 dB needs to be made to account for the variability due to local atmospheric and weather conditions. For the equivalent earth-bound system an allowance of more than 30 dB may be necessary.

The high level of reliability of the current technology is such that the service lifetime of a satellite is now in the order of 15 years. When compared with terrestrial microwave relays and undersea cables, and taking into consideration the fact that, in some areas of the world, satellites are the only possible way of providing radio communications, then the system becomes very cost-effective.

Communication satellites occupy either *equatorial*, *elliptical*, or *polar* orbits as shown in Fig. 31.1(a). *Geosynchronous* satellites are those whose period of rotation is synchronised to that of the earth or some multiple of it. The *geostationary* orbit is a unique geosynchronous one, located over the equator. The satellite has a height and velocity such that it appears stationary to earth-bound observation. It is the earth's period of rotation relative to the fixed stars in space (the *sidereal* time) that is important. This is slightly less than the solar period and is approximately 23 h 56 min 4.1 s (23.9345 h). The height above the earth's surface of 35 765 km and velocity of 3.073 km/s, can be calculated as follows:

The gravitational force of attraction between two bodies is proportional to the product of their masses M, and inversely proportional to the square of the distance

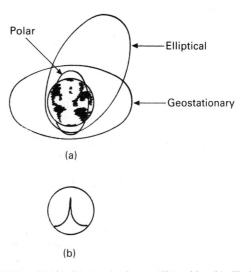

Figure 31.1(a) Communication satellite orbits; (b) elliptical orbit tracking.

r, between their centres of mass. The constant of proportionality is G, the universal gravitational constant of 6.67×10^{-11} Nm^2/kg^2. Thus the gravitational force acting between a geostationary satellite and the earth is given by:

$$F_g = GM_eM_s/r^2 \text{ Newtons} \qquad (31.1)$$

where M_e and M_s are the masses of the earth and satellite, respectively. ($M_e = 5.98 \times 10^{24}$ kg.)

The force acting on a body, constrained to circular motion, is proportional to its mass, the radius of the circular path and the square of the angular velocity. Thus the force acing on a geostationary satellite against gravity is given by:

$$F_c = M_s r \omega^2 \text{ Newtons} \qquad (31.2)$$

where r is the radius of the circular path and ω is the angular velocity.

The satellite continues in circular orbit around the earth when their separation and velocity are such that equations (31.1) and (31.2) are equal, so that:

$$GM_eM_s/r^2 = M_sr\omega^2$$

i.e.

$$r^3 = GM_e/\omega^2$$

Now since the satellite is geostationary, its angular velocity ω is the same as that of the earth, 2π radians per sidereal day:

$$= 2\pi/(23.9345 \times 60 \times 60)\,\text{rad/s}$$
$$= 7.30 \times 10^{-5}\,\text{rad/s (approx.)}$$

thus

$$r^3 = (6.67 \times 10^{-11} \times 5.98 \times 10^{24})/(7.3^2 \times 10^{-10})$$
$$r = 42143\,\text{km (approx.)}$$

Assuming a mean earth radius of 6378 km, the satellite average height above the equator is 35 765 km.

$$\text{Satellite velocity} = 2\pi(42143 \times 10^3)/(23.9345 \times 60 \times 60)$$
$$= 3.073\,\text{km/s (approx.)}$$

These values were presented by engineer and science fiction writer, Arthur C. Clarke in an article first published in the October 1945 issue of *Wireless World*. This effectively set out the ground rules for satellite communications, so that the geostationary orbit is now named for Clarke in recognition of this work. The *Clarke orbit* is used to provide worldwide *point-to-point* (narrowcasting) and *point-to-multi-point* (broadcasting) services.

The elliptical orbiting satellites operate with the earth's centre as one of the two focus points of an ellipse, that is inclined at a suitable angle to polar axis. Such an orbit has the advantage that the launch is achieved for a lower expenditure of energy.

Satellites in elliptical orbits obey Kepler's laws of planetary motion. Rephrased for satellite applications these state:

(1) The satellite moves in an elliptical orbit with the centre of the earth at one focus.
(2) The radius vector sweeps out equal areas in equal time.
(3) The square of the period of revolution is proportional to the cube of the semi-major axis, giving rise to the equation:

$$T = 2\pi(a^3/GM_e)^{0.5} \qquad (31.3)$$

where

T = orbit period in seconds
a = semi-major axis in metres
$M_e = 5.98 \times 10^{24}\,\text{kg}$
$G = 6.67 \times 10^{-11}\,\text{Nm}^2/\text{kg}^2$.

A typical satellite has an *apogee* (highest point) of about 35 600 km and a *perigee* (lowest point) of about 3960 km above the earth's surface and a period of slightly less than 12 h. Around the apogee, the satellite appears *pseudo-stationary*, remaining within a beamwidth of less than $\pm15°$ for more than 8 h over the service area. Three satellites in such an orbit can therefore provide 24 h per day service. These orbits sometimes described as *super-synchronous* are also known as *Molniya* orbits after the Russian communication satellites that used them. The track of such a satellite, over its service area, is shown in Fig. 31.1(b).

The low earth orbiters (LEOs) in polar orbits have a typical altitude of about 850 km and a period of around 100 min. These are used extensively for navigation and weather forecasting applications. LEOs traverse the North and South Poles covering the earth's surface in a series of strips, to scan the conditions that exist around the world.

Radio propagation is such that frequencies below about 30 MHz are trapped within the ionosphere by reflection and refraction, a critical frequency that varies during the 11-year sun-spot cycle and with atmospheric conditions. For this reason, only frequencies above about 100 MHz are used for space communications.

Each satellite transponder acts as a frequency changer, receiving the up-link signal on one frequency and transmitting the down-link signal on another. To the end user, apart from the propagation delay, the transponder is virtually transparent.

In general, the up-link service is provided in a higher frequency sub-band and the corresponding down-link in a lower one. There are some economic and technical advantages in using this arrangement. The gain of an antenna is proportional to the operating frequency and inversely proportional to its beamwidth. If the high band were used for the down-link, the receiver antenna would have a narrower beamwidth, increasing the problems of maintaining alignment in windy conditions and as the satellite drifts around in space about its mean position. This may introduce the need to add a servo-controlled steering system to provide compensation. By using the high band for the up-link, the extra gain can be usefully employed to make up for the high path length attenuation. In this way, the overall system S/N ratio will be better managed. The large dish, ground transmitting station will in any case require a servo-controlled tracking system, so that there is no added cost in this case.

The methods used to regulate the exploitation of the frequency spectrum for terrestrial systems provided unfair advantage to those nations first in the field of radio communications. The result in some cases, of almost haphazard development, can be heard in nearly all wavebands. In certain parts of the spectrum, the level of mutual interference produced by transmitters is fast becoming intolerable. In a restricted sense, the frequency spectrum is a non-renewable resource. When a frequency is occupied by one transmitter, it is of little use to any other within the same coverage area. World Administrative Radio Conferences (WARCs) are regularly convened in order to manage the spectrum for space communications and minimise the effects of terrestrial systems. The problem of avoiding interference in space-based communications systems has a significant effect on the way in which geostationary satellites are managed. In these cases, mutual interference can be minimised by controlled allocation of orbit position, the use of frequency division (FDMA), time division (TDMA) or code division multiple access (CDMA) techniques, and the use of various forms of signal polarisation.

The number of satellites that the Clarke orbit will support depends upon the antenna pointing accuracy that can be achieved and maintained. Satellite operational lifetime is to some extent dependent upon the amount of fuel it can carry for station keeping, the accuracy required being within an arc of $\pm 0.1°$. A simple calculation shows that such a satellite can wander about its mean position, within a sphere of about 62.5 km. Taking all these points into consideration, it is unlikely that spacings closer than about $2°$ can be tolerated for satellites using the same frequency bands.

Frequency division multiple access (FDMA) is a very common technique in use. Each transmitting station is allocated different carrier frequencies, as indicated in Fig. 31.2(a). Many stations may use the same transponder amplifiers simultaneously within the limits of the total bandwidth. Capacity allocation is simple and the system requires no complex timing, or synchronism. The baseband signal is easily recovered using relatively simple and inexpensive equipment. The major disadvantage lies in the nonlinearity of the transponder's high power amplifier. Because several carriers are present simultaneously, intermodulation (IM) can arise. This is normally countered by *backing-off* the power output, thus reducing the signal level received by the ground station.

Time division multiple access (TDMA) allows each earth station to be assigned a time-slot for transmission using the entire transponder bandwidth as illustrated in Fig. 31.2(b). Each ground station transmitter uses the same carrier frequency within a particular transponder and transmits in bursts. As each burst must carry a

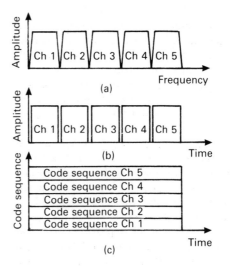

Figure 31.2 (a) Frequency division multiplex; (b) time division multiplex; (c) code division multiplex.

destination address, the flexibility of interconnection is improved. Since only one carrier is present at any one time, the transponder amplifiers can be operated near to saturation, for greater power output. There is no intermodulation or adjacent channel interference from the same satellite. This leads to an improved S/N ratio at the ground receiver. The costs of these improvements lie in the extra complexity of the equipment. However, the concept is very compatible with digital signal processing techniques. A ground control station is necessary to maintain control and synchronisation of the system network. The throughput of data is better than FDMA by a factor of about two.

Code division multiple access is a method well suited to digital transmission. All users operate at the same carrier frequency and simultaneously use the whole channel bandwidth. A unique digital code is added to each digital transmission sequence and the intended receiving station equipped with the same *key* (see Fig. 31.2(c)). By using *correlation detection*, the receiver extracts the wanted signal from the noise of all the other transmitters. As the number of users in these systems increase, the system performance degrades gracefully, unlike normal digital systems whose performance *crashes* when overloaded. When the system is under-used, there is an automatic improvement in the signal error margin. Signal polarisation variation can permit *frequency re-use* to extend the occupancy of a waveband and provided that the services are separated by a sufficient ground distance, there will be minimum mutual interference. '*x*' and '*y*' linear and left- and right-hand circular polarisation can be used in the same way as vertical, horizontal and circular polarisations are used for terrestrial systems.

By using all these concepts, an equitable share of the frequency spectrum should be available to all who need to use it. In addition, there should be much less mutual interference between users than has been the case with earth-bound communication systems.

From the above, it can be seen that the transponder characteristics may be power or bandwidth limited. In certain cases *on-board* demodulation, noise reduction/error correction and remodulation can be usefully employed to improve the ground station's received signal quality. However, this makes further demands on the limited power available. The down-link antenna beams are specifically shaped to cover the required service area. Global coverage requires a beamwidth of about 17.5° to cover a little over one-third of the earth's surface. Reducing the beamwidth to provide a *spot beam* with a width of about 4° is equivalent to increasing the gain over the smaller service area by about 13 dB. Therefore, for the same signal level at the ground receiver, the global beam makes a much greater demand on the available power and limits the services available through the satellite. Higher effective radiated down-link power leads to smaller and lower cost receiver antennas.

The transponder high power stages commonly use *travelling wave tube* (TWT) amplifiers. These have the advantage of relatively low weight and high power efficiency. When operated near to saturation, the nonlinearity and *group delay* (rate of change of the phase response) will seriously distort amplitude modulated signals, and cause intermodulation when several carriers are present simultaneously. Such operation requires the TWT operating conditions to be backed-off, reducing the effective output power. Using frequency modulation however, the TWT can be operated nearer to saturation for higher efficiency.

31.1 Satellite services

Aeronautical mobile system

This developing service operates by using the satellite as a relay link between the aircraft and a ground station. This in turn provides the interconnection to the terrestrial PSTN system. The service which operates in the band 1.67 to 1.8 GHz provides both passengers and aircrew with an almost global voice and data communications link. Two digital voice channels are provided at 9.6 Kbit/s for passenger use, a single channel at 4.8 Kbit/s for crew use, and 600 bit/s for a digital data messaging service. Due to the high aircraft speeds, Doppler shift could be a problem. The RF, modulation and coding schemes are devised so that provided the Doppler shift does not exceed ±100 Hz, the system can provide a high quality of service. The aircraft antennas,

which are encased in an aerofoil shaped housing, are mounted externally and take on the appearance of *shark fins*. To provide all-round coverage and maintain constant contact with the satellite, the antennas are of the steerable, planar, phased array types.

The voice signal is PCM coded by sampling at 8 kHz to produce 8-bit samples every 125 μs. These are processed by a microprocessor and the bits are clocked out at 1536 kHz. The 8-bit bursts at 1536 kHz are then converted into a continuous stream at 9.6 Kbit/s. This output is then multiplexed with a dual tone multi-frequency dialling tone, to generate a gross bit rate of 10.368 Kbit/s. After the addition of FEC bits to counter the effects of multipath fading, the bit rate has doubled to 21 Kbit/s. This serial bit stream is then modulated QPSK on to the final RF carrier.

The system bandwidth is wide enough to accommodate additionally, engineering, navigational and other data, which will help to improve the safe and economic operation of the aircraft.

Business services (see also LASS, p. 272 and SSSO, p. 280).

In the main, the major owners of the satellite space segments are the consortiums such as INTELSAT (International Telecommunications Satellite Organisation), EUTELSAT (European Telecommunications Satellite service), an arm of the European Space Agency, and PASAT (Pan American Satellite Inc). These then lease the day-to-day operation of the network to other suitably qualified organisations, who are often but not always, the national PTTs. These operate ground stations with antennas that may range in diameter from 3.5 to 13 m. The transponder channels may then be sub-leased to other end users. The services provided include video, voice and data and these are often linked into the national PSTN or ISDN systems. Apart from the propagation delay, the high grade of service makes these satellite channels virtually transparent to the end users.

Briefcase satellite newsgathering (SNG) terminal

This small, light, go-anywhere system was developed by the cooperation between Columbia Broadcasting System (CBS), Teleglobe Canada and Skywave Electronics Ltd. Weighing only 14.5 kg, it is capable of providing good quality speech and still pictures plus data, into any PSTN (public switched telephone network) over a satellite channel. It will operate in the full duplex mode over INMARSAT-C links of 1.5 to 1.6 GHz to a hub earth station which itself operates in the 4 and 6 GHz band.

The antennas consist of two flat patch microwave arrays, providing circular polarisation with a gain of 17 dBi. Together with a 20 W RF power amplifier stage, the system produces an EIRP of 27 dBW in a 5 kHz bandwidth. This allows two terminals to operate simultaneously over a single 25 kHz INMARSAT channel. The antennas are fixed to the lid top which is orientated in azimuth and elevation using locking hinges. During deployment, the antenna is aligned using the satellite beacon signal. Speech transmissions employ *amplitude companded single sideband* (ACSSB) and the digital modulation scheme uses DMSK at 2400 bit/s.

Power for the terminal may be provided by locally rechargeable NiCd batteries or by 110/230 V 50/60 Hz mains supplies. It is anticipated that operation over a higher power satellite would allow an even smaller terminal to be produced.

When deployed, the terminal can also receive calls, but with the distant operator dialling first the hub terminal code, followed by the briefcase terminal code.

Environmental satellite services

Because the data is supplied from low earth polar orbiters equipped with scanning systems, these have much in common with the meteorological services. Typically, the orbits are at an altitude of about 800 km with a period of 100 min. The sensors are chosen to collect the data required by scanning the earth's surface using a wavelength range varying from about 500 nm to 900 nm. This represents colours ranging from green to near infra-red. High resolution images can be generated from the received data that is accurate to less than 20 m.

LANDSAT. This US series of satellites is operated by Earth Observation Satellite Corporation (EOSAT) and is intended to be financially self-supporting through the sale of the remote sensing information. The on-board sensors are designed to provide images that can be used for a wide range of operations. These include land use and mapping, soil erosion, crop management, volcanic activity, mineral deposits and environmental disasters. The sensors have the following earth resolutions:

visual images, 30 m
cartographic images, 15 m
infra-red thermal images 120 m

The data that is collected on-board is stored using high density magnetic tape recorders. To save wear and tear on the recorder mechanism and space craft power, the tape machines transmit the stored data in reverse. This avoids wasting time and power to rewind before replay. The data is relayed to the ground stations either directly on command from a control station, or via the tracking and data relay satellite system (TDRSS). For the relay of the stored data, modulated X band carriers are used,

with data rates of 15 Mbit/s for visual and IR images and 85 Mbit/s for cartographic data.

SPOT (Systéme Probatoiré d'Observation de la Terre). This series of French satellites is owned by a public company of which MATRA and Centre National d'Etudes Spatiale (CNES) are major shareholders. The on-board sensors provide typical environmental information with a resolution of 10 m for colour images and 20 m for the wide band coverage data. A unique feature involves the ability to provide 3D images from the received data. The collected data is stored on a high density magnetic tape recorder using *three-position* modulation, a technique that provides three data bits for every two flux reversals. The relayed data is PCM encoded at 25 Mbit/s onto 20 W carriers of either 8.025 or 8.40 GHz.

ERS-1 and ERS-2 (European remote sensing). This series of satellites is owned and managed by the European Space Agency (ESA) and is basically intended to provide information about the marine environment. The features being monitored include, mean sea height to an accuracy within 10 cm, wave movement, sea and cloud temperatures, ice flows and icebergs, the ozone layer and climatology in general.

FleetNET system

This network operates over the INMARSAT system and through type C terminals. It allows individual companies to communicate with all their associated mobiles in a broadcast manner, or individually by using a unique addressing technique. In addition, navigation sensors can be added to the mobile terminals which allows them to automatically transmit regular position reports to the home base station.

Global Maritime Distress and Safety System (GMDSS)

The meteorological and other polar orbiting satellites often perform a dual role by providing a relay for emergency and navigation signals. These satellites are sun-synchronous, in that they pass over the same point on the earth's surface at the same time each day. The operating frequencies being used are all in the VHF or low UHF bands, 121.5, 243, 401, and 406 MHz, which includes the common International Distress Frequencies. The services are typically international, but are generally within the SARSAT/COSPAS agreements. (SARSAT = Search And Rescue SATellite, COSPAS = Cosmicheskaya Systema Poyska Avarinich Sudov, which roughly translates to: Space System for the Search of Ships in Distress.)

This service which is expected to be globally operational by 1999, could be made mandatory for all sea-going vessels under the SOLAS (Safety Of Life At

Sea) Convention. It already has the support of the United Nations via the work of the International Telecommunications Union (ITU) and the International Maritime Organisation (IMO), two of the organisations that have been responsible for the development of the technical standards involved.

The important emergency part of this system is based on a small beacon transmitter that, when activated, either manually or automatically, continuously emits an AM signal at a power output of 100 mW. This signal is picked up by a polar orbiting satellite and repeated in real time to be received by a local user terminal (LUT) ground station. The beacon transmitters are generally known as emergency position indicating radio beacons (EPIRB) or search and rescue beacon equipment (SARBE). The position of the emergency is determined by the LUT using the Doppler shift affecting the beacon signal received at the satellite. The satellite processes the beacon identification, emergency and user category codes, adds a time code and then retransmits the message. Each LUT is capable of providing coverage over an area of about 2500 km radius, and depending upon conditions, produces a position accurate to 5 to 10 km.

Iridium system (element 77 in the atomic table)
This proposal, originally made by Motorola Corp., is being considered by a consortium that additionally includes British Aerospace, Lockheed Aircraft Corp. and INMARSAT, as a possible global network that could be in operation as early as 1996.

The initial proposal was based on 77 low earth polar orbiters at an altitude of about 660 km, the satellites to be positioned within seven orbits, 11 to an orbit, and each equipped with inter-satellite communications links.

With improvements in reliability and technology the system will probably now deploy 66 satellites in six orbits.

Calls will be made from handsets similar to the current cellular telephones, switched between satellites, then ultimately to a ground station to be patched into the terrestrial telephone system.

Various frequency bands have been allocated to mobile operations within the range of about 1.48 GHz to 2.7 GHz. These separately cover land, air and maritime mobile services and Iridium could operate in the 2 GHz to 2.7 GHz segment.

For low population densities and remote rural areas, this service, which is intended to be compatible with and complementary to INMARSAT, should prove to be more economic than extending the present terrestrial telephone system.

Large aperture satellite systems (LASS)
The wide range of antenna sizes in operation allows for a correspondingly wide range of LNAs/LNBs to be used to meet the system's overall G/T ratios. The parameters and standards for the complete ground stations, including up-link, down-link and footprints (specified down to a minimum elevation angle of $10°$) are closely specified.

Power budget calculations and a comparison of the relative merits of TDMA versus SCPC are explained by reference to Table 31.1.

Services are provided in both C and Ku bands and the latter is probably most popular outside of the USA. The transponder bandwidths are in the order of 36 or 72 MHz and these may be occupied by either analogue or digital signals. All forms of analogue modulation are catered for but FDMA/SCPC is most commonly used.

Table 31.1   *Down-link power budget*

	TDMA	SCPC
(a) Transponder saturated EIRP (−3 dB contour)	42 dBW	42 dBW
(b) Back-off for multi-carriers	0	6 dB
(c) Operating EIRP (a−b)	42 dBW	36 dBW
(d) Transponder power per carrier	42 dBW	16 dBW
(e) Free space attenuation	208 dB	208 dB
(f) Receiver antenna gain	47 dB	47 dB
(g) Received signal power (d−e+f)	−119 dBW	−145 dBW
(h) Receiver noise temperature	25 dBK	25 dBK
(i) Down-link C/T ratio (g−h)	−144 dBW/K	−170 dBW/K
(j) Boltzmann's constant	−229 dBW/HzK	−229 dBW/HzK
(k) C/N_o ratio (i−j)	85 dBHz	59 dBHz
(l) Transmission rate	74 dBHz	48 dBHz
(m) E_b/N_o (k−1)	11 dBHz	11 dBHz
(n) Degradation margin	4 dB	4 dB
(o) E_b/N_o (theoretical (m−n))	7 dB	7 dB

The bulk of the digital services operate with QPSK/2-4PSK modulation with TDMA/SCPC protocols. However, in the interests of efficient frequency spectrum management, the use of higher levels such as 64QAM will become progressively more important.

The bandwidths for these services are usually allocated in contiguous blocks of 64 Kbit/s depending upon the user requirement, up to a maximum of 8.448 Mbit/s. This, however, can be extended up to 24.576 Mbit/s for special cases using differential encoding and demand assignment (DA). Forward error correction (FEC) using 1/2 or 3/4 rate convolutional coding plus Viterbi decoding is commonly employed to produce a bit error rate better than 10^{-3}. A form of spread spectrum transmission is commonly employed using scrambling to act as a form of energy dispersal to obtain full channel occupancy. Even so, it is still allowed for the originator to include encryption within his own domain.

Meteorological services

These services were set up under the auspices of the World Meteorological Organization (WMO), to provide a permanent World Weather Watch (WWW) and carry out experiments within the Global Atmospheric Research Programme (GARP). The total system at any one time, involves the services of six geostationary satellites and at least four polar orbiters.

Of the geostationary satellites, two are operated by the USA and are known as Geostationary Operational Environmental Satellites (GOES); one is operated by Japan, the Geostationary Meteorological Satellite (GMS); one is operated by the USSR, and referred to as Geostationary Operational Meteorological Satellite (GOMS); one is operated by the Indian Satellite Research Organisation (ISRO) and known as INSAT; while that operated by the European Space Agency (ESA) belongs to the Meteosat series.

The geostationary satellites are spaced about 70° longitude apart, in order to provide a wide coverage up to about 75° latitude. The actual orbit positions are:

GOES-F at 135°W (Pacific Ocean region W) (POR)
GOES-H at 75°W (Atlantic Ocean region) (AOR)
METEOSAT at 0° (Greenwich meridian)
GOMS at 70°E
INSAT at 74°E (Indian Ocean region) (IOR)
GMS at 140°E (Pacific Ocean region E) (POR)

The view from these satellites takes in a full earth disc, that covers about one-quarter of the surface, although, near to the horizon, the view is too oblique to be of much value. These satellites are supported by about 9000 earth stations, 7000 ships and 850 balloons, from more than 150 countries. Unlike the other satellites,

GOES-H also monitors the Earth's magnetic field, the intensity of solar winds and radiation belts surrounding the earth. It also acts as a relay link in the 406 MHz band for the International search and rescue service, COSPAS-SARSAT.

The polar orbiting satellites provided by the USA are of the TIROS and NIMBUS series and are operated by the National Oceanographic and Atmospheric Administration (NOAA). The satellites provided by the USSR are of the Meteor series. All these have a period of about 100 min, at an altitude of between 700 and 1500 km. Due to the earth's rotation, each orbit crosses the equator about 25° longitude further west than the previous one. The on-board instruments see the earth surface as a series of strips, which are scanned from side to side. A particular location is thus viewed at least twice a day, once during a north to south pass and again, about 12 h later, during a south to north pass. Because of the width of each scan, there is an overlap of tracks, so that some areas are viewed on successive orbits. The two systems together provide a complete worldwide coverage, to detect weather patterns as they develop.

A radiometer is carried on each satellite to measure the magnitude of earth radiation (heat, light, etc.), which sensors convert into electrical signals for transmission. To provide the most useful images for meteorological purposes, the radiation waveband is restricted by the use of filters. The Meteosat data that is transmitted covers the following bands or channels:

0.4 to 1.1 μm – visible range (VIS)
5.7 to 7.1 μm – water absorption range (WV)
10.5 to 12.5 μm – infra-red range (IR)

to provide three distinct images.

Visual range (VIS). These images show only the reflected light from the earth and so are only available for the daylit hemisphere. In the reproduced image, space appears black, and snow and clouds white. Other areas are shown as varying shades of grey.

Water vapour (WV). These images do not provide an earth view and only represent the upper atmosphere. White areas represent low temperatures, which can be equated with high humidity. The darker areas represent a lower level of atmospheric humidity.

Infra-red (IR). Infra-red radiation is proportional to temperature and so these images are available both during daylight and darkness. Since space and high clouds are cold, these regions are depicted as white. Hot deserts thus appear very dark. The grey scale is thus temperature dependent.

Each geostationary meteorological satellite spins on its axis at 100 rpm, and its radiometer scans the earth in the east to west direction. After each revolution, the on-board sensors are deflected in small steps in the south to north direction. This dual action causes the earth to be scanned in a series of lines, a full image being generated at half-hourly intervals. Earth image data is gathered during 1/20 of each revolution, which represents a one line sampling period of the earth radiation. Each sample contains 2560, 32-bit words, each of which contain visual, infra-red and possibly water vapour data.

The raw data for each line of the image is stored in memory and then transmitted in somewhat less than 570 ms, at 166.7 Kbit/s using digital split phase modulation (SP-L), which is a version of the Manchester code, in which a transition occurs at each bit cell centre, effectively $1 = 10$ and $0 = 01$. On Meteosat, the carrier frequency is 1686.833 MHz, linear polarised, at an EIRP of 19 dBW per transponder. If the line store fails, provision is made to by-pass the 'stretched mode' and transmit the raw data in a 'burst mode' at the same rate that it was gathered (2.73 Mbit/s).

The data for each image is gathered in 25 min and a further 5 min is allowed to reset the vertical scanning mechanism, thus providing a new image every 30 min. The maximum earth resolution is either 2.5 or 5 km, depending upon the image mode. After the raw data has been received and stored at the earth station meteorological centre, it is processed, before being retransmitted through the satellite, in both analogue and digital form, for dissemination to the various users. Processing is necessary to remove the defects in the radiometer, sensors, and filter characteristics, to correct any inaccuracies in synchronism and to ensure registration of the three views. Processing also allows other derived images to be transmitted, such as cloud motion vectors and cloud top heights.

Apart from earth imaging, the geostationary satellites provide two further services to meteorology, the dissemination of processed raw data to other areas of the world and data collection via data collection platforms (DCPs). These are small automatic or semi-automatic units that monitor environmental data and use the satellite to transmit this back to a central ground station. These platforms may be carried in ships or aircraft, or be fixed on land sites. Some are self-timed to transmit their data by an internal clock. Some may transmit their data upon interrogation from a central station. 'Alert' DCPs automatically transmit a message if one of its measuring parameters exceeds a pre-set level. This gives early warning of risks of serious disturbances. The DCPs are interrogated over one of two channels and transmit within the frequency range 402 to 402.2 MHz.

The DCPs are supported by a system of three geostationary satellites that are known as tracking and data relay satellites (TDRS). In addition, the ARGOS system is used to locate, collect and disseminate marine environmental data from fixed or mobile DCPs.

The geostationary satellite stations fall into two classes, primary data users stations (PDUS), which receive the processed data in digital form for high resolution images, and secondary data users stations (SDUS) which receive the data in analogue form to the WEFAX standard. Down-link frequencies for both Meteosat PDUS and SDUS are either 1691 MHz or 1694.5 MHz, linear polarised. In general the two channels lie in the range 1690 to 1697 MHz.

The PDU stations in general, are much the more complex of the two. SDU stations provide images that are easily displayed on CRTs or FAX machines, using a facility known as automatic picture transmission (APT), which is also used by the polar orbiters where the image data is being continuously transmitted. With the latter satellites, reception is continuous while it is more than about 5° above the horizon. Typically such data can then be received over a circular area of about 2500 km radius from the antenna.

Normally the polar orbiting satellites work in pairs, separated by 90° longitude, so that each earth point is scanned every 6 h. The TIROS-N series of satellites carry advanced very high resolution radiometers (AVHRR) whose outputs are focused on to five sensors to gather data covering the following channels:

0.58 to 0.68 μm
0.725 to 1.1 μm
3.58 to 3.93 μm
10.3 to 11.3 μm
11.5 to 12.5 μm

The data is digitised by an on-board computer, partly to linearise each line scan to compensate for the curvature of the earth and partly because the data will be received with fairly simple ground stations, without processing capability. After processing, the data from two channels is selected and synchronising signals added, that will identify the image edges. This composite signal is then converted back into analogue form and then used to amplitude modulate a 2.4 kHz sub-carrier. This DSB AM sub-carrier in turn, is used to frequency modulate a final RF carrier that is typically in the range, 137.15 to 137.62 MHz. The transmitter power output is approximately 39 dBm (5 W) and the transmission is right-hand circular (RHC) polarised. APT facility is accomplished by continuous transmission at 120 lines per minute, to give an earth resolution of 4 km.

Later TIROS-N satellites also include a high resolution picture transmission (HRPT) facility. This is a digital service that provides for an earth resolution of 1.1 km. The outputs of all five AVHRR data channels are time multiplexed for transmission by split-phase (SP-L) modulation at a bit rate of 665.4 Kbit/s on a carrier of 1698 or 1707 MHz. RHC polarisation is again used, at a level of 39 dBm.

Micro-satellites

These small satellites are the result of cooperation between Amateur Radio Satellite Corp. (USA), the University of Surrey (UK) and Surrey Satellite Technology Ltd (UK), and form part of the PACSAT Communications Experiment (PCE). They occupy a low earth polar orbit with a period of 90 to 100 min, so that any earth station is covered at least three times every 12 h. The satellite carries a full scale *store and forward* messaging system, with messages and data being stored in an on-board 4 Mbyte RAM disk. The current satellite carries an on-board *charge-coupled-device* (CCD) image sensor, but other devices may be used. The stored messages and data are transmitted using a packet technique for distribution to the many scattered outposts, the system software being IBM-PC compatible. Due to their relatively very low cost, micro-satellites form a useful concept for testing new ideas under actual space conditions.

Mobile services

INMARSAT standard B and C systems. The standard A system is somewhat limited in the facilities offered and to the size of the vessel on which it can be installed. To overcome some of these problems, the standard B system was developed. This provides a telephony service using digital speech processing and coding plus QPSK modulation. One of two digital communications data rates (9.6 and 16 Kbit/s) are user selective. The system uses adaptive predictive coding with FEC (forward error correction) and a half-rate convolution code, plus Viterbi soft decoding at the receiver. The telephony mode is voice operated to allow faster and simpler operation of this lower cost SES.

The standard C system uses the same transmission frequencies as standard A and was designed to extend digital services to smaller vessels and land vehicles over the INMARSAT system. To keep the total power requirement as low as possible required that the transmission output power should be restricted to about 15 to 20 W (13 dBW). The usual Shannon–Nyquist signal to noise trade-off thus implies that operation at a low C/N ratio means that the system has to operate at a low bit rate and a narrow bandwidth. The resulting design is a low cost, flexible, go anywhere, satellite communications system that is capable of interconnecting a mobile terminal to practically any PSTN system.

The basic data rate is 600 bit/s, but this can be doubled in the redundant mode, when each message is sent twice. The coding method uses a half-rate Viterbi code which can be interleaved for added error protection to give a bit error rate (BER) better than 10^{-3}. The modulation of the RF carriers is by binary PSK (BPSK). The system frequency tuning is synthesiser control in steps of 5 kHz.

The mobile unit uses a non-stabilised, omni-directional antenna and the specification calls for a G/T figure of merit of −22.8 dB/K. The transmission coding and system protocol is based on the CCITT X25/X400 standards, with the data organised into packets with 8.64 s duration frames.

The service also provides facilities for the mobile terminal to transmit position and data reports. These consist of up to 32 bytes contained within three packets. Both position and data reports can be initiated either by polling or interrogation from a base station. Poll commands instruct the terminal how and when it should respond. There are three types of polling signals available, either for individual, group or area responses. In the latter two cases, the polling terminal arranges for queueing control. An *enhanced group call* service is also available for messages fed into the standard C system via a terrestrial PTT network. This allows messages to be routed to mobiles according to a priority. A single message can be simultaneously received by a number of mobiles anywhere within global range, or just to mobiles within a specific region.

Since the standard C terminals require very little power and can operate in an automatic mode, they are sometimes used for the remote sensing and reporting of terrestrial situations. This supervisory control and data acquisition (SCADA) mode allows the terminals to be positioned to gather data from such situations as volcanic, geothermal, pressure ridges and plate movement, thus providing advanced warning of developing environmental disasters.

Navigation services (see also Radar and navigation systems, p. 255)

Transit system. Specialised receivers tuned to the transmissions of polar orbiting satellites of the US Navy Navigational Satellite System can provide very accurate navigational positions or fixes. Although the system was originally intended for military purposes, it is available for civilian applications, by both water- and land-based mobile vehicles.

The space segment of this system consists of five satellites, at an altitude of about 1100 km, in polar orbits, with periods of approximately 107 min. The RF radiated power is of the order of 1 W. The transmitted data includes a time code and precise details of the orbit parameters. A fix is obtained by decoding the data and measuring the Doppler shift in the received carrier frequency. The information is then evaluated in a computer-controlled receiver.

The satellites' precise orbit is not circular, but affected by minor perturbations. These would lead to errors in the calculated positions. To ensure that the data is as accurate as possible, ground control stations monitor each satellite transmission and use a computer to establish the true orbit. Updated, corrected information is then transmitted back to the satellite, twice in each 24 h period.

A satellite continuously transmits messages of precisely 2 min duration on frequencies of 150 and 400 MHz, the latter being commonly used for civilian navigation purposes. Each message consists of 156, 39-bit words, plus one 19-bit word, making a total of 6103 bits at an equivalent rate of 50.858 bit/s. The first three words define universal time (GMT) and the next 25 words, used for navigation, define the minor perturbations and the fixed parameters of the current and future orbits. The remaining words are for military purposes and are not decoded. The primary bit pattern (see Fig. 31.3) has a duration of 19.7 ms. This pattern is used to provide a PSK of $\pm 60°$ of the 399.968 MHz carrier. This particular form of modulation is chosen so that the Doppler shift in the received carrier frequency can be easily and continuously measured.

The received frequency will be different to that transmitted due to the Doppler effect, which depends upon the relative velocities of the satellite, the receiver and the axial velocity of the earth. A simple mathematical formula for the observed change of frequency is given by:

$$f_t - f_r = f_t(2V/c)$$

where

f_t = transmitted frequency
f_r = received frequency
V = relative velocity of source and receiver
$c = 3 \times 10^8$ m/s.

As the satellite rises towards its zenith, approaching the receiver, the received frequency will be higher than that transmitted. At the zenith, the relative velocities are momentarily zero, so that received and transmitted frequencies are the same. As the satellite recedes, the received frequency falls. To avoid the problem of dealing with positive and negative Doppler shifts, the satellite navigation receiver compares the received frequency with a highly accurate and stable local frequency of 400 MHz (see Fig. 31.4). The difference frequency thus varies around 32 kHz during each satellite pass.

During each acceptable satellite pass, the Doppler shift is measured and counted over periods of 30 s and stored in the computer memory. By the end of each pass, the Doppler count has been used to compute a curve that uniquely relates the receiver position relative to the orbit. The computer then uses this together with the data from each 2-min message to compute and display the position. Between such fixes, data from the ship's log (speed) and compass (bearing) are input to the computer which then continually updates the position by dead reckoning (DR).

Orbits that produce angles of elevation between about 15° and 75° can produce accurate fixes that even under worst-case conditions can be within 500 m. Orbits outside of this range are likely to produce unacceptable errors due to the non-spherical nature of the surface of the earth.

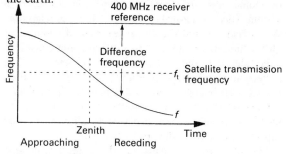

Figure 31.4 Doppler shift relationship.

Global positioning system (GPS)

NAVSTAR. This highly complex navigation system, initially developed by the US Department of Defense for military purposes, can also be used in a restricted form by civilian operators. Unlike the TRANSIT system that uses the Doppler shift of UHF signals as the measuring domain, NAVSTAR uses the propagation

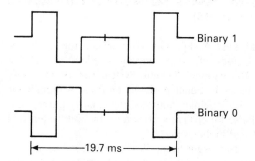

Figure 31.3 Primary 'transit' bit code pattern.

delay of the signals transmitted in L(S) band by each satellite. The space segment will ultimately consist of 18 operational satellites plus six spares. These will be positioned in six circular and inclined orbits at an altitude of about 20 170 km and with a period of 12 h. The angles of inclination are such that any point on the earth's surface can see at least four satellites at any given time. Each satellite carries a very accurate atomic clock that transmits digitally coded time signals on a frequency of 1227.60 MHz. Other transmission frequencies are 1381.05 MHz and 1575.42 MHz, but only the latter is used for navigation purposes (see Fig. 31.5(a)).

Each satellite transmits on the same carrier frequency of 1575.42 MHz but is identified by its unique Gold code that provides BPSK modulation and direct sequence-spread spectrum (DS-SS) transmission. The transmission power level at −163 dBW is well below the system general noise level.

Each Gold code is 1023 bits long and repeated 1000 times per second, producing a code chip rate of 1.023 Mbit/s. Modulation raises the nominally 100 Hz wide baseband up to a transmission bandwidth of about

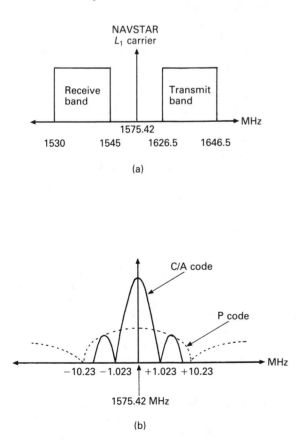

(a)

(b)

Figure 31.5 (a) NAVSTAR L_1 frequency relative to INMARSAT-C frequencies; (b) spectra of NAVSTAR C/A and P codes.

2 MHz. This code, referred to as the coarse acquisition code (C/A), provides the standard positioning service (SPS) for civilian applications.

A second code, referred to as the P code, at a rate of 10.23 Mbit/s is available for military purposes and provides the precise positioning service (PPS) (see Fig. 31.5(b)). This code is about 6 Mbits long and would repeat every 267 days. However, this sequence is reset weekly and with a different starting point for each satellite. As an indication of the accuracy of positioning, 1 bit of the SPS code is equivalent to 300 m, while that for the PPS code is about 1/10 of this. Significantly better accuracy can be achieved with both codes by using a phase measuring technique in the receiver.

Because of the variations in earth geometry and the vagaries of propagation, each satellite additionally transmits data about the parameters of the present and future orbits of all the satellites. This *almanac* allows the receivers to calculate accurate positional fixes and to be prepared with the correct code as a new satellite appears over the horizon. Information for the almanac is collected by ground stations for processing in the system central computer at the master control station in the USA. Four other ground stations ensure that any satellite is never out of contact with a ground station for more than 2 h, the regular updating period. To allow a significant margin of error, the almanac remains valid for up to 4 h.

In order to obtain a fix in three dimensions, it is necessary to obtain data from three satellites. However, positional accuracy will not be uniform across the earth's surface, due in the main to:

(1) changing propagation conditions in the ionosphere and the use of a minimal number of satellites;
(2) gravitational effects and *relativity*. The latter feature causes the earth-bound clocks to appear to run slow;
(3) the effects of multipath reception can give rise to noise and phase disturbances in the demodulation stages.

To overcome these problems, a fourth satellite is always used to provide a cross-reference. Many current receivers use five satellite channels simultaneously: four to provide an accurate fix and the fifth to provide quick access to the correct codes as a new satellite appears above the horizon. The mobile antenna needs to have a hemispherical response to signals with an elevation as low as 5° above the horizon.

To descramble/despread the received signal, the receiver has to generate a copy of the satellite code and multiply this with the incoming code. When the code patterns agree, the correlation detector ensures that the energy spread throughout the 2 MHz bandwidth becomes concentrated into the original 100 Hz baseband, with the significant coding gain. The recovered data is then automatically entered into the mobile terminal computer. This then solves the necessary four simultaneous equations in an iterative manner, before presenting the results to the user on a suitable display.

Because of the very low level of input signal, the use of a double-superhet receiver is practically mandatory. Conversion to a 1st IF in the range 100–200 MHz gives good image frequency rejection, while high gain can readily be established in a 2nd IF of 5–20 MHz. Since all the satellite frequencies are multiples of the basic chip rate of 1.023 MHz, it is useful to use the same concept in the receiver. Thus a carrier frequency of 1540×1.023 MHz $= 1575.42$ MHz, and 1st IF of 120×1.023 MHz $= 122.76$ MHz or 160×1.023 MHz $= 163.68$ MHz requires a local oscillator frequency (LO) of 1452.66 MHz or 1411.74 MHz. By using such related frequencies, the locally generated signals can be synchronously locked. This avoids spurious beats and harmonics that could cause interference problems, leading to a reduction in processing accuracy.

Differential GPS

This technique greatly enhances the accuracy of fixes without affecting the integrity of the military part of the system. It requires the additional use of a fixed GPS receiver located to within a surveyed positional accuracy of better than 1 m. The computer in this reference receiver compares its accurately known position with that obtained from GPS so that it can calculate a correction factor. This is then transmitted to all suitably equipped mobile GPS receivers, where it can be applied to the raw GPS fix to calculate a position, accurate to within 5 m. This extra feature requires that the GPS receiver must be equipped to receive the correction factor transmissions either via satellite or over a conventional radio link.

GLONASS (GLObal NAvigation Satellite System) (USSR)

This system operates in a very similar manner to NAVSTAR, in that it employs spread spectrum ranging techniques, but uses multiple L(S) band carriers in the range 1602.5625 MHz to 1615.5 MHz. The system will eventually employ 24 operational satellites in three inclined orbits. Again, two levels of location accuracy can be achieved using a C/A code of 511 Kbits and a P code of 5.11 Mbits. The code repetition rate is also 1000 times per second but with a basic bit rate of 50 bit/s.

PRODAT data service

The European Space Agency (ESA) is developing a system known as PROSAT, to be available for land, air and maritime mobile operations. The latter service is referred to as PROMAR, while a low cost digital data

only service that is included in the scheme is known as PRODAT.

This system operates through existing ESA ground stations and INMARSAT communications satellites and provides a low speed, low power and narrow bandwidth service. It is initially intended to work alongside the more common HF communications radio systems currently used by aircraft. Since this mode suffers from the vagaries of variable propagation, the satellite system will provide a more consistent and reliable service. The flexible system design allows for communications mobile to ground, mobile to mobile, and ground to mobile. In addition, it provides for extension to both land and maritime operations.

The ground station transmits to the satellite in the 6 GHz band using time division multiple access (TDMA) with differential binary phase shift keying (DBPSK). Each TDM frame is 1.024 s long and is divided into 64-bit slots, each slot representing a channel sending data to a user at 47 bit/s. The other bits in each slot are used for coding and control. This signal is then relayed to the mobile by a transponder of 24 dBW output using a 1.5 GHz link. A lower power output is envisaged but using fewer slots per frame. PRODAT uses a two-dimensional Reed–Solomon (RS) interleaved code that is capable of correcting both random and burst errors. In addition, if the burst errors are too long for correction, an automatic request repeat (ARQ) for retransmission can be generated, acknowledgement of accurate reception being given during the transmission of the following block. Because of the delay over the satellite channel, each block must be numbered.

Quadrifilar helix antennas, with 0 dBi gain and a hemispherical radiation pattern, are mounted externally to the aircraft and housed within a low drag aerofoil casing. The aircraft transmissions use the 1.6 GHz band to the satellite, which in turn relays the signal to ground using a 6 GHz carrier. The down-link uses code division multiple access (CDMA) with suitable collision detection. The *spread spectrum* signal is generated by the addition of a PRBS (pseudo-random binary sequence) to the channel signal. The system allows for 32 simultaneous channels using the same carrier frequency and occupying a total bandwidth of 650 kHz. At the ground station, the CDMA signal is decoded using correlation detection and the appropriate PRBS code, before being directed to the user terminal.

Radio determination satellite service (RDSS)

This commercial system is used to report accurate information about the location of mobile units back to transport operators' headquarters. It is suitable for use on road vehicles, railway trains, aircraft or boats and can provide positional fixes accurate to less than 10 m.

In addition, provision is also made to allow for short, two-way messages to be passed between headquarters and the mobile user. The system operates through either low earth orbiters (LEO) or geostationary satellites, using either the standard LORAN-C (LOng Range Aid to Navigation), Transit or GPS positioning services.

In 1986 the US GEOSTAR Corp. was granted a licence and the first system became operational in 1987. In the same year, a European consortium LOCSTAR signed a licensing agreement as the first step to a global expansion of the service. Since that time, many more countries including Australia, Brazil, China and India have adopted the concept.

Development is still progressing in stages. The original System 1 operated through spare transponders of the ARGOS system LEOs known also as NOAA 9 and 10. This provided basic positional reporting and allowed for only a one-way data service. System 2 went through a number of enhancement stages during 1988 and 1989, providing a more accurate location service plus two-way messaging. System 3, which provides for satellite ranging and navigation, came into service during 1991 and provides a positional fix accurate to less than 7 m, plus two-way messaging. System 4 is due to come into operation during 1993/94 and provides for a worldwide network that allows *roaming* and *visiting*, with many millions of users.

The 1987 World Administrative Radio Conference for Mobile Services (known as MobWARC'87) agreed to the International Standards jointly proposed by the International Telecommunications Union (ITU), the Federal Communications Commission (FCC), the International Radio Consultative Committee (CCIR) and the International Civil Aviation Organisation (ICAO). Such was the demand for this service that almost 100 MHz of spectrum was allocated in the following frequency bands:

1610–1626.5 MHz, L band user up-link
2483.5–2500 MHz, S band user down-link
5150–5216 MHz, C band satellite to ground station link
6525–6541.5 MHz, C band ground station to satellite link

The system uses two radio channels, one for data and messages from the mobile to the central control station (in-bound) and the other in the reverse direction (out-bound). Each message sent over the system carries the unique identity code (ID) of the user, thus ensuring that only the addressee receives that message. Because RDSS uses spread spectrum, burst transmission techniques, many users can communicate simultaneously with the geostationary satellite. The Gold spreading codes form the basic protection from data collision in this CDMA

controlled system. The GEOSTAR control centre transmits interrogation signals through the satellite to all users many times per second. When received by a mobile, it replies with its own unique ID together with the digital data that it has waiting. The mobile transmitter then automatically shuts down while the control centre stores the messages in its computer. Once stored, the control centre transmits an acknowledge signal and then forwards the message to a prearranged destination.

The basic packet structure consists of about 130 000 time slots or chips. Each packet has a maximum length of 128 bytes of which 105 are allocated to messages. A packet consists of header, address, data or text, CRC error check, and trailer, with small variations between in-bound and out-bound messages. Messages that cannot be contained within a single packet require the addition of a sequence number.

The user terminal automatically counts from the start of a frame and if prompted, it transmits a prearranged chip sequence. This prevents other users from accessing the same slot. Generally three-chip separation is allowed between all users existing in the same frame. If any user terminal is blocked in any given frame, it has a better than 99% chance of access within the next frame. A tier of priority is provided in the protocol, so that a terminal with an emergency can obtain rapid access.

Binary PSK is used for the primary coding format at a basic rate of 15.6 Kbit/s. The use of a half-rate convolution code, with constraint length 7, doubles this to a transmission rate of 31.2 Kbit/s. Burst lengths vary between 20 ms and 80 ms, and the use of a Gold spreading code expands the bandwidth to 16.5 MHz. The typical transmit power level is in the order of 50 W (17 dBW).

As an example of the value of RDSS to transportation generally, a fully operational System 3 installation can handle 4 million out-bound, and more than 20 million in-bound messages per hour. For added security and privacy, the system also provides for message encryption.

A typical user terminal consists of a main receiver/processor/transmitter, a keyboard/display unit, receive/transmit antennas, plus the option of a hard-copy printer. During the few years of system development, the size and power requirement of the terminal has reduced markedly and hand-held units are now a possibility.

The antennas used vary from a 42 cm monopole for LORAN reception to microwave patch and helical devices for L and S band.

SafetyNET system

A further INMARSAT network that is exclusively available to the maritime user. This is used for the transmission of safety related information such as weather warnings for specific areas.

Small system satellite services (S⁴)

The technical developments of the last decade have had a marked effect on both satellite-borne equipment and ground station characteristics. Improvements in semiconductor technology now provide amplifier devices that operate up to higher frequencies, with higher gain and lower noise factors. Improvements to high power amplifiers (HPA) for ground stations and the travelling wave tube (TWT) and solid-state RF power amplifiers for satellite transmissions, contribute to the need for smaller antennas. Due to the availability of suitable integrated circuits (ICs), the efficient modulation schemes and coding techniques devised over a decade ago have now become a practical reality. The traditional system design trade-offs using Shannon, Hartley and Nyquist criteria still apply, but with all these improvements, adequate signal to noise ratios, particularly with digital processing, allow the use of low power transmissions using small antennas, capable of providing significantly high data rates with acceptably low bit error rates. As an example of the extent of development, hand held radio-paging devices have now been demonstrated to be practical.

Specialised satellite services operators (SSSO)

In the UK a number of operators have been licensed to operate fixed and mobile ground stations through most satellites. Of these organisations, Satellite Information Services (SISLink) was probably one of the first operators to lease transponders on a permanent basis. SISLink provides satellite newsgathering services and news feeds on an international basis; a private racing network of video and sound distribution to bookmakers throughout Europe; plus videoconferencing and other corporate communication services. Although the racing network uses a B-MAC system developed by Scientific Atlanta Inc. for vision with Dolby ADM for sound, all other forms of modulation and television formats can be handled.

Television services

These broadband services, operating in either C or Ku band, provide for a number of specific purposes:

- direct broadcast by satellite (DBS);
- head end feeds to large area networks;
- satellite master antenna TV (SMATV) feeds to small local networks;
- international programme distribution;
- satellite newsgathering (SNG) feeds and news distribution.

There are two different types of satellite distribution employed. The fixed satellite service (FSS) originally intended for telecommunications purposes and cable system feeds, providing a receiver site EIRP of up to 50 dBW and the higher power transponder satellites intended for the direct to home (DTH) service providing an EIRP of around 54 dBW. The transponder bandwidth is typically in the order of 54 or 72 MHz and this may be allocated into an 18 or 27 MHz channel as necessary.

For a DTH service, the received S/N ratio needs to be in the order of at least 43 dB, a level that may be achieved with antennas as small as 40 cm diameter. However, rebroadcast systems such as cable head ends feeds require a S/N ratio of at least 50 dB and this may require the use of 3 m diameter antennas. In the interests of adequate image quality, frequency modulation is invariably employed.

The peaky nature of the TV signal spectrum, particularly during the sync pulse periods, can cause interference with terrestrial services. To minimise this effect and spread the spectral energy, a low amplitude triangular waveform at field frequency is added to the composite TV signal. The effect of this *energy dispersal* waveform is to prevent a constant carrier frequency being radiated during sync pulse periods. Provision has also been made for the future transmission of high definition television (HDTV) signals in the 21.4 to 22 GHz segment of the Ka band. In the analogue form, these signals will need at least 54 MHz of bandwidth. However, it is anticipated that a digital format will be developed that, using image compression, will occupy a bandwidth nearer to that currently used for NTSC, PAL and SECAM transmissions.

Tracking and data relay satellites (TDRS)

Because a geostationary satellite can *see* a polar orbiter for longer periods than any ground station, the TDRS system of three satellites was designed by NASA to provide links between LANDSAT, the space shuttles, C band satellites and up to 19 LEOs. In addition, this network can also act as relays for the data collection platforms of the meteorological services.

The TDRS system operates on several carrier frequencies, using S band for data rates up to 250 Kbit/s and Ku band for higher rates. The downlink to earth station is by Ku band (13.5 to 15.2 GHz), using a spread spectrum technique that provides a multiplex for up to 30 spacecraft channels simultaneously.

TSAT

A third generation VSAT system, so called because it has been designed to operate at the CCITT T1/E1 data rates of 1.544 Mbit/s (North America) or 2.048 Mbit/s (Europe). Unlike the VSAT that normally operates in a *star* mode of connection, the TSAT is capable of operating in a *mesh* configuration, connecting each location to all others without the need of a hub station. Not only are these systems faster with a higher data capacity, they are also significantly cheaper. The networks, originally conceived by SPAR Aerospace of Canada, are ISDN compatible in that they provide both basic rate access of 2B + D (64 Kbit/s) and primary rate access of 23B + D (1.544 Mbit/s) or 30B + D (2.048 Mbit/s) (B channel used for voice/data services and D channel for signalling purposes).

Very small aperture terminals (VSAT)

VSAT systems are usually considered to be those designed to operate with antennas that are the equivalent of 1.8 m diameter or less. The relatively low gain provided by these has to be countered by very careful design. Reference to the rules of communications (Shannon, Hartley, Nyquist), will show that systems operated in this way suffer from a reduced signal to noise (S/N) ratio and hence an increase in error rate. However, an acceptable error rate can be achieved by reducing the information transmission rate. VSAT systems are therefore essentially designed to provide digital services at a reduced data rate, typically 9.6 Kbit/s, a value that can rise to 1.544 or 2.048 Mbit/s under special circumstances.

Second-generation VSAT terminals operate as transceivers, working through a satellite and a higher powered ground or hub station equipped with an antenna of 3.5 m diameter or more. The large antenna and greater power level is necessary to maintain an adequate overall S/N ratio. The hub station provides the link to other network users, either via a second satellite link, or the terrestrial PSTN or other suitable network. Due to improving technology, third generation

VSATs are able to operate directly in the VSAT to VSAT mode, without the intervention of a hub station.

VSAT networks form a valuable global link or bridge between isolated terrestrial local area networks (LAN) with which they have much in common and can easily be integrated. They thus add considerably to the communications capability of international corporate users, even to the extent of providing private video/telephony conferencing.

VSAT systems operate both in C and Ku band, but the latter probably carries the greater part of these services. The systems may use any of the techniques of modulation, access and control that apply to satellites and LANs.

Terminals are divided into two basic sections (see Fig. 31.6). An outdoor section comprising low noise block convertor (LNC) and high power amplifier (HPA) are usually contained within a single casing and mounted at the antenna focal point. The HPA, which is a semiconductor device, commonly produces an RF power of some 2 to 3 W. This assembly also carries the orthomode transducer and waveguide filters. These are necessary to separate the transmit and receive signals. Both transmitter and receiver function on the dual conversion, double superhet principle. The local oscillator is normally a crystal controlled phase lock loop, with frequency synthesis to select the correct frequencies for up- and down-link operation. The typical 1st IFs are around 1 GHz for Ku band and 600 MHz for C band, while the 2nd IF is commonly either 70 or 140 MHz.

The typical figure of merit (G/T) for the front end is in the order of 18 to 20 dB/K, depending upon the combination of antenna size and low noise amplifier characteristics.

The indoor section contains the second stage of frequency conversion, plus the modem, codec and muldex (multiplexer/demultiplexer), these latter elements providing complementary functions in the

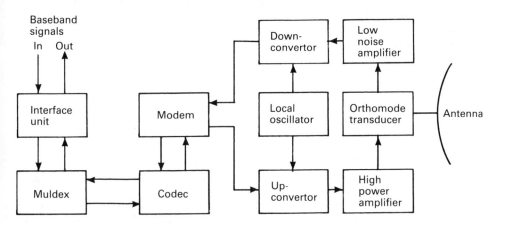

Figure 31.6 Typical VSAT earth station.

transmit and receive modes. The interface unit processes the baseband signals and matches the various input devices to the digital processing stages. The muldex combines/distributes a given transmission channel between several inputs. The codec and modem stages handle the forward error correction (convolution code), and convert the source code, usually pure binary, into the channel code, either BPSK or QPSK, the latter being preferred because it doubles the channel capacity without increasing the bandwidth requirement.

A basic network provides 20×9.6 Kbit/s SCPC trunks from VSAT to hub and a single TDM frame of 20×9.6 Kbit/s timeslots in the reverse direction. The VSAT to hub channels consist of one random access signalling channel and 19 SCPC DAMA (demand assigned multiple access) message channels. The reverse direction timeslots are similarly allocated; one for signalling and 19 DAMA for messages.

The system control operates on a modified SPADE protocol (see Access protocols, p. 282). A VSAT accesses the hub via its signalling channel f_1 requesting connection. The hub informs the calling station of the assigned channel via its signalling timeslot t_1 of the TDM frame. The VSAT then transmits on frequency f_x and receives on timeslot t_x. For VSAT to VSAT links, two DAMA channels need to be allocated by the hub, f_x and t_x as described above and f_y and t_y for the relay between the hub and the called VSAT. The system link power budget can be calculated in the manner shown in Table 31.1.

31.2 Miscellaneous terms

Access protocols. Generally, the traffic loading on a business data terminal tends to be very variable, so that these systems tend to be bandwidth rather than power limited. The access problems are very similar to those of local area networks (LANs) and therefore the choice of access protocol, which can be critical, needs to be carefully chosen bearing in mind the following points:

- average acceptable transmission delay
- range of data rates involved;
- acceptable message failure rate;
- message format (fixed or variable length).

ACK/NACK protocol (acknowledged/not acknowledged). When the receiving station decodes a packet without error, it transmits an acknowledgement signal. If the transmitting station does not receive this acknowledgement, it decides that an error has occurred and automatically retransmits the data packet.

ALOHA (system originally devised by University of Hawaii for interworking of computer networks). With the pure ALOHA protocol, terminals transmit as and when they have data ready for transmission and also continually monitor the network. If two terminals transmit together, they detect a collision of data and both back off for a random time period before trying again. This is basically a collision detect–multiple access (CDMA) system.

Reservation TDMA. Each terminal is allocated a particular time slot of the time frame for its transmissions.

Selective spectrum ALOHA. An ALOHA protocol reserved for use with spread spectrum transmissions, when packets of data will not be affected by collisions.

Slotted ALOHA. As in pure ALOHA but each terminal transmission is restricted to allocated time slots.

SPADE (single-channel per-carrier demand assigned multiple access). A control station monitors the channel frequencies in use, then answers a request for interconnection by allocating a free channel to the two stations.

Reservation TDMA is particularly useful when relatively long transmissions are required. Slotted ALOHA has an advantage over pure ALOHA in that once a terminal has accessed a time slot, it has sole use of this until the end of transmission. Selective spectrum ALOHA has the advantage that data collisions are not problematic but the system has to include some form of forward error correction (FEC). ACK/NACK system is not very useful over satellite links. Due to the overall time delay of about 250 ms, the following block may be transmitted before the lack of acknowledgement has been recognised and this will produce a data collision.

Azimuth and elevation angle calculations. A calculator with trigonometrical functions and this algorithm can quickly produce the azimuth/elevation angles required for geostationary satellite reception.

For the northern hemisphere, the convention is to use positive angles for satellites and ground stations west of the Greenwich Meridian (easterly locations use corresponding negative values).

(1) Find the angle WEST (the longitude difference). Satellite °W − Station °W.
(2) Find $X = (\cos \text{WEST}° \times \cos(\text{Station latitude}°))$.
(3) AZIMUTH$= \tan^{-1} (\tan \text{WEST}/\sin \text{Station latitude})$ add 180° if satellite is west of Greenwich Meridian.

(4) Calculate $Y = \sqrt{1 + K^2 - 2KX}$, where $K = 6.608$, the distance between the satellite and the earth centre in terms of earth radii.

(5) ELEVATION $= \cos^{-1}((1 - KX)/Y) - 90°$.

(6) Calculate range $Z = 6378 \times Y$ km.

Note. For latitudes greater than $81°$, $X < 0.15$ and this indicates that the satellite is below the horizon.

Eclipse. During the equinox periods (March and September) the earth, sun and satellite are in alignment with the result that the earth's shadow eclipses the satellite solar cells. This interrupts the battery recharging process and the transmitters may lose power. Low power satellites can usually continue operation using battery back-up power, but high power satellites are shut down for all but essential services. The eclipse effect is noticeable for periods of about 4 weeks and the peak duration lasts for about 1.25 h.

Footprint. Jargon term used to describe the equi-power flux contours provided by a satellite transmission antenna. Unlike terrestrial transmitters' field strength contour maps, footprints are usually drawn for either equal power flux density (dBW/m^2) or effective isotropic radiated power (dBW). These contours thus define a particular transponder's service area.

LIDAR (LIght Detection And Ranging). A technique that can be used for measuring wind speed and the distribution of airborne pollution. A frequency stable pulsed laser is used to generate a light beam that is aimed to pass through the atmosphere of interest. The radiation becomes scattered due to discontinuities in the gas molecules and some is reflected back towards the transmitter. The receiver detects the Doppler shift in the returned beam and this is interpreted as an accurate measurement of the wind speed.

Molniya orbit. An elliptical orbit with a 12 h period and named after the Russian satellites that originally used such an orbit.

QUASAT (QUAsar SATellite). This satellite is primarily designed for the study of the high energy radio sources commonly known as *quasars* (quasi-stellar objects). The satellite occupies a highly elliptical orbit with an *apogee* (highest altitude) of 36 000 km and a *perigee* (lowest altitude) of 5000 km, inclined at an angle of 30°. The observation frequencies are within the radio astronomy bands of 1.66 to 1.67 GHz, 4.8 to 5 GHz and 22.21 to 22.5 GHz. In addition, a small segment at 327 MHz has been reserved for the study of *pulsars*.

Sun blinding or outage. A complementary eclipse effect that occurs when a satellite is in alignment with the sun and an earth station. In this case, the ground station antenna which is pointing directly at the satellite is also aimed at the sun which acts as a high level noise source to interfere with normal reception.

Tundra orbits. An elliptical orbit with a 24 h period and named after the Russian satellites that originally used these.

Useful references

Ackroyd B. (1990) *World Satellite Communications and Earth Station Design*, Butterworth-Heinemann, Oxford.

Fthenakis E. (1984) *Manual of Satellite Communications*, McGraw-Hill, New York.

Ha T.T. (1986) *Digital Satellite Communications*, McGraw-Hill, New York.

Lewis G.E. (1992) *Communication Services via Satellite*, 2nd edition. Butterworth-Heinemann, Oxford.

Mattos P. (1992/3) G.P.S. *Electronics World and Wireless World.* December 1992, 982; January 1993, 29; February 1993, 146; March 1993, 210; April 1993, 296; May 1993.

Rainger P. *et al.* (1985) *Satellite Broadcasting*, Wiley, London.

Stephenson D.J. (1991) *Newnes Guide to Satellite TV*, 2nd edition, Butterworth-Heinemann, Oxford.

32 Semiconductor devices and technology

Without a semiconductor industry, which is still less than 50 years old, telecommunications as known today would not exist. Devices range from single discrete active or passive components such as diodes, transistors, resistors and capacitors, to highly complex integrated circuits (ICs) that can virtually represent systems on a single *chip*. The term transistor is derived from a contraction of *transfer resistor* which, to a limited extent, explains its operation. The manufacturing technology is likewise very wide-ranging and complex, both in the methods of fabrication and the materials used. Because contaminants are responsible for device inefficiencies and premature failures, a very high degree of material purity is demanded by this industry. In turn, this has driven the chemical and allied industries to unprecedented levels of purification of the solid, liquid and gaseous materials employed. Material purity levels in the order of 99.99999% with production clean room atmospheres containing no more than 10 particles greater than $0.2\,\mu m$ per cubic foot are common.

The fabrication part of the industry represents a highly automated production system backed up by computer-aided design (CAD) and extensive research and development (R&D). All this only becomes cost-effective due to the accurate and relatively low cost repeatability of the product lines.

The industry probably represents the most rapidly developing and changing section of the communications business. Although it has a solid base of well-established devices and fabrication technology, changes occur at such a rate that the only real source of up to date information is that obtained from learned papers presented at various conferences.

The level of device standardisation relative to that which obtained for thermionics is low. In-house standards and second sourcing manufacturing agreements have produced a degree of interchangeability, but in spite of the work of such organisations as the US Joint Electronic Device Engineering Council (JEDEC), there is still a surfeit of similar devices with differing parameters.

The future of the communications industry is closely bound to the developments in IC production. These have developed from small scale integration (SSI) with less than 100 components per chip, through medium scale integration (MSI) with between 100 and 1000 components per chip, large scale integration with between 1000 and 10 000 components per chip, to very large scale integration (VLSI) with greater than 10 000 components per chip. Currently the density record exceeds 1 000 000 components per chip. The stage has been reached when if an IC can sell in sufficient quantities, then many systems can be completely fabricated on a single device. Many advantages accrue from using ICs and these include:

(1) increased reliability due to a construction that leads to fewer interconnections that can develop *dry joints*;
(2) reduced size compared with their discrete component counterparts;
(3) reduced costs through volume production;
(4) faster operation or better high frequency response due to shorter signal path lengths;
(5) the fabrication methods give a better control over the spread or variation of device parameters;
(6) reduced assembly time due to fewer soldered joints;
(7) the overall simpler circuit design leads to improved performance.

The basic materials that are employed have a conductivity about halfway between that of conductors and insulators, hence the term semiconductor. These include:

germanium (Ge), silicon (Si) and the Group III/V elements, aluminium (Al), gallium (Ga), indium (In) / arsenic (As), antimony (Sb)

Germanium is now less used than silicon because of its poorer temperature characteristics (typical operating temperature 75°C compared with 200°C for silicon).

32.1 Behaviour of PN junctions

Just as controlled conduction can be achieved through doping *intrinsically* pure tetravalent semiconductor elements (I type material) with trivalent or pentavalent impurities in a precisely controlled way, pseudo-semiconduction can be produced by the careful combination of the Group III (trivalent) and Group V (pentavalent) elements to produce a new class of device with even more useful characteristics. By adding pentavalent impurity atoms to I type material, additional electrons are donated to the crystal structure. This forms an N type semiconductor material in which electrons form the majority current carriers.

In a similar way, the addition of trivalent impurities leaves spaces or *holes* in the crystal structure which represent energy levels where electrons could exist. This P type material thus has the positive charges of the holes as majority carriers. A hole therefore has an equal but opposite polarity charge to that of an electron. In both P and N type materials there will always be some thermally generated charges of the opposite polarity and these are known as minority carriers.

In certain applications, the P and N type impurity layers may be doped to a relatively higher or lower level than normal. By convention, these regions would be indicated as P+ or N+; or P− or N−, respectively.

If a surface of a mono-crystalline P type material is heated to about 1100°C and exposed long enough to a gaseous atmosphere of N type impurity, the surface will be converted into N type. Therefore, somewhere below the surface, there will exist a junction on an atomic scale, between the P and N type materials. Holes and electrons will migrate across this junction to produce an electric field described as the *barrier potential*. When this reaches some particular level (about 100 mV for Ge and 600 mV for Si), the region becomes depleted of free current carriers to produce a *barrier* or *depletion region* as indicated in Fig. 32.1. If an external voltage greater than the barrier potential, is now applied to this junction with a positive polarity on the P region (forward bias), current will flow through the junction and in the external circuit. If the applied voltage is reversed, the barrier potential is reinforced and very little current flows (reverse bias). This current would in fact fall to zero if it were not for the thermally generated hole–electron pairs that act as minority carriers. Operated in this way, the PN junction thus has diode characteristics which, in many cases, can conveniently be considered simply as a *piecewise* linear device, with a relatively low

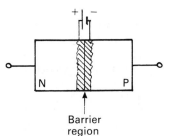

Figure 32.1 PN junction.

forward resistance and very high or even infinite reverse resistance. However, the total voltage/current characteristic is more precisely described by the following equations, as logarithmic:

$$I = I_o \exp \frac{qV}{nkT} \text{ or } V = \frac{(nKT)}{q} \ln \frac{I}{I_o}$$

where I = forward current,
I_o = reverse saturation current
q = electron charge (1.602×10^{-19} coulombs)
k = Boltzmann's constant (1.38×10^{-23} J/K)
V = voltage across junction
T = absolute temperature
n = constant typically close to unity.

These equations hold true for any PN junction and have been shown to be accurate to within 1% over more than eight decades of current.

32.1.1 Bi-polar transistors

Figure 32.2 shows the cross-section through one type of transistor. The surface of a piece of N type silicon which forms the substrate is coated with a thin insulating layer of silicon dioxide (SiO_2). This operation is often performed by an evaporation process at a temperature of 300 to 400°C, and at low pressure in the presence of either silicon or silicon monoxide and oxygen.

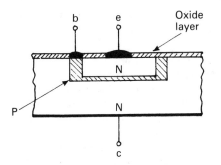

Figure 32.2 Planar bi-polar transistor.

A small area of the SiO₂ is then etched away using a photographic and masking technique to leave a window in which the P type base region will be formed. When this area is exposed to an atmosphere of a P type gas such as boron, the exposed region will be converted into P type semiconductor. The surface is again exposed to the SiO₂ process so that the insulating layer is reformed. A smaller window is then etched into this layer and the area exposed to an N type impurity atmosphere such as phosphorus. This converts the exposed region back into N type semiconductor. A further insulating and etching operation leaves small windows for the metallisation of the electrode connection areas. Because the transistor is built up layer by layer, the process is referred to as *planar*. These transistors are described as bi-polar because both holes and electrons form the majority carriers in the various regions. These are respectively referred to as emitter (e), base (b), and collector (c). Although an NPN device has been described, complementary PNP transistors are also manufactured.

This structure can be regarded as two back-to-back diodes sharing a common P type region. In normal amplifier configuration, the base to emitter junction is forward biased while the base to collector junction is reverse biased. Because the base region is very thin, any electrons injected into it will be influenced by the collector potential and swept out as collector current. Because about 99% of the emitter current reaches the collector, practically the same level of current flows from the low resistance junction into the high resistance one (transfer-resistor), thus representing a significant voltage and power gain.

Transistors may also be fabricated as PNP devices. However, these tend to have a poorer high frequency response because the majority current carriers are holes and these have a lower mobility than electrons.

32.1.2 Field effect transistors (FET)

The conduction through these devices is controlled via an electric field and not by the transfer resistance principle. They should therefore be more accurately described as field effect devices. However, they are generally erroneously known as FETs and operate by one of two modes.

Depletion mode FET. These are fabricated in the manner indicated by the cross-section in Fig. 32.3 using a similar technology to that described above. Two N type regions are formed in a P type substrate and these are linked by an N type channel that is only a few microns thick.

A thin layer of SiO₂ is formed over the channel and metallised areas are provided for the terminations. These are defined as source (s), drain (d), and gate (g) as

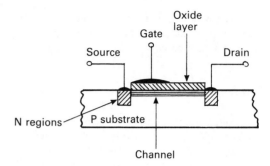

Figure 32.3 Depletion mode FET.

indicated. When correctly biased, current flows from the source, through the channel to the drain. If a voltage negative with respect to the substrate is applied to the gate, the resulting electric field will penetrate into the channel to increase its resistance and lower the drain current. The channel thus becomes depleted of current carriers. If the gate voltage is now made positive, the electric field attracts electrons into the channel from the substrate, with the result that the drain current is enhanced. P channel field effect devices are also fabricated in a similar way.

Enhancement mode FET. As shown by the cross-section in Fig. 32.4, these are made by a technology similar to that of the depletion mode devices but without a conduction channel. With correctly biased source and drain potentials and with zero gate voltage, the current flow through the device is practically zero. If, however, the gate is made positive with respect to the substrate, the electric field generated under the insulating layer will cause electrons from the substrate to form an inversion layer that bridges the source and drain regions to form a conduction channel. Increasing the positive gate voltage thus increases or enhances the source to drain current. Again, devices are manufactured with the complementary induced P channel conduction.

Hybrid circuits (see Thick film circuits, p. 287).

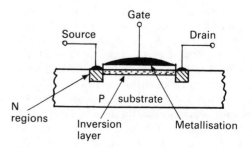

Figure 32.4 Enhancement mode FET.

32.1.3 Integrated circuits

These devices range from the general purpose to the application and user-specific (ASIC and USIC). Monolithic (Greek for single stone) ICs are formed by diffusing impurity elements into single crystal I type material and the smallest chip or die size used is 0.05 in square. ICs are usually fabricated using an *epitaxial* or planar process, in the form of four basic layers. The basic substrate forms the first layer upon which is grown an epitaxial extension layer of less than 25 microns thick. All of the passive and active components of the circuit will be formed in this layer by a diffusion process. The third layer, consisting of SiO_2, is used to protect the chip from impurity penetration and contamination. Windows are formed in this layer so that the fourth layer of metallic interconnections can provide the necessary links between the individual circuit components. The number of lead-outs can vary from three or four, up to a matrix of a few hundred pins in recent VLSI devices.

The minimum line width or chip feature is less than 1 micron (sub-micron technology), but will be greater for higher power devices.

Resistors are formed as a strip of P or N type semiconductor, with values ranging from about $10\,\Omega$ up to $30\,k\Omega$. Thin film capacitors are formed by using SiO_2 as the dielectric between a metallic layer (aluminium) and a heavily doped N^+ region. This feature is described as a metal-oxide-semiconductor (MOS) device. Values will rarely exceed about 200 pF. The actual component values for both resistors and capacitors tend to be temperature and voltage dependent and with a tolerance of around ±10%. However, the component value ratio on a given chip can be as close as ±1%.

Inductors are usually attached externally to the circuit but small values may be achieved by using a gyrator to convert a capacitance into an equivalent inductance.

32.1.4 Thick film circuits

These may consist of several active and passive discrete components, plus monolithic ICs in the unmounted state, all attached to the same ceramic substrate which carries the circuit pin connections. The inter-component connections are provided by direct wire bonding or via screen printed conductor strips using a surface mounting technique. Electrical isolation in these hybrid devices is provided by physical separation. The ceramic substrate provides good thermal conductivity and heat dissipation in a relatively compact circuit. Hybrids are used in applications where the power dissipation cannot be conveniently handled by an IC and only relatively few components are involved.

32.2 Devices

32.2.1 Diodes and junctions

Abrupt junction

A PN junction in which the carrier concentration across the barrier region changes abruptly. The junction capacitance varies as the square root of the applied voltage, and even with zero voltage the junction capacitance is still determined by the barrier potential. A forward bias voltage increases the capacitance quite considerably but the low forward resistance prevents its use. Used in reverse biased mode, the diode finds application as a voltage dependent capacitance (varactor or varicap diode).

Avalanche diode

See Breakdown diode, below.

Backward diode

A semiconductor diode whose V/I characteristic is practically the reverse of that for a conventional diode. These can be constructed by pulse bonding a gallium plated gold wire to a chip of lightly doped N type germanium to form a special type of tunnel diode, but without the negative resistance region. Because of the small cross-sectional area of the wire, the junction capacitance is correspondingly very small. Due to this and the high sensitivity at zero bias, backward diodes are very suitable for use as microwave mixers or detectors.

Breakdown diode

These are designed with adequate power rating so that they may be operated in the reverse breakdown region, which is achieved by using heavily doped PN junctions. These devices find applications as voltage reference diodes or constant voltage sources due to a characteristic in which a large change of current is accompanied by a very small change in the volt drop across the diode. There are two physical mechanisms involved, avalanche breakdown that occurs above about 6 V and Zener breakdown below this level. However, both types are often referred to as Zener diodes.

Avalanche breakdown. In the reverse biased mode, a few thermally generated current carriers may acquire sufficient energy to produce hole–electron pairs through collision with the crystal ions. In turn, these new carriers may well generate additional hole–electron pairs through further collisions, to give rise to an avalanche effect. Silicon avalanche diodes are available with ratings of several hundred volts and power dissipation of more than 100 W.

Zener breakdown. The applied reverse voltage may be sufficient to strip some electrons away from their parent atoms to generate carriers that aid reverse conduction.

Breakover diode

This device, designed as a transient suppressor on telephony or computer network transmission lines, is capable of protecting line equipment against the effects produced from such as lightning strikes. It consists of a four-layer PNPN thyristor type semiconductor sandwich which automatically develops a very low ON resistance whenever the applied voltage exceeds some threshold level. As soon as the transient has passed, the device current falls below a holding level and it then reverts to its original high OFF resistance state. Because of the symmetrical four-layer construction, the device will also react in the same way to a transient of opposite polarity.

ESAKI diode

See Tunnel diode, p. 290.

Fast recovery diode.

The transition period for switching from reverse to forward bias (forward recovery) is relatively small and usually unimportant. By comparison, the reverse recovery period can create problems. When the conventional diode is forward biased, the minority carrier concentration at the junction is high. If the voltage is suddenly reversed, it can take a significant time for these to disperse against the diode resistance while the junction capacitance charges through the load resistance. Fast recovery diodes are therefore usually formed from lightly doped PN junctions.

Gold bonded junction

These are point contact diodes formed by pulse bonding a gold wire whisker to a chip of N type germanium. This forms a PN junction with a very low forward resistance just beneath the wire tip.

Graded junction

A PN junction in which the carrier concentration varies, usually linearly, across its width. This makes the junction capacitance proportional to the cube root of the applied voltage and hence less sensitive to voltage changes.

Gunn device

Although often described as a diode, this two terminal oscillator device operates on a transit effect principle. A layer of N type gallium arsenide (GaAs) is grown as an epitaxial layer on a low resistivity substrate which forms the device anode. A gold contact lead is bonded to the epitaxial layer and this forms the cathode connection.

The device thus has a diode like structure. When the applied voltage across the device generates an electric field in excess of about 35 V/cm, a high level electric field domain is produced in the cathode region and this drifts rapidly towards the anode. As the domain passes out at the anode, the electric field falls rapidly, only to restart almost immediately. A new domain forms when the field rises above the critical level, and the cycle repeats.

The equivalent circuit of the device is a negative resistance of about -5Ω in series with a capacitance in the order of 0.2 pF. To make the circuit oscillate, the load connected to the device should have:

(a) a series reactance that tunes the device to resonance;
(b) a series load resistance less than the modulus of the negative resistance;
(c) at frequencies other than resonance, a load resistance greater than the modulus of the negative resistance.

Provided that these conditions are met, a variable reactance can be used to tune the circuit over at least an octave frequency range.

Hot carrier diode

A hot carrier is an electron (or hole) that has an energy level higher than that of the crystal structure, i.e. several kT above the Fermi level. A rectifying junction can be formed between a metal and a semiconductor (Schottky diode). If forward biased, majority carriers (electrons in N type semiconductor) with an energy level greater than the Schottky barrier — the hot carriers — can cross the barrier and produce a current flow. The number of hot carriers increase with the forward bias so that the forward current rises rapidly. This feature considerably increases the forward sensitivity of the diode relative to that of the conventional PN junction. The reverse biased current is very small and the diode can switch state very rapidly. These properties make the device valuable for handling very low amplitude, very high frequency signals.

IMPATT diode (IMPact Avalanche and Transit Time)

Constructed from two heavily doped P and N type regions separated either by a lightly doped N type region or an I type region, this diode exhibits a negative resistance region in its V/I characteristic. When this is placed in parallel with the positive resistance of a resonant circuit or cavity, oscillations at microwave frequencies occur.

The origin of the negative resistance is the transfer of electrons from the normal conduction band into a satellite band, which depends on the magnitude of an applied electric field. The result is a pulse of current that flows through the device from anode to cathode.

Josephson junction

This junction relies upon superconducting properties at very low temperatures. One practical implementation uses niobium electrodes separated by a thin layer of insulator cooled to the temperature of liquid helium (4.2 K). Superconducting electrons tunnel through the installation barrier in the absence of an applied voltage to create a supercurrent. If this current exceeds about 1 mA, the junction voltage suddenly jumps from zero to about 3 mV. This is described as the dc Josephson effect.

An ac Josephson effect occurs when a small dc voltage is applied across the junction to produce an ac current through the device.

The junction operates with low voltage and current and is therefore a low power device that can switch state in a few picoseconds. The junction may be used as a cryogenic memory cell or in laboratory standards measurements applications.

Light emitting diode (LED)

When holes and electrons combine energy is released. In germanium and silicon, this energy is released as heat into crystal structure. In material such as gallium arsenide and gallium phosphide, the energy is released as light, with different semiconductor compounds releasing light of different wavelengths (or colour). The LED is constructed using a planar diode moulded into a plastic tube that acts as a light pipe to radiate light from its one domed end. When forward biased beyond about 1.6 V, the current rises rapidly and light is radiated from the junction. The forward current needs to be limited to a value rather less than 40 mA.

MOTT diode

This diode, designed for microwave frequency operation, is fabricated on a relatively thick gallium arsenide substrate using planar techniques. Over the normal operating range, the junction capacitance is practically constant and independent of the applied voltage. The device is often used as a low noise mixer diode for signal frequencies up to 110 GHz.

PIN diode

This type of diode has a thin I type or lightly doped N or P layer sandwiched between highly doped P and N type regions. The device has a normal reverse biased characteristic with high OFF resistance. When forward biased, however, it exhibits an almost pure voltage-dependent, linear, variable resistance with a value ranging from about 1 Ω to 10 kΩ, thus making it useful as a variable signal attentuation for frequencies up to around 40 GHz. Because of the suitable forward and reverse resistance values, the device also finds applications as a high speed switch for antenna change-over or RF band switching.

Photodiode

Photodiodes are formed from one very thin junction region and usually equipped with a lens so that light energy can be directed into the depletion or barrier region. The light energy creates hole electron pairs that increase the diode's conductivity. Many diodes have a peak response in the infra-red region, but still have a very useful response to visible light. Under static conditions, these diodes are operated either reverse biased or very lightly forward biased so that no current flows. Typical *dark current* for germanium types may be as low as 2 nA, rising to about 100 μA in bright light. Modern PIN silicon versions may have a peak power dissipation greater than 100 mW. The increase in current due to incident light is practically linear. The response time to a pulse of light is very short so that these diodes find applications in high-speed switching circuits. Generally the small output and relatively high impedance requires the use of a buffer amplifier.

READ diode

This diode, which finds applications in microwave circuits, consists of a four layer sandwich of N^+PIP^+ semiconductor material, with a P–I section providing the space charge region. Diode operation depends upon a combination of transit time effects and avalanche breakdown. This produces a phase difference of 180° between the applied voltage and resulting current, equivalent to a negative resistance. When the electric field across the diode exceeds about 5 kV/cm, avalanche breakdown occurs and the forward current rises rapidly.

Schottky diode

A contact between a metal and a semiconductor produces a rectifying junction with diode like characteristics. Schottky diodes are therefore formed by fusing an aluminium lead to a chip of N type silicon. The junction capacitance is very small so that the charge storage is negligible and this gives the diode a high switching speed. With a forward biased threshold level of only 300 mV, the diode characteristics are much closer to the ideal. These diodes are often fabricated into ICs to act as protective shunts to the transient sensitive gate electrodes.

Solar cells

Based on the photodiode, these cells are formed from heavily doped silicon PN junctions. A cell of about 4 cm² is capable of providing 0.6 V on open circuit, with a short-circuit current of up to 100 mA in bright light. The cells may be linked in a series–parallel configuration to provide much higher levels of power supply.

Switching diode

The low forward and high reverse resistances in a diode characteristic enables these to be used as a switch. However, if these are used in very high frequency circuits, the diode capacitance will provide a bypass path for the signal. Switching diodes are therefore chosen for a very low junction capacitance.

Tunnel or ESAKI diode

For the conventional diode, a carrier needs an energy level at least equal to that of the barrier before conduction can take place. The tunnel diode consists of a heavily doped germanium or gallium arsenide PN junction but with a very thin barrier region. In this case, carriers with a lower energy level can produce conduction, creating the effect of tunnelling through the barrier. This gives the device a negative resistance region and, if the diode is suitably biased and placed in parallel with the small positive resistance of a resonant circuit, results in sustained oscillations.

Varicap diode (see Abrupt junctions, p. 287)

These devices are formed from abrupt junctions that produce a significant voltage dependent capacitance. The range of a typical device varies from about 200 pF to 20 pF as the reverse bias changes from zero to about −15 V. The capacitance has an effective shunt resistance of about 1 MΩ with a series ohmic contact resistance of less than 10 Ω.

Voltage reference diode

See Breakdown diode, p. 287.

32.2.2 Programmable logic devices (PLD)

Any combinational logic function can be converted into *sums of products* terms using De Morgan's theorem. This then allows many complex functions to be generated with an array of OR (sum) and AND (product) gates.

PLDs are ICs that are based on the programmable read only memory (PROM) device (see Memories, p. 167) and this was originally designed as a look-up table type of memory. The ability to program and reprogram ROMs in the field, plus the recognition that ROMs could be programmed to act as logical devices, has led to the development of many variants of the basic concept that have similar titles. The term PLD tends to be used generically to cover all these devices but Generic Array Logic (GAL) is a trademark of National Semiconductor Corp. Further, Programmable Electrically Erasable Logic (PEEL) and Electrically Erasable Programmable Logic Device (EEPLD) are trademarks of International CMOS Technology Inc. and Altera Inc., respectively. Other terms in general use include, programmable array logic (PAL), programmable logic array (PLA), field programmable logic array (FPLA), and Programmable Logic Element (PLE) which is a trademark of Monolithic Memories Inc. PLDs can be programmed to behave as either combinational or sequential logic systems. In the latter case, the devices may be referred to as programmable logic sequencers (PLS).

Table 32.1 gives an indication of the degree of flexibility that can be achieved with these devices.

Table 32.1 *Programmable logic devices*

Device	AND array	OR array	Output
PROM	Fixed	Program	T/S. O/C
FPLA	Program	Program	T/S. O/C
PAL	Program	Fixed	T/S. Registered
GAL	Program	Fixed	User defined Polarity programmable

T/S = tri-state, O/C = open collector. Polarity programmable = output active high or low. Fixed = hard-wired interconnections.

The conversion of any logical operational function can easily be performed using a suitable computer program, the programming of the PLD can then be performed using a suitable micro-controlled programmer.

The basic principle of the PLD is indicated by Fig. 32.5. The suitably converted input data is passed through AND and OR gates to provide the output control signals.

Figure 32.6 shows a number of circuit elements that are commonly associated with PLDs. Two types of fuse links are in use. First, the metallic fuse, which consists typically of an aluminium deposition linking all the cells. An unprogrammed PLD has all its fuses intact which thus represent logic 1. When a fuse link is blown during programming, this is converted into a logic 0. Alternatively an *anti-fuse* device may be used. In this case, the metallic tracks are initially insulated with a layer of silicon dioxide. Programming this involves applying a voltage between two tracks that is high enough to rupture the insulator and fuse the tracks together. The input buffer amplifier shown in Fig. 32.6(b) is designed to provide both true and false outputs. The inverted term simplifies the logic function and reduces the number of logic elements needed on the chip. To improve the clarity of circuit diagrams, the conventional logic AND and OR element symbols are modified as indicated in Fig. 32.6(c). Although the inputs A, B, and C are shown connected to a single input line, the diagram is intended to represent a gate with three separate inputs.

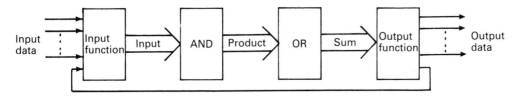

Figure 32.5 Forming sum of products in PLD.

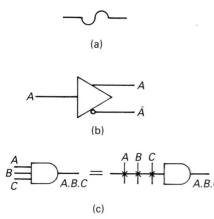

Figure 32.6 PLD symbols: (a) fuselink; (b) input buffer; (c) 3-input 'AND' gate.

The general architecture of a PLD is shown in Fig. 32.5. A set of data inputs are coupled to the AND gate array via buffer amplifiers that provide both true and complement outputs, and a matrix of cross-connections. This latter may be fully decodable, selectively hardwired during fabrication, or open circuit. In the particular example shown, the AND gates would each have three inputs. The AND gate outputs are similarly matrixed to the OR gate array and again the cross-connections may be hardwired or programmable as indicated by Table 32.1. The outputs therefore represent logical sums of products.

Bi-directional I/O
By using tri-state buffers in the output stages, device pin connections can be arranged to function as either inputs or outputs. A dedicated AND gate is used to control the tri-state device, while the same pin is also connected to the AND array as an input. Such a pin may be used either as a:

dedicated input — all fuses intact
dedicated output — all fuses blown
controlled output — programmed by control logic, or
output with feedback — output signal applied to the AND array.

Grow-back
See Metal fuse regrowth, p. 291.

Hard array logic (HAL) device
A PAL device that is programmed during fabrication and all interconnections are hard-wired. A standard chip is produced as far as the metallisation stage. A custom mask is then used to fabricate all the necessary hard-wired links in place of the programmable fuses.

Metal fuse regrowth
This term refers to the ability of blown fuses to develop a self-healing effect and corrupt the programmed logic pattern. This may arise because the metal in very thin films can migrate under the influence of electric fields. Programmable fuses need to be blown with a sufficient current density to ensure a clean rupture. If the programming current is too low, a ragged break results and this promotes the possibility of grow-back. In this respect, it is important to ensure that device pin contacts with the programmer are of low resistance.

Programmable output polarity
To minimise the logic function without resorting to the use of external discrete components, some PLAs include an inbuilt programmable output inverter. This takes the form of a two input Ex-Or gate with one input fuse programmable.

Registered output
Sequential logic systems provide an output that depends not only on the present input, but also the previous output, thus providing a need for some form of memory. This usually takes the form of an edge triggered D type flip-flop. By comparison, an unregistered output is usually provided via a tri-state buffer.

Security fuse
After the circuit has been programmed and verified, an additional fuse may be blown to disable the verification logic. This makes it very difficult to copy the device program.

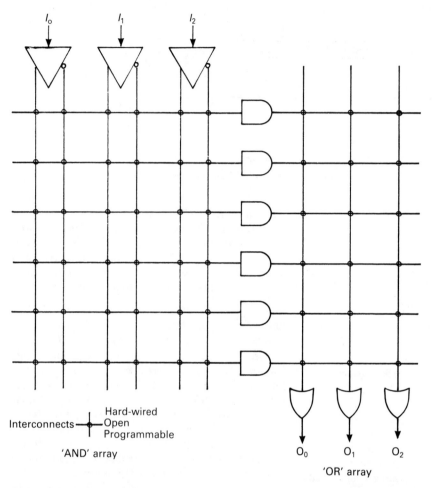

Figure 32.7   Architecture of PLD.

32.2.3 Transistors and other devices

Charge coupled or transfer devices (CCD or CTD)

These devices consist of a linear series of capacitor cells or potential wells, fabricated by the planar process between either a P type or an N type substrate, and a series of metallic gate electrodes with an oxide layer as the dielectric.

CCDs may be used as: a serial shift register for both analogue and digital signals, an analogue signal delay line, a simple serial memory, or a dynamic filter.

Because the signal energy that is stored in each well is moved from well to well under the influence of a series of overlapping clock pulses applied to the gate electrodes, the CCD is often referred to by the jargon term *bucket brigade device*. The signal transfer efficiency is very high, often exceeding 99.999%.

The signal charge may be injected either from a diffused region at one end of the line of wells, or by the exposure of the whole array to radiant energy. In this latter form, the device is often used as a television camera image sensor, when each well represents one pixel. These devices are fabricated in a matrix form to produce more than two million pixels.

Gate turn off thyristor (GTO) (see also Silicon controlled rectifier, p. 293)

During design and fabrication, the gains of the two internal transistors are carefully adjusted. This allows a negative gate voltage to divert a fraction of the ON state current out at the gate electrode. This in turn, causes the circuit to unlatch so that the device turns off. When operated in the GTO mode, the thyristor current must be limited to a level below that of the maximum rated value.

High electron mobility transistor (HEMT)

These are based on very thin heterojunctions of GaAlAs and GaAs. The epitaxially grown layer is only a few electrons thick, so that the interface is practically only two-dimensional. The electrons that collect in this region due to an applied electric field behave in a different manner because they now only have two degrees of freedom of movement. This results in an electron mobility that is a factor of 3 better than that for silicon. HEMT devices can therefore operate at microwave frequencies with very high gain and low noise levels.

Insulated gate bi-polar transistor (IGBT)

These devices are a hybrid of MOSFET and bi-polar technology, with the transistor base being controlled via an insulated gate in the manner of the FET. This produces a device that is more robust than a bi-polar transistor and not so susceptible to gate–source breakdown as MOSFET. IGBTs have a lower ON resistance than many FETs and a transition frequency higher than an equivalent bi-polar device. Maximum operating parameters include:

Power rating 100 W
Peak voltage 1500 V
Transition frequency 50 kHz.

Low dimensional structures (see HEMT, p. 292)

Resonance-tunnelling hot electron transistor (RHET)

In bipolar transistors, barrier potentials are created between different semiconductor materials. Varying the applied voltage at the base, then controls the current flowing between emitter and collector by varying the level of the barrier. In quantum effect devices, the characteristic discrete energy levels of the different materials act as a barrier to control the current flow. When the modulated base voltage exactly matches these energy levels, electrons resonate (hot electrons) and tunnel across the base region to control the current flow. The output voltage rises in phase with the input voltage up to the point where the hot electrons start tunnelling out of the device. Circuits constructed from these transistors are capable of operating at frequencies in excess of 120 GHz with very high gains.

Silicon controlled rectifier ((SCR) or thyristor)

These devices are formed from a four layer PNPN sandwich in which the outer sections form the anode and cathode, respectively. A control gate electrode is attached to the P region closest to the cathode. If an alternating voltage is applied between cathode and anode with the gate at zero, no current will flow through the structure because one PN junction will always be reverse biased. If, however, the gate voltage is held at a small positive voltage, then current will flow whenever the anode is positive with respect to the

cathode. Once conduction has started, the dual transistor (equivalent circuit shown in Fig. 32.8(b)) will latch up and current will continue to flow even if the gate voltage is removed. Conduction will now only cease if the anode–cathode voltage is reduced below the near zero holding level. Devices are available to pass peak currents and withstand peak voltages well in excess of 100 A and 1000 V, respectively.

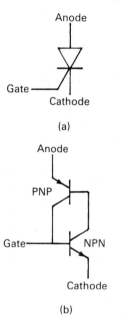

Figure 32.8 Thyristor: (a) symbol; (b) equivalent circuit.

Triac (bilateral triode switch or silicon controlled switch).

These devices effectively consist of two SCRs coupled in inverse parallel, with a common trigger electrode. Current can flow in either direction, provided that conduction is initiated by the presence of a suitable polarity trigger pulse.

Tetrode transistor

This is an FET device designed to operate at very high frequencies (typically 1 GHz upwards). It is constructed with a short channel overlaid with two gate electrodes, each capable of controlling the source to drain current. As indicated in Fig. 32.9, it operates in the manner of the cascode amplifier where gate 1 provides the signal input. The dc bias on gate 2 may be used to provide gain control. Alternatively gate 2 may be driven with a second modulated signal, so that the device functions as a mixer stage.

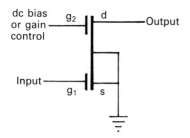

Figure 32.9   Tetrode transistor.

VMOS (vertical MOS device) (see also Amplifiers, p. 22)

The conventional MOS device has a lateral form with source and drain being fabricated close to the semiconductor surface. By comparison, the VMOS device has a vertical structure. The drain electrode is fabricated below a V shaped channel to ensure good heat transfer to the substrate. The gate is formed in the groove that is etched into the silicon. Current flows vertically between the source and the drain through the parallel paths formed by the two sides of the V. This type of structure produces a low resistance, high current and high power device capable of operating at upwards of 50 W power output.

Velocity modulation transistor (VMT)

An FET structure with a double GaAs/AlAs channel sandwiched between barrier layers of AlAs. Unlike the FET, the drain current is modulated by changes in electron velocity and since the electron density under the gate does not change when the device changes state, its switching rate (typically around 1 ps) is not transit time limited.

32.3 Technology

Unlike semiconductor devices, fabrication technology tends to change at a very much slower and evolutionary pace. Innovative technologies are introduced for specific reasons and these are progressively refined over a period of years.

32.3.1 Comparisons

Device types

Bi-polar-CMOS (BiCMOS). An IC that contains both bipolar and complementary (CMOS) transistors, a type of structure that exploits the advantages of both processes. Bi-polar transistors operate faster than the

equivalent CMOS devices and commonly have a better set of analogue operating parameters. CMOS devices are preferred for their lower power dissipation and this leads to a greater component density on a given chip area. The technique is particularly valuable in the production of mixed signal ICs.

Bi-polar device. An IC based on bi-polar transistor technology. Devices are formed as planar structures by oxide growth, photolithographic etching and the diffusion of the appropriate impurities in turn, as epitaxial layers. The substrate commonly forms the collector electrodes in order to improve heat dissipation. Individual components are isolated by deep diffusion regions that effectively form reverse biased junctions of high resistance. The interconnection between components is provided by metallic aluminium diffusions. The bi-polar device has a higher transconductance than the equivalent CMOS device and provides a higher output drive level, properties and parameters that are most useful in analogue circuits.

CMOS device. An IC formed by a planar process of metal oxide on semiconductor (silicon), but using complementary transistors with either P or N type channels, incorporating circuits similar to that of the inverter shown in Fig. 32.10. If the gate input is at logic 1 (+5 V), the lower channel will be conductive while the upper one is cut off. The output will thus be at logic 0 (0 V). If the input changes state to logic 0, the conduction states of the two channels change over and the output goes to logic 1, but this time, with the lower channel cut off. It will be apparent that whatever binary state is taken up by the input, one of the channels will always be non-conducting so that practically no power is

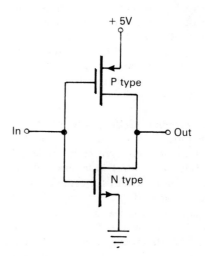

Figure 32.10   CMOS inverter stage.

being drawn from the power supply. With CMOS ICs it is important that any spare inputs are permanently wired to ground or the positive supply. If any such inputs are left floating, the gate will charge due to leakage currents and latch-up. This creates an excessive current drain that can lead to device failure. In addition to the obvious digital applications, CMOS devices may also be used as linear amplifiers or switches in analogue circuits.

Fabrication materials

Ceramics. Barium titanate, the ceramic material commonly used for the production of positive and negative temperature coefficient (PTC and NTC) thermistors, develops semiconducting properties when doped with suitable amounts of strontium and lanthanum. This produces a device whose resistance increases by a factor of more than 10^5 between 50°C and 100°C with almost logic-like switching characteristics.

Diamond. Synthetic diamond which is unconditionally stable in most environments, finds applications as a substrate at the very high operating temperatures of around 600°C, for the following reasons:

- a dielectric constant about half that of GaAs;
- breakdown voltage about 50 times that of GaAs;
- power capability about 2500 times that of GaAs;
- electron mobility greater than that of silicon and about twice that of GaAs, with even greater hole mobility, and thermal conductivity about four times that of copper.

In addition, diamond is much more radiation resistant and more area efficient than any other semiconductor IC material.

Germanium (Ge). This relatively rare element found in nature as an oxide, requires considerable processing to obtain an acceptable degree of purity. Although germanium has a lower intrinsic resistance and higher dielectric constant than silicon, it finds few applications in modern semiconductor technology. Like silicon, the reverse biased leakage current of a PN junction approximately doubles for every 11°C rise in temperature. Since germanium has a significantly higher leakage at the same temperature, the operating range is restricted to about 75°C maximum. The PN junction threshold level is about 150 mV.

Group III/V materials. Certain compounds of Groups III and V elements may be produced that have P and N type semiconductor properties. The most important elements include gallium (Ga), indium (In) and aluminium (Al) from Group III, with arsenic (As) and phosphorus (P) from Group V. The compounds most extensively used are GaAs (gallium arsenide) and GaAlAs (gallium aluminium arsenide). GaAs has a resistivity similar to that of silicon and an electron mobility significantly higher. This provides for a higher speed of operation with a lower power consumption. The PN junction barrier potential is in the order of 0.75 V. Gallium arsenide devices have a better radiation resistance and a higher maximum operating temperature range than similar silicon devices. These features are valuable in high speed signal processing, and light emitting and detecting circuits. GaAs technology is thus compatible with optical fibre systems.

Silicon (Si). This second most common earthly element is found as a silicate (sand). Being produced in vast quantities, silicon devices form the workhorse of the electronics industry. The PN junction threshold level is typically about 600 mV and silicon transistors have a high maximum operating temperature (typically in the order of 250°C). Although rather slower than equivalent GaAs devices, when transistors are fabricated with a silicon/germanium alloy base region, the speed difference is significantly reduced.

32.3.2 Other features

Application-specific ICs (ASICs)
This term applies to any IC designed to perform a function or set of functions, specific to a circuit or system and to meet a user's specific need. In general, the architecture includes PLDs and other cell-based structures. The former is based on a matrix of logic circuits while the latter may be fabricated from a set of any predefinable circuit blocks. The devices are produced as general purpose and are only specifically programmed during the final stages of metallisation. The design of ASICs, which are produced using either bipolar, CMOS or GaAs technologies, is very flexible. The concept, which is valuable in reducing costs for large-run production systems, leads to reduced size and improved reliability and power consumption.

Chemical vapour deposition (CVD)
See Vapour phase epitaxy, p. 297.

Diffusion
One of the processes used for doping impurities into intrinsic semiconductor materials. A slice of semiconductor material is exposed in a heated and slightly pressurised chamber, to a gas carrying the impurities. At a temperature that may be as high as 1000°C, the impurity atoms transfer from a region of high density into one of a lower density by penetrating the surface of

the cooler slice. The depth of penetration depends upon the temperature, pressure, duration of exposure, and the nature of the surface. Areas of unwanted diffusion can be protected by using an oxide layer which inhibits the process. Using the known factors about diffusion, a very accurate control of the fabrication process can be achieved.

E-Beam

A computer-assisted design technique that is used in the fast development of ASICs. Silicon slices in the fully metallised state are directly written with the required pattern using an electron beam controlled by the computer and program. The pattern details are stored in a computer file for future use, either for modification after prototype testing or for production runs of the final device. The technique therefore significantly reduces costs and the time span between initial design and final production due to the debugging capability.

Ion implementation

A technique that is used as an alternative to diffusion, to implant impurity ions into a semiconductor material by surface bombardment. The dopant molecules are first ionised and then accelerated by an electric field to produce a velocity high enough to penetrate the water surface. Typically this requires a beam current and power of about 25 mA and 4 kw, respectively. However, lower levels can be used if the process time is extended. A mask of aluminium or photoresist is used to define the implantation area, and a deeper penetration can be achieved if bombardment takes place along a preferred crystal axis. The chief advantages of this technique are the relatively low process temperature and the sharply defined doped regions.

Isolation techniques

The electrical isolation between circuit components of an IC which is needed to avoid unwanted interconnections can be achieved by a number of different methods.

Dielectric isolation. A passive layer of silicon monoxide, silicon dioxide, ruby or glazed ceramic can be used to surround small groups of components. This layer is made thick enough to minimise the parasitic capacitance that naturally results, but this reduces the number of components that can be accommodated on a given chip size. However, this problem is minimised using a derivative known as trench isolation.

Diffusion isolation. In the example shown in Fig. 32.11, a P type substrate is more heavily doped to provide a P$^+$ region. Deep N type regions are then formed which will carry the components; either diodes, resistors or transistors. In operation, the NP$^+$N regions thus form

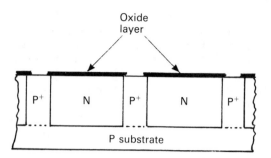

Figure 32.11 Diffusion isolation.

back to back reverse biased junctions of very high resistance to provide the isolation regions. While this produces an adequate dc isolation, the parasitic capacitances of the junctions thus formed will provide unwanted coupling at higher frequencies to limit the device performance.

Trench isolation. This is a development of dielectric isolation in which the circuit elements are formed within a vertical trench of insulator that extends down into the substrate. The technique produces a greater component count on a given chip area.

Lithography

The elemental components of an IC are built up from layers of silicon, insulator (usually silicon dioxide) and conductor (usually aluminium). To produce the desired pattern, the wafer surface is coated with a particular material. This surface is then completely covered with a photoresist (PR) or light sensitive chemical. When this is exposed to light through an accurately made mask and photographically developed, the desired pattern appears in PR. When this etched and the PR dissolved by solvent, the pattern has been transferred to the silicon surface. Ultraviolet or mercury vapour lamps are often used for the photographic exposure stage, for which the shortest wavelength available is around 365 nm. As transistor and insulator features have shrunk in size in order to increase the chip density, the diffraction at the mask edges creates a distortion that blurs the pattern and generates a density limit. One solution uses soft X-rays with a very much shorter wavelength for the light source.

MISFET/MESFET

Metal insulator semiconductor FETs have a significant negative resistance region in their characteristic, a feature particularly useful in certain analogue circuits such as frequency multipliers. They are based on a heterojunction of GaAs/GaAlAs, behave in a very similar way to HEMT devices and have a simpler structure than MESFETs.

Metallising

The interconnections between components of an IC are formed by the deposition of metallic tracks of aluminium or gold. These metals are chosen because of low resistivity and good adhesion to the semiconductor materials. Aluminium is most commonly used because gold has a poor adhesion to silicon dioxide and requires an intermediate layer of chromium. Aluminium suffers from ion migration under the influence of high density current flow. The track width therefore has to be wide enough to avoid this problem. If high current densities are unavoidable, then gold/chromium or titanium provides a suitable alternative. Metallisation of complex ICs can extend to four levels, each layer of metal being separated by oxide layers. This multi-layer approach is described as buried metal and provides a means of producing chips with a very high component density. Post metal programming (PMP) uses a high energy ion implant technique to program the on-chip ROM (read only memory) of a microcontroller after the final metallising stage.

Molecular beam epitaxy (MBE)

See Vapour phase epitaxy, below.

MOS technology variations

At least three variations of the basic MOS technology are in use. These generally relate to modifications of the gate structure and result in devices with characteristics of higher speed, lower power consumption or both.

These are commonly referred to as: metal nitride oxide semiconductor (MNOS); silicon nitride oxide semiconductor (SNOS); or silicon oxide nitride oxide semiconductor (SONOS) technologies.

Multi-chip modules

A technique devised to allow large chip sets to be mounted into a very small space. Each IC, bare of any packaging, is equipped with a series of *bumps* at each connecting point. These bumps then make contact with mounting pads on a silicon or ceramic substrate.

Radiation hard

For semiconductor devices, space is a hostile environment and silicon ICs fail when bombarded by cosmic radiation. Radiation hardened devices are expected to be able to withstand a total dose of 1 Mrad (1 rad is the unit of absorbed dose of ionising radiation). By comparison, CMOS devices, which are more resistant than bi-polar ones, will only withstand a maximum of 100 krad. Silicon on insulator (SOI) or silicon on sapphire (SOS) devices offer the best performance against such radiation. Transistors are formed by depositing silicon layers on a sapphire substrate so that components are built upwards. Since very little silicon is used, there is also

very little to be affected by radiation. Furthermore, a layer of silicon dioxide is used as an overlay to complete the protection. Circuit components formed in this way are well isolated from each other so that latch-up effects are also avoided. While the process is expensive and produces ICs with a relatively low component count, the devices have very low parasitic capacitances and therefore operate at high speed with very low power dissipation.

Reactive ion etching (RIE)

A fabrication technique used to etch very narrow slots (less than $1 \mu m$). The etchant is a stream of ionised reactive gas molecules accelerated to a high energy level. When these strike a surface, molecular disassociation occurs to produce spaces.

Silicon on sapphire (SOS)

See Radiation hard, above.

Sputtering

This term refers to one technique that is used in the electro-deposition of materials as thin films. Sputtering is the ejection of atoms from the surface of a material when bombarded by high energy positive ions obtained from a gaseous plasma. Many materials that are difficult to evaporate, such as refractory metal alloys, semiconductors and dielectrics, can all be deposited by sputtering. The energies of sputtered atoms is significantly higher than those obtained by evaporation so that a sputtered film adheres more firmly to a given base. There are many ways of achieving deposition by sputtering, but one common method uses a cathode of the material to be deposited with the target used as the anode. These are contained in a chamber with non-reactive argon at low pressure and a voltage between $1 kV$ and $10 kV$ is used to create a glow discharge through which the transfer occurs. If a reactive gas is used, then it is possible to deposit compounds of the cathode material.

User-specific IC (USIC)

See Application-specific ICs, p. 295.

Vapour phase epitaxy (VPE)

A growth process whereby silicon atoms from a carrier gas (vapour phase) are deposited onto a substrate so that the single crystal structure can be extended. The grown layer may differ from the substrate in impurity concentration. PN junctions can be formed using this technique, or alternatively, highly resistive layers may be grown onto low resistive substrates and vice versa.

Wafer scale integration (WSI)

ICs are produced *en masse* by fabricating many identical devices simultaneously on typically a 6 in diameter semiconductor slice. The slice is then diced into small squares to provide the individual chips for final preparation and packaging. Wafer scale technology takes the integration process a stage further by recognising that a slice contains many fully operative chips, plus a few defectives. Instead of dividing the slice into individual chips, interconnections are provided on the slice to integrate all the chips into a single large circuit. The slice incorporates a logic module that can be configured in the manner of a programmable ROM that will provide a contiguous data path through the slice, avoiding the defective cells. Having a memory, this circuit will ensure that the same path is in operation at each power up. The technique is particularly valuable for high density, large semiconductor memories where a single defect in one memory would result in a completely defective chip.

32.4 Miscellaneous terms

Amorphous. Non-crystalline and without regular form.

Biosensors. These are devices that are formed from a thin organic film deposited onto a semiconductor substrate. They may be used as electrical sensors for the measurement of the concentration of certain biochemicals.

Black death. The formation of a blackened region in an IC that results from metal ion migration with diffusion of gold into an aluminium area. This gradually spreads through the metallisation and ultimately causes high resistance links or even open circuits. The basic solution to this problem is to use either all gold or all aluminium for metallisation or to use a pre-deposition of chromium under the gold.

Hall effect devices. If a current-carrying conductor is placed in a magnetic field and restrained from movement, the electrons (current carriers) in the wire are forced to one side. This sets up an emf across the conductor width which is referred to as the Hall voltage. In normal conductors this effect is very small, but in semiconductors it can reach a significant value. Reversing either the magnetic field or the direction of the current, reverses the polarity of the Hall voltage. Hall effect devices can therefore be used in many applications including:

- measurement of the strength of a magnetic field;

- measurement of the intensity of a current flowing in a conductor without breaking the circuit;
- provision of contactless switches.

HARP (high gain avalanche-rushing amorphous photo-conductor). A light sensitive semiconductor sensor used in solid-state imaging circuits (CRT replacement).

Heterojunction. A junction formed between two dissimilar semiconductors. The properties of the junction are improved if there is an interface with different band gap energies between the two single crystal semiconductors. These junctions are classified as either, iso, N–N, or P–P types. In the iso-type, the minority carriers are confined to a small active region due to a potential barrier within the structure. The injection efficiency of either holes or electrons is improved with an aniso-type junction with a sufficiently large band gap difference.

Hole. An empty energy level in the valency band of a semiconductor due to the loss of an electron either due to thermal effects or the attraction produced by an acceptor impurity. Has a charge equal to that of an electron but of opposite polarity. The mobility of a hole is not usually the same as that of an electron from the same semiconductor element.

Homojunction. A junction fabricated from a single crystal semiconductor material.

Infra-red LED radiation. Devices made from Group III/Group V elements are particularly useful in optical fibre systems. Diodes manufactured from arsenic (As), antimony (Sb), gallium (Ga), indium (In) and phosphorus (P) exhibit the following radiation properties:

Combination	Wavelength	Bandwidth
GaSb	1.8 μm	360 nm
InGaAsSb/GaSb	2.0 μm	134 nm
InAs	3.8 μm	640 nm
InAsSbP/InAs	4.2 μm	450 nm

Moore's law (Dr Gordon Moore, co-founder of Intel Corporation). States that the number of transistors that can be fabricated on a given area of silicon doubles every year. Data has been plotted since 1965 and even during the 1990s the growth is still fairly accurate although the doubling period is now nearer to 18 months.

Polysilicon (polycrystalline silicon). This material is used in the fabrication of gate electrodes for FET devices. The silicon is highly doped so that it exhibits almost metallic (conductor) like properties.

Purple plague. The formation of purple coloured areas at the bonds between gold wires and aluminium pads of an IC. The welds become brittle and are prone to break. The phenomenon arises at temperatures above about 300°C in the presence of silicon that appears to act as a catalyst.

Silicon compiler. A computer program that allows an engineer to describe a target system to be configured on a PLD or ASIC chip, in a very high level language. A series of instructions are entered and the compiler software converts these into a chip design. Such compilers are knowledge based; that is, they are capable of learning from previous design runs.

Transferred electron device (TED). A device based on the Schottky barrier gate Gunn effect that exhibits two distinct stages in its characteristic that impart very high speed binary logic properties. Below some threshold level, the device has a simple resistive impedance, but above this, high electric field domains exist which are self-sustaining.

VHDL or VHSIC. Hardware design languages for ASICs or PLDs (see Silicon compilers, p. 299).

Vias. Holes etched through dielectric layers. These are filled with metal at points where the interconnections between the various layers of metal need to be made to provide continuity.

32.5 Useful references

The most recent information on this topic will be obtained from the learned papers of the International Solid State Circuits Conference (ISSCC), held annually under the auspices of the Institute of Electrical and Electronic Engineers (IEEE), USA.

Berry J. (1992) *Application Specific Integrated Circuits (ASICs)*, Butterworth-Heinemann, Oxford.
Bostock G. (1987) *Programmable Logic Handbook*, Butterworth-Heinemann, Oxford.
Millman J. and Grabel A. (1987) *Microelectronics*, 2nd edition, McGraw-Hill, New York.
Pease R. (1993) Understanding diodes and their problems. Identifying and avoiding transistor problems. *Electronics World and Wireless World*, February, 130 and March, 218.

33 Signals

Communication systems function by transmitting information as time varying electrical quantities of voltages or currents, the signals, which may be of either analogue or digital format. The study of signals is related to the way in which these are processed through a network or system.

By comparison, the study of networks or systems can be described in terms of analysis, synthesis and excitation. Analysis is used to determine the response of a given network to a given excitation. Synthesis is used to design a network to produce a given reponse from a given excitation. Both excitation and response signals can be expressed in terms of their time varying characteristics or alternatively, in terms of amplitude, frequency and phase. As examples of this, it might be pointed out that an oscilloscope and a spectrum analyser provide windows in the signal's time and frequency domains, respectively. In a similar way, for ease of analysis/synthesis or other advantage, signals may be transformed from one domain into another. The general principle of this is shown in Fig. 33.1, where the problem and solution domains are shown as α and β. The solution may be obtained directly in domain α or, if more convenient, via a transform into domain β where an alternative solution can be obtained. Applying a suitable inverse transform then returns the solution to the original problem domain. As a simple example of this process, the addition of logarithms is often used instead of the multiplication of numbers.

33.1 Basic signal elements

It is generally convenient to consider, in both electrical and mathematical terms, that the lowest common denominator of all complex signals is the sinusoid as indicated in Fig. 33.2(a). In the time domain, this can be represented mathematically by:

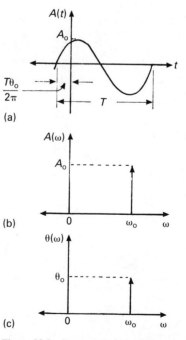

(a)

(b)

(c)

Figure 33.2 Representation of sine wave signal: (a) sinusoid (time domain); (b) amplitude response (frequency domain); (c) phase response (frequency domain).

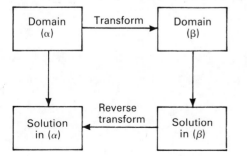

Figure 33.1 Problem solving.

$$A(t) = A_0 \sin(\omega_0 t + \theta_0)t$$

where $A(t)$ represents the general time varying amplitude, A_0 represents the peak amplitude, ω_0 represents the angular velocity in radians per second (rad/s), $\omega = 2\pi f$ where f represents the frequency in cycles per second or Hertz (Hz), and θ_0 represents the phase relationship at time $t = 0$. The periodic time $T = 1/f$ seconds (s).

In the frequency domain, the frequency and phase relationship is represented by the line spectra as shown in Fig. 33.2(a) and (b).

The sinusoid can also be obtained from the rotating phasor or vector as shown in Fig. 33.3, where the instantaneous amplitude

$$A(t) = \sqrt{(r\sin\theta)^2 + (r\cos\theta)^2}$$

and the phase angle

$$\theta = \tan^{-1}\left(\frac{r\sin\theta}{r\cos\theta}\right)$$

From this it is also evident that one rotation or one cycle represents 2π radians, confirming that $f = \omega/2\pi$.

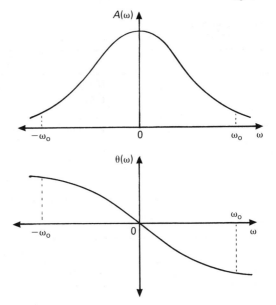

Figure 33.4 Continuous spectra.

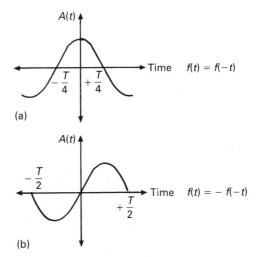

Figure 33.5 (a) Even function (cosine terms only); (b) odd function (sine terms only).

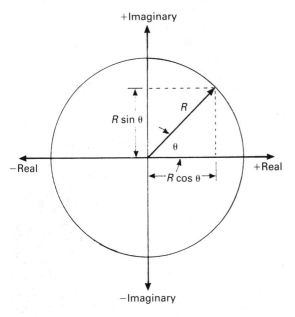

Figure 33.3 Rotating phasor (vector).

Complex waves represent the summation of many sinusoids, each representable by a line spectrum but of varying amplitude. As the wave complexity increases, the spectral line spacing decreases and the gaps tend towards zero so that individual discrimination is lost. The spectrum then becomes continuous in both frequency and phase as shown in Fig. 33.4.

For convenience, waveforms may be characterised as either *odd* or *even* by reference to the type of symmetry that exists at $T = 0$. This is demonstrated by the use of sine waves in Fig. 33.5.

Table 33.1 *Bessel functions*
Extract from Bessel function tables for index 1 to 6.

n	$J_n(1)$	$J_n(2)$	$J_n(3)$	$J_n(4)$	$J_n(5)$	$J_n(6)$
0	0.7652	0.2239	−0.2601	−0.3971	−0.1776	0.1506
1	0.4401	0.5767	0.3391	−0.0660	−0.3276	−0.2767
2	0.1149	0.3528	0.4861	0.3641	0.0466	−0.2429
3	0.0196	0.1289	0.3091	0.4302	0.3648	0.1148
4	–	0.0340	0.1320	0.2811	0.3912	0.3576
5	–	–	0.0430	0.1321	0.2611	0.3621
6	–	–	0.0114	0.0491	0.1310	0.2458
7	–	–	–	0.0152	0.0534	0.1296

Bessel function values for index less than unity.

n	$J_n(0.2)$	$J_n(0.4)$	$J_n(0.6)$	$J_n(0.8)$
0	0.9900	0.9604	0.9120	0.8463
1	0.0995	0.1960	0.2867	0.3688
2	–	0.0197	0.0437	0.0758
3	–	–	–	0.0102

33.1.1 Bessel functions

The integral

$$J_n(x) = \frac{1}{2\pi} \int_0^{2\pi} \cos(n\theta - x\sin\theta)\,\mathrm{d}\theta$$

was deduced by Bessel in 1824 as the solution to Kepler's interplanetary perturbation problem. Although the integral has no straightforward solution, the results are tabulated in the manner of the extract shown in Table 33.1. Bessel functions provide the solutions to

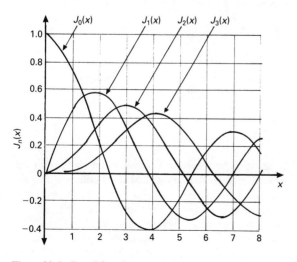

Figure 33.6 Bessel function curves.

many communications problems including frequency modulation (FM), when expansion of the integral as a series, reveals the sideband structure (1). Furthermore, the integral can be used to evaluate the side lobe response of antenna systems (2).

When the tabulated values are plotted graphically, a set of curves as shown in Fig. 33.6 is obtained. In spite of the appearance, the curves are neither sine nor cosine shaped because the periodicity is continually changing. The growth and decay rate of the wave envelope is also non-exponential.

33.1.2 Complex frequency and numbers

Complex numbers represent a valuable tool for the analysis of many communications problems. Consider the phasor or vector S_1 in Fig. 33.7(a), at rest, along the horizontal *real* axis. If this is rotated anticlockwise through 180° it will now have a value of $-a$ and the operation was the equivalent of multiplying by −1. If the vector is further multiplied by −1, this is equivalent to a complete rotation of 360°. If the phasor is rotated through 180° in two 90° stages, this is equivalent to multiplying by $\sqrt{-1}$ twice. For convenience sake $\sqrt{-1} = \mathrm{j}$ and since this has no *real* value, it is described as being *imaginary*. The phasor S_1 in Fig. 33.7(a) can now be shown to consist of two parts, a real part a and an imaginary part $\mathrm{j}b$, so that $S_1 = a + \mathrm{j}b$ is a complex number. Since both a and $\mathrm{j}b$ can lie in any of the four quadrants, the generalised complex number is represented by $z = \pm a \pm \mathrm{j}b$ on the so-called Argand diagram.

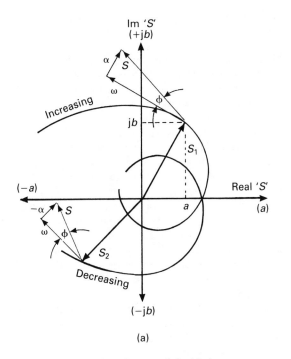

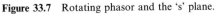

(a)

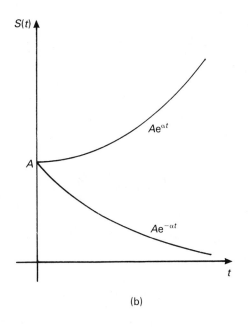

(b)

Figure 33.7 Rotating phasor and the 's' plane.

From Fig. 33.7(a) it can be seen that the magnitude of $S_1 = \sqrt{a^2 + b^2}$ and its phase angle $\phi = \tan^{-1} b/a$ so that a phasor can be represented either in *polar coordinates* (length and angle) or in the complex domain. Furthermore, multiplying a phasor by j is equivalent to phase shifting by $90°$.

The complex conjugate of $z = a + jb$ is a second complex number $a - jb$ and the product of a complex number with its conjugate always produces a completely real number.

In a similar way, a frequency variable can be expressed as $s = \alpha \pm j\omega$ having real and imaginary parts. α represents the growth or decay in amplitude while ω represents the angular frequency. This is indicated by the two exponential spirals that represent the trajectories of phasors S_1 and S_2 in Fig. 33.7(a). The angular velocity ω represents the velocity at right angles to the tip of the phasor, while the quantity $\pm\alpha$ parallel to the phasor represents its growth or decay. s, which represents the velocity at some angle ϕ, can be seen to be complex in nature. From this it can be deduced that the real part of $S = A \exp(\pm\alpha t)\cos\omega t$ and the imaginary part is equal to $A \exp(\pm\alpha t)\sin\omega t$. As shown in Fig. 33.7(b), which is described as the S-plane representation, the positive and negative exponents indicate increasing and decreasing amplitudes respectively.

In general, $S(t) = A \exp(st)$, where $s = \alpha + j\omega$.

33.2 Specific transforms

When using a transform pair to transpose a signal from one domain to another, it sometimes happens that a finite signal in one produces an infinite equivalent in the other. For example, the transform of a square pulse may be a spectrum of infinite bandwidth. This implies that to generate an ideal square wave requires the use of circuitry with infinite bandwidth. The fact that neither conditions are practically achievable prevents such a transform pair from becoming a paradoxical lie. That the use of such transforms provides the right answers to practical problems within acceptable limitations, indicates the value of such mathematical tools in a practical environment.

33.2.1 Discrete cosine transform (DCT) (see Image processing, Transform coding, p. 140; and Television systems, Digital processing, p. 357).

The DCT transform pair are closely related to the Fourier transforms but only retain the *real* or cosine part of the signal. It is commonly used for the non-analytical applications of image processing and may be applied to a matrix of practically any dimensions. The forward and reverse transform pair for an $N \times N$ matrix are given by:

$$F(u, v) = \frac{4C(u)C(v)}{N^2} \sum_{i=0}^{N-1} \sum_{j=0}^{N-1} f(i,j) \cos A . \cos B$$

$$f(i,j) = \sum_{u=0}^{N-1} \sum_{v=0}^{N-1} C(u).C(v).F(u, v) \cos A . \cos B$$

where

$$A = \frac{(2i + 1)u\pi}{2N} \text{ and } B = \frac{(2j + 1)v\pi}{2N}$$

$$C(u) = C(v) = 1/\sqrt{2} \text{ for } u = v = 0$$

or

$$C(u) = C(v) = 1 \text{ for } u = v \neq 0$$

The factor of 4 is often included so that application of forward and reverse transforms in series results in an overall unity gain. The application of the forward transform maps an image space into a frequency space. The mapping produces a set of coefficients that represent amplitude variations in the image domain, where large- and small-scale features represent low and high frequency components, respectively.

The major advantage of this process lies in the fact image features do not normally change very fast so that many of the DCT coefficients are either zero or very small. This makes it possible to store and process a digitised image with a significantly compressed bit rate requirement.

33.2.2 Fourier transforms

As stated above, any periodic waveform can be synthesised by the summation of sinusoids with suitable amplitude and phase relationships. Each frequency component in the complex wave will be a simple integer multiple of the frequency of the original wave. Mathematically this can be expressed as:

$$f(t) = a_o + a_1 \cos \omega t + a_2 \cos 2\omega t ...$$
$$+ b_1 \sin \omega t + b_2 \sin 2\omega t ...$$

where a_o represents a dc component and a_1, a_2, ... and b_1, b_2, ... are the Fourier coefficients of the harmonics ω, 2ω, 3ω, etc; or alternatively as:

$$f(t) = \sum_{n=0}^{\infty} a_n \cos(\omega nt) + b_n \sin(\omega nt)$$

Table 33.2 lists the important equations associated with the calculations for these coefficients for both the single and doubled sided frequency spectra, together with the basic transform pair equations for non-periodic waveforms.

Figure 33.8 shows some of the more important transform pairs in a diagrammatic way. This points out some of the similarities and complementary features that arise between a signal and its frequency spectra. The sine and cosine waves shown at (a) and (b) are basically the same except for a phase or time shift. Therefore it is logical that both should have similar spectra. Note also that the Gaussian pulse and spectrum (d) both have the same exponential shape. The two pulses shown at (c) and (d) are particularly valuable ones because the pulse energy is concentrated within narrow bandwidths. The complementary transforms shown at (e) and (f) are part of the paradoxical problem; how is it possible to generate a pulse now when its origin was some time in the past $(-\infty)$? At (g) and (h) the pulses only differ in amplitude and dc component, therefore it is logical that both should have similar spectra. Although the unit impulse function and its transforms shown at (i) and (j) are a mathematical anomaly, they are important in the solution of sampling problems. This impulse, also known as the *delta function*, represents a pulse of zero width, infinite amplitude and unity area. Such a pulse generated at any time must therefore have the spectrum of white noise, that is, it consists of all frequencies from dc to infinity with equal amplitude. The spectrum at (j) represents a single frequency with an infinite amplitude spectral line. Applying the inverse transform thus yields a dc level of unity.

Table 33.3, which shows some of the important Fourier transform properties, should be read in conjunction with Table 33.2 and Fig. 33.8.

Convolution

As indicated in Table 33.3, multiplication of waveforms/signals in one domain is equivalent to *convolution* in the other. The convolution or *folding* integral is defined as:

$$g_1(\omega) g_2(\omega) = f_1(t) \otimes f_2(t) = \int_0^t f_1(t - T)f_2(T)dT$$

The solution is achieved by time reversing one of the component envelopes, multiplying by the second and then finding the area under the resulting curve by integration between the limits 0 and t.

Table 33.2 *Fourier representation of signals*

Fourier series
Representation of a repetitive signal with period
$T = 2\pi/\omega_s$ and a finite number of finite
discontinuities, in terms of its *single-sided frequency
spectrum* with real components $d_0, d_1,..., d_n$.

$$f(t) = d_0 + \sum_1^\infty d_n \cos(n\omega_s t - \phi_n)$$

where

$$d_n = \sqrt{a_n^2 + b_n^2}$$

and

$$\phi_n = \tan^{-1} \frac{b_n}{a_n}$$

$$d_0 = a_0 = \frac{1}{T} \int_{-T/2}^{T/2} f(t) \cdot dt$$

$$a_n = \frac{2}{T} \int_{-T/2}^{T/2} f(t) \cos(n\omega_s t) \cdot dt$$

$$b_n = \frac{2}{T} \int_{-T/2}^{T/2} f(t) \sin(n\omega_s t) \cdot dt$$

Complex form of Fourier series
Representation of a signal, $f(t)$, as above but in
terms of its *double-sided frequency spectrum*, with
complex components c_n.

$$f(t) = \sum_{-\infty}^{+\infty} C_n \cdot \exp(jn\omega_s t)$$

$$C_n = \frac{1}{T} \int_{-T/2}^{T/2} f(t) \cdot \exp(-jn\omega_s t) \cdot dt$$

$-\infty < n < +\infty$ with integer values.

$$\left[\mid C_n \mid = \mid C_{-n} \mid = \frac{1}{2} d_n \right]$$

Fourier transform pairs
Representation of a non-periodic random signal,
$f(t)$, in terms of its *continuous frequency spectrum*
$g(\omega)$, in general, a complex function of ω.

$$f(t) = \frac{1}{2\pi} \int_{-\infty}^{\infty} g(\omega) \cdot \exp(j\omega t) \cdot d\omega$$

$$g(\omega) = \int_{-\infty}^{\infty} f(t) \cdot \exp(-j\omega t) \cdot dt$$

Correlation

This integral provides a way of evaluating the degree of
similarity between two waveforms during the period t_1
to t_2 and is expressed as:

$$\int_{t_1}^{t_2} f_1(t) f_2(t) \, dt$$

A positive value indicates a degree of similarity and the
higher the value, the greater the similarity. If the result is
negative, then $f_1(t)$ is more closely related to $-f_2(t)$. It is
possible to introduce a time delay into the comparison
so that

$$R_{12}(T) = \int_{t_1}^{t_2} f_1(t) f_2(t + T) \, dt$$

when this is described as the *cross-correlation* function
between the two signals.

Signal *f(t)*   Spectrum *g(ω)*

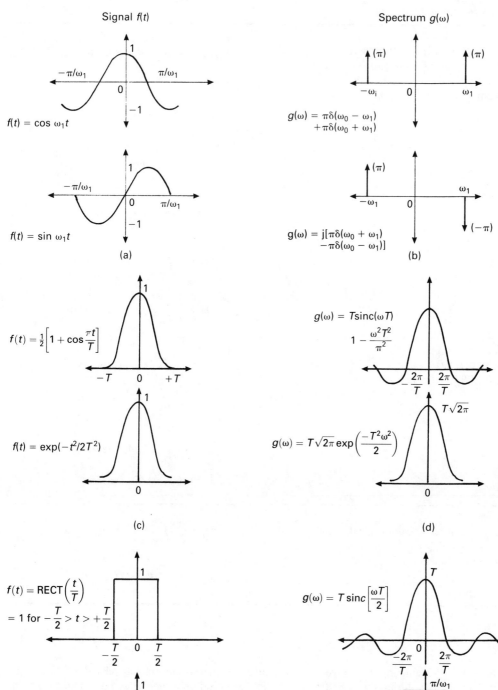

$f(t) = \cos \omega_1 t$

$g(\omega) = \pi\delta(\omega_0 - \omega_1)$
$\quad + \pi\delta(\omega_0 + \omega_1)$

$f(t) = \sin \omega_1 t$

$g(\omega) = j[\pi\delta(\omega_0 + \omega_1)$
$\quad - \pi\delta(\omega_0 - \omega_1)]$

(a)   (b)

$f(t) = \frac{1}{2}\left[1 + \cos\frac{\pi t}{T}\right]$

$g(\omega) = T\mathrm{sinc}(\omega T)$
$\dfrac{1 - \dfrac{\omega^2 T^2}{\pi^2}}{}$

$f(t) = \exp(-t^2/2T^2)$

$g(\omega) = T\sqrt{2\pi}\exp\left(\frac{-T^2\omega^2}{2}\right)$

$T\sqrt{2\pi}$

(c)   (d)

$f(t) = \mathrm{RECT}\left(\dfrac{t}{T}\right)$
$= 1 \text{ for } -\dfrac{T}{2} > t > +\dfrac{T}{2}$

$g(\omega) = T \, \mathrm{sinc}\left[\dfrac{\omega T}{2}\right]$

$f(t) = \mathrm{sinc}\,(\omega_1 t)$

$g(\omega) = \dfrac{\pi}{\omega_1}\mathrm{RECT}\left[\dfrac{\omega_0}{2\omega_1}\right]$

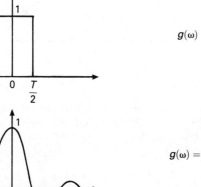

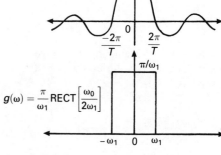

(e)   (f)

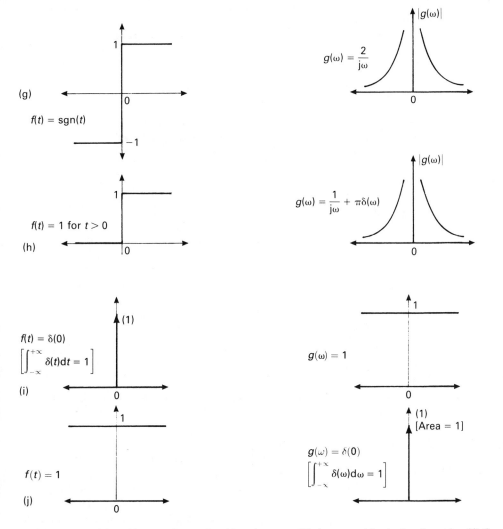

Figure 33.8 Useful Fourier transform pairs: (a) cosine wave; (b) sine wave; (c) raised cosine pulse; (d) Gaussian pulse; (e) rectangular pulse (RECT); (f) sinc pulse ($\frac{\sin x}{x}$); (g) signum function (sgn); (h) unit step function; (i) unit impulse function; (j) constant (dc).

The *auto-correlation* function refers to the similarity between a signal and its delayed version. The integral now operates on $[f_1(t)]^2$ so that the solution represents the total energy in the wave. Dividing this solution by the period T, yields the average power in the wave.

33. 2. 3 Discrete Fourier transform (DFT)

This transform can be used with discrete values as found in sampled data. If a waveform is converted into a digital format as a set of N consecutive samples, this

may be regarded as one cycle of a new periodic signal. The DFT for N samples can be computed from:

$$X_k = \sum_{n=0}^{N-1} x_n \exp\left(\frac{-j2\pi nK}{N}\right)$$

where both X_k and x_n may be complex and $k = 0, 1, 2, ..., N-1$.

The transform is most conveniently performed using digital signal processing (DSP) devices. However, using this direct application, the computation of which

Table 33.3 *Some important Fourier transform properties*

Operation	Signal	Spectrum
1. Transform	$f(t)$	$\int_{-\infty}^{\infty} f(t)\exp(-j\omega t)dt$
2. Inverse transform	$1/2\pi \int_{-\infty}^{\infty} g(\omega)\exp(j\omega t)d\omega$	$g(\omega)$
3. Complex conjugate	$f^*(t)$	$g^*(-\omega)$
(real signal)	$f^*(t) = f(t)$	$g^*(\omega) = g(-\omega)$
4. Symmetry	$f(t)$ real and even	$g(\omega)$ real and even
	$f(t)$ real and odd	$g(\omega)$ imag and odd
	$f'(t)$ real	$\mid g(\omega)\mid$ always even
5. Interchange	$g(t)$	$2\pi f(-\omega)$
6. Amplitude scaling	$Af(t)$	$Ag(\omega)$
7. Superposition	$Af_1(t) + Bf_2(t)$	$Ag_1(\omega) + Bg_2(\omega)$
8. Time shift (delay)	$f(t-T)$	$g(\omega)\exp(-j\omega T)$
9. Level shift	$A + f(t)$	$2\pi A\delta(\omega) + g(\omega)$
10. Frequency shift (translation)	$f(t).\exp(j\omega_0 t)$	$g(\omega - \omega_0)$
11. Time reversal	$f(-t)$	$g(-\omega)$
12. Differentiation	$\dfrac{d^n}{dt^n}[f(t)]$	$(j\omega)^n g(\omega)$
13. Product of signals	$f_1(t)f_2(t)$	$1/2\pi[g_1(\omega) \otimes g_2(\omega)]$
14. Product of spectra	$f_1(t) \otimes f_2(t)$	$g_1(\omega) g_2(\omega)$
15. Correlation	$\int_{-\infty}^{\infty} f_1(t)f_2^*(t-T)dt$	$g_1(\omega) g_2^*(\omega)$

[$\otimes$ = convolution]

Table 33.4 *Comparison of DFT and FFT*

N	DFT	FFT
8	64	24
16	256	64
32	1024	160
64	4096	384
128	16384	896
256	65536	2048

requires N^2 complex multiplications and additions, tends to be relatively slow.

33.2 4 Fast Fourier transform (FFT)

This transform uses algorithms designed to remove much of the redundancy in the calculation process of the DFT. This can be achieved by recognising that multiplication by sine and cosine produces many repeating values but only a relatively few unique ones. The discarded values have very little effect on the accuracy of the end result. Any algorithm that achieves this is described as a *fast* transform. Using such an algorithm it is possible to reduce the number multiplication and addition operations to $N \log_2 N$. As an indication of the improvement available, Table 33.4 makes comparison with the DFT.

The processing is carried out using dedicated DSP integrated circuits that are constructed around multiplying accumulator (MAC) stages. The design of these devices is well described in the references. Figure. 33.9 shows how part of such a device may be organised to perform four multiply and six addition operations on two complex inputs. The switching action gives rise to the descriptive title of butterfly operation.

33.2.5 Laplace transform

Many analytical time domain problems in communications engineering involve the solution of differential equations. By using a Laplace transform, the problems

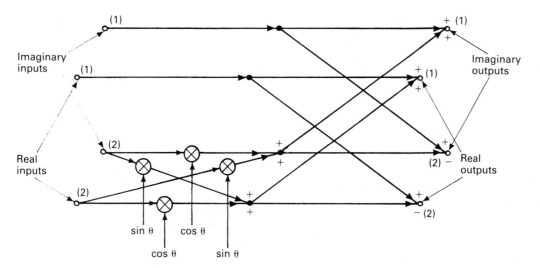

Figure 33.9 Computational 'butterfly'.

Table 33.5 Laplace transforms

$f(t)$	$F(T)$
a	a/s
$\exp(at)$	$1/(s-a)$
$\exp(-at)$	$1/(s+a)$
t^n	$n!/s^{n+1}$
$t^{n-1}/(n-1)!$	$1/s^n$
$t\exp(-at)$	$1/(s+a)^2$
$\sin \omega t$	$\omega/(s^2+\omega^2)$
$\cos \omega t$	$s/(s^2+\omega^2)$
$\sinh \omega t$	$\omega/(s^2-\omega^2)$
$\cosh \omega t$	$s/(s^2-\omega^2)$
$\exp(-at)\sin \omega t$	$\omega/[(s+a)^2+\omega^2]$
$\exp(-at)\cos \omega t$	$(s+a)/[(s+a)^2+\omega^2]$
$\mathcal{L}[af_1(t)+bf_2(t)]$	$a\mathcal{L}[f_1(t)]+b\mathcal{L}[f_2(t)]$
$\mathcal{L}[f'(t)]$	$s\mathcal{L}[f(t)]-f(0)$
$\mathcal{L}[f''(t)]$	$s^2\mathcal{L}[f(t)]-sf(0)-f'(0)$
$\mathcal{L}[f^n(t)]$	$s^n\mathcal{L}[f(t)]-s^{n-1}f(0)$
	$-s^{n-2}f'(0)...$
	$...-f^{n-1}(0)$

where $f(0)$ is the value of the function when $t=0$ and $f',f''...f^n$ are the 1st, 2nd ... nth derivatives.

can be converted into an algebraic form in a new domain where the solution is much simpler. Using the inverse Laplace transform then provides the solution in the time domain. The transform is obtained by multiplying $f(t)$ by $\exp(-st)$ and integrating the product with respect to time between the limits $t=0$ and $t=\infty$. Thus

$$\mathcal{L}f(t) = \int_0^\infty \exp(-st)f(t)\,\mathrm{d}t = F(s)$$

produces a new function in the s domain. If this is compared with the Fourier transform

$$g(\omega) = \int_{-\infty}^\infty f(t)\exp(-\mathrm{j}\omega t)\,\mathrm{d}t$$

the transforms are identical if $s=\mathrm{j}\omega$ and the integration is restricted to a period beginning at $t=0$, i.e. positive values of t only.

Due to the way in which the Laplace transforms are generated, these exist for many signals that have no Fourier transform. Table 33.5 lists the most important transforms in standard form and these may be used in reverse to provide the inverse transform.

33.2.6 Lapped transforms

Unlike conventional block transforms, lapped transforms (LT) allow consecutive information blocks to overlap before calculating the transform coefficients (see Fig. 33.10). LTs are therefore closely allied to filter banks. For continuous signals, the rectangular window (see Filters, p. 130) provides a superior resolution but is

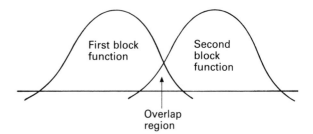

Figure 33.10 Lapped transform concept.

only reliable if some *a priori* information is available about the signal outside of the window. For image and audio signals this is not generally available. If the relative phase information can be retained, then a series of overlapping windows can be used with a fast transform so that the signal may be stored and later reconstructed with the minimum of error. The good complexity/performance trade-off of LTs can prove useful for image and audio signal processing.

33.3 Smith's charts (see also Transmission lines and waveguides, p. 378)

The major applications for the Smith's chart are in the solution of transmission line problems and the design of impedance matching networks. Although this graphical method yields less accurate results than computer-aided design, within the limits of typical component value tolerance, it is accurate enough for most practical applications.

33.4 *z* and chirp-*z* transforms

These are techniques that have been developed for handling sampled data using digital signal processing (DSP) and computer-aided system evaluation. The Laplace transform of a train of impulses is given by:

$$\mathcal{L}[f(t)] = A_0 + A_1 \exp(-sT) + A_2 \exp(-2sT) \dots + A_n \exp(-nsT)$$

where A represents the pulse amplitudes and T the time duration between pulses. If the substitution $z = \exp(sT)$ is made, then the expression becomes:

$$\mathcal{L}[f(t)] = A_0 + \frac{A_1}{z} + \frac{A_2}{z^2} \dots + \frac{A_n}{z^n}$$

which is described as the z transform of $f(t)$.

Figure 33.11 shows how a set of points M, representing sampled values in the s-plane are mapped into the z-plane. In general, straight lines in the s-plane map into spirals in the z-plane using the following transform equations:

$$z_K = A W^{-1} \qquad K = 0, 1, 2 \dots M - 1$$
$$W = W_0 \exp(j2\pi\phi_0)$$
$$A = A_0 \exp(j2\pi\theta_0)$$

The start coordinates of the spiral are defined by A_0 and θ_0, with the rate of decay to or from the centre and the angular spacing between samples by W_0 and ϕ_0. These parameters can be varied under software control to define the start and end frequencies. This provides the facility to zoom in or out on frequencies of particular interest and so display peaks and troughs in a waveform that might otherwise be masked.

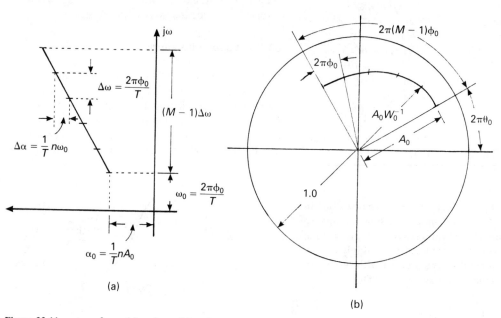

(a)

(b)

Figure 33.11 *z* transform: (a) *s*-plane; (b) *z*-plane.

The interior of the *z*-plane unit circle corresponds to the left-hand half of the *s*-plane, while the exterior corresponds to the right-hand half.

The title chirp-*z* derives from the fact that *W* represents a complex sinusoid of increasing frequency, a waveform similar to that used in *chirp* radar systems.

33.5 Miscellaneous terms

Aperiodic. Non-periodic; the pulse pattern does not repeat after a given period.

Causal. A relationship between cause and effect. A system is described as causal if its response cannot be anticipated.

Crest factor. Defined as the ratio of peak to rms value of a periodic wave.

dc value (RMS value). That value of a waveform that produces the same power or heating effect as an equivalent dc source.

Discontinuous. A waveform whose level changes with a jump at a given instant. At this time the wave has an infinity of level values.

Duty cycle. The ratio of the positive duration of a waveform to its periodic time.

Orthogonal. Two waveforms are said to be orthogonal if they contain no common components. Mathematically, the condition is defined from the following integral:

$$\int_{t_1}^{t_2} f_m(t) f_n(t) \, dt = 0 \text{ for } m \neq n$$

and non-zero for $m = n$.

Periodic. A waveform that repeats over a given period of time.

Time constant. Although the overall response curve of a *CR* or *LR* circuit is exponential, the initial rate of change of voltage/current is practically linear. If this rate of change continued, then the voltage or current would have reached the maximum or zero value in a time equal to *CR* or L/R seconds. However, due to the exponential shape of the response, the level obtained in this time is only 63% or 37% of the final or initial values respectively.

Time invariant. If a network or system produces a particular response output due to an input at time t_1 and then produces an identical output for the same stimulus at time t_2, the system is said to be time invariant.

33.6 Useful references

Coates R.F.W. (1985) *Modern Communications Systems*, 2nd edition, Macmillan, London.

Connor F.R. (1975) *SIGNALS – Introductory Topics in Electronics and Telecommunications*, Edward Arnold, London.

Karwoski R.J. (1980) *Introduction to Digital Spectrum Analysis and High Speed FFT Processor Design*, TRW Products Inc. El Segundo, California.

Kuo F.F. (1966) *Network Analysis and Synthesis*, Wiley, New York.

Malvar H.S. (1992) *Signal Processing with Lapped Transforms*, Artech House, London.

Oran B.E. (1974) *The Fast Fourier Transform*, Prentice Hall, New Jersey.

Schwartz M. (1980) *Information Transmission, Modulation and Noise*, 3rd edition, McGraw-Hill, Tokyo.

Slater J.N. and Trinogga L.A. (1986) *Satellite Broadcasting Systems*, Ellis Horwood, Chichester.

Stroud K.A. (1978) *Laplace Transforms*, Stanley Thornes, Cheltenham.

34 Spread spectrum techniques

A spread spectrum (SS) communications system is one in which the transmitted frequency spectrum is much wider than absolutely necessary. Amplitude modulation (AM) is usually considered to be a minimum bandwidth technique, thus by comparison, wideband frequency modulation (FM) represents an analogue example of a spread spectrum system. FM can be shown to have a signal to noise (S/N) ratio advantage over AM of $3\beta^2 F$, where β is the deviation ratio and F is the ratio of peak deviation to baseband width.

Similarly the transmission bandwidth of a digital signal can be expanded so that many bit periods are necessary to represent one bit at baseband. Such a system has a processing gain expressed as the ratio of transmission code rate to original information bit rate, which leads to a reduction in transmission errors, which is in turn, equivalent to an improvement in S/N ratio.

This trade-off of increased bandwidth for improved S/N ratio, is a good example of the applications of the Hartley/Shannon laws of information theory, to combat the effects of noise, interference, fading and other variations in propagation conditions.

34.1 Spread spectrum systems

Digital SS systems are used extensively in military (covert operations) and satellite systems for high security reasons. A low level wideband SS signal, can easily be hidden within the same spectrum as a high power television signal where each signal appears to be noise to the other.

Code division multiple access (CDMA) techniques can be adopted to allow several transmissions to occupy the same channel simultaneously. This in turn leads to an improvement in the spectrum utilisation.

The technique can also be applied to communications over mains electrical power wiring and local area network (LAN) systems. Historically, frequency and time division multiplex (FDM and TDM) systems have been used in order to achieve maximum use of the frequency spectrum. For the future, code division multiplex (CDM) systems employing spread spectrum techniques will further improve the utilisation of this limited resource.

At the heart of all digital spread spectrum systems there is a pseduo-random binary sequence (PRBS) or similar derived code. There are two basic ways in which these are employed, with several variations of each plus hybrid versions. For the *direct sequence* SS systems, either the original baseband bit stream is multiplied by the PRBS to produce a new bit stream, or the PRBS and its inverse are used to represent logic 1 and 0, respectively. For *frequency agile* or *hopping* systems the PRBS is used to make the RF carrier frequency hop around within the spread spectrum bandwidth. In both cases, only those receivers equipped with the correct PRBS can decode the original message. While longer sequences provide greater security, it takes much longer to obtain synchronism at the receiver. By using a second PRBS to select truncated versions of the first, the security of longer codes can be achieved with shorter synchronism times.

Basically, these techniques cause the energy in the baseband signal to spread across a much wider transmission bandwidth as indicated in Fig. 34.1. At the receiver, the low level wideband signal will be accompanied by noise, and by using a suitable detector/demodulator, this signal can be squeezed back into the original narrow baseband. Because noise is completely random and uncorrelated, the wanted signal can easily be extracted.

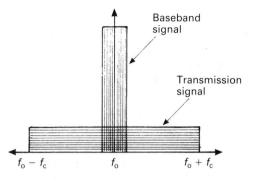

Figure 34.1 Signal spectra.

34.1.1 Direct sequence spread spectrum (DSSS) systems

The smallest duration of the code sequence, the PRBS clock period, is usually referred to as a *chip*. Each data bit is normally an integral number of chips in length so

that the data transitions all coincide with clock pulse edges.

With a carrier frequency of f_0, a chip duration of t_c, a chip frequency of $1/t_c$ and a sequence L chips long, the resulting spectrum is centred on f_0 and consists of discrete spectral lines at $1/Lt_c$ Hz intervals. The envelope distribution is given by $E(f) = (t_c/2)$ $(\operatorname{sinc} x)^2$, where $x = \pi f t_c$ and $\operatorname{sinc} x = (\sin x)/x$. Theoretically, more than 90% of the transmitted energy lies within the band $f_0 - f_c$ and $f_0 + f_c$.

For a carrier frequency in the low gigahertz range and modulated BPSK or QPSK (bi or quadri phase shift keying), f_c may be as high as 10 MHz. Figure 34.2(a) shows the basic arrangements made for coding and modulation at the transmitter.

At the receiver, the received signal is multiplied by the correct PRBS. This concentrates the energy back into the original baseband so that the demodulator's output S/N ratio is much higher than that at the input due to the processing gain. For code sequences that occupy a single data bit period, this gain is equal to L.

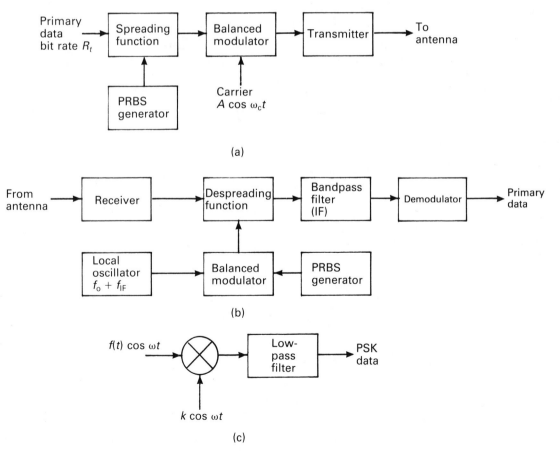

Figure 34.2 Direct sequence spread spectrum system: (a) transmitter; (b) receiver; (c) PSK demodulator.

Figure 34.2(b) shows how the received signal is processed. The simplest receiver, shown at Fig. 34.2(c), compares all the possible phase relationships between the received signal and the correct PRBS by allowing the clock signals to slip until a match is obtained. This technique is described as correlation detection or code tracking.

34.1.2 Frequency hopping spread spectrum (FHSS)

Figure 34.3 indicates how a frequency synthesiser is used to select the various carrier frequencies under the control of a PRBS generator at both transmitter and receiver.

If Δf is the baseband width and the separation between N discrete frequencies within the allotted bandwidth, the processing gain is the ratio of RF bandwidth to the message baseband width and equal to N. The RF bandwidth is equal to $f_n - f_1$ plus twice the baseband width.

FHSS systems are basically simpler to implement than DSSS ones but generally have a slower transmission code rate (about 200 Kbit/s) due to the relatively slow frequency synthesiser, but with a faster acquisition time.

34.1.3 Hybrid and alternative systems

Hybrid systems are the result of compromises that are used in order to combine the various merits of DSSS and FHSS systems and to achieve a simpler system implementation. For maximum processing gain, DSSS systems must have a very high chip rate, while FHSS systems must have a large number of channels. The overall gain of the hybrid system is then the sum of the DSSS and FHSS gains.

Also employed are time hopping systems using rapid bursts of carrier frequency, with each burst having a PRBS duration, and pulsed FM or chirp systems where the carrier frequency is swept over a wide frequency range.

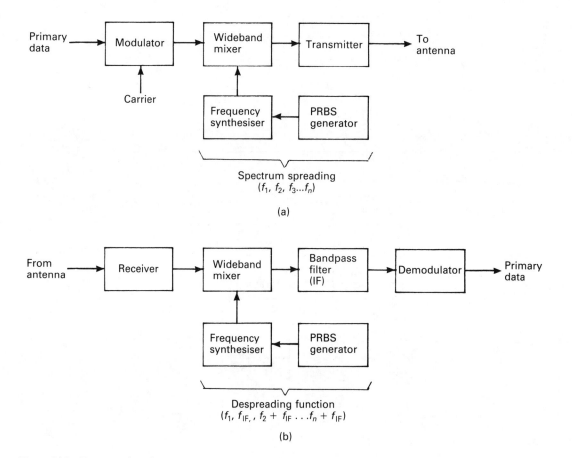

Figure 34.3   Frequency hopping spread spectrum system: (a) transmitter; (b) receiver.

In one example, a low frequency baseband is sub-divided into 10 channels and these are alternately allocated five for logic 1 and five for logic 0. To send a single bit, the system transmits on each of the five appropriate frequencies for a fixed time period. The receiver uses 10 filters from which the output is accumulated at the end of each bit period. The receiver then detects a logic 0 or 1 by the transmission of the five appropriate frequencies plus the absence of a signal in the other five channels. This low bit rate, wideband, low error rate system allows the use of low power, with good noise and interference rejection properties. The system is capable of accurately receiving data at signal strengths below the general noise level.

34.2 Demodulation (see also Modulation and demodulation (digital), p. 197)

The PRBS, which is the key to accurate demodulation, must be sent to all authorised users before the message. This can be handled in the same way as the keys for the various encryption systems.

34.2.1 Correlation detection

Table 34.1 indicates the manner in which the received bit stream and the PRBS are scanned in order to detect synchronism using the correlation function.

Correlation detection is achieved by replacing logic 0 with −1 and then multiplying the bit stream and PRBS bit-by-bit as shown. The bit sum at the end of each step is only maximum when the two signals are correlated.

34.2.2 Costas loops (after J.P. Costas)

Figure 34.4(a) indicates how the use of a phase lock loop (PLL) and three multipliers are used for BPSK and QPSK demodulation in the Costas loop. This diagram also shows how the mathematical equations develop at each stage and the effects of the low-pass filters in removing the unwanted harmonic components. The output signal $\pm A/2(\cos \phi) = \pm A/2$ when $\phi = 0$ and this represents the bi-polar signal. As shown in this diagram, the Costas loop will also function as an analogue FM demodulator. The main disadvantage of this demodulator is that it does not provide an *amplitude* output to drive a *signal present* or *in lock* detector or a *coherent* AGC system within the receiver. This problem can be overcome by adding the circuit shown in Fig. 34.4(b) in parallel with the loop. The output signal
$$A^2/4[\cos(0 + 2\phi) + \cos 90°] = A^2/4[\cos 2\phi].$$
When the system is locked, $\cos 2\phi = 1$, and the output amplitude is simply the dc level $a^2/4$. As the receiver drifts out of lock, $\cos 2\phi$ rapidly approaches zero as does the output dc level.

34.3 Miscellaneous terms

Gold code sequences. These represent a family of pseudo-random sequences that were defined by R. Gold in 1967 as having a low cross-correlation between each other. As such they can occupy the same signalling channel without producing destructive interference. Because of these properties, Gold codes find applications in code division multiple access (CDMA) or spread spectrum multiple access (SSMA) systems. Each code sequence is produced from a pair of PRBS generators each with n stages by modulo-2 addition. If all the $2^n + 1$ Gold sequences of period $2^n - 1$ are concatenated, the resulting sequence period becomes $(2^n + 1)(2^n - 1) = 2^{2n} - 1$, the same as would be generated by a shift register with $2n$ cells.

Pseudo-noise. An alternative title for a pseudo-random binary sequence.

Table 34.1 *Correlation detection*

Bit stream	−1	−1	−1	1	1	−1	1	1	Left shift
PRBS	1	−1	−1	−1	1	1	−1	1	
Product	−1	1	1	−1	1	−1	−1	1	Sum = 0
Bit stream	−1	−1	−1	1	1	−1	1	1	Synchronised
PRBS	−1	−1	−1	1	1	−1	1	1	
Product	1	1	1	1	1	1	1	1	Sum = 8
Bit stream	−1	−1	−1	1	1	−1	1	1	Right shift
PRBS	−1	−1	1	1	−1	1	1	−1	
Product	1	1	−1	1	−1	−1	1	−1	Sum = 0

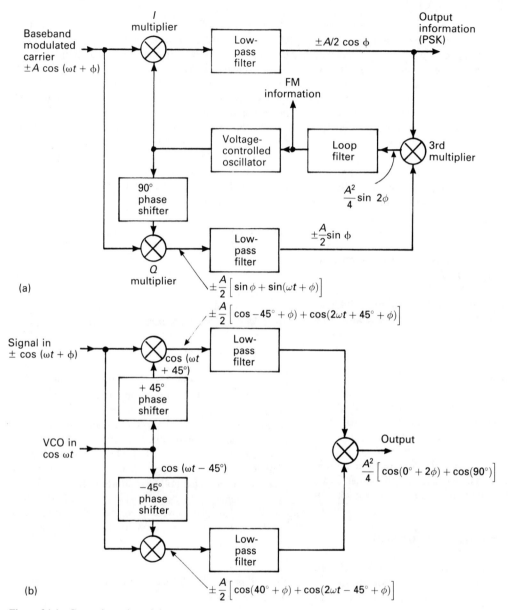

Figure 34.4 Costas loop demodulator.

RADAS (random access discrete address system). An early digital development in individual addressable receivers. Each receiver within the system was allocated a unique binary code that served as the address and this was used in the manner of the later DSSS systems. The message originator or operator was provided with a look-up table that contained the code for any particular receiver on the system.

Self-noise (of system). Noise-like characteristics generated by other transmissions within the spread spectrum system.

34.4 Useful references

Dixon R.C. (1976) *Spread Spectrum Systems*, Wiley, New York.
Gardner F.M. (1979) *Phaselock Techniques*, Wiley, New York.

35 Standards organisations and associated bodies

In the technical sense, the specification, of standards are important and are designed to:

- ensure the interconnection between different users;
- allow the portability of equipment between different applications and in different areas;
- ensure that equipment purchased from one vendor can be interfaced with that from another.

In addition, standards can also be important to the health and wellbeing of both users and operators.

Basically there are two types of standards; those that have been developed through lengthy international discussions, and those known as *de facto* standards. These latter very often develop locally because no previous standard exists. As the systems to which these apply become more widely accepted and the technology spreads, the manufacturers' standard becomes an acceptable alternative. At a later stage, these limited agreement standards might well become the basis of an international standard.

Historically within the telecommunications industry, each national post, telegraph and telephone (PTT) authority had a monopoly and tended to set local or national standards. With the widespread need for global interoperability, however, it was quickly disovered that international standards were an economic necessity. To allow for some flexibility, the international bodies now tend to produce *base standards* from which a number of variants or functional standards can develop to meet the rapidly changing needs of industry.

Telecommunications standards organisations are now well over 100 years old. In 1865, the Union Telegraphique was formed to coordinate and standardise the telegraph industry. In 1947 this became the International Telecommunications Union (ITU), as a special agency of the United Nations charged with the task of managing the standards for the whole field of communications. Today, the ITU consists of more than 165 member nations.

In 1926, the International Standards Association was formed and this converted to the International Standards Organisation (ISO) in 1954. The International Electrotechnical Commission (IEC) was formed in 1904. Both the ISO and the IEC are world, private, voluntary, non-governmental and scientific organisations, affiliated to each other and the United Nations Organisations (UNO). Together they represent at least 80% of the world population. The principal members of the ISO and IEC are the national standards institutions such as BSI (UK), DIN (Germany), ANSI (USA) and JISC (Japan). All the ISO/IEC standards are published as being international but it is the responsibility of the national organisations to apply these to the local needs, no member state being forced to adopt and enforce a recommendation.

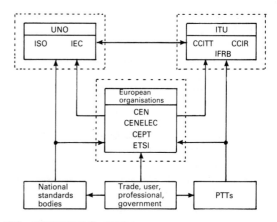

UNO = United Nations Organisation
IEC = International Electrotechnical Commission
ISO = International Standards Organisation
ITU = International Telecommunications Union
IFRB = International Frequency Registration Board
CCITT = Consultative Committee for International Telegraph and Telephones
CCIR = Consultative Committee for International Radio
CEN = European Committee for Standardisation
CENELEC = European Committee for Electrotechnical Standardisation
CEPT = Conference of European Posts and Telecommunications
ETSI = European Telecommunications Standards Institute
PTTs = Posts, Telegraph and Telephones (Ministry for)

Figure 35.1 Standards-making hierarchy (Europe).

In the main, the ISO is responsible for information technology, while the IEC looks after the interests of electrical and electronic engineering. Due to the obvious overlap of interests, both bodies work in close liaison. The official languages of the ISO/IEC are English, French and Russian.

The ITU has four permanent associated bodies, a General Secretariat, CCITT, CCIR and IFRB (see Fig. 35.1). These control such conferences as WARC (World Administrative Radio Conference) and WATTC (World Administrative Telephone and Telegraph Conferences).

In Europe, organisations such as CEN, CENELEC, CEPT, ETSI, etc., all liaise with the national standards bodies, the agencies representing the trade, user groups, professional, governmental and PTTs. Figure 35.1 gives an indication of this very close liaison and interworking.

35.1 Organisations

ABSOC (Advanced Broadcasting Systems of Canada) or SYRPOC (SYstèmes de Radiodiffusion de POinte du Canada)

This body was founded in 1990 to coordinate the introduction and testing of advanced broadcasting systems into Canada. It is composed of members from all sides of the industry, including cable and satellite interests and government agencies. It replaces and builds on the work and expertise of the Canadian Advanced Broadcasting Systems Committee that was created in 1988 with a similar remit.

ABSOC, 280 Albert Street, 10th Floor, Ottawa, Ontario K1P 5G8, Canada.

Tel. 613 236 5850. Fax. 613 236 9241.

ANSI (American National Standards Institute)

ANSI was founded in 1918 as a privately funded and non-profitmaking organisation, by five professional/technical societies and three government agencies. It was charged with the coordination and approval by consensus, of voluntary standards prepared by such bodies as EIA, IEEE and SMPTE, etc. within the United States. Until 1946 when ANSI became a member of the ISO, most of the work had been associated with the electrotechnical field. ANSI is now one of the leading American standard organisations and member of both the ISO and IEC.

American National Standards Institute, 1430 Broadway, New York, NY 10018, USA.

Tel. 212 642 4995.

AT&T Bell Laboratories (American Telephone & Telegraph Co.)

The prominent role in research and development (R&D) of this international organisation has ensured that many of its products form the basis of industry *de facto* standards. As a result of this, Bell Labs provides a very significant input to the national and international standards-making bodies.

AT&T Bell Laboratories, 600 Mountain Ave, Murray Hill, New Jersey, 07974, USA.

Tel. 201 582 3000.

BNSC (British National Space Centre)

BNSC was formed in 1985 to direct UK civil space activity and develop future policy. It coordinates the civil space interests of the government, including the Department of Trade and Industry, Ministry of Defence, Scientific and Engineering Research Council and the Natural Environmental Research Council. Within this remit, it makes representations to the related national standards-making bodies.

BNSC, Millbank Tower, Millbank, London SW1P 4QU, UK.

Tel. 017 211 3000. Fax. 071 821 5387.

BSI (British Standards Institution)

BSI is the national standards body which provides technical specifications for products and processes which are used throughout industry as a basis for economic production. BSI also represents the UK viewpoint at international and European discussions on standards. It is an independent body operating under Royal Charter.

BSI also offers extensive services in testing, quality assurance and technical advice to exporters.

In telecommunications, BSI works for the government in preparing British Standards intended to be designated under the Telecommunications Act 1984. These standards are used as a basis for testing apparatus which, following approval by the appropriate authority, can be connected to the telecommunications network or service.

Internationally, BSI represents the UK at discussions in the European Electrotechnical Standards Bodies, CEN and CENELEC, which are also concerned with the safety aspects of telecommunications standardisation. On behalf of the Department of Trade and Industry, BSI provides a forum for discussion to establish the UK view to be presented in CEPT/ETSI on draft European Telecommunications Standards known as NETS.

BSI, 2 Park Street, London W1A 2BS, UK.

Tel. 071 629 9000. Fax. 071 629 0506.

BTA (Broadcasting Technology Association)

BTA was formed by a small number of broadcasters and broadcast equipment manufacturers in 1985 in order to coordinate the R&D work of the industry for the national benefit. Today, the organisation has about 50 members and provides the industry's input to the Japanese standards-making organisations such as EIAJ and JISC.

B.T.A. (Japan), Daini Doi Building, 4F 2-8-12 Nishi-shinbashi Minato-ku, Tokyo 105, Japan.

Tel. 03-503-2531. Fax. 03-503-2533.

CCETT (Centre Commun d'Etudes de Télédiffusion et Télécommunications)

The CCETT is a centre for applied research in the wider field of audio-visual communication services. It has been set up as a result of cooperation between CNET, the national research centre of France Telecom and TDF, the French broadcasting organisation. The main areas of research include:

- broadband services associated with HDTV and cable and telematic networks;
- multi-media services via the ISDN systems;
- broadband networks for future services;
- the development of high performance modulation and coding systems;
- the development of encryption and security systems.

CCETT thus presents the manufacturers' and users' viewpoints to the national and European standards bodies.

C.C.E.T.T., 4 rue du clos Courtel BP 59, 35512 Cesson-Sevigne, Cedex – France.

Tel. +33 99 02 41 11. Fax. +33 99 02 40 98.

CCIR (Comité Consultatif International des Radiocommunications. International Radio Consultative Committee)

The CCIR is the organ of the ITU which is concerned with the establishment of internationally agreed system and technical characteristics for the compatible inter-working of radio communications and allied services and the efficient use of the radio frequency spectrum. The services include sound and television broadcasting, fixed and mobile radio, standard frequencies and time signals, and many other associated services. The resulting conclusions of the CCIR are therefore essentially performance specifications for the services, including bandwidth, signal and modulation character-istic, rather than manufacturing and similar specifica-tions and standards.

The conclusions of the CCIR are reached at Plenary Assemblies which are held at about 4-yearly intervals.

The adopted texts that represent Recommendations to member nations are published in Volumes after the close of each Plenary.

Due to the continuous convergence of the interests and technology of radio and telecommunications, the work of the CCIR and the CCITT is to be progressively merged. It is expected that the CCIR will have been phased out by 1995.

CCIR, 2 Rue de Varembe, 1211 Geneve 20, Switzerland. Tel. +41 22 99 51 11. Fax. +41 22 33 72 56.

CCITT (Comité Consultatif International Téléphonique et Télégraphique. Consultative Committee for International Telephone and Telegraph)

The CCITT is the organ of ITU that is charged with the task of establishing the global standards for telecommu-nications. These are produced as Recommendations to be implemented by member countries. The work proceeds in 4-yearly cycles known as Study Periods and each ends with a Plenary Assembly of the whole CCITT. The results of these deliberations are then published in *coloured* books; Red book for 1984 and Blue book for 1989.

The Recommendations produced by CCITT are organised into a number of Series of Recommenda-tions, with each series covering some major topic. For example:

- E Series Recommendations cover international tele-phone services;
- I Series Recommendations cover ISDN; and
- V Series Recommendations cover data communica-tions over telephone networks.

A new Study Period starts at the end of each Plenary when a set of questions is allocated to each of the various Study Groups. The answers should result in a set of Proposed Recommendations for the following Plenary Assembly. After suitable discussion and test-ing, these Recommendations are published.

CCITT, Place des Nations 1201, Geneve, Switzerland. Tel. 022 99 52 85. Telex. 421 000.

CEN-CENELEC (Comité Européen de Normalisation–Comité Européen de Normalisation Electrotechnique)

Both CEN and CENELEC are standards writing bodies rather than certifying agencies. CENELEC is chiefly responsible for electrotechnical matters while CEN covers all other areas. The organisations function separately except in the area of information technology when they combine to form JESI (Joint European Standards Institute).

CEN and CENELEC have direct links to the ISO and EIC, respectively, and their outputs are referred to as Europaische Norm (EN) or European Standards. The official languages of both bodies are English, French and German.

A separate arm CECC (CENELEC Electronic Components Committee) has been set up to look after the interests of the electronic components industry.
European Committee for Electrotechnical Standardisation, Rue Brederode 2, Bte. 5-1000 Bruxelles, Belgium. Tel. 025 11 79 32. Telex. 26257 CENELEC B.

CEPT (Conference Européene des Administrations des Postes et Télécommunications)

This group of PTT operators from 26 member countries including the European Community, the European Free Trade Area, Turkey and Yugoslavia was set up in 1959 to establish closer links between administrations and to improve technical cooperation. Through its various sub-committees it became heavily involved in the Pan-European digital cellular radio network, the broadband EURONET network and mandatory standards. Until its amalgamation with ETSI in 1989/90, CEPT made representations to both the CCIR and CCITT.
CEPT Liaison Office, Case Postale 1283, CH-3001, Berne, Switzerland.

CIE (Commission Internationale de L'Eclairage, International Commission on Illumination)

This technical, scientific and non-profitmaking organisation is devoted to international cooperation and the exchange of information relating to all matters associated with the science and art of lighting. Its objectives are:

- to develop basic standards and procedures of metrology;
- to provide guidance in the principles of the development of applicable international standards;
- to prepare and publish standards, reports and publications associated with the science and art of both indoor and outdoor lighting;
- through the provision of an international forum, maintain liaison with other international organisations that are concerned with the art and science of light and lighting.

These objectives embrace such fundamental subjects as vision, photometry and colorimetry, involving natural and man-made radiations ranging from infra-red through to ultraviolet wavelengths.

The management is vested in about 40 National Committees which are expected to be fully cognisant with the standards of regional, national and international bodies to which the CIE provides an appropriate input.
CIE Central Bureau, PO Box 169, Kegelgasse 27, A-1033 Wien, Austria.
Tel. 0222 75 31 87/88. Telex. 111 151 cie a.

DIN (Deutsches Institut für Normung)

This German National standards body represents manufacturing, consumers, service industries, scientific and technical inspection, quality assurance and the federal and local authorities.

The standards are voluntary and DIN endeavours to restrict the number to a minimum. It also ensures that the standards reflect what is achievable in reality and that no two standards have conflicting conditions.

With its wide ranging viewpoint, DIN provides inputs to all the European and International standards-making bodies.

The main source for reference to the DIN technical regulations and standards in Germany is Beuth Verlag, Postfach 1145, Berlin 30, Germany.
Deutsches Institut fur Normung, Burggrafenstrasse 6, Postfach 1107, D-1000, Berlin 30, Germany.
Tel. 030 26 01 600. Telex. 185 269 ditr d.

EAN (Association Internationale de Numerotation des Articles. International Article Numbering Association)

This organisation, known as EAN, was developed from the European Article Numbering Association formed in 1977. The title was changed in 1981, to reflect the international status that was acquired due to an expanding membership from other countries and continents. The EAN abbreviation was retained to identify the numbering and symbol marking system. As a non-profitmaking organisation, EAN sets out to promote the development of a worldwide system of identification of goods by a number which can be expressed in machine readable symbols, together with harmonisation of the applications in which this number is used. EAN standards also ensure that the rules for defining such numbers are applied in a consistent manner.

The association represents its members at international associations and institutions, and provides a forum for the liaison and negotiation of the extension of its aims. It also plays a role in the development of information technology and networking concepts as applied to sales and check-out operations. Apart from this, EAN has no direct links with the international standards setting bodies of telecommunications. The EAN system is compatible with Universal Product Code (UPC) used in North America.

The UK representative organisation is the ANA (Article Numbering Association), 6 Catherine Street, London WC2B 5JJ, UK.
Tel. 071 836 2460. Fax. 071 836 0580.
Association Internationale de Numerotation des Articles, Rue des Colonies 54 – Bte 8 – B-1000 Bruxelles, Belgium. Tel. 02 218 76 74. Fax. 02 218 75 85.

EBU (European Broadcasting Union)
The EBU, which was founded in 1950, owes its origins to the International Broadcasting Union formed in 1925. It was formed to promote international cooperation and developments in broadcasting, by coordinating European interests and expertise. As an independent, non-governmental, non-profitmaking body, it works closely with the principal international bodies such as the United Nations, the Council of Europe, the International Telecommunications Union (ITU) and its Consultative Committees, the United International Bureaux for the Protection of Literary and Artistic Property (BIRPI), and the various satellite communications agencies.

Although consisting essentially of all the European states, plus those countries bordering the Mediterranean, the EBU has more than 40 associated members from countries around the world.

The EBU is responsible for the management of the Eurovision network for programme distribution and also a number of working party groups engaged on technical studies. These latter are involved in R&D associated with sound and television broadcasting, recording (both film and magnetic), high definition television and the development of technical training. In many cases, the proposals from these groups are submitted to the international standards bodies with a view to worldwide adoption. Due to the political changes that are occurring in Europe, the services provided by OIRT, are being progressively merged with those of the EBU.

Both the secretariat and the technical division of the EBU are now located in Geneva.
E.B.U., 17a Ancienne Route, Case Postale 67, CH-1218 Grand-Saconnex, Geneve, Switzerland.
Tel. 41 22 798 77 66. Fax. 41 22 798 58 97.

EIA (Electronic Industries Association)
The EIA, which was formed in 1924, functions essentially as a trade association looking after the wide-ranging interests of the North American electronics industry. It is organised in groups to manage the specific interests of components, consumer and industrial electronics, information technology, government services and marketing. A significant engineering support group assists with the development of standards and specifications for the industry in general, with a particular interest in the problems that arise from interchangeability. The EIA is managed via a Board of Governors through various group committees and annually publishes a valuable listing of all Recommended Standards (RS).
E.I.A., 2001 Eye Street NW, Washington DC, 20006, USA.
Tel. 202 457 4966. Fax. 202 457 4985.

ESA (European Space Agency)
The membership of ESA consist of the majority of European states plus a number of non-European associates. The purpose of the organisation is to support and promote, for exclusively peaceful purposes, cooperation among European States in space research, technology and applications.

At regular intervals the member states propose plans for the future direction of European space research. ESA also plays a coordinating role by attempting to integrate the national programmes of members into its own programmes.

ESA is essentially an R&D organisation but does not develop or manufacture its own spacecraft. The definition stage of a programme is carried out by ESA engineers and scientists, but the manufacturing element is contracted out to suitable European industrial organisations. In this respect, ESA has a direct interest in developing manufacturing standards for space-craft as well as those for telecommunications.

ESA activities are: either mandatory, where all member states contribute to the costs in proportion to the average national income; or optional, when the members' contributions reflect their interest in a particular field of activity.

ESA consists of four major segments; a Headquarters in Paris, a Research Centre in Noordwijk, an Operations Centre in Darmstadt and a Data Retrieval Centre at Frascati. In addition, the Ariane Spacecraft launch base is located in Kourou in French Guiana.
ESA Headquarters, 8–10 Mario-Niki -75738, Cedex 15, Paris, France. Tel. 331 42 73 76 54.
European Space Operations Centre (ESOC), Robert-Bosch-Strasse 5, 6100 Darmstadt, Germany.
Tel. 06151 88 66 02. Telex. 419453.
European Space Research and Technology Centre (ESTEC), Keplerlaan 1, Postbus 299, 2200 AG Noordwijk, Netherlands. Tel. 01719 86555.

ETSI (European Telecommunications Standards Institute)
ETSI was founded in 1988 as a result of an initiative of the European Commission. It was established to produce telecommunications standards by democratic

means, for users, manufacturers, suppliers, administrations, and PTTs. ETSI's main aim is the unrestricted communication between all the member states by the provision of essential European standards.

It is now an independent, self-funding organisation that has progressively merged since 1990, with both CEPT and the EBU. The Joint Technical Committee (JTC) therefore presents a common viewpoint to the CEN–CENELEC standards bodies

E.T.S.I. Route des Lucioles Sophia-Antipolis, B.P. 152–06561 Valbonne Cedex, France.

Tel. +33 92 94 42 00. Fax. +33 93 65 47 16.

EUMETSAT (EUropean METeorological SATellite Organisation)

This organisation operates as the European contribution to the World Weather Watch (WWW) and World Meteorological Organisation (WMO), which in turn operates under the auspices of the Coordination of Geostationary Meteorological Satellites (CGMS) body. This, considering the political situation in 1993, must represent a prime example of international cooperation.

EUMETSAT is an independent, inter-governmental organisation that was established by agreement with most European countries in 1986. Its aim is to establish, maintain and exploit European systems of operational meteorological satellites. A prime task is to ensure the continuation of Meteosat, the European meteorological satellite system developed by ESA in 1977. The Meteosat series of satellites is operated for EUMETSAT by the European Space Operations Centre (ESOC), an ESA establishment.

EUMETSAT, Am Elfengrund 45, D-6100 Darmstadt-Eberstadt, Germany.

Tel. +49 6151 5392-0. Fax. +49 6151 5392-25

EUREKA

This is basically an enabling organisation set up by the Commission of the European Communities (CEC) in 1985 to promote and fund research, development and cooperation for the implementation of advanced technologies. Already, this has resulted in a significant number of new and improved standards within the broad field of communications.

The EUREKA Secretariat, 19H Avenue des Arts, Bte 3, B-1040 Bruxelles, Belgium.

Tel. 322 217 00 30. Fax. 322 218 79 06.

EUTELSAT (EUropean TELecommunications SATellite Organisation)

As the result of an initiative of CEPT, this organisation was set up by 17 European countries during 1977 to establish and operate a telecommunications satellite system. The initial agreement provided for two space segments: one for fixed services and the other for maritime mobile services. This latter segment was later transferred to INMARSAT.

ESA is responsible for the procurement, launch and in-orbit maintenance of the satellite system. EUTELSAT either owns or leases the space segments that it uses to provide telephone, television and business data services.

Membership of EUTELSAT is open to any European states that also belong to the ITU and CEPT. Orbital locations and operational frequencies are allocated by the IFRB.

EUTELSAT, Tour Maine-Montparnasse, 33 Avenue du Maine, 75755 Paris, Cedex 15, France.

Tel. 331 45 38 47 47. Fax. 331 45 38 37 00.

FCC (Federal Communications Commission)

The FCC is an independent United States government agency, responsible directly to Congress. The Radio Act of 1927 created the Federal Radio Commission with regulatory powers over radio services.

By the Communications Act 1934, this became the FCC which was charged with regulating interstate and international communications by radio, television, wire, satellite and cable.

The commission is organised by function and operates four bureaux: Mass media, Common carrier, Field operations and Private radio. Its functions include:

- monitoring the radio spectrum to check that technical requirements are being met;
- inspecting stations of all types;
- conducting operator examinations;
- issuing permits or licences to qualified persons;
- locating and closing down unauthorised transmitters;
- providing radio bearings for aircraft or ships in distress;
- locating sources of interference and suggesting remedial measures;
- carrying out special engineering work for other government agencies;
- obtaining and analysing technical data for Commission use.

F.C.C., 1919 Main Street NW, Washington DC, 20554, USA.

Tel. 202 653 8102.

Hayes Microcomputer Products Inc.

During 1978, this company was formed to develop, manufacture and market modems for use with personal computers. At that time, this was a new development and no suitable standards existed. Hayes therefore produce their own standards of interconnection, command signals and operating protocols. The rapid expansion of the concept led to the acceptance of the Hayes standards by other manufacturers and so these became an industry *de facto* standard.

Hayes modems have a built in intelligence due to the use of a microprocessor which allows the system to be controlled by software from the computer keyboard. This replaced the previous alternative of setting switches or jumper links to select the wanted system parameters. The operating protocol is known as the AT Command Set.

Hayes Microcomputer Products Ltd, 1 Roundwood Avenue, Stockley Park, Uxbridge, Middlesex UB11 1AH, UK.

Tel. 081 848 1858. Fax. 081 848 0224.

Hayes Microcomputer Products Inc, PO Box 105203 Norcross, Atlanta, Georgia, USA 30348.

Tel. 404 449 8791. Telex. 703500.

IEC (International Electrotechnical Commission) or CEI (Commission Electrotechnique Internationale)

The IEC is an organisation which was formed as a result of the resolution of the Chamber of Government Delegates at the International Electrical Congress of St Louis (USA) in September 1904. The first statutes which were drawn up at the Preliminary Meeting held in London in 1906 were adopted in 1908.

It was resolved that steps should be taken to secure the cooperation of the technical societies of the world, by the appointment of a representative Commission, to consider the question of the standardisation of the Nomenclature and Ratings of Electrical Apparatus and Machinery.

The object of the Commission is to promote international cooperation on all questions of standardisation and related matters in the fields of electrical and electronic engineering and thus to promote international understanding. This object *inter alia* is achieved by issuing publications including recommendations in the form of international standards, which the National Committees are expected to use for their work on national standards, in so far as national conditions will permit.

I.E.C., 3 Rue de Varembe, PO Box 131, 1211 Geneve 20, Switzerland.

Tel. +41 22 340 150. Telex. 28872 CEIEC CH.

IEEE (Institute of Electrical and Electronic Engineers)

The IEEE was formed in 1963 by the amalgamation of the American Institute of Electrical Engineers (AIEE-1884) and the Institute of Radio Engineers (IRE-1912). It operates primarily as an international learned society

with almost 300 000 members worldwide. The objectives centre on advancing the theory and practice of electrical, electronics and computer engineering and computer science. To meet these objectives, it sponsors conferences and meetings, publishes a wide range of professional papers and provides educational programmes. In addition, the institute works to advance the professional standing of its members.

As a leading authority in all areas of communications and associated fields, it also provides a forum for the development and introduction of national and international telecommunication and allied standards.

IEEE Inc, 345 East 47th Street, New York, NY, USA 10017.

Tel. 212 705 7910.

IFRB (International Frequency Registration Board)

This is the arm of the ITU that has been charged with the task of managing and controlling the international allocations of the radio frequency spectrum. To ensure an equitable share for all nations, their requirements are openly discussed at World Administrative Radio Conferences (WARCs).

I.F.R.B., Place des Nations, CH-1211 Geneve 20, Switzerland.

Tel. +41 22 99 51 11. Fax. +41 22 33 72 56.

INMARSAT (INternational MARitime SATellite Organisation)

This organisation was formed in 1982 to provide communications services for all sea-going vessels and off-shore oil industry installations. It can also provide emergency transportable communications systems for use in disaster areas. As the service has developed, provision is now additionally made for global land and air mobile operations, the latter providing for both operational flight data as well as passenger telephone and fax facilities.

INMARSAT, which is now well into the second generation stage of development, leases MARECS satellites from ESA, MARISAT satellites from COMSAT General of the USA and maritime communications sub-systems from INTELSAT.

The coast earth stations which form the link between the satellites and the international telecommunications networks, are generally owned and operated by the signatory organisations of the countries that form INMARSAT. To date, there are now almost 60 member countries.

INMARSAT, 40 Melton Street, London NW1 2EQ, UK.

Tel. 071 387 9089. Fax. 971 382 2115.

INTELSAT (INternational TELecommunications SATellite Organisation)

This international organisation of more than 100 country members, owns and operates a global commercial communications satellite system, primarily dedicated to domestic operations. INTELSAT, which was formed in 1964, now has a network that (including in-orbit spares) consists of almost 30 satellites. During its long period of operations, INTELSAT has been responsible for the development of a significant number earth station standards (A, B, C, D, E, F, G and Z) for operation with antennas ranging from 30 m down to 0.8 m in diameter.

The services provided include telephone, television, fax and ISDN compatible data. INTELSAT business service (IBS) provides for voice telephony, low and high speed data, packet switched networks, video-conferencing and electronic mail. These are used for financial services, oil and gas exploitation, environmental data collection and distribution, plus interactive services, all using small terminals operating through a large hub earth station. VISTA and SuperVISTA systems provide remote and rural areas with a telecommunications service, again using small earth stations.

INTELSAT, 3400 International Drive NW, Washington DC, USA 20008-3098.

Tel. 202 944 6800. Telex. 89–2707.

ISO (International Standards Organisation)

The ISO is a specialised international standards agency, comprising more than 90 national standards bodies. Its work provides industries with several types of standards, including specifications, performance requirements, testing methods, definitions and symbols.

The International Standards are produced as a result of the work carried out by almost 170 technical committees. The major part of the representation is through the ISO member bodies, and each committee maintains liaisons with other international bodies which have an interest in standardisation, in order to reduce duplicated efforts and ensure the widest awareness of the international work.

ISO's scope covers standardisation in all fields except electrical and electronic engineering, which is the responsibility of the International Electrotechnical Commission (IEC). However, ISO and IEC work jointly in the field of information technology. The official body engaged in this work is the ISO/IEC Joint Technical Committee (Information Technology) and this has developed close links with the ITU through the CCITT and the CCIR.

I.S.O., 1 Rue de Varembe, Case Postale 56, CH-1211 Geneve 20, Switzerland.

Tel. +41 22 34 12 40. Fax. +41 22 33 34 30.

ITU (International Telecommunications Union)

This organisation owes its origins to Union Telegraphique which was formed in 1865, with the specific aims of developing standards for the telegraph industry. In 1947, under a United Nations charter, it was reformed as the ITU. This body has three main aims:

- To maintain and extend international cooperation for the improvement and interconnectivity of equipment and systems, through the establishment of technical standards.
- To promote the development of the technical and natural facilities (the spectrum) for most efficient applications.
- To harmonise the actions of national standards bodies to attain these common aims. In particular, to encourage the growth of communications facilities in developing countries.

Due to the rapid growth of the telecommunications industry, it was necessary to set up the International Consultative Committees (CCITT, CCIR and IFRB) within the ITU's jurisdiction in order to adequately manage this expansion. The aims are achieved by organising international conferences and meetings, by sponsoring technical cooperation, and by publishing information and promoting world exhibitions. Currently, the ITU has about 170 member nations.

I.T.U., Place des Nations, CH-1211 Geneve 20, Switzerland.

Tel. +41 22 99 51 11. Fax. +41 22 33 72 56.

JEDEC (Joint Electronic Device Engineering Council)

This council forms the standards division of the EIA for solid state devices. In the main, it registers device numbering, component outlines and packaging.

E.I.A., 2001 Eye Street NW, Washington DC, USA 20006.

Tel. 202 457 4966. Fax. 202 457 4985.

JISC (Japanese Industrial Standards Committee)

This is the major segment of the Japanese national standards body.

J.I.S.C., c/o Standards Dept, Agency of Industrial Science and Technology, Ministry of International Trade and Industry, 1-3-1 Kasumigaseki, Chyoda-Ku, Tokyo 100, Japan.

NAB (National Association of Broadcasters)

Although this is basically a trade association looking after the interests of the North American broadcasters, there are almost 50 international associated members.

NAB was formed in 1922 and now represents over 5000 radio stations, 900 television stations and 1200 associated members. It files pleadings with the FCC and other federal government agencies and represents broadcasters' interests before Congress.

NAB is represented on many industry standards developing committees including EIA (Electronic Industries Association), SMPTE (Society of Motion Pictures and Television Engineers), IEEE (Institute of Electrical and Electronic Engineers), ANSI (American National Standards Institute), ATSC (Advanced Television Systems Committee), and NRSC (National Radio Systems Committee).

N.A.B., 1771 North Street NW, Washington DC, USA 20036.

Tel. 202 429 5340.

NBS (National Bureau of Standards)

This national organisation was established by act of Congress in 1901. The aims and objectives are to strengthen and advance the nation's science and technology and facilitate their effective application for the public benefit. To this end, NBS conducts research and thus provides a basis for:

- the national measurement system;
- scientific and technological services to industry and government;
- a technical basis for equity in trade;
- technical services to promote public safety.

The Bureau's technical work is carried out by the National Measurement Laboratory, the National Engineering Laboratory, the Institute for Computer Sciences and Technology, and the Institute for Materials Science and Engineering.

U.S. Department of Commerce, National Bureau of Standards, Gaithersburg, Maryland, USA 20899.

Tel. 301 975 2000.

OIRT (Organisation Internationale de Radiodiffusion et Télévision)

For more than 40 years, OIRT has served the broadcasting interests of the so-called European Eastern Bloc and associated countries. It owes its origins to the International Radio Organisation formed in Europe during 1946. During 1950, OIRT formed a separate split-off organisation. Due to the current political and cultural changes taking place in Europe, OIRT is due to merge with the EBU by 1994.

O.I.R.T., Liebknechtova 15, Prague 5, Czechoslovakia.

Tel. 34 14 67. Telex. 12 15 58.

SITA (Société Internationale de Télécommunications Aéronautiques Société Coopérative)

This is a cooperative body that was initially founded in 1949 by 11 airlines. The basic aims were to provide fast, reliable and economical solutions to their communications problems. Today it is an international organisation with over 300 member airlines.

It now has in operation the world's largest specialised communications network, which operates as a complement to the Aeronautical Fixed Telecommunication Network (AFTN) that operates under the auspices of the International Civil Aviation Organisation (ICAO).

S.I.T.A., 112 Avenue Charles-de-Gaulle, 92522 Neuilly, Cedex France.

Tel. 1 47 38 50 00. Fax. 1 47 47 15 42.

SMPTE (Society of Motion Picture and Television Engineers)

Founded in 1916 as a non-profitmaking learned society the SMPTE now has a worldwide membership serving both the motion picture and television industries. It meets the needs of the engineering fraternity by publishing learned papers and journals, and sponsoring conferences.

The society has always been involved with standards-making at both the national and international levels.

The SMPTE Technology committees develop standards; recommend practices and engineering guidelines; review existing documents to ensure that they are compatible with international standards; recommend and develop test specifications, methods, and materials; prepare tutorial material and institute any action necessary to ensure the compatibility and interchangeability of products, within the scope of the society.

A Standards committee reviews the work of the Technology committees and examines their recommendations in relation to the overall planning.

S.M.P.T.E., 595 W Hartsdale Avenue, White Plains, New York, USA 10607.

Tel. 914 761 1100. Fax. 914 761 3115.

35.2 Useful standards database references

Canadian Standards Association, 178 Rexdale Blvd, Rexdale, Toronto, Ontario, Canada M9W 1R3. Tel. 416 747 4044. Fax. 416 747 4149.

Elsevier, Advanced Technology Group, PO Box 2400, 1000 CK Amsterdam, Netherlands.

Elsevier BEP Data Services, Chiltern House, 146 Midland Road, Luton, Bedfordshire, UK. Tel. 0582 421981. Fax. 0582 25473.

Eurodata Foundation, Empire House, 175 Piccadilly, London W1V 9DB, UK. Tel. 071 629 1143. Fax. 071 583 0516.

Infonorme London Information (ILI), Index House, Ascot, Berkshire SL5 7EU, UK. Tel. 0990 23377. Fax. 0990 291 194.

INSPEC Marketing, I.E.E. Station House, Nightingale Road, Hitchin, Hertfordshire SG5 1RJ, UK. Tel. 0462 53331. Fax. 0462 59122.

Verband Deutscher Elektrotechniker, VDE-Verlag GmbH, Merianstrasse 29, D-605 Offenbach, A.M. Germany. Tel. 0611 83 061. Telex. 41 52 678 VLG D.

36 *Telephony and associated systems*

Historically, telephony refers to the transmission of analogue voice signals over open wire or twisted pair unshielded copper cables. The bandwidth of such a system is restricted to 300 to 3400 Hz, which is just sufficient to provide readily recognisable speech of adequate quality.

The interconnections between subscribers is provided via exchanges that select routes or circuits through the system for each telephone call, this circuit switching being achieved by the use of relays and solenoids in the early Strowger type exchanges. Later in the 1960s, this gave way to the use of cross-bar and reed relay switched exchanges to achieve a more flexible means of control. Since that period, there have been significant developments in solid-state electronics and a convergence with the telecommunications industry's needs. The stage has now been reached when it is expected that the telephone system will be almost completely digital in the very near future.

The developments in the telecommunications industry have been forced by technology and the high capital cost of the equipment necessary to maintain the local, national and international networks. Such costs have to be amortised over many years to provide an economical service. Fortunately, the technological developments have aided the expansion necessary to maintain an acceptable subscriber cost level. Current digital techniques now allow the use of completely electronic exchanges that are capable of handling several orders of magnitude more calls than any previous similarly costing network. Today's digital network now also provides for data, video and computer links, in addition to the voice service.

36.1 Analogue systems and techniques

Many of the services originally provided by analogue means are now in the process of being transferred to the digital environment. This area of telecommunications must therefore be considered to be almost obsolescent and reaching the end of its economic lifetime. However, much of the infrastructure of the system has a valuable role to play even in a digital era.

36.1.1 Basic network structures

Each subscriber is connected to a local or junction exchange which provides the power supply for the terminal and the means of switching the subscriber's communications needs. Each local exchange is connected via *trunk* lines into group switching centres that manage the needs of a number of local exchanges. The total local, plus trunk networks form the national network. Within each country there exists a number of international exchanges which act as switching centres to the international network. The international links may be provided either via land-lines, submarine cables or satellites.

In the trunk network, two twin wire pairs are used for transmit and receive (go and return) – known as four-wire operation. By comparison the subscriber's connection to the local exchange is via a single wire pair that performs both functions, the go and return signals being separated by a special type of transformer known as a *hybrid*. This device, shown in Fig. 36.1, provides two-wire to four-wire conversion as follows. A signal from the four-wire side at input 1 will drive a current through L_1

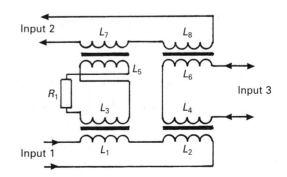

Figure 36.1 Hybrid transformer.

and L_2. This will induce equal voltages in L_3 and L_4 and also L_5 and L_6. Because the latter inductances are connected in series opposition, the voltages induced in L_7 and L_8 will be self-cancelling. Consequently none of the signal at input 1 will appear at input 2 and vice versa. At the same time however, a transmission path exists to the two-wire side of the hybrid. The resistor R_1 is included to balance the line impedances. Minor unbalances in impedance matching will create some cross-talk (interference) between the go and return lines via the hybrid and this can give rise to echo effects; hence the need to include echo cancelling circuits in the system. Provision is also made for a small side-tone leak between the subscriber's transmitter and receiver. This avoids the effect of talking to an apparent dead telephone. Because of the attenuation of the cabling system, signal amplifiers or repeaters are installed at strategic points along the network, the amplifier power supply being provided over the same cables.

The subscriber's terminal is dc coupled to the local exchange to similarly provide a power supply, but is also ac coupled for the signals, this dual function being controlled via a transmission bridge.

36.1.2 Basic exchange structures

In the interests of economy, exchange lines and equipment can be shared by subscribers by using a multiplexing technique. Using the assumption that not all subscribers will need to use the system at any given time, the technique shown in Fig. 36.2 may be used. Every 1000 input lines may be fed in via a concentrator to 100 exchange distribution lines. Then to ensure that every subscriber can still be connected to every other, a distributor routes these lines through the exchange to a number of expandors, connecting all its 100 lines to all 1000 subscribers' lines. If more than 100 calls are presented to such a system, it is obvious that the excess calls will be *blocked* for a period of time. So that the exchange can also play its part in the group switching concept, the distributor stage has to handle incoming and outgoing trunk connections.

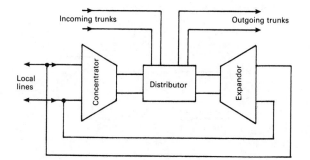

Figure 36.2 Exchange distribution.

36.1.3 Centrex services

Centrex is the generic title of a number of services that offer the large user (more than about 100 extension lines) the facilities of a PABX (private automatic branch exchange), within certain major exchanges. This avoids the user cost of installing and operating an on-site private exchange. The concept also allows for different user sites within the same call charge area to share the same equivalent PABX.

The extension phones are given short dialling codes within the exchange which can be reached by direct dialling in the manner of normal PABX extension lines.

Although the concept will function over the analogue public switched telephone network (PSTN), it is particularly well suited for operation over an ISDN system.

36.1.4 Frequency division multiplex (FDM) hierarchy

To maximise the frequency spectrum available over trunk cables and international links, the subscribers' base band voice signals covering from 300 to 3400 Hz are translated using single sideband (SSB) modulation to a higher frequency range suitable for propagation over coaxial cables and radio links.

Twelve basic channels are modulated on to carriers in the range 64 to 108 kHz and spaced 4 kHz apart. When the lower sideband (LSB) is selected, these form a *group* with a bandwidth of 48 kHz, extending from 60 to 108 kHz. Five groups are then modulated in similar manner on to carriers spaced at 48 kHz intervals from 420 to 612 kHz to form a *supergroup*.

Sixteen supergroups are then LSB-SSB modulated on to carriers spaced by 248 kHz from 1116 kHz upwards. This results in a band of frequencies from 564 kHz upwards.

To utilise the range below 564 kHz, a supergroup is modulated on to a 612 kHz carrier which after selection of LSB is reduced to a band between 60 and 300 kHz. The band between 300 and 564 kHz is filled with another supergroup in basic form (312 to 552 kHz).

This hierarchy, referred to as a *master* or *hypergroup*, provides a multiplex (including frequency gaps or *guardbands* to cater for the characteristics of practical filters), with an upper frequency of close to 4 MHz which is easily carried over a coaxial cable.

36.1.5 Personal communications services (analogue)

Cellular radio concept

By using either frequency, time or code division multiple access techniques with low power radiation, and suitably located antennas, the carrier frequencies can be reused many times over without creating cross-channel interference. Thus by suitably selecting a power level, radiating frequency and antenna system, a complete network can be built up to provide continuous coverage over a wide area. Such cellular radio telephone systems use power levels ranging from around 10 mW up to 20 W and frequencies from around 27 MHz up to 1.8 GHz to provide communications areas ranging from picocells of about 200 m diameter, through micro cells, to macrocells of about 5 km diameter. By using frequency agile mobile transmitter/receivers, it is possible for a subscriber to roam from cell to cell and still maintain communications contact.

CT1 (cordless telephone, first generation)

This first step towards cordless operation was applied to the domestic telephone and involved replacing the normal cord (wiring) between the handset and instrument with a radio link. These telephones operate in 1 of 4 channels within the 27 MHz band and provide a cell of about 200 m diameter. In addition to providing this remote link, the system also usually provides an intercom between the handset and the instrument base station.

CT1+ (cordless telephone first generation development)

This standardised system moved cordless operation into the European business field. The system operates using a trunked version of CT1 with frequency division multiple access (FDMA) to provide 80 channels within the band 885 to 887 MHz. Because two channels are needed to provide for a single duplex speech communications link, a maximum of 40 simultaneous calls are possible within each cell. Within the UK, 40 duplex channels are provided within the bands 914 to 915 MHz and 958 to 959 MHz. The mobile and base station transmitters are allocated to the lower and higher frequency range respectively to provide a maximum range of about 400 metres.

NMT is a Nordic version operating within the 450 MHz band and C-NET is an equivalent German system.

Private mobile radio (PMR)

Trunked PMRs are basically private communications systems operating in the UK within VHF Band III that was previously used by 405 line television services. Several variants are in use and operate within the terms of MPT1327/MPT1343. The service allows roaming between cells and even connections to the international networks.

Radiopaging

Essentially this low cost type of service is intended to alert a subscriber that he is required to respond by calling back to base over the national telephone network. There are three basic techniques in use:

- a simple single or multiple audible tone emitted from the pocket receiver which requires the user to call directly back to a controlling agency;
- a numeric pager that displays a limited number of digits to indicate which of several possible callers need attention;
- an alphanumeric display receiver that can transmit simple but detailed messages.

TACS/ETACS (Total Access Communications Systems/Extended TACS)

This UK cellular telephone system operates within two paired 15 MHz bands, 890 to 905 MHz and 935 to 950 MHz, with the mobile transmitter using the lower frequency. The system is a derivative of the American AMPS (Advanced Mobile Phone Service) system that operates in two segments of the 800 MHz band, but modified to operate with 25 kHz channel spacing. The cells are typically of 1 to 2 km diameter. As the service expanded and congestion occurred, two further sub-bands were added (872 to 888 MHz and 917 to 933 MHz) and this gave rise to the Extended TACS concept.

36.1.6 Signalling

Signals are needed within a telephone network to establish line acquisition at the start, call routeing through the network, monitoring the call progress and detecting line clearance. In addition, it is necessary to provide for such services as billing, quality of service records and general network management. When such signalling is carried over the communication lines it is referred to as *channel associated signalling*. Since this technique is not compatible with stored program control, it is progressively being replaced by *common channel signalling* where a separated dedicated channel is reserved for this function.

A system known as the *loop disconnect* dialling system, where the line is pulsed at 10 p/s via interrupter contacts, has been commonly used in the UK. This is progressively giving way to a multi-frequency dialling technique in which each key or pushbutton selects a unique combination of two tones of different frequency for transmission over the voice channel. These *dual tone multi-frequency* (DTMF) tones (see Digital communication systems, p. 89) provide more reliable signalling over analogue lines than do the digital pulses of the loop disconnect system. At the exchange decoder, the signals are filtered into low and high frequencies and processed to provide addresses in a ROM (read only memory) look-up table. This then outputs each dialled digit as a 4-bit binary code.

36.1.7 Stored program control (SPC)

As networks have expanded and the range of services offered has increased, system control using computers and microprocessors has become more efficient and economical. Using SPC, all the necessary operational instructions are held in programmable read only memories (PROMs). Due to the high degree of flexibility that this provides, systems of widely differing size and architectures can be made to operate together in a compatible manner.

The principal motives for this development include:

- economy and flexibility;
- speed of operation;
- the ability to reconfigure the systems to meet developing needs;
- the ease with which new subscribers and subscribers services can be added;
- improved network management and control of fault conditions.

36.1.8 Switching

There are basically two concepts used in telephony switching systems. *Circuit switching*, whereby an actual path is set up between subscribers and operates in real time, and *message switching*, which involves storing the signals in memory. This latter, which is restricted to digital systems, is also known as *store and forward*. Overloading produces call blocking and *queueing* in circuit and message switched systems, respectively.

The obsolescent Strowger automatic exchanges utilised a number of dual motion electro-mechanical selector switches coupled in series. Each selector consists of a bank of 10 semicircular rotary switch elements each carrying 10 output contacts. Any particular contact on the switch is then selected by a vertical motion to locate the wanted bank followed by a rotary motion to select the required contact. Thus any input line can be connected to any one of 100 output lines. By using three such selectors in series it is possible to service any telephone system based on six-figure numbers.

For modern exchange applications, electro-mechanical switching is much too slow and unreliable. The first replacement consisted of a matrix of cross-point switches used to interconnect input and output lines. This is now giving way to *common control switching* whereby the incoming directory number is stored in a digital memory and this is then used to find the best route through the switching matrix. In large installations, switching matrices are coupled in series and this is

referred to as a *space divided network*. To maximise the use of exchange equipment and utilise computer control, these are operated on the statistical basis that only a relatively few interconnections will be needed at any one time.

36.1.9 Traffic parameters

The utilisation and provisioning of exchange equipment within a telephone system is based on the measurements of traffic flow through the network. The basic unit is the *Erlang* which is defined as follows:

One permanently engaged circuit has a traffic flow of 1 Erlang, a dimensionless unit. The volume of traffic is measured as the average traffic intensity of 1 Erlang for 1 hour or 1 Erlang hour.

If a total of n calls are made during a period of T seconds and each has a duration h_1, h_2, h_3, ..., h_n seconds, then the total use made of the system is given by:

$$\sum_{i=1}^{i=n} h_i \text{ call seconds (traffic volume for period } T)$$

$$\text{Average traffic } E = \frac{\sum_{i=1}^{i=n} h_i}{T} \text{ Erlangs}$$

The average time for which calls occupy the equipment, the *mean holding time*

$$h = \frac{\sum_{n=1}^{i=n} h_i}{n} \text{ seconds}$$

The average traffic E for the period T

$$= nh/T \text{ Erlangs}$$

The average rate at which new calls are made for period T, (a calls/s) $= n/T$, so that

$$E = ah \text{ Erlangs}$$

If during the busy hour, C_A calls are attempted and C_B of these are blocked, then the probability of blocking $P_B = C_B/C_A$.

In practice, the average traffic will always be less than the number of equipment stages available to handle the service. For example, 45 trunks may carry 36 Erlangs of traffic so that the mean traffic $= 36/45 = 0.8$ Erlangs. In spite of this, the number of calls being handled at any one time can vary between 0 and 45.

Telephony and associated systems 331

36.1.10 Trunking and scanning

These terms are usually applied to telephony channels used over radio networks.

The frequency spectrum of a trunked network is divided into channels which are managed by a central controller. A caller obtains access to the network via the controller which allocates a free channel for the duration of the call. Once completed, the channel is returned to the control pool for future re-use.

Scanning or self-trunking is an extension of this technique. The equipment of the user making a call, scans the available spectrum to locate a free channel. The user then calls the other party either over a calling channel or the free one, to establish communications over a nominated channel. Without the central control, such a system is likely to crash under overload condition, rather than degrade gracefully like the trunked network.

36.2 Digital systems and technologies (see also ISDN, p. 86 and PCM, p. 91).

As telephone usage has expanded, it has become economically necessary to maximise the use of the analogue equipment and system. By using the time sharing concept of *time division multiplex* (TDM) considerably more subscribers can be catered for without completely re-engineering the analogue network.

The effects of aliasing, and interference with the 4 kHz spaced carriers of the analogue system, can be avoided by sampling the audio channel at the internationally agreed rate of 8 kHz, with quantisation at eight bits per sample and using relatively simple filter banks.

Internationally there are two digital systems in operation. The American Bell/A.T&T with 24 time slots (channels) and the European 32 time slots (channels) per frame, respectively. Both of the systems use time frames of 125 µs duration. The Bell/AT&T system uses 24 time slots each with eight bits capacity for message plus one extra bit for frame synchronism. This gives a total of 193 bits per 125 µs or a gross bit rate of 1.544 Mbit/s and is referred to as the D1 or T1 rate.

By comparison, the European system uses time slot 0 (TS-0) for synchronism and TS-16 for signalling and control. This leaves time for 30 audio channels. The bit rate in this case is 32×8 bits per 125 µs or 2.048 Mbit/s and this is referred to as the D2 or E1 rate.

Interworking between the two rates with the same time frame duration is readily achieved using microprocessor control and semiconductor memories.

The trunk network may be structured around copper coaxial or optical fibre cables. In the latter case, light wavelengths around 1300 and 1500 nm are used for carriers, with the speech modulated signals being organised as a wavelength division multiplex (WDM) hierarchy.

Paired wires for the subscriber's *drop cables* or *final loop* are generally retained because sampling and quantisation is more economically carried out at the local exchange.

36.2.1 Cordless communications and telephones

CT2 plus CAI (second generation CT, plus common air interface)

This system which was originally devised for UK use is now an ETSI (European Telecommunications Standards Institute) standard. CT2 is basically the portable cordless handset and the CAI provides a highly flexible way to interconnect handsets from different manufacturers and sub-systems into the PSTN or ISDN. The concept is variously known under a number of proprietary names such as Pointel and Zonephone, but covered by the generic title of *telepoint*.

The handsets may either be used in business, domestic or public access modes, with the latter normally only being able to originate a call. The digital approach provides a speech quality rather better than CT1, together with a limited degree of security and a capability of carrying data signals. The system operates in the band 864.1 to 868.1 MHz using FDMA (frequency division multiple access) with time division duplex (TDD) to provide a full duplex service over a single channel. Within the CAI, each channel is subdivided into three segments: a B-channel for 32 Kbit/s speech or data, a D-channel for in-band signalling and a SYN-channel for synchronism purposes.

The 4 MHz bandwidth allows for 40 channels with 100 kHz spacing, using ADPCM (adaptive delta pulse code modulation) at a bit rate of 32 Kbit/s. Digital speech packets of 2 ms duration are time compressed into 1 ms and these are transmitted alternately in opposite directions for full duplex operation. An additional 6 Kbit/s is needed for signalling and control purposes, plus a further overhead for duplex control, giving a gross data rate of 72 Kbit/s. A handset maximum power output of 10 mW provides a maximum cell range of about 200 m.

Where permitted, calls are set up from either a base station or a handset, by selecting a free channel. A base station transmits a burst of 32 bits of D-channel information at 16 Kbit/s, followed by 36 bits of SYN-channel data at 17 Kbit/s. The latter carries a special marker bit pattern that is used by the handset for bit and burst synchronism. This burst provides a polling sequence which carries the identities (ID) of the base station and target handset. When a handset recognises its ID, it responds.

Because the base station has to manage a large number of handset links, each calling handset transmits an extended burst of 720 bits. This is divided into 5×144 bit sequences each of which are further subdivided into 4×36 sub-multiplexes. Each sub-multiplex carries indentical information that consists of a preamble and data. The remaining 144 bits provide the SYN-channel. Each burst carries data to identify the target base and particular handset. The handset stops transmitting the extended burst as soon as it detects the SYN bit pattern during its listening cycle and this confirms that the communication link has been established.

CT3 or DCT900 (third generation, Digital Cordless Telephone 900 MHz)

This development of CT2 technology utilises similar operating frequencies and the time division duplex (TDD) technique but exploits the advantages of time division multiple access (TDMA). As indicated by Fig. 36.3, 16 ms *time frames* are divided into 16 *time slots* each of 1 ms duration. The latter are alternately used for transmission in opposite directions (TDD) to provide a full duplex service, with a gross bit rate of 640 Kbit/s. By using TDMA, each time slot is capable of supporting eight speech channels or 32 Kbit/s of digital data. A further advantage of TDMA accrues as a mobile handset moves from one cell to another; the change of base station and operating channels (hand-off) is virtually transparent to the user. Figure 36.3 also shows how each time slot is divided into segments for synchronism, indent, data and error correction purposes.

DCS 1800 (Digital Cellular System, 1800 MHz)

Apart from the operating frequency which is in the 1.8 GHz band, this developing system is compatible with the specification for GSM 900, but is expected to support cells that are less than about 7 km radius. This concept forms one of the UK commitments to PCN (personal communications networks).

DECT (Digital European Cordless Telephone)

This developing European standard that uses TDMA/TDD has much in common with CT3 but with a frame duration of 10 ms, divided into 24 time slots to accommodate 12 full duplex time division channels. Operation falls in the 1880 to 1900 MHz band and this provides enough spectrum for up to 120 simultaneous calls. The concept is designed for low power operation and this restricts the cells to a radius of about 200 m.

GSM-900 (Group Speciale Mobile-900 MHz, also Global System for Mobile Communications)

This cordless, cellular and frequency agile system represents a pan-European development backed up with ETSI standards. It operates in the paired bands,

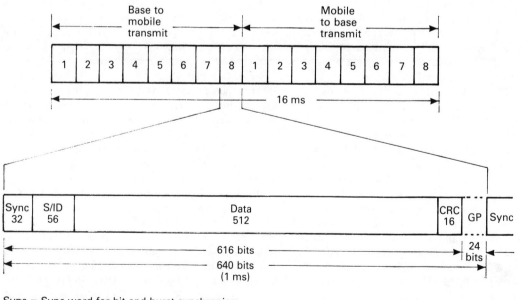

Sync = Sync word for bit and burst synchronism
S/ID = Signalling and identification
Data = Speech allocation
CRC = Cyclical redundancy check
GP = Guard period

Figure 36.3   Frame and time slot structure for CT3/DCT900 systems.

890 to 915 MHz and 935 to 960 MHz with 200 kHz spaced carriers. As is common with other cellular concepts, GSM uses the lower frequency band for transmissions from the low power mobile handset. The transmission system uses TDMA with TDD to provide *seamless hand-off* and a high frequency reuse factor, typically allowing up to 12 simultaneous communications links. With eight time slots per carrier, this allows 992 simultaneous users per cell.

The broadcast system uses a complex encryption technique, with interleaving and convolution coding to provide a high degree of system security and bit error control. Compared with an analogue system which requires a carrier to noise (C/N) ratio in the order of 30 to 35 dB for acceptable quality, GSM can offer a similar service with a C/N as low as 12 dB.

Signal modulation of the RF carrier uses Gaussian minimum frequency shift keying (GMSK) with a modulation index of 0.3 (MSK is a continuous phase modulation (CPM) version of FSK). The use of Gaussian shaped pulses maximises the bandwidth/spectral efficiency trade-off and improves the resilience to co-channel interference.

Each speech channel is sampled at 8 kHz and quantised to a resolution of 13 bits. This is then compressed by a factor of 8 in the *full rate* vocoder (voice coder) to give a bit rate of 13.4 Kbit/s. It is expected that half-rate coders will eventually reduce this to 6.7 Kbit/s. As indicated by the handset block diagram shown as Fig. 36.4, signal processing utilises both speech

compression and synthesis. The vocoders are designed around high speed digital signal processors (DSPs) and use a form of *linear predictive coding* (LPC) referred to as *residual excited* LPC. The algorithms for LPC attempt to model the human vocal chords in order to produce realistic synthetic speech with the minimum of memory.

An additional provision is made for an extra digital data channel with a capacity of 9.6 Kbit/s.

Transmission power levels range from 20 W maximum for a base station down to 0.8 W minimum for a handset. A power saving provision is also made for the power levels to be reduced by a maximum of 15, 2 dB steps depending on a signal level that gives an adequate bit error rate (BER). Typically, a cell range varies from about 3 km to 35 km depending upon the local environment.

The specification provides for vehicle speeds up to 250 km/h (156 mph) and at maximum range this represents a time delay of around 15 μs. Because Doppler shift and multipath echoes can affect the received signal quality, each TDMA burst includes a number of *training bits*. The receiver compares this with a known training pattern and from this deduces the transfer function of the propagation path. An adaptive filter is then made to perform the inverse function, thus cancelling any unacceptable distortion.

Each 4.62 ms TDMA frame is divided into eight time slots with 557 μs of user capacity. These are interleaved so that the handset does not have to receive and transmit within the same time slot. Each block of 456 data bits is

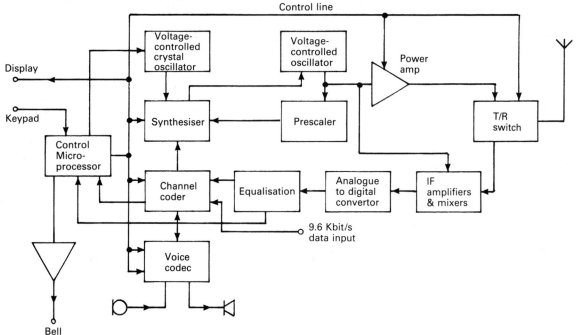

Figure 36.4 GSM-900 handset.

divided into four groups of 144 bits and these are loaded into alternate time slots, each with 20 training bits inserted between sub-groups of 72 data bits. The data packets are switched to a new carrier at about 217 Hz (frame rate). Altogether this produces a gross bit rate in the order of 270 Kbit/s.

Since each handset transmits for only 1/8 of the time (1/16 in the case of the half-rate vocoder), the average power consumption is very low.

GSM systems are structured in a hierarchical manner and installed on *intelligent networks* (IN); that is, networks with a significant element of computer control. Each base station (BS) serves a limited number of cordless terminals simultaneously and can switch up to 12 users to the trunks of a public telephone network (PTN). A switching sub-system (SSS) consists of a mobile service centre (MSC) plus the necessary database. The MSC provides the connections to the PTN and other MSCs. Each subscriber belongs to one *public land mobile network* (PLMN) and the relevant data is stored in a *home location register* (HLR). If a subscriber is currently outside his home PLMN, then the necessary data is held in a *visitor location register* (VLR). It is this feature that permits *roaming*. Each BS consists of a station controller and the necessary transcoding equipment, with the controller operating as an air interface for the mobile stations.

The operation and control of a GSM network is very largely managed by computer software. Unlike the conventional PTN where control is established at the beginning and end of each call period, GSM systems provide control and management signalling throughout the communications period. The following represents a brief explanation of the protocol as a MS tries to initiate a call.

- Phase 1. The mobile requests to set up a call which should result in the allocation of a control channel.
- Phase 2. The MSC checks the identity of MS and if authenticated it issues a new cipher key. If this fails, a disconnection release is initiated.
- Phase 3. The MSC agrees the ciphering with MS and provides the necessary synchronisation.
- Phase 4. This represents the actual call set-up part of the message, including the called party's number.
- Phase 5. A traffic channel is assigned if available. If not, then the call is placed in a queue. If the queueing capacity is exceeded, the calling MS is advised of the blocking situation.
- Phase 6. The radio link is established after the issue of *call connect* and *call acknowledge* signals.

Hand-off or hand-over occurs when MS moves from one cell to another. BSC periodically (every 480 ms) monitors the signal strength from a BS. It then employs a complex algorithm that considers the signal level relative to all neighbouring cells to make the hand-off decision. If the decision is made, then hand-off occurs at the next time frame. The new BSC will then make a similar decision to check that the correct hand-off was made.

36.2.2 Developing digital technologies

ACD (automatic call distribution)

This represents a concept that was designed for the optimum management of a communications system where the traffic consists largely of very many incoming calls. Each call is distributed to a free *agent* as it arrives without the intervention of an operator, typical applications being credit card verification, air line booking facilities and similar situations that require the back-up of a large computer-held database. Such a system provides better queue control with fewer lost calls. Generally, if the system overloads, then provision is made for automatic call back by the first free agent.

ACD differs from the conventional PABX (private automatic branch exchange) in the following ways:

- Because an incoming call is directed to the first free agent rather than being picked up by an operator, less time is lost.
- The agent's terminals are more sophisticated than the conventional telephone terminal. They are normally equipped with a full keyboard and visual display screen, with a direct link to the corporate mainframe computer.
- ACD is capable of being able to provide a great deal of statistical management information, such as call queue lengths, periods of overload and other lost time features.
- The greater cost of the ACD system is offset by the greater throughput of calls, reduced lost time and calls and by the use of fewer operators.
- In some cases, the use of a DTMF telephone allows callers to make direct access to sections of the computer data while waiting for a free agent. When the call is eventually picked up, the necessary customer information can be displayed on the agent's screen immediately to reduce lost time.
- ACDs can be networked over the PSTN (public switched telephone network) so that several remote common databases can be directly linked. This allows sharing of databases under the loading variation that is bound to occur throughout the day. It also allows for the shut down of some ACDs during off peak periods.
- Developing systems also allow for the use of voice recognition techniques to further reduce a queueing problem.

ATM (asynchronous transfer mode)

Conventional networks carry data in a synchronous manner and because empty slots are circulating even when the link is not needed, network capacity is wasted. The ATM concept, which has been developed for use in broadband metropolitan area networks (MAN) and optical fibre based systems, is supported by both CCITT and ANSI standards, and can also be interfaced to SONET (Synchronous Optical NETwork). ATM automatically adjusts the network capacity to meet the system needs and can handle data, voice, video and television signals. These are transferred in a sequence of fixed length data units called cells.

Common standards definitions are provided for both private and public networks so that ATM systems can be interfaced to either or both.

ATM is therefore a wideband, low delay, packet-like switching and multiplexing concept that allows flexible use of the transmission bandwidth and is capable of working at data rates as high as 622.08 Mbit/s.

Each data packet consists of five bytes of header field plus 48 bytes for user data. The header contains data that identifies the related cell, a logical address that identifies the routeing, forward error correction (FEC) bits, plus bits for priority handling and network management functions. FEC applies only to the header as it is assumed that the network medium will not degrade the error rate below an acceptable level. All the cells of a *virtual connection* (VC) follow the same path through the network that was determined during call set-up. There are no fixed time slots in the system so that any user can access the transmission medium whenever an empty cell is available. ATM is capable of operating at bit rates of 155.52 and 622.08 Mbit/s and the cell stream is continuous and without gaps.

The position of the cells associated with a particular VC is random and depends upon the activity of the network. Cells produced by different streams to the ATM multiplex are stored in queues awaiting cell assignment.

Since a call is only accepted when the necessary bandwidth is available, there is a probability of queue overflow. Cell loss due to this forms one ATM impairment. However, this can be minimised through the use of statistical multiplexers. Bit errors in the header which are beyond the FEC capability can lead to mis-routeing.

Cell relay or switching (see also ATM, above)

This is the generic title for a member of the packet switching family in which data is loaded into short packets or cells for asynchronous transmission.

Centrex (see also p. 328)

Within Europe, digital Centrex services are ISDN compatible and meet the CCITT 2.048 Mbit/s PCM standard. These are designed to fit the PSTN's call progress tones, ringing cadences, numbering plans, billing and trunk signalling. The system is well suited to the needs of large corporate users who may also be members of a particular *closed user group* (CUG). As in the analogue case, the network supplier provides PABX facilities within a local exchange and this saves the subscriber space, equipment costs and operating staff expenses. In addition, it is very easy for the network provider to update the facilities. The services provided by Centrex include call transfer, call forward, call pick-up, call waiting, teleconferencing, short code dialling, call parking or holding, call splitting, automatic recall, and the display of the called or calling numbers.

Cordless PABX

The common air interface (CAI) of the CT2 specification represents a radio link access standard and this allows for cordless extensions to the standard PABX. Retro-fitting the CAI allows handsets to roam within the limits of the micro-cell inside a building. The concept was developed in order to reduce lost calls and time spent locating missing users. CT3 systems are also amenable to this conversion and allows roaming over an even greater area. The technique permits the use of cordless data modems, which in turn gives handheld computers and portable Fax machines access to the telephone network.

The introduction of a fully cordless PABX also removes the need to employ extensive in-building wiring.

E-Mail (electronic messaging) (see also X-400 services, p. 341)

E-Mail is a text-based messaging service designed for simple access over analogue type telephone lines and using relatively low speed data modems (up to 9.6 Kbit/s). The subscriber has access to certain public domain databases, together with the facility to exchange messages with other users through personally allocated sections of the system memory. The messages essentially consist of files, spread-sheet data, word processed text or memos transferred through a series of *mail-boxes*. Although initially established to provide a domestic service through linking personal and home computers, E-Mail has expanded into the corporate market. In this latter case, the technique is often referred to as *electronic data interchange* (EDI). The system now provides for tele-shopping and banking and provides mobile sales representatives with a quick access to corporate databases. Bar code readers linked via data modems to a corporate network are also a feature of some EPOS (electronic point of sales) terminals.

Frame relay

Although initially intended to provide a simplified high speed technique for data transport, frame relay can also be used for voice services. The concept is ISDN compatible and can handle data speeds as high as 200 Mbit/s. Each frame consists of a flag byte, two header bytes, a field carrying five to 8000 data bytes, two bytes CRC (cyclic redundancy check) frame check and an end flag byte. The header consists of a *data link connection identifier* (DLCI) that allows the network to route each frame from end to end through a series of nodes. This virtual path (VP) is defined either on call set-up or at subscription time. The two delimiting flags and the two CRC bytes are identical to those used by HDLC (high-level data link control) or SDLC (synchronous data link control) based interfaces.

When user A requires to send a file to user B, A initiates the process by sending a request to establish a transport layer link. This is passed through the ISDN over a D signalling channel. As this is routed through the network, it is used to define the VP and the calling parameters that will be used at the data transfer stage. Once the call and VP have been established, the data is passed from A to B by node hopping under the guidance of the DLCI in the frame header.

Frame relay achieves its high speed because there is very little processing carried out within each node, the responsibility for error control being delegated to the first and last nodes only. Because of this, frame relay relies on the use of an error-free carrier medium such as optical fibre. Due to its variable length data field and the use of dynamic bandwidth allocation, frame relay is well suited to *bursty* type data channels.

MDNS (managed data network services)

Both large private and public networks require continuous monitoring and control to ensure efficient operation and a high quality of service. This is particularly true for networks using leased and dedicated PSTN lines.

Network management architectures are designed to provide these functions and to maintain a statistical database that can be used to analyse the quality of service being provided.

ONP (open network protocol)

This represents the extension of the ISO Open System Interconnect (OSI) concept for applications within wide area or global networks.

SDH (synchronous digital hierarchy)

Normally, the high capacity trunks and the local networks are not fully synchronised so that the interfaces, particularly those for high speed operation, find this a problem. With the SDH concept, the traffic in all parts of the network is transferred in synchronised packets. This permits higher data speeds and better utilisation of the network capacity. For example, an SDH system operating at 155 Mbit/s can provide for 1800 simultaneous telephone calls over each optical fibre. The service also allows for the integration of data, voice, video and television traffic.

The SDH is defined by CCITT recommendations G707, G708 and G709, with the lowest multiplex data rate of 155.52 Mbit/s being described as STM-1 (synchronous transfer module-1). The multiplex is described in general terms as STM-N where N takes on the values 3, 4, 6, 8 and 16. Thus the upper rate is $16 \times 155.52 = 2488.32$ Mbit/s. The value 155.52 was chosen because it represents the convergence of different multiples of the CCITT and Bell/AT&T standard rates.

Each SDH frame can be represented as a 9 row by 270 column matrix as indicated in Fig. 36.5. The first or top left-hand corner byte, known as the *framing byte*, provides bit rate clock synchronism and represents the start of a frame. Transmission then follows on a row-by-row basis. The 81 bytes described as section overhead which are interleaved with the data, provide byte synchronism, system monitoring and frame delimiters.

The useful payload of each SDH frame is contained within the 9 rows of 261 bytes. The overall bit rate is therefore given by:

$$270 \times 9 = 2430 \text{ bytes/frame}$$
$$8 \times 2430 = 19440 \text{ bits/frame}$$

At 8000 frames/s this gives $8000 \times 19440 = 155.52$ Mbit/s. The actual user channel capacity is given by:

$$261 \times 9 \times 8 \times 8000 = 150.336 \text{ Mbit/s}$$

Although initially designed for use over an optical fibre network, STM-1 can be handled over a microwave link. Using either 32-QAM or 64-QAM, this yields a bandwidth between 30 and 40 MHz.

SNMP (simple network management protocol)

As networks expand it becomes more important to maintain a database holding certain statistical and operation data about the system and the quality of service that it is providing. SNMP is one of the *de facto* management schemes that has developed to meet this need. Management information is stored at each network element and this is available for remote access. Each element holds the addresses of the other elements that have access to it and the status of that access, either *read only* or *read/write*. Encryption techniques may be added to this simple protocol to prevent piratical access to this information.

TCP/IP (transmission control protocol/Internet protocol)

This represents a proven alternative to the ISO Open Systems Interconnect (OSI) concept to allow different types of networks (Internet) to be interconnected.

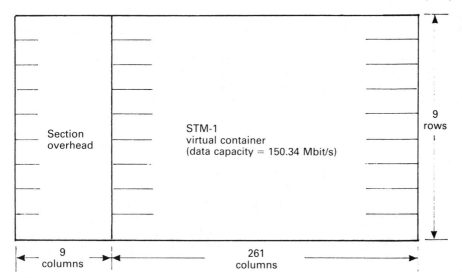

Figure 36.5 SDH STM-1 frame structure.

Although originally developed in the USA for government, research and academic applications, it has expanded into the commercial field. As TCP/IP has developed from an operational need, it has become one of the industry *de facto* standards.

VANs (value added networks)

This term is used to describe the additional value of services that may be provided over telecommunication networks in addition to the basic voice and data services. Thus videotex, which provides subscribers with access to certain public or corporate databases, is a value added service. The users of VANs include banks, mail order suppliers and market researchers. Voice messaging services may also be provided in the manner of E-Mail.

Videophones (see also Image processing, DCT coding, p. 138 and Television signal processing and systems, p. 357)

Simple video cameras and small screen displays may be added to the basic telephone service to provide dial up face-to-face communications. The image compression technique that is required to squeeze the video signal into an ISDN 64 Kbit/s or the analogue telephone channels is covered by CCITT H.320 standards.

36.2.3 Packet switching and fast packets (see also ATM, p. 335)

The term packet switching refers to the technique of passing or switching fixed length blocks (packets) of data around a network. The packet sequence, routeing through the network, error control and mode of access, is fixed by the network protocol.

At the originating network terminal, the data is organised into packets of pre-determined length, together with the necessary control and error correction bits by a PAD (packet assembler/disassembler). To ensure delivery to the correct node, each packet contains the originating and terminating addresses. In cases where the packets may reach the final node by various routes, each packet also carries a sequence number so that messages may be reassembled in the correct order. Each data field of a packet may also carry a CRC (cyclic redundancy check) pattern for error correction.

A packet switching system functions by a receiving node, checking each data field for errors, making the necessary corrections, duplicating the packet and retransmitting forward to the next node. A receiving node may also make a backward acknowledgement that a packet has been received.

Fast packet networks (ATM) are designed to provide a high throughput of data with very short delays. These networks are characterised by high speed trunks, hardware switching, virtual circuit routeing, short packets and simple protocols. When used on conventional PCM channels (E1/T1) rates fast packet networks can make use of the silence periods that occur in speech.

A fast packet network circulates frames that are either empty or contain data. If a frame must be dropped due to overload, the network can distinguish between these two types and drop only the empty packet, a feature that is not possible with a circuit switched network.

36.2.4 Paging

Paging systems provide a low cost, one-way mobile, cellular radio type service that alerts the subscriber that messages are to be delivered.

Calls for paged users are initiated by a PABX or message control centre (Bureau) and the subscriber responds to the alert either by calling back to the centre or by making a direct response to the instruction.

The service can be linked to personal computers (PCs) or similar terminals so that messages may be sent in batches to reduce the *air time*. Using the PC, the service may be also linked to the E-mail system. The more sophisticated mobile receivers may have an in-built printer to provide a hard copy of any messages.

The service may operate in either the VHF or UHF bands, typically at 153.275 MHz, 169.425 to 169.80 MHz and in the 466 MHz band.

Provision is made for three basic types of messaging:

- A tone-only service (bleepers) in which one or more tones may be transmitted to signify the unique number that has to be called back.
- A numeric pager service in which up to 24 or more digits may be displayed to indicate directly the numbers to call back.
- An alphanumeric display system to convey more explicit messages. This type can be equipped with a memory to display messages that would normally overflow the display capacity.

As a comparison of spectrum efficiency, the alphanumeric type uses about 15 times the capacity of the simple bleeper system.

ERMES system

ERMES (European Radio MEssage System) supported by ETSI standards represents a pan-European approach to paging. Unlike the POCSAG code system, the ERMES standard defines the whole system and is intended to meet the European needs for the foreseeable future.

The system operates with 16 channels spaced by 25 kHz within the band 169.425 to 169.80 MHz and frequency agile receivers scan all channels. This latter feature improves the system roaming characteristics and makes for flexible use of the frequency spectrum. The basic data rate is 6.25 Kbit/s, but this reduces to 3.125 Kbit/s due to the use of four-level pulse amplitude modulated frequency modulation (PAM/FM/FSK). A forward error correction (30, 18) code is used to minimise bit errors and this is supplemented by the interleaving of message bits to provide for burst error protection.

Offset superhet receiver

The provision of suitable RF selectivity and tracking for small portable receivers of the type used for paging presents a significant problem. The front-end circuits are therefore equipped with antenna matching networks, simple filters and RF low noise amplifiers.

The crystal controlled local oscillator frequency is offset from the carrier by half the signal bandwidth or deviation. The resulting very low frequency IF (intermediate frequency) is therefore half the deviation (typically 2.25 kHz). As a consequence of this, image frequency interference does not arise because this signal becomes part of the wanted signal spectrum.

POCSAG systems

The term POCSAG (Post Office Code Standardisation Advisory Group) refers only to a coding technique which is also known as the CCIR Radiopaging Code No 1. It was designed for transmission at either 512 bauds (tone pagers) or 1200 bauds (alphanumeric pagers) (up to 2400 bauds for other suitable applications) and utilises a single carrier frequency of 153.275 MHz radiated at sufficient strength to penetrate most buildings.

This radiopaging code which is based on a primary non-return-to-zero (NRZ) format and frequency shift keying (FSK), is constructed according to the following rules:

A transmission starts with a preamble of at least 576 alternate 1s and 0s to provide clock synchronism and this is followed by a number of *batch blocks*. Each batch contains a further 4-byte synchronisation code word and eight frames (frames 0 to 7). Each frame consists of 32 bits which form either an address, a message or an idle code word, an idle code word being inserted either to fill an empty batch or to separate messages. An address frame starts with a leading 0, while a message frame starts with a 1. Each frame contains a 10-bit CRC (cyclic redundancy check) and ends with one even-parity bit. Of the remaining 20 bits in each frame, all are used for message information, while the address frame is divided into an 18-bit address plus two function bits. These latter are used to distinguish between one of four different types of call to a particular user.

36.2.5 Personal and private systems

INMARSAT aeronautical mobile system (APC and ACARS)

This developing system, which is also referred to as aeronautical public correspondence (APC) and addressing communications and reporting system (ACARS), operates within the aeronautical mobile bands of 1545 to 1555 MHz and 1646.5 to 1656.5 MHz.

The service operates by using a satellite as a relay between the aircraft and a ground station, which in turn provides the connection to a terrestrial PSTN system. Two digital voice channels are provided at 9.6 Kbit/s for passenger use, a single channel at 4.8 Kbit/s for crew use, and 600 bit/s for a digital data messaging service. The modulation and coding scheme is designed to offset the effects of Doppler shift due to aircraft speed. The antennas are encased in an aerofoil housing and when mounted externally, these take on the appearance of *shark fins*. These antennas are of the planar, steerable, phased array type and are designed to provide all-round coverage. The voice signal is PCM coded by sampling at 8 kHz to produce 8-bit samples every 125 μs. These are processed via a microprocessor and the bits are clocked out at 1536 kHz. The 8-bit bursts are then converted into a continuous 9.6 Kbit/s data stream. This output is then mulitplexed with a dual tone multi-frequency dialling tone, to generate a gross bit rate of 10.368 Kbit/s. After the addition of FEC bits to combat the effects of multipath fading, the bit rate has doubled to 21 Kbit/s. This bit stream is then modulated QPSK (quadrature phase shift keying) on to the final RF carrier. The system provides enough spectrum to accommodate the transmission of engineering, navigational and other aircraft management data.

Iridium
See Satellite systems, p. 273.

PARS (private advanced radio service), SRR (short range radio)
This developing short range (about 10 km) system operates in the 888 to 890 MHz and 933 to 935 MHz bands to provide both voice and data services. In the UK, provision is made for 80, 25 kHz channels each supporting a digital bit rate of 16 Kbit/s. The system utilises a polling protocol so that it becomes self-trunking, thus avoiding the need for a central control station. Each handset therefore contains the necessary circuitry to select a free channel and make or receive a call.

36.2.6 Signalling

Whereas analogue systems utilise a common channel for both communications and signalling purposes, digital systems use a dedicated channel for this latter function. Of the many digital signalling methods in use, only two are particularly international in their applications.

CCITT No. 6 (Common Channel Signalling System-6. CCSS-6)
This signalling specification was designed for operation over a dedicated voice band analogue circuit to handle the requirements of many simultaneous callers. The system also permits the use of low speed digital modems operating at 1200 bauds using QPSK and a carrier frequency of 1800 Hz over the same audio channel. In the USA, a variation known as *common channel inter-*

office signalling (CCIS) procedure is adopted by the Bell/AT&T system.

CCITT No. 7
Although the Common Channel Signalling System (CCSS) No. 7 was initially developed to manage call connection and disconnection, it has expanded to provide many other service functions. The specification which is based on the ISO 7 layer model, is ISDN compatible and suitable for use over a satellite link.

As indicated in Fig. 36.6(a) the message transfer section covers the lower three OSI levels. The data link level is a full duplex, 64 Kbit/s link dedicated to signalling. Because the system may have to handle the needs of many thousands of callers simultaneously, it is most important that this link be error free. This level ensures that blocks of data are delivered in the correct order, are not transmitted at a rate too high for the receiver and make sure that data is not duplicated. Figure 36.6(b) shows the extensive nature of this control feature. The actual data transfers through the interfaces and switching system are managed by the upper levels of the protocol. CCSS-7 has a large degree of inbuilt flexibility that allows new services to be added without changes to the network or protocol.

36.2.7 Switching (digital systems)

In order to obtain a high degree of flexibility with low cost, a *time divided* (T) switching concept is often used with the TDM (time division multiplex) transmissions used over the ISDN. This functions by utilising input and output buses which are switch interconnected on a cyclic basis. A memory is used in a *time slot interchange* (TSI) to delay a PCM code word from the input bus until it is needed at the output in synchronism with the previous output frame, this action being obtained via a pair of digital switches. The functions of multiplexing/demultiplexing and TSI are usually incorporated into a device described as a *codec* (coder/decoder). This simple type of time divided switching can easily overload and so the concept is generally limited to use on small PABXs. The problem is resolved by introducing a *space divided* (S) switch concept into the structure. Large exchanges are usually managed as a cascade of space–time–space (S–T–S) or time–space–time (T–S–T) switching or even multiples such as S–S–T–S–S or T–S–S–S–S–T. In this way, an exchange can provide services for many thousands of subscribers.

36.2.8 X series services

X25
CCITT Recommendation X25, which is a close relation to frame relay, defines a packet switching network

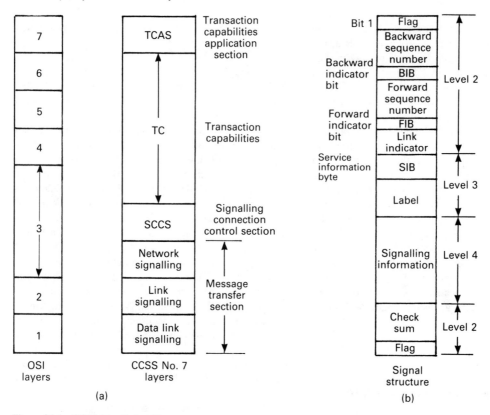

Figure 36.6   CCSS No. 7 signalling.

interface rather than the complete network. Therefore any device that matches this requirement can be connected to an X25 network. The protocol at each X25 node error checks each received frame, makes the necessary corrections and retransmits the packet to the next node which may or may not be the final destination.

The node also makes a backwards acknowledgement through the network to confirm that it has received the packet. X25 networks originally operated on 64 Kbit/s leased lines, but can now handle data rates as high as 2.048 Mbit/s.

X50, X50bis and X51

The basic parameters of the PCM frame structure and multiplex are contained in these CCITT Recommendations. These create a standard for the interworking of synchronous data networks. The PCM frame time slot structure provides for blocks of 64 Kbit/s data streams but to accommodate lower bit rates, a further multiplexing scheme needs to be applied.

In this process, the customer serial data stream is divided into 6-bit blocks to which leading and trailing bits are added to form an 8-bit (6 + 2) *envelope*. The leading bit provides *alignment* while the trailing *status*

bit is used for network signalling and call control. In some countries, a 10-bit envelope is used in a similar way, but with eight data bits sandwiched between the alignment and status bits (8 + 2). The basic multiplex consists of 80, 8-bit envelopes interleaved in blocks of five, onto the basic 64 Kbit/s channel. Each block of five envelopes is referred to as a *phase*. Discounting the housekeeping bits, the actual subscriber capacity is 48 Kbit/s and this can be employed in various combinations, ranging from 1×9.6 Kbit/s to 16×600 bit/s of user data per phase.

If the 80 envelope structure is considered as a 16×5 matrix (16 envelopes, 5 phases), the leading bit in every tenth byte is used for housekeeping purposes (eight bits) instead of alignment.

An alternative multiplex scheme is defined in X50 and described as Division 3. In this case the structure consists of 20, 8-bit envelopes which allows multiplexing at channel rates of 12.8, 6.4 and 3.2 Kbit/s. A 20-bit overhead is distributed across the multiplex in the manner described above, but in this case 19 bits are used for framing, leaving just one bit for housekeeping purposes.

For interworking between 8- and 10-bit envelopes, X50 defines a Division 4 structure, the interchange being effected by loading the data stream into a four envelope by 8- or 10-bit matrix and unloading in the appropriate pattern.

X400

This CCITT Recommendation, which was developed to overcome the incompatibility problems of many E-mail systems, provides a standard for *message handling systems* (MHS) that is recognised by the ISO. It therefore provides a very flexible interface between disparate terminals in a transparent manner. This allows the extension of such services as computer-to-computer communications, the interchange of telex, facsimile, videotex and teletex messages, electronic data interchange (EDI) and fund transfers. The transferred files may include text, data, graphics or even digitised voice signals.

The system is built around a message handling centre (MHC) which provides a message transfer agent (MTA), a message store server (MSS) and a directory agent. The MTA switches messages to other MTAs and to the store and access section. It also provides all the routeing and transmission facilities to control the receipt, store and dispatch functions. Each MSS provides the memory and control for handling stored messages. The directory agent holds information about each user and the services to which each subscribes. Basic system security is provided via the use of a password or PIN (personal identity number), plus the facility for adding encryption.

X500

The Directory standard specified in CCITT Recommendation X500 and supported by the ISO is central to the success and expansion of the Open System Interconnect (OSI) concept. This aims to make possible the interconnection of different directory systems on a global scale and is probably the most far reaching and open concept that the CCITT has ever considered. Such a directory would include information about such entities as people, organisations and resources. The system as proposed supports and extends the possibilities associated with X400 message handling systems.

The standard specifies how the data must be organised for search by tree and branch techniques. This is most important particularly if the global concept should ever develop, because searching could then become a very costly and time consuming business.

36.3 Miscellaneous terms

Abbreviated dialling. The full address or number of the required subscriber is held in memory. *Short code* or *speed dialling* then only requires the use of the access number of the memory location.

Absent extension diversion. Any incoming PABX calls may be diverted on a temporary basis to some other extension.

Activity factor. A USA term that is used to define the expected occupancy of a circuit during busy hour. Normally expressed as a decimal fraction of an hour.

Add-on conference or third-party mode. A PABX technique that allows several extensions to be simultaneously linked to provide conference facilities.

Auto-answer. An electronic circuit within a modem that automatically switches the system to receive upon the reception of an incoming signal.

Auto-dialler. An electronic circuit within a modem that is memory driven and automatically carries out the dialling operation.

Call congestion ratio. Also known as the *blocking ratio*. The ratio of the number of calls blocked to the number of call attempts made during a specific period.

CINDI. A combined computer and telephone with hard disk drive storage. The system provides for voice mail, telephone memo pad for callers, answer-phone and recorded delivery facility, and automatic forward to an alternative number including a car phone. The system is also capable of adding services to a local area network.

Cross-talk. Cross-talk between lines occurs as interference due to capacitive pick-up and system impedance mis-matching. When necessary this is divided into two regions, near end cross-talk (NECT or NEXT) and far end cross-talk (FECT or FEXT).

Digital filling or bit stuffing. The addition of extra bits to a time slot made in order to pad out the bit rate to match the requirements of a particular multiplex frequency.

Duplex operation. Full duplex systems provide for simultaneous two-way communications. By comparison, half-duplex systems provide for two-way communications but only one way at any one time.

Echo cancelling/suppression. A feedback network that is added to cancel interference due to echo effects. The signals in the go and return paths are compared so that a replica of the echo component can be generated. This is then added in anti-phase to produce a cancellation.

Future public land mobile telecommunications systems (FPLMTS) frequency allocations. For these developing services, the following allocations have been made: 1.885 to 2.025 GHz and 2.11 to 2.2 GHz.

Frame relay assembler/disassembler (FRAD). These devices perform similar functions to the PAD of X25 systems, but with a much greater degree of flexibility. FRADs are designed to allow a network with non-frame relay protocols to access private and public frame relay services. They support the network integration of disparate data streams and have many of the characteristics of statistical multiplexers.

High data-rate subscribers line (HDSL). This transparent replacement system for E1/T1 repeatered circuits uses standard 3B1Q coding giving a four level pulse amplitude modulation scheme to achieve a high data rate on a two pair cable system. This allows a total of 2.048 Mbit/s to be carried over either 4000 yards (approx. 3700 m) of 24 gauge or 3000 yards (approx. 2700 m) of 26 gauge copper cable without repeaters. The data and control bits are mapped into two data streams for the transceivers employed at each end. The total bit rate of 2.048 Mbit/s is shared equally across the two pairs to provide full duplex operation.

Hundred call seconds (HCS). Also known as *unit call* or *cent call seconds.* A unit of traffic intensity related to the Erlang, and equal to the number of calls per hour, assuming an average holding time of 100 s. Thus 1 Erlang = 36 HCS.

Leased lines. These may be allocated to special subscribers by the PTTs to provide an immediate and direct link between two terminals. This provides a valuable and secure way of interconnecting two separate local area networks (LANs).

Minitel. The French PTT's teletex (videotex) service. The system carries several thousand different services, including data base information, telephone directory, E-mail, business directory (Yellow Pages), etc.

Pocsagger. A unit devised to convert ASCII (American Standard Code for Information Interchange) code into the POCSAG (Post Office Code Standardisation Advisory Group) code. This allows communications between a host computer and a paging receiver.

Queued packet synchronous exchange (QPSX). A metropolitan area network (MAN) concept based on the DS3 rate of 45 Mbit/s. The network provides for a mix of voice, data, video and graphics with the bandwidth being dynamically allocated. The system multiplex is capable of delivering data to a subscriber's premises at up to 155 Mbit/s.

Ringing equivalence number (REN). The number of equivalent telephone devices that can be added to a subscriber's drop cable. If the line loading is increased beyond about REN 3, then the line current will not be able to support the necessary ringing current.

Routeing information. The housekeeping bits that are added to a data packet to ensure that the message is delivered to the wanted address and over an appropriate route.

Simplex operation. A system that provides communications over a given path but only in one direction.

Statistical multiplexer. A device that allocates time slots in a PCM system only to active terminals. If the channel capacity is likely to overload, then the multiplexer stores or buffers the data until such times as capacity is available.

Teleinformatics/telematics. Generic terms used to describe the non-voice facilities of a telecommunications service.

Telemetry. The process of transmitting variable measured quantities at some remote location over a communications link for processing with a central computer.

Teletex. A service complementary to telex, that allows subscribers to exchange textual information between computers over a telecommunications network. The data is usually transferred in a page format.

Teletext. A domestic broadcast service for text and simple graphics, in which digital data is transmitted within the normal television signal.

TRANSPAC. The French PTT's public data network designed to include facsimile and telex services.

Wavelength and time division multiplex (WTDM). Baseband signals from several sources are digitally combined into time slots of a time division multiplex (TDM) hierarchy. Several similar networks can then be wavelength division multiplexed (WDM) onto a single optical fibre using either 1300 or 1500 nm lasers. These light sources may provide channels separated by as little as 4 nm so that 16 different wavelengths can provide a total capacity in excess of 40 Gbit/s.

36.4 Useful references

Calhoun G. (1988) *Digital Cellular Radio*, Artech House, London.

Coates R.F.W. (1985) *Modern Communications Systems*, 2nd edition, Macmillan, London.

Mazda ·F. (1989) *Electronic Engineers' Reference Book*, 6th edition, Butterworth-Heinemann, Oxford.

Smoll G., Mamer M.P.R. and Hills M.T. (1981) *Telecommunications – A Systems Approach*, Allen & Unwin, London.

37 Television systems and signals

Analysis of light reveals three components:

- brightness, which is electrically equivalent to the amplitude of the light;
- hue, the colour which is determined by its wavelength; and
- saturation, which is the variation in the depth of colour, i.e. the variation between a pastel and a deep shade.

White light is the summation of the total radiation within the visible spectrum and can also be simulated, as shown later.

A saturated colour that is diluted with white light becomes desaturated. Thus white is the ultimate desaturated colour. An object in nature appears to be coloured because it reflects certain parts of the spectrum only. Most of the colours that occur in nature can be simulated by the addition of suitable intensities of red, green and blue light, or by the addition of voltages that represent different levels of intensity of these primary colours. This effect can be explained with the aid of Fig. 37.1. The colours at the apexes represent saturated levels

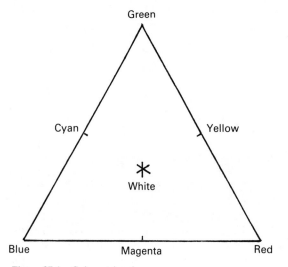

Figure 37.1 Colour triangle.

and can be considered as the colour centres of three light beams. Along the edges, the colours merge and gradually change in hue. The colours halfway along each edge, cyan, yellow and magenta, are known as secondary or complementary colours. These are formed as follows:

cyan = blue + green   yellow = green + red
magenta = red + blue

These are described as complementary for the following reason. If two edge colours connected by a line drawn through the triangle centre are added, the resultant is white. Following any line from the edge towards the centre generates desaturated colours. Thus:

white = red + green + blue (in suitable proportions)

Any colour can therefore be described in terms of its brightness, hue and saturation.

The visible spectrum can be represented as in the Commission Internationale de l'Eclairage (CIE) (1931) diagram shown in Fig. 37.2. All the real colours are contained within the horseshoe shaped boundary and the individual wavelengths can be located from the marked points. The lower edge of the boundary is unmarked because this represents the non-spectral colours, the purples. Figure 37.2 also shows the locations of the primary and secondary colours in relationship to their xy coordinates. The values typically chosen for television, to match the phosphorescence and emission properties of the elements used in TV display tubes, are listed in Table 37.1.

D65 white is one of the standard whites defined for colour TV. It represents the colour that would be produced by a full radiating body at the equivalent colour temperature of 6500 K.

Colours may also be represented by the CIE three-dimensional model shown in Fig. 37.3. This is described as the CIELUV model, where the U and V axes represents the red-green and the yellow-blue components, respectively. The vectors C_{UV}, H_{UV}, therefore

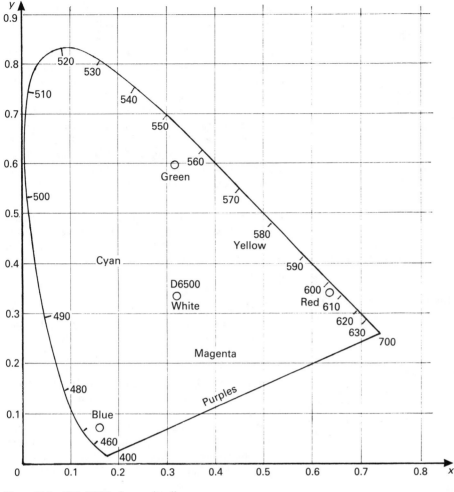

Figure 37.2 CIE (1931) chromacity diagram.

Table 37.1 *TV xy colour values*

Colour	x	y
Red	0.630	0.340
Green	0.310	0.595
Blue	0.155	0.070
White (D65)	0.3127	0.3290

represent the chrominance or saturation and hue, respectively. The brightness relative to a reference white is represented by the *L* axis. Thus any colour can be represented by *L*, *U* and *V* values.

In order to produce a TV image, the scene has to be scanned in a series of very narrow strips or lines. The greater the number of lines, then the better will be the definition of the image. The complete set of lines then form a frame or raster. Frames are displayed at a rate high enough to avoid a flicker effect and this is usually achieved by interlacing or interleaving two fields that make up each frame. The image is thus scanned in two directions (horizontally and vertically) simultaneously. To obtain a satisfactory image definition along the horizontal axis, each line can be considered to be divided into a number of picture elements or pixels. The television signal therefore represents the serial variation of pixel information organised on a line by line basis.

Because the TV display tube has a nonlinear light output/voltage input characteristic, the signal has to be pre-distorted in a complementary manner, in the camera channel before transmission. The display tube characteristic has the form $B = \alpha + \beta V^{\gamma}$, where B is the brightness, V is the applied drive voltage, α and β are constants and γ, which describes the curvature of the characteristic, has a value within the range 2.2 to 2.8.

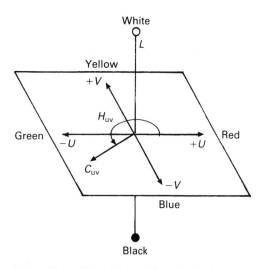

Figure 37.3 'CIELUV' three-dimensional chromacity model.

The gamma correction coefficient $(1/\gamma)$ applied before transmission, therefore lies between 0.45 and 0.35.

37.1 Camera and lens system

The lens is the first element in the system between the scene to be televised and the viewer. It is therefore critically important that this component has practically zero imperfections.

37.1.1 Lens fundamentals

Angle of view. This represents the angular range that can be focused within the image size. It is usually expressed as the angle subtended from the principal point of the lens to the horizontal, vertical and diagonal extremes. It is thus related to the image size and focal length.

Depth of field and focus. Although the focal point is clearly defined in theory, in practice an image will appear to be in focus over a range of distances in front and behind the point on which the lens is focused. By comparison, the range over which points within the depth of field produce focused images is described as the depth of focus. As shown in Fig. 37.4, these parameters are related to the *permissible circle of confusion*, which to a certain extent for the TV camera lens is determined by the width of the raster lines. The depth of field is greater for large F numbers, shorter focal lengths and greater subject distances. The depth of field is also usually greater behind the subject than in front of it.

Effective aperture. This dimension is somewhat smaller than the physical diameter of the lens. It is equal to the diameter of the image of the diaphragm or iris falling on the lens. To a limited extent, this depends upon the angle of view.

F numbers. These represent the brightness of an image formed by a lens. For a lens with a focal length f and effective aperture D, the F number F_N is given by f/D. Lower values of F_N produce brighter images. The lens stop ring is marked with a sequence of values for F_N in the ratio $\sqrt{2}$ (1.4, 2, 2.8, 4, 5.6, 8, 11, 16, etc.). Since the amount of light entering a lens is proportional to the cross-sectional area of the luminous flux (proportional to the square of the diameter), the image brightness will be inversely proportional to the square of F_N. Increasing the setting by one stop decreases the brightness by a half.

Focal length. Parallel light rays passing through a convex lens converge to a point along the optical axis. This is measured from a point at the centre of the lens as indicated by Fig. 37.4. Although a typical system consists of several lenses the principle still applies. The whole system behaves as a single lens but with an imaginary *principal point* located on the optical axis. A

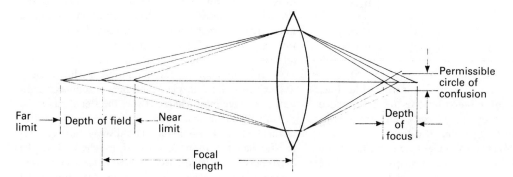

Figure 37.4 Relationship between depth of field and depth of focus.

lens has two focal points, a primary one located on the object side and a secondary one on the image side. The focal length is used to calculate the magnification of the lens and the image position.

Image size. The image formed by a lens is normally circular, while that required for a TV system is a rectangular area just touching the image circle. Hence the major reason for describing the size of a TV image tube or sensor by its diagonal dimension.

Minimum object distance (MOD). This represents the shortest distance between a subject and the front face of the lens for which a focused image can be produced. A single convex lens, a close-up lens, can be added in front of the camera to reduce the MOD, but at the expense of a reduced depth of field.

Modulation transfer function (MTF) or optical transfer function (OTF). This refers to the power of a lens to resolve fine image detail or high spatial frequencies. MTF can be equated with the cut-off frequency of a low-pass system whose output falls with rising frequency. For the lens and camera systems, the input is provided by a grating of alternate black and white bars of reducing width. As the spacing decreases, the system fails to transmit the higher repetition rate of contrast changes and a point is reached where the image becomes uniformly grey with no contrast. The MTF characteristic can be shown on a graph with the contrast reproducibility on the vertical axis and the spatial frequencies in lines/mm horizontally. This then allows comparisons between various lenses to be made.

T numbers. The F numbers express the speed of a lens based on the assumption of 100% transmittance of the incident light. In practice, the transmittance can vary between lenses with the same F number. The T number T_N resolves this problem by taking both the diaphragm diameter and the transmittance into account. The two number scales are related as follows:

$$T_N = \frac{F_N \times 10}{\sqrt{\text{transmittance}\,(100\%)}}$$

Two lenses with the same T number always produce the same image brightness.

37.1.2 Lens imperfections (aberrations and distortion)

Chromatic aberration

Two types of colouration distortions arise because of the wavelength dependence of the refractive indices of glass.

Lateral chromatic aberration arises because the image magnification of a lens changes with wavelength and hence colour. This creates registration errors between the red, green and blue pixels in the image.

Longitudinal aberration causes the different colours to be focused on different image planets. Thus a white spot would be displayed with coloured fringing.

Seidel's aberrations

These five basic aberrations are named after Seidel, the researcher who classified them. The first four affect image sharpness and degrade the MTF. Each can be mitigated to a certain extent, by adjusting the lens to give greater depth of focus.

Astigmatism. Points off the optical axis may not all focus on to the same image plane. Concentric circles come into focus on one plane while radial lines focus on another. Therefore this feature is checked by using a resolving power chart that consists of concentric circles and radial lines.

Coma. The title is descriptive of the aberration effect, images being formed with a comet-like tail. This arises because rays that are incident at an angle to the optical axis of the lens may not focus to a point in the image plane.

Curvature of field. This occurs when the lens fails to focus a plane object as a plane image. If the centre is in focus, the edges will be out of focus. Refocusing on the outer edges simply defocuses the centre.

Spherical aberration. Parallel rays that are incident on to points away from the centre of the lens tend to focus to different points along the optical axis. Thus an object point will be resolved as a disc image.

Distortions. Whereas aberrations are associated with focus problems, distortions affect the image shape. A rectangular object may appear either barrel (negative distortion) or pin-cushion (positive distortion) shaped.

37.1.3 Lens coatings

As light rays pass through the interface between a lens surface and air, spurious reflections occur that represent a loss of light. For a multi-lens system, this loss could be very significant. The effect is considerably reduced by coating each lens with a thin transparent layer, usually of magnesium fluoride (MgF_2). For television purposes, the coating is usually built up in layers (multi-layer) of alternate materials using a vacuum evaporation process.

37.1.4 Prisms – colour separation

The separation of light into the red, green and blue colour channels is carried out with the aid of a prism structure similar to that shown in Fig. 37.5. Trimming filters are added at the exits to each prism in order to eliminate unwanted light and improve the colour rendition. This structure makes use of the reflective and colour selective properties of dichroic layers to separate the incident light into its three primary channels (red, green and blue). A dichroic layer is formed by the vacuum evaporation of between 10 and 20 layers of materials with alternating high and low refractive indices. The choice of a particular material and the layer thickness determines the colour separation process. As indicated by Fig. 37.5, the blue and red components are extracted by reflection, leaving the green light to pass straight through the prism assembly. At the exits, three images are formed in the primary colours; either on the face plates of special cathode ray tubes described as camera pick-up tubes, or on the surface of a solid-state sensor of the charge coupled device (CCD) type. In either case, these light images are scanned in raster fashion to provide the video signal. In order to reduce the effect known as after-image, a bias light is normally used. This evenly illuminates the image plane on the front face of the prism assembly at a fixed level. The light is often introduced by edge lighting a thin glass plate mounted in front of the prism structure.

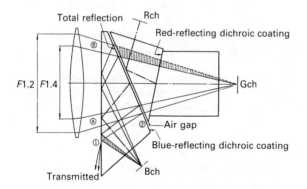

Figure 37.5 *F*1.2 lens fitted to a three-piece *F*1.4 prism (courtesy of Canon (UK) Ltd).

37.1.5 Zoom lens

A zoom lens system is one in which the focal length can be continuously varied without losing focus. If the position of a lens placed between a viewer and a distant object is varied, the size of the image changes, as does the point of focus. Thus a change of magnification must be accompanied by a simultaneous change of focus. If two lenses are used and their movements coordinated, then magnification can be varied without loss of focus. Therefore a lens assembly designed to provide the zoom function, consists of multiple lenses, some convex (converging) and some concave (diverging). The lenses are arranged in four groups in the manner shown in Fig. 37.6. The two inner groups provide adjustment for magnification (the variator) and the maintenance of focus (the compensator). The first fixed group provides overall focus while the fourth fixed group acts as a relay lens, matching the transmittance characteristic to those of the following prism. To ensure that the variator and compensator move in precise unison, their movement is controlled via a barrel cam mechanism. This has both linear and curved paths to control the motion whose complexity can be seen in Fig. 37.6 (a) and (b).

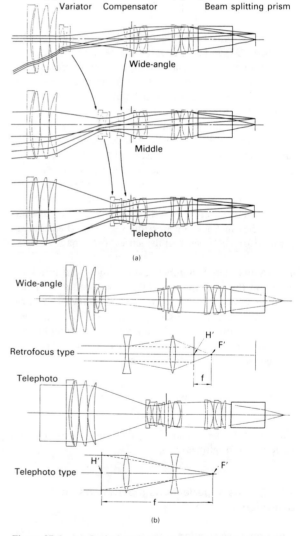

Figure 37.6 (a) Optical path of hand-held zoom lens; (b) lens positions at wide-angle and telephoto ends of zoom (courtesy of Canon (UK) Ltd).

37.1.6 Semiconductor image sensors (CCD)

Until very recently, the CRT was the main image pick-up device. With the developments in semiconductor technology, these are progressively giving way to solid-state, light sensitive devices based on the charge coupled concept (see Semiconductor devices and technology, p. 292). For many applications, CCDs with their low power requirements, light weight and high sensitivity can provide images equal to those of the CRT.

The $\frac{1}{2}$ in or $\frac{2}{3}$ in diagonal image pixel area is fabricated as a rectangular matrix. This provides a resolution ranging from 756×485 pixels for standard 525 line television, up to more than 1920×1036 pixels for HDTV applications.

Although there are many minor variations between types from different manufacturers, CCDs in general operate with the following basic principles.

Interline transfer (IT). In these devices, the vertical columns of pixel cells alternate between the function of image sensor and storage element. During the horizontal blanking interval (HBI), the accumulated charges due to the illumination are transferred vertically from the sensors to the storage cells. These are then read out horizontally during the active line period. High levels of light may penetrate into the substrate region and create excessive charges. These may leak into the storage registers to give rise to vertical smearing. One of the various methods adopted to minimise this effect and at the same time improve the vertical resolution, is to offset the pixel sites by half a pixel pitch between odd and even fields.

Frame transfer (FT). These devices can achieve a high pixel density due to the way in which the image and storage areas are organised. A high speed shutter (typically $1/2000\,\text{s}$) allows the image to integrate over a field period. The shutter then closes so that the image charges for the complete field can be moved to the storage area. From here, the charges are read out line by line during the following field period, to produce the video signal.

Frame interline transfer (FIT). These devices utilise a variable speed shutter to maximise the exposure and charge transfer periods to improve performance. An intermediate gate area is positioned between the image and transfer regions so that the effects of over-exposure can be better controlled.

Hole accumulation diode (HAD) sensors. Under conditions of darkness, random electrons can be created within the light sensitive region to give rise to a *dark noise* component. To overcome this, the Sony Inc.

developed sensor carries a parallel diode with a heavily doped P type region of just two or three molecules thickness to act as a trap.

Hyper-HAD. This further Sony development of the HAD sensor produces a doubling in sensitivity which is equivalent to an extra F-stop to the camera. This is achieved by fabricating a micro lens over each pixel to concentrate the light energy into the photo-sensitive region.

37.2 Current analogue systems

All the current systems, NTSC (National Television Standards Committee), SECAM (Sequential Coleur à Mémoire), and PAL (Phase Alternation Line-by-line) are compatible with earlier monochrome systems.

Each variant of these systems is based on the concept of the brightness or luminance (Y) component signal of the monochrome systems, plus chrominance (colour) information. In all three systems, the Y component is formed from the weighted addition of red (R), green (G) and blue (B) gamma corrected camera voltages, that represent the three primary colour signals, i.e.

$$Y' = 0.299R' + 0.587G' + 0.114B'$$

Three colour difference signals are needed at the receiver, $(R' - Y')$, $(G' - Y')$ and $(B' - Y')$. When the Y' component is added to each, a colour signal voltage is regenerated that represents the original colours.

Of the three colour difference signals, $(G' - Y')$ is always of the lowest amplitude and will thus be affected the most by noise in the transmission channel. Since the Y' component contains a portion of all three colours, $(G' - Y')$ can be regenerated at the receiver from the Y', $(R' - Y')$ and $(B' - Y')$ components. These latter three components are therefore all that is necessary for the transmission of colour TV signals. Thus the two colour difference signal components to be transmitted are of the form $(B' - Y')$ and $(R' - Y')$. These are then modulated on to the sub-carrier frequencies. In order to conserve bandwidth, the spectral components of modulation can be interleaved with those of the luminance signal. This can be arranged because the luminance signal spectrum consists of spectral lines separated by $25\,\text{Hz}$ (or $30\,\text{Hz}$), which are displaced in packets that are, in turn, spaced at line frequency as indicated by Fig. 37.7. In a limited way, this interleaving influences the choice of the actual sub-carrier frequency which has to be chosen, not only for reasons of frequency interleaving, but to avoid patterning due to beat frequencies between the luminance and chrominance signal components. Furthermore, the

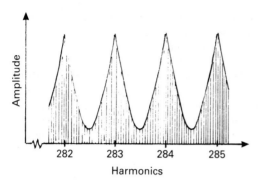

Figure 37.7   TV luminance signal spectra.

$$\begin{bmatrix} Y \\ I \\ Q \end{bmatrix} = \begin{bmatrix} 0.299 & 0.587 & 0.114 \\ 0.596 & -0.275 & -0.322 \\ 0.211 & -0.523 & -0.312 \end{bmatrix} \cdot \begin{bmatrix} R' \\ G' \\ B' \end{bmatrix}$$

or alternatively as:

$$I = 0.596R' - 0.275G' - 0.322B'$$
$$Q = 0.211R' - 0.523G' - 0.312B'$$

As indicated in Fig. 37.8, it can be deduced that the I and Q components contain a fraction of both colour difference signals expressed as follows:

$$I = 0.736(R' - Y') - 0.269(B' - Y')$$
$$Q = 0.478(R' - Y') + 0.414(B' - Y')$$

This was arranged to combat the effects of nonlinearity, where amplitude distortion might cause a change of colour saturation. More importantly however, phase distortion would produce an actual change of hue (colour).

Consistent with the technology then available, the NTSC system provided very adequate colour images and maintained compatibility with the previous monochrome service.

The later developing European colour TV industry recognised these possible problems and devised the SECAM system in France and the PAL system in West Germany, almost in parallel. Due to technological advances made during the intervening time, both of these systems provided better quality pictures and so became competing European Standards.

technique can lead to a useful bandwidth reduction. If, for example, it is required to transmit the luminance signal with an image resolution of 6 MHz, then it would be necessary to use three 6 MHz channels for each of the red, green and blue signals. Since relative to the luminance, the eye is only about half as sensitive to colour level variation, the two colour difference channels need only 3 MHz of bandwidth each, thus reducing the bandwidth requirement from 18 MHz to 12 MHz.

The major differences between the systems lie in the way in which the chrominance information is modulated on to the sub-carriers.

37.2.1 NTSC system

The NTSC system of America, was the first system to enter operational service. This was made compatible with the earlier monochrome service that was based on a 525 line image format. The line and field rates were nominally 15.75 kHz and 60 Hz, respectively. To accommodate the 3.579545 MHz sub-carrier and minimise the interference beats between this and the sound carrier, these frequencies were marginally offset to 15.734254 kHz and 59.94 Hz, respectively. (These changes were insufficient to cause a monochrome receiver to lose timebase lock.) The colour sub-carrier therefore lies exactly halfway between the 227th and 228th harmonic of the line timebase frequency. The chosen colour difference components amplitude modulate quadrature (QAM) versions of the same carrier frequency, to produce I (in phase) and Q (quadrature) components, with the double side band, suppressed carrier (DSBSC) mode being employed. Like the Y' component, both I and Q signals are formed by weighted addition of the R', G' and B' primary components as indicated by the following matrix:

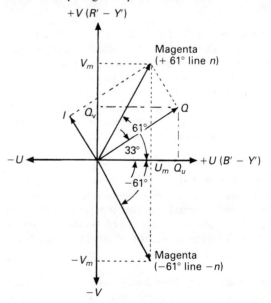

Figure 37.8   Comparison of NTSC and PAL chrominance phasors.

37.2.2 PAL system

For the PAL system, the three signal components can be derived from the following matrix:

$$\begin{bmatrix} Y \\ U \\ V \end{bmatrix} = \begin{bmatrix} 0.299 & 0.587 & 0.114 \\ -0.169 & -0.3316 & 0.500 \\ 0.500 & -0.4186 & -0.813 \end{bmatrix} \cdot \begin{bmatrix} R' \\ G' \\ B' \end{bmatrix}$$

The two colour difference signals are scaled or weighted so that the total signal of luminance plus chrominance does not produce over-modulation. The two modulating colour difference components are thus represented by:

$$U = 0.493(B' - Y') \text{ and } V = 0.877(R' - Y')$$

Like the NTSC system, PAL utilises QAM, DSBSC, but uses an alternating phase inversion, line-by-line, of the $V(R' - Y')$ component. In this way, phase errors on successive lines tend to average out and cancel, thus improving the colour performance. Figure 37.8 shows how the NTSC and PAL chrominance components differ. Both the I and the Q signals contain an element of each colour difference signal due to the rotation of the reference phase. The diagram compares the phasors for a magenta hue that forms the basis of the two chrominance signals.

The relationship between the sub-carrier and line timebase frequencies is much more complex due to the problem of *line crawl*. The precise value of 4.43361875 MHz represents 283.75 times line frequency plus an off-set of 25 Hz. This results in an eight-field sequence before the colour sub-carrier phase repeats itself.

37.2.3 SECAM system

The SECAM system adopted the technique of transmitting the two colour difference signals sequentially, on alternate lines, using frequency modulation of two different sub-carrier frequencies, thus ensuring that there would be no cross-talk between these two components. However, the two frequencies needed to be very carefully chosen, to minimise the effect on monochrome areas of the image. Because, unlike both NTSC and PAL, that both use DSBSC, the carrier amplitudes do not fall to zero, when both the $(R' - Y')$ and $(B' - Y')$ components are absent, as in the grey picture areas. The two colour difference signals of SECAM are designated as $D_R = -1.902(R' - Y')$ and $D_B = 1.505(B' - Y')$ which are pre-emphasised before being used to frequency modulate the two sub-carriers of 4.40626 MHz and 4.25 MHz, respectively. The magnitude of the coefficients is chosen so that total deviation is restricted to 3.9 to 4.75 MHz for both sub-

carriers. The negative coefficient of D_R results in the two difference signals deviating their carriers in opposite directions. This was adopted in order to minimise the possible effects of differential phase errors, but practice has shown this to be unnecessary. The later specifications for SECAM therefore use negative coefficients for both D_R and D_B. The sub-carriers are maintained to an accuracy within ± 2 kHz and represent the 282nd and 272nd harmonics of the line timebase frequency.

37.2.4 Sampling problems — aliasing

The increased size and brightness of the modern picture tube causes system impairments (*artifacts*) to become apparent. These include such as *cross-colour/luminance* effects, that arise due to the imperfect separation of the interleaved spectra. Luminance information can reach the chrominance channel and create false colours, and in a similar way chrominance information can create high frequency patterning in the receiver luminance channel. These effects can be minimised by the use of modern comb filter techniques or by digital processing of the composite video signal.

However, the sampled nature of the image signal creates aliasing problems that are not so easily solved. The response of the image can be expressed in terms of three kinds of frequency, as indicated by Fig. 37.9(a).

Considering the horizontal frequency f_x as a continuous function of time and hence analogue, the remaining two frequencies f_v and f_t represent the two-dimensional vertical (line) and temporal (field) sampling frequencies, respectively. Figure 37.9(b) represents this spatio-temporal spectra for a period when $f_x = 0$.

A sampling operation always generates repeated sideband pairs, related to the baseband and disposed around multiples of the sampling frequency. Figure 37.9(b) translates such spectra into two dimensions simultaneously, where the *quincunx* areas (four corner and centre points of a square) represent the repeating spectra. The diagram is scaled for a 625 line system using 575 active lines per frame, to give a vertical resolution of $575/2 = 287.5$ c/ph (cycles/picture height), with a 25 Hz temporal or frame frequency. Even if the spectral groups do not overlap to produce aliasing, the human eye will act as an imperfect filter. The approximate response of the eye is enclosed by the dotted line. This includes some of the repeat spectra, giving rise to an aliasing effect.

Referring to Fig. 37.9(b), area A is responsible for *large area flicker*, which is essentially a peripheral vision effect. It is not only related to the signal, but also to the display viewing angle. Area B represents the *vertical aliasing*, which is responsible for the visibility of the line structure and the *Kell effect* which describes the loss of vertical resolution relative to that in the horizontal plane. Area C, the *temporal aliasing*, only exists in

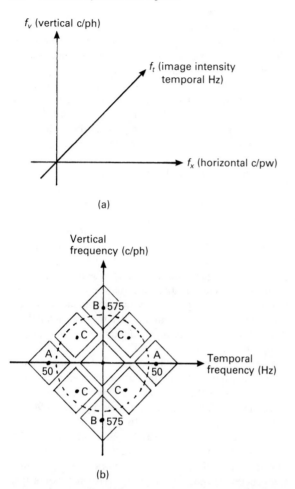

(a)

(b)

Figure 37.9 (a) Three-dimensional representation of image response; (b) two-dimensional spectra for $f_x = 0$.

interlaced raster structures and is responsible for the apparent inter-line flicker or crawl. This effect also causes wheels to appear to rotate in reverse. Increasing the frame rate will only increase the frequency at which this occurs. Further technological improvements, in either of the current analogue systems, are unlikely to remove these alias-based artifacts completely.

37.3 Current analogue/digital systems

In addition to the cross-colour/luminance artifacts produced by the current analogue systems, a noise problem arises when these signals are transmitted through a satellite link using frequency modulation (FM). Figure 37.10 shows the disposition of the chrominance components within the luminance spectrum, together with the superimposed triangular noise spectrum for FM. As can be seen, the chrominance components lie in the noisier region. Demodulation

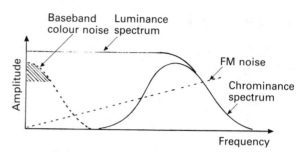

Figure 37.10 Video signal spectra and FM noise effect.

returns the chrominance components to baseband together with the noise element. The colour response of the human eye is approximately triangular and complementary to the FM noise spectrum. Therefore the colour noise appears in the most annoying region, being particularly troublesome in the highly saturated colour areas of the image.

It was to combat these problems and to improve the image quality that the following analogue/digital systems were developed.

37.3.1 Multiple sub-Nyquist sampling encoding system (MUSE or Hi-vision)

This system, developed by the Japan Broadcasting Corporation (NHK), Technical Research Laboratories, was intended from the outset to be capable of providing a high definition service. The basic parameters of the system are shown in Table 37.2.

Table 37.2 *The MUSE system parameters*

Scanning lines per frame	1125
Active lines per frame	1035
Pixels/line	1920
Field frequency	60 Hz
Interlace	2:1
Aspect ratio	16:9
Luminance bandwidth	20 MHz
Chrominance bandwidth	
Wideband C_W	5 MHz
Narrow band C_N	4 MHz

The signal components are formed using the following matrix:

$$\begin{bmatrix} Y \\ C_W \\ C_N \end{bmatrix} = \begin{bmatrix} 0.3 & 0.59 & 0.11 \\ 0.63 & -0.47 & -0.16 \\ 0.03 & -0.38 & 0.41 \end{bmatrix} \cdot \begin{bmatrix} R' \\ G' \\ B' \end{bmatrix}$$

By using a technique known as *time compression integration* (TCI) and multiple sub-sampling, this signal can be squeezed into a baseband of just 8.1 MHz.

To achieve time compression, the chrominance components are sampled at one rate, stored in memory and then clocked out at a higher rate. This process automatically increases the bandwidth requirement. Two compression factors ($\times 4$ and $\times 5$) are used so that the processed signals C_W and C_N occupy the same 20 MHz bandwidth as the uncompressed luminance.

Two different systems are defined where either the two chrominance components may be multiplexed in sequence into about 19% of each line period, or the whole 19% period allocated to C_W and C_N on alternate lines.

The luminance component which transmitted in interlaced form is uncompressed, and time multiplexed along with control and sync information in the manner indicated by Fig. 37.11. The three intervals shown represent 12, 94 and 374 pixel points at the 16.2 MHz sampling rate.

Bandwidth reduction is achieved by spatial and temporal filtering, sampling and sub-sampling. The video signal is sampled at 64.8 MHz, which produces 1920 sample points per line. These are stored in a four-field memory which is then resampled at the lower rate of 16.2 MHz. This results in the selection of every fourth pixel data. The field sub-sampling pattern and sequence is shown in Fig. 37.12.

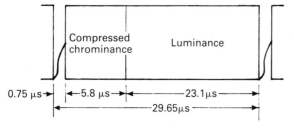

Figure 37.11 Line multiplex format.

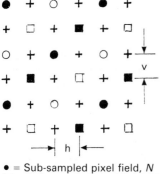

- ● = Sub-sampled pixel field, *N*
- □ = Sub-sampled pixel field, *N* + 1
- ○ = Sub-sampled pixel field, *N* + 2
- ■ = Sub-sampled pixel field, *N* + 3
- + = Interpolated pixel
- v = Vertical sampling spacing
- h = Horizontal sampling spacing

Figure 37.12 Raster sub-sampling pattern.

At the same time the image data is compared, inter-frame and intra-frame, to detect minor movement within a frame. From this, a motion compensation signal is generated. In addition, camera operations such as pan and zoom can create whole image movements which can be compensated for by the application of signals known as *motion vectors*. The motion compensating signals are then multiplexed into the vertical blanking interval for decoder control purposes.

The stereo sound channel is sampled at 32 kHz and the data bits transmitted as four-phase DPSK, within the vertical blanking interval of the time multiplex. At this stage, the composite signal has a baseband width of 8.1 MHz.

37.3.2 Multiplexed analogue component systems (MAC)

In the UK, the Independent Broadcasting Authority (IBA), now National Transcommunications Ltd (NTL) developed an alternative solution. The *Y* and *U,V* components of video are each time compressed for time division multiplex transmission which automatically resolves the cross-luminance/chrominance problem. This concept, known as multiplexed analogue components (MAC), has given rise to a series of variants, each with some different properties. These have largely been developed within the framework of the European Broadcasting Union (EBU) and the European Community's (EC) Eureka scheme. The format of the video line time multiplex, which is shown for one of the variants in Fig. 37.13, is typical of almost all versions. The active line period for the current systems is about 52 μs. For MAC systems, the luminance signal is time compressed by a ratio of 3:2 so that it occupies about 35 μs. The two colour difference signals are compressed by a ratio of 3:1, with a corresponding duration of approximately 17.5 μs. By transmitting the compressed *Y* signal on each line, with one of the compressed *U, V* signals alternately, the active line period is still 52 μs. As the action of the conventional line sync pulse can be replaced by a digital control, there is a period of about 10 μs which is available for digitised sound, data, and digital control signalling.

A-MAC. This original version employed the time compressed analogue *Y,U,V* multiplex but with the sound channel modulated onto a separate carrier spaced 7.1 MHz from the vision carrier. Since this was likely to lead to intermodulation problems, A-MAC has not been adopted.

B-MAC. This version time multiplexes the digital sound and data information at baseband, into the 10 μs horizontal blanking period. The video bandwidth is

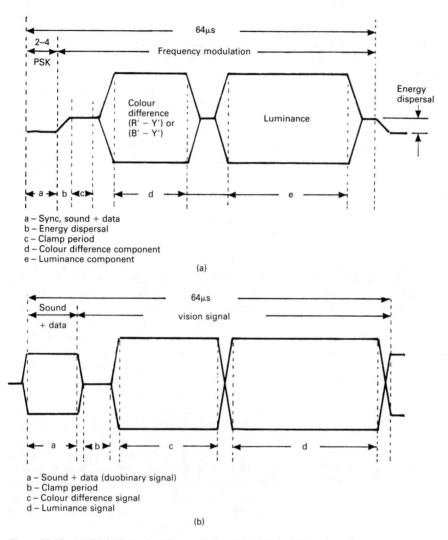

a – Sync, sound + data
b – Energy dispersal
c – Clamp period
d – Colour difference component
e – Luminance component

(a)

a – Sound + data (duobinary signal)
b – Clamp period
c – Colour difference signal
d – Luminance signal

(b)

Figure 37.13   (a) C-MAC (packet) line multiplex; (b) D/D2-MAC (packet) systems line multiplex.

limited to about 6 MHz and so is suitable for distribution via both satellite and cable systems.

C-MAC. This uses an RF time division multiplex. The carrier is frequency modulated by the analogue vision signal components for 52 μs and then digitally modulated during a further 10 μs by a multiplex of the sound channel and data.

C-MAC/packet. This is a derivative of C-MAC and forms one of the acceptable standards for DBS in Europe. The variation applies only to the organisation of the digital information into *packets* each of 751 bits.

D-MAC/packet. This member of the family, as originally specified, used an RF multiplex for the vision and sound plus data signals. The two components are carried on the separate carriers, typically spaced by 10.5 MHz. These carriers could either be frequency modulated for transmission through a satellite link, or amplitude modulated (vestigial sideband) for a cable network. The current specification calls for the sound and data to be time multiplexed into the 10 μs line blanking period, with all components being modulated on to the same carrier. While the vision signal is frequency modulated in the conventional MAC manner, the digital components use *duo-binary* digital modulation at a sampling rate of 20.25 MHz.

D2-MAC. This variant of D-MAC has been adopted as a standard by the EBU, chiefly for use in France and Germany, the major change being a reduction in the sound and data bit by a factor of two (10.125 MHz).

Further variations have been developed for special purposes, and Table 37.3 shows the standard MAC parameters adopted for the 625 line services.

Table 37.3 *Standard MAC parameters*

Frame frequency	25 Hz
Line frequency	15.625 kHz
Interlace	2:1
Aspect ratio	4:3
Y compression ratio	3:2
U, V compression ratio	3:1
Luminance bandwidth	5.6 MHz
Chrominance bandwidth	2.8 MHz
Transmission baseband	8.4 MHz (Luminance, $5.6 \times 3/2 = 8.4$) (Chrominance, $2.8 \times 3 = 8.4$)
Sound channel	40 to 15000 Hz
Sampling frequency	32 kHz
Dynamic range	> 80 dB

All the MAC versions eliminate cross-colour/luminance and reduce the colour noise. In addition, provision is made for sound and data channels that include multiple stereo, bilingual sound, teletext and controlled access.

While Fig. 37.13(a) shows the general arrangement of the time multiplex used for the MAC systems, the periods a, b, and c specifically refer to the C-MAC/packet variant. Figure 37.13(b) shows the corresponding line multiplex for the D/D2-MAC/packet variants.

The digital components within period a, include line and demodulator sync, sound and data, which are organised into packet multiplexes of 751 bits. Each packet contains a header and data section. Each header contains a 10-bit address, to identify any one of 1024 different services, two continuity bits to link successive packets of the same service and 11 protection or parity bits for a Golay 23,12 cyclic error correcting code. This can correct any three errors in up to 23 bits. Packet 0 is permanently allocated to the service indentification system.

The useful data area contains one 8-bit byte to indicate the packet type (PT). For instance, the sound decoder needs to know whether the coding law is linear or *near-instantaneously companded* and also whether the sound signal is full bandwidth (high quality), or reduced bandwidth. The remaining 720 bits (90 bytes) may contain an *interpretation block* (BI), with the necessary instructions for setting up the sound decoder; or an actual sound data coding block (BC).

A triangular energy dispersal signal of 25 Hz and at an amplitude suitable to provide a peak to peak deviation of 600 kHz is added. This is gated to zero during sound, sync and data periods. The clamp period allows the use of a receiver clamping circuit to remove this waveform at demodulation.

Figure 37.14 shows the structure of the C-MAC/packet frame multiplex. Luminance information is carried on lines 24 to 310 and 336 to 622. Chrominance is additionally carried on lines 23 and 335, when the luminance period is set to black level. Line 624 can be used for carrier recovery and the setting of the receiver AGC system. Line 625 is completely reserved for frame sync and service identification. The field blanking period thus occupies 25 line periods.

Line 625 carries the maximum of 1296 bits, and with a duration of 64 μs this represents a bit rate of 20.25 Mbit/s. On the remaining lines, the first seven bits are used for demodulator and line sync, while the remaining 198 bits are used for sound and data purposes. There are thus $(623 \times 198) + 8 = 123\,362$ bits/frame, which are organised into 164 non-contiguous packets, each of 751 bits, within the two sub-frames as shown in Fig. 37.14. Thus there are a significant number of bits that are unused and available for future developments.

Figure 37.15 shows the C-MAC/packet signal as displayed without expansion and remultiplexing, on a standard receiver. The digital data, chrominance and luminance components can be clearly recognised. In particular, the chrominance section shows the sequential nature of the colour difference signals.

37.4 High definition TV (HDTV) systems (see also image compression, p. 357)

High definition systems are defined as those capable of providing an image resolution of greater than 1000 lines per picture height (lines/ph) and with an aspect ratio of 16:9. To provide such images, a video base bandwidth in the order of 30 MHz is required. Since there is virtually no part of the frequency spectrum available to accommodate such a signal using the modulation techniques of current systems, some form of image compression/processing is absolutely necessary.

37.4.1 Analogue/digital HDTV systems

The characteristics and parameters of the Japanese MUSE/Hi-vision system introduced in Section 37.3.1 describe one HDTV system already in commercial use.

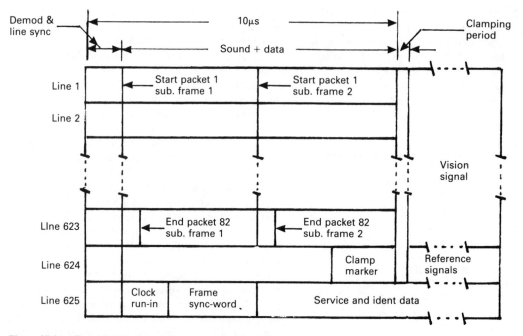

Figure 37.14   C and D-MAC (packet) systems frame multiplex.

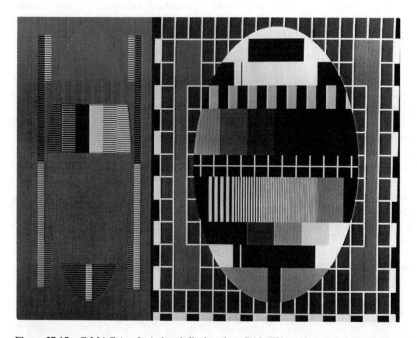

Figure 37.15   C-MAC (packet) signal displayed on PAL TV receiver (courtesy Independent Broadcasting Authority).

In Europe, much of the development work in this area has been based on the extension of the MAC concept.

HDB-MAC

In the USA, extensive development work has been carried out by Scientific Atlanta Inc. to expand the capabilities of B-MAC. As a result, HDB-MAC is a wide screen (16:9) system based on B-MAC encoding/ decoding but with extended bandwidth filters (10.5 MHz) to cater for the high definition input signals, together with a pre-processor stage. Using horizontal and vertical filtering, this stage produces spectrum folding at 7 MHz. The encoder stage then compresses this signal and adds other components to provide a combined transmission bandwidth of 10.5 MHz.

At the receiver, a standard B-MAC decoder discards the folded components, selects the central 4:3 aspect ratio section and then converts the HD signal into a standard NTSC signal. For the HD receiver, the HDB-MAC signal is decoded in the wide aspect ratio format, with 525 lines sequentially scanned at the 59.94 Hz field rate. A high definition image is then formed by using a field store and line by line interpolation.

HD-MAC

The HD-MAC signal (1250/50/2:1) encoding starts by reducing the number of lines by a factor of 2, to ensure that the signal fits into a standard 6 MHz, 625 line MAC channel and is compatible with a standard MAC receiver. Then by including a digital assistance signal (DATV), that is transmitted in the field blanking interval, the HD-MAC receiver is able to regenerate the missing lines.

Squeezing the 25 MHz bandwidth vision signal into a standard MAC channel, means that the bandwidth has to be compressed by a factor of 4. At the encoder, this is achieved for the luminance signal by using ÷4 sub-sampling. Both the static and moving area digital signals are filtered before sub-sampling. Two interpolators are then used to reconstruct the original signal from the reduced sample sets. Both versions are then compared with the original input to detect which produces the least error. As indicated in Fig. 37.16, this is then gated to the output, while the minimum error component is used to generate the DATV signal.

Thus the encoding is based on an adaptive process that depends upon the motional changes in the image. To take into account motion at different rates, information is derived over 4, 2 or 1 field periods to represent stationary, slow or fast changes, respectively. At the receiver, line-by-line interpolation regenerates the 1250 line format and the DATV signal component is used to indicate not only the direction of motion but also its rate of change.

37.4.2 Digital HDTV systems

On any broadcast network, transmission frequencies can only be reused if mutual adjacent channel interference is avoided, a feature that gives rise to the term *taboo* channels. This interference arises primarily due to the power distribution throughout each channel, being particularly high in the region of the carrier and sub-carrier frequencies. By using digital processing and time multiplexing of the luminance, chrominance and sound signals, all these components can be modulated onto a single carrier. Since digital signals can invariably be transmitted at a lower power level than that required for analogue signals, a digital system becomes more taboo channel friendly. To transmit an HDTV signal with a 30 MHz base bandwidth within a 6 or 8 MHz channel using analogue methods, requires bandwidth compression techniques, which to date have not been developed. Therefore the only alternative is to use digital signal processing (DSP).

The bandwidth required to process the luminance component of a 625/25/2:1 image using a 4:3 aspect ratio can be calculated as follows:

$$625 \times 625 \times 4/3 \times 1/2 \times 25 = 6.51 \text{ MHz}$$

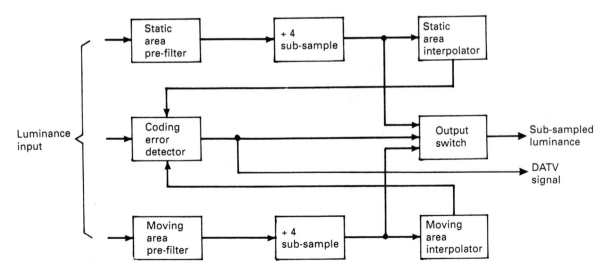

Figure 37.16 HD-MAC encoding.

Using a sampling frequency of 13.5 MHz and a resolution of 8 bits per sample yields a bit rate of 108 Mbit/s. To include the two chrominance components, each with a resolution of 4 bits per sample, adds a further 108 Mbit/s to give a total bit rate of 216 Mbit/s. It can be shown in a similar way that the total bit rate for an HDTV image using the European 1250/25/2:1 format and with a 16:9 ratio, yields a total bit rate of almost 1.2 Gbit/s.

DSP provides the key to significant bit rate reductions which can be carried out in several stages:

- by making use of the redundancy that occurs in most images where large areas do not change significantly from frame to frame. This allows a form of differential PCM (DPCM) to be employed;
- by manipulating the image pixel data using a two-dimensional (2D) transform coding technique, followed by matrix scanning in a manner that produces long runs of near zero values;
- by using adaptive quantisation or rounding of the matrix coefficients to reduce the number of values that need to be transmitted;
- by adopting a form of run-length coding where for example, the values 7,7,7,7 would be coded 7,4 (i.e. 7 four times);
- by using a modulation scheme that allows each transmitted symbol to represent several binary digits. For example, digital 16 quadrature amplitude modulation (16 QAM) based on the PSK system but using eight vectors spaced by 45°. Since each may occupy one of two levels this provides 16 unique code vectors to generate 4 bits per transmitted symbol. However, relative to bi-phase PSK, reducing the vector spacing from 180° to 45° represents a noise penalty of 6 dB. In the future, it might be possible to use 256 QAM which will allow each transmitted vector to represent 8 bits per symbol.

Because digital signals are generally more robust than analogue ones under noise conditions, a lower transmission power level can be used. It is generally agreed that a C/N ratio of 45 dB is necessary to achieve good image quality in an analogue system. With current digital technology, similar image quality can be achieved with a C/N ratio of around 20 dB, and with a more uniform spectral power distribution.

37.4.3 DigiCipher system (General Instrument Corp., USA)

This all-digital system has been designed for transmission within a 6 MHz channel and is therefore suitable for terrestrial as well as satellite distribution. It can operate on relatively low power using small antennas, and a C/N

ratio of less than 20 dB will produce error-free reception. It is therefore also taboo channel friendly.

The coding system generates a digital data stream multiplex that includes vision, sound and teletext type data. Therefore neither a sound carrier nor chrominance sub-carrier is necessary for its transmission. Using digital 16-QAM with four carrier phases each with four permitted amplitudes, produces a peak to average power variation of only 5 dB.

The source image is obtained from R, G, B inputs that are matrixed to produce Y, U and V components with luminance and chrominance bandwidths of 22 MHz and 5.5 MHz, respectively. The image format is based on 1050 lines at a field rate of 59.94 Hz, with 2:1 interlace and 16:9 aspect ratio. The sampling rate is 51.8 MHz with 8-bit resolution for both luma and chroma components. This video bit stream is then compressed to provide a bit rate of 13.83 Mbit/s.

Four audio channels of compact disc quality are provided for by sampling at 44.05 kHz to give a total audio bit rate of 1.76 Mbit/s. These bit streams are then multiplexed with data and text (126 Kbit/s) and system control data (126 Kbit/s) to give a total effective bit rate of 15.84 Mbit/s. When this is processed through the forward error correction (FEC) circuit using a Reed–Solomon (154,130) code, the final transmission bit rate rises to 19.43 MBit/s. Using 16-QAM with its 4 bits per symbol, the final baseband signal becomes 4.86 MHz.

The video signal processing is carried out in five stages:

(1) chrominance processing;
(2) discrete cosine transform (DCT) (see Image processing, p. 139);
(3) adaptive quantisation;
(4) variable length coding (Huffman);
(5) motion compensation and estimation.

Chrominance information is compressed by the use of decimation filters that averages pixels in groups, four horizontally and two vertically. The luma (Y) signal bypasses this stage before being multiplexed with the processed chroma component one block (8×8 pixels) at a time. This multiplex is then passed through the DCT stage with forward (encoder) and reverse (receiver decoder) transformations as follows:

$$F(u,v) = \frac{4C(u)C(v)}{N^2} \sum_{i=0}^{N-1} \sum_{j=0}^{N-1} f(i,j) \cos A \cos B$$

$$f(i,j) = \sum_{u=0}^{N-1} \sum_{v=0}^{N-1} C(u)C(v)F(u,v) \cos A \cos B$$

where

$$A = \frac{(2i+1)u\pi}{2N} \qquad B = \frac{(2j+1)v\pi}{2N}$$

$$C(u) = C(v) = 1/\sqrt{2} \qquad \text{for } u = v = 0$$
$$C(u) = C(v) = 1 \qquad \text{for } u = v \neq 0$$

i, j and $u, v = 0, 1, 2, ..., N-1$, where i, j are the spatial coordinates in the image plane, and u, v are the corresponding coordinates in the transform plane. N is the horizontal and vertical block size.

This DCT coder operates with a form of DPCM input as indicated by points (1) and (4) in Fig. 37.17. The information for the next frame is predicted and compared with the current image data and only the differences DCT coded. A good predictor for this is simply the previous frame. The output forms an 8×8 matrix of coefficients of which the element 0,0 has a value that is twice the average of the 64 elements. It thus represents the dc energy in the block. The horizontal elements represent increasing horizontal frequencies, and the vertical elements represent increasing vertical frequencies present in the image. The elements along the diagonals thus represent the energy in the diagonal frequencies. Generally the magnitude of these higher order elements rapidly tends towards very small or zero values, since images only rarely contain significant diagonal information. Hence the matrix is scanned in the manner shown in Fig. 37.18.

The following stage of quantisation or rounding sets all the small values to zero and truncates certain coefficients in an adaptive manner according to a look-up table stored in ROM. The matrix of elements is then scanned in a manner that leads to long runs of similar values so that an amplitude/run-length coding can be applied. This takes the form of a modified Huffman code using a two-dimensional code book. If it is detected that a block ends with a long run of zeros, an *end-of-block* code can be appended after the last non-zero value to further compress the data stream.

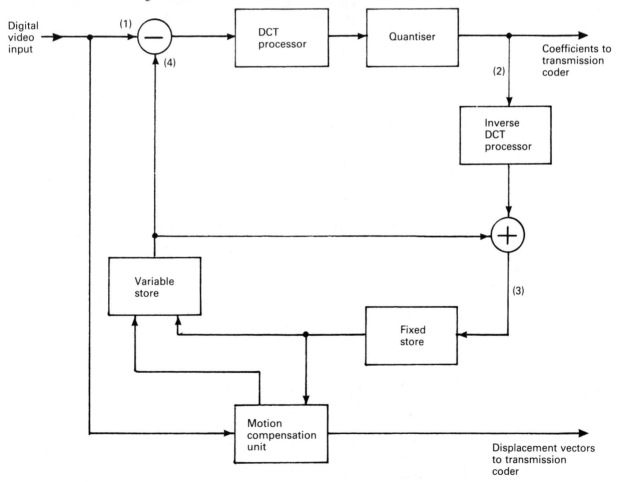

Figure 37.17 DigiCipher image processing.

The listing in Table 37.4 shows how the processing of an arbitrary pixel block produces this run of zero or near zero values. At the same time, the table also shows how the application of the inverse transform reproduces pixel values with very little error.

Motion estimation/compensation is applied to compensate for the errors introduced by temporal image compression. This is achieved by comparing images on a frame-by-frame basis to detect moving areas and predict how this will appear in the next frame. From this data, motion vectors can be derived for transmission to the decoder. In order to reduce the information needed to describe the motion, estimation is performed on a block matching basis, using blocks of 32×16 pixels (a superblock). Since this dimension is compatible with the 4-times horizontal and 2-times vertical sub-sampling

of the chroma signal, a single vector can be used to describe the movement of both components. In this way, the bit rate overhead needed to describe motion is just 9 bits per superblock or about 0.018 bits per pixel.

The variable bit rate output from the Huffman coder needs to be matched to the constant bit rate required for transmission. This is achieved by the use of a buffer memory in both encoder and decoder. This store is capable of holding data for 1 frame ± 1 field variation. The memory level, which is continually monitored, is used to control the adaptive quantisation process to avoid under and overflow of the memory.

The 15 kHz baseband audio signal is pre-emphasised, sampled at 44.05 kHz and quantised to 15 bits resolution. This is then instantaneously μ-law companded to 10 bits. The transmission multiplex allows for

Table 37.4 *Forward and reverse DCT processing of 8×8 pixel block (courtesy of General Instruments Corp., VideoCipher Division)*

139	144	149	153	155	155	155	155	
144	151	153	156	159	156	156	156	
150	155	160	163	158	156	156	156	
159	161	162	160	160	159	159	159	Original 8×8 block
159	160	161	162	162	155	155	155	
161	161	161	161	160	157	157	157	
162	162	161	163	162	157	157	157	
162	162	161	161	163	158	158	158	
314.91	−0.26	−3.02	−1.30	0.53	−0.42	−0.68	0.33	
−5.65	−4.37	−1.56	−0.79	−0.71	−0.02	0.11	−0.30	
−2.74	−2.32	−0.39	0.38	0.05	−0.24	−0.14	−0.02	DCT processed block
−1.77	−0.48	0.06	0.36	0.22	−0.02	−0.01	0.08	
−0.16	−0.21	0.37	0.39	−0.03	−0.17	0.15	0.32	
0.44	−0.05	0.41	−0.09	−0.19	0.37	0.26	−0.25	
−0.32	−0.09	−0.08	−0.37	−0.12	0.43	0.27	−0.19	
−0.65	0.39	−0.94	−0.46	0.47	0.30	−0.14	−0.11	
315	0	−3	−1	1	0	−1	0	
−6	−4	−2	−1	−1	0	0	0	
−3	−2	0	0	0	0	0	0	Quantised coefficients
−2	0	0	0	0	0	0	0	
0	0	0	0	0	0	0	0	
0	0	0	0	0	0	0	0	
0	0	0	0	0	0	0	0	
−1	0	−1	0	0	0	0	0	
139	145	150	154	154	153	154	153	
145	150	154	157	157	155	156	156	
150	155	158	161	160	157	157	155	Reconstructed 8×8
159	161	161	163	161	158	159	158	block
159	160	161	163	161	157	156	155	
163	162	160	162	161	157	157	158	
162	161	159	162	161	157	157	157	
164	162	160	163	162	158	159	160	

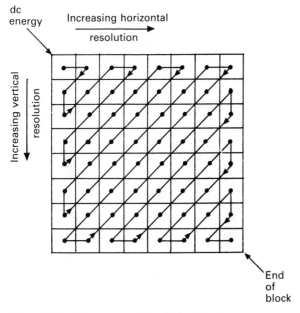

dc energy

Increasing horizontal resolution →

↑ Increasing vertical resolution

End of block

Figure 37.18 Zig-zag scan of 8×8 pixel block.

four such audio channels. Provision is also made to include text type data in 4×9.6 kbaud channels.

The control, data, audio, video and sync bit streams are multiplexed in the manner shown in Fig. 37.19. Video lines 2 through to 1050 contain four control bits, four data bits, 56 audio bits and 440 video bits. For line 1, which is chiefly associated with system synchronism, the last 48 bits are allocated to frame sync, system control and next macro-block position (NMP) (macro-block $= 256 \times 16$ pixels). Once the receiver decoder clock has synchronised, the 24 bits in line 1 provide frame sync. The last 16 bits (NMP) are used to support user channel changes and system recovery from error conditions. This is achieved by using this data to signal the number of bits from the end of the NMP field to the beginning of the next macro-block.

The digital multiplex is then FEC coded using a Reed–Solomon (154,130) code before being applied to the QAM transmission modulation system that

reduces the 19.42 Mbit/s bit rate to a bandwidth of 4.86 MHz.

37.4.4 Digital Spectrum Compatible HDTV System (Zenith Electronics Corp./American Telephone & Telegraph Corp. (AT&T))

The image input source for this system is gamma corrected R, G, B signals that are matrixed to provide Y, U, and V components. These in turn are bandwidth limited to 34 MHz for luma and 17 MHz for chroma. The image format is based on 787.5 lines per frame and 59.94 frames per second, displayed as 1575 lines progressive (or sequentially) scanned every 1/29.97 s, with an aspect ratio of 16:9. This produces a line scan frequency of about 47.20275 kHz, or exactly three times that of standard NTSC. The digital sampling frequency is 1596 times line frequency or about 75.3356 MHz. This image structure provides a high Kell factor in the order of 0.9. Conversion of this image format to standard NTSC only requires interpolation of 4:3 horizontally and 3:2 vertically.

Motion compensated transform coding is used to remove the temporal (inter-frame) and spatial (intra-frame) redundancy. Temporal redundancy is removed by estimating the motion of objects from frame to frame using a block matching technique, while motion vectors derived from the luminance signal only, are used to remove the spatial redundancy.

The principle of encoding is shown in Fig. 37.20. Essentially this predicts how the next frame will appear and, using a decoder similar to that in the receiver, compares the prediction with the actual next frame. Any difference is then applied to the transmitted frame data stream to minimise the error.

The motion estimator compares two successive frames and generates motion vectors for the next frame. These are compressed and stored in the channel buffer for transmission. Each frame is analysed before processing to determine the perceptual importance of each coefficient. Motion vectors and control parameters resulting from forward estimation are then input to the encoder loop. This outputs the compressed prediction

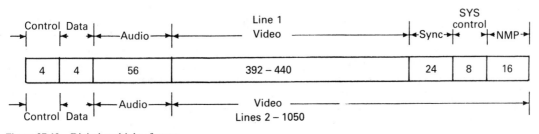

Figure 37.19 Digital multiplex format.

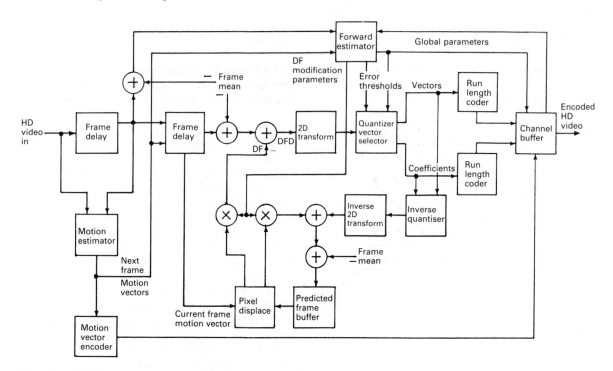

Figure 37.20   Digital spectrum compatible HDTV system encoder.

error to the channel buffer memory, the control parameters being weighted according to the buffer state, which is in turn monitored by the forward estimator.

In the predictive loop, the data for the new image and the predicted image are coded using a two-dimensional adaptive transform, the parameters for this coding stage being in part controlled by the forward estimator. Before processing in the prediction loop, the input frame has its mean value subtracted in order to produce a *zero-mean* input to maximise the efficiency of the transform stage.

The encoded information that is to be transmitted is also decoded as in the receiver. The pixels of the previously decoded image are displaced (motion compensated) and the result modified by control parameters from the forward estimator. This displaced frame (DF) is subtracted from the input zero-mean frame to produce the displaced frame difference (DFD) and this data stream is then 2D transformed. The coefficient data is quantised and run length coded before being passed to the channel buffer, the luma and chroma difference pixels being coded separately. The total data is then formed into a packet and Reed–Solomon FEC coded (167,147), for transmission. In spite of the variable data rate at its input, the channel buffer output is maintained at about 17 Mbit/s by a control parameter obtained by monitoring the buffer level.

In case of transmission errors or viewer channel changes, the decoder automatically switches to a special mode which allows the image to build up quickly to full quality.

When sound, control and other data is added to the video data stream, the total bit rate rises to 21.5 Mbit/s. This is then used to pulse amplitude modulate the carrier with four discrete levels corresponding to 2 bits/symbol, reducing the symbol rate to about 10.8 Msymbols/s or 5.4 MHz. The transmission uses suppressed carrier, vestigial sideband (VSBSC), with the carrier being positioned half- way down the lower band edge of the 6 MHz channel, 4-VSB being the term being used to describe this form of modulation.

37.5 TV sound channels

For the current TV broadcasting services, the main monophonic (mono) sound signal is conveyed by FM on a separate carrier, displaced by either 5.5, 6 or 6.5 MHz from the vision carrier. By comparison, a variety of techniques are used to provide the extra information needed for stereophonic (stereo) systems. Essentially the stereo signal is contained within two sub-channels, left (L) and right (R). For the mono receiver, the sum signal $L + R$ is the only component necessary. If the extra component is transmitted as a difference signal ($L - R$), the L and R channels can be recovered simply by adding and subtracting the sum and difference signals. Therefore stereo operation can be achieved by transmitting just the difference signal on a separate carrier (the German system). An alternative arrangement employs a frequency multiplex using a 38 kHz sub-carrier to carry the difference signal (the Zenith GE Pilot tone system). It has been found that the allocated sound spectrum can support several FM signals, provided that the modulation index of the sub-carriers is restricted to less than about 0.18, and with a channel separation of 180 kHz. These channels may be used to convey either stereo sound or completely unrelated audio or digital data signals (the Wegener system, sometimes described as FM–FM or FM^2).

Various companding systems are employed, particularly over satellite links, in order to maximise the system signal to noise ratio. However, due to the system flexibility and the improved quality available, the future of sound broadcasting in general is likely to be a digital one.

37.5.1 Dolby adaptive delta modulation digital audio system

Delta modulation (DM) uses only one bit per sample to indicate whether the analogue signal is increasing or decreasing in amplitude. This effective bit rate reduction technique allows the use of a higher sampling frequency, which in turn leads to a simpler filter arrangement in the decoder, without the risk of aliasing. Unlike PCM, a single bit in error produces the same signal effect wherever it occurs. When a bit error is detected in a DM system, the introduction of an opposite polarity bit will reduce the audible effect to practically zero. The only major disadvantage is that an overload can arise when the signal amplitude being sampled changes by more than the step size. However, even this can be overcome, if the sampling frequency is high enough.

Dolby Laboratories Inc. have devised an adaptive delta modulation (ADM) system that has been adapted for use with the B-MAC and other systems. This uses a variable step size to overcome the overloading and variable pre-emphasis to further improve the overall S/N ratio.

At the encoder of the ADM system, a pre-emphasis section analyses the frequency spectrum of the audio signal, to determine the optimum pre-emphasis characteristic. After pre-emphasis, the signal is passed to a step size section, which continually evaluates the signal slope to select the step size. The pre-emphasis and step size information is then coded into two low bit rate control signals. The main audio signal is then digitised and delayed by an extra 10 ms. This allows the control signals to reach the decoder in time to process the audio signal in a complementary manner.

The digital data is formatted into blocks for transmission, when provision is made for synchronisation. Two types of format are provided for: one for bursty systems such as sound-in-sync or B-MAC, and the other for continuous channels.

The basic function of the ADM decoder can be explained by the use of Fig. 37.21. After demodulation, the signal is filtered to separate out the components. The audio data for each channel typically runs at a bit rate of 200 to 300 Kbit/s and the control data at 7.8 Kbit/s (half TV line rate). The audio data is clocked into a multiplier stage as a bi-polar signal, with the step size data acting as the multiplying constant. It is then converted into analogue format using a leaky integrator. The de-emphasis control signal functions in a similar way, but instead of being used as a gain varying element, this amplifier stage functions as a variable, single pole frequency, de-emphasis network. The decoder, which is available in integrated circuit format, is simple and relatively insensitive to component tolerances.

37.5.2 NICAM-728 (near instantaneous companded audio multiplex — 728 Kbit/s)

This UK developed (British Broadcasting Corporation, BBC, and National Transcommunications Ltd, NTL) audio system uses a second sub-carrier at a level of −20 dB relative to the peak vision carrier and 6.552 MHz (6.552 MHz = 9 × 728 kHz) above the vision carrier. (As modified for NORDIC TV, NICAM-728 uses a sub-carrier of 5.85 MHz.)

The sub-carrier is differentially encoded with the digital signal for both channels of the stereo pair. The present 6 MHz (5.5 MHz) FM sound channel is still retained in the interests of compatibility with mono receivers. The digital sound carrier is quadrature (four phase) PSK modulated, where each resting carrier phase represents two bits of data, thus halving the bandwidth requirement. Because the data is differentially encoded (DQPSK), it is only the phase changes that have to be detected at the receiver, the bits to phase change relationship being as follows:

$$0\,0 = -0° \text{ phase change}$$
$$0\,1 = -90° \text{ phase change}$$
$$1\,0 = -270° \text{ phase change}$$
$$1\,1 = -180° \text{ phase change}$$

Pre-emphasis/de-emphasis to CCITT Recommendation J.17 is applied to the sound signal either when in the analogue format, or using digital filters while in the digital domain.

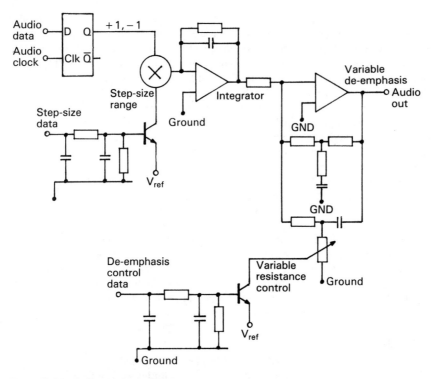

Figure 37.21   Dolby ADM decoder.

The left- and right-hand channels are sampled simultaneously at 32 kHz, coded and quantised separately to 14-bit resolution and transmitted alternately, at a frame rate of 728 bits per millisecond or 728 Kbit/s.

The NICAM compander processes the 14-bit samples in the manner shown in Fig. 37.22. The rule for discarding bits can be summarized as follows:

The most significant bit (MSB) is retained and the four following bits are dropped only if they are of the same consecutive value as the MSB.

If this leaves a word longer than 10 bits, then the excess bits are dropped from the least significant bit (LSB) region.

A single even parity bit is added to check the six most significant bits in each word. The data stream is then organised into blocks of 32, 11-bit words in the 2's complement form. A 3-bit compression scaling factor is determined from the magnitude of the largest sample in each block. This is then encoded into the parity bits for that block. At the receiver, the scale factor can be extracted using a majority decision logic circuit. At the same time, this process restores the original parity bit pattern.

Two blocks of data are then interleaved in a 16 × 44 (704 bits) matrix, to minimise the effects of burst errors. Adjacent bits in the original data stream are now 16 bits

apart. A transmission frame multiplex is then organised in the manner of Fig. 37.23, the additional bits being used as follows:

8 bits are used as a frame sync word (framing word), 5 control bits are used to select the mode of operation, either $(C_0 - C_4)$
- stereo signal composed of alternate channel A and B samples,
- 2 independent mono signals, transmitted in alternate frames,
- 1 mono signal plus one 352 Kbit/s data channel on alternate frames,
- 1 × 704 Kbit/s data channel. Plus other concepts so far undefined;

11 additional data bits are entirely reserved for future developments.

Following the interleaving of the 704 sound data bits (64 × 11-bit samples), the complete frame, except the framing word, is scrambled for energy dispersal, by adding Ex-Or, a PRBS of length $2^9 - 1$. The PRBS generator is reset on receipt of the framing word.

To limit the bandwidth, the data stream is passed through a spectrum shaping filter that removes much of the harmonic content of the data pulses. This, combined with the action of a similar filter in the receiver,

MSB   LSB

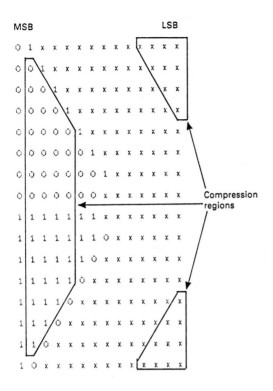

X = Don't care

Figure 37.22 NICAM-728 companding coding scheme.

produces an overall response that is described as having a full or 100% cosine roll-off.

The data stream is finally divided into bit pairs to drive the DQPSK modulator of the 6.552 MHz sub-carrier.

Decoding the NICAM-728 signal (Fig. 37.24). The secondary sound channel sub-carrier appears at either 32.948 MHz or 6.552 MHz, depending upon the method adopted for processing the receiver sound IF channel.

The spectrum shaping filter forms part of the system overall pulse shaping and has an important effect upon the noise immunity. The overall filtering ensures that most of the pulse energy lies below a frequency of 364 kHz (half bit rate).

The QPSK decoder recovers the data stream and the framing word detector scans this to locate the start of each frame and reset the PRBS generator. This sequence is then added Ex-Or to the data for de-scrambling. The de-interleaving process is also synchronised by the arrival of the framing word.

Error control follows standard procedures, but since this is usually buried within an IC, the process is transparent. The operating mode detector searches for the control bits $C_0 - C_4$, to automatically set up the data and audio stage switches, the data outputs being those for the 352 or 704 Kbit/s data channel options.

The NICAM expansion circuit functions in a complementary manner to the compressor, but using the scaling factor to expand the 10-bit data words into 14-bit samples.

Finally, the data stream is converted back into analogue form for delivery to the audio amplifier stages. These need to be designed to a very high standard, because NICAM-728 has an audio quality equivalent to that of compact disc systems.

The DQPSK decoder. This complex stage, Fig. 37.25, is available in IC form. The two main sections are associated with the recovery of the carrier and bit rate clock. The first section relies upon a voltage controlled crystal oscillator running at 6.552 MHz and two phase detectors to regenerate the parallel bit pairs, referred to as the I and Q signals (in-phase and quadrature). A second similar circuit, but locked to the bit rate of 728 kHz, is used to synchronise and recover the data stream.

Parallel adaptive data slicers and differential logic are used to square up the data pulses and decode the DQPSK signals. The bit pairs are then converted into serial form. The practical decoder carries a third phase

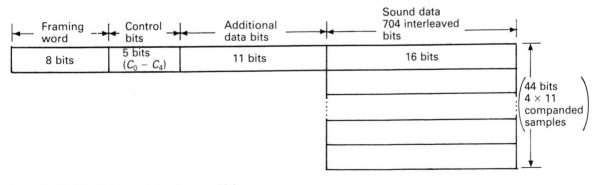

Figure 37.23 NICAM transmission frame multiplex.

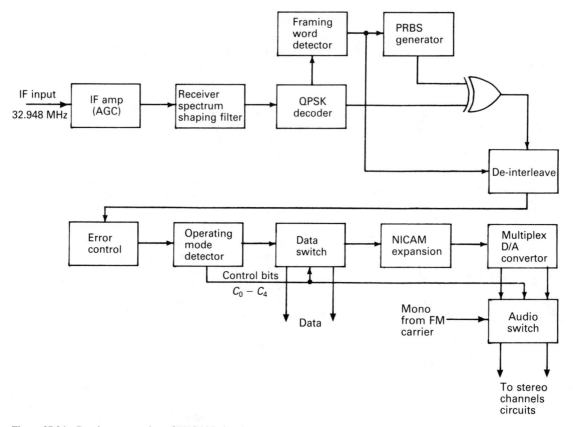

Figure 37.24   Receiver processing of NICAM signal.

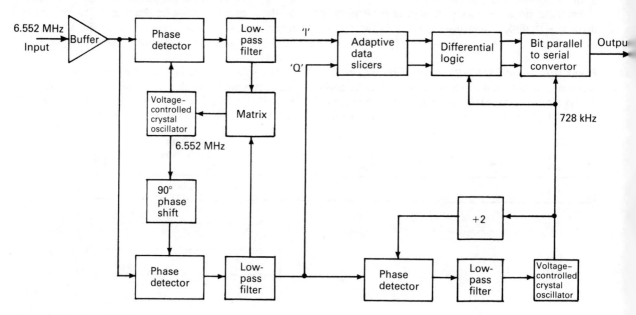

Figure 37.25   The NICAM decoder.

detector circuit driven from the Q chain. This circuit is used as an amplitude detector to generate a muting signal if the 6.552 MHz sub-carrier is absent or fails. This is then used to switch the audio system to the 6 MHz FM mono sound signal.

Apart from Europe where NICAM-728 has been adopted as a stereo standard, the system is being installed in many areas around the world. It is also capable of modification to suit the 525 line NTSC services.

Because of the commonality the system shows with the MAC/packet systems, it has become economical to produce dual standard decoding chips.

37.6 Scrambling and access control

Scrambling in the amplitude and time domains provides the two basic methods of denying the user of a video signal its entertainment value. The most elementary method that has been used, without success, is the suppression of line or field sync pulses. This fails chiefly because modern TV receivers, with their flywheel sync/PLL type timebases, require only little modification to produce a locked picture, direct from the luminance signal.

There are two levels of service that have to be provided for. In some countries, including the UK, a *must carry* rule applies to the cable network provider. This means that the system must distribute national programmes without restriction. If scrambling is applied to all channels, then a descrambling key has to be freely available. Therefore for premium services such as subscription, or pay-per-view television, an extra level of security is needed.

Under conditions of adjacent and co-channel interference, a scrambled picture can have a 2 dB *S/N* ratio advantage. This arises because such interference produces patterning which would apply to the scrambled picture. Descrambling then breaks up the patterning to make it much less noticeable.

37.6.1 Access control

The European Broadcasting Union (EBU) devised a control standard for conditional access to the MAC/packet systems. This, however, can be adapted for other standards. It uses an extension of the DES, three-key algorithm and its general principle is shown in Fig. 37.26. A PRBS (psuedo-random binary sequence) is defined by a *control key* that, for free access to must carry signals, is fixed and publicised, so that a receiver can automatically descramble such programmes. Where access is controlled, the control key is encrypted by an *authorisation key* which is in turn encrypted by a *distribution key*.

Decryption of the authorisation key is effected by the use of the distribution key, which may be transmitted over the air, or input via a smart card. The encrypted version of the control key is decrypted by the use of the combination of authorisation and distribution keys. This then enables the selection of the correct PRBS to descramble the signal.

Subscriber billing can also be built into the system and this is managed via the system control computer, using reprogrammable data on the smart card, or through the distribution key.

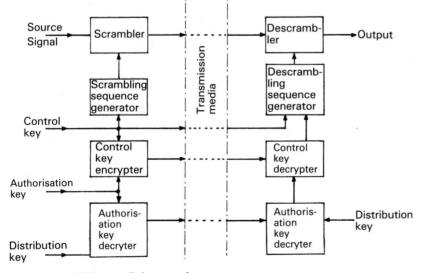

Figure 37.26 EBU controlled access scheme.

37.6.2 Basic principles of scrambling

Inversion of video signal

Simple video signal inversion, which is not very secure, can be successfully combined with a PRBS to provide a very high degree of security. This system inverts alternative groups of lines in a continuously variable manner under the control of the PRBS. It is effectively a three-key system, with digital data regarding the scrambling sequence being transmitted with the signal and changing with every programme. This is used in conjunction with a unique key held in each descrambling/decoder unit and the subscriber's smart card. The decoder key is hidden within the decoder electronics and the user smart card electronics can be reprogrammed on a monthly basis. The card is programmed to work only with one decoder and this alone provides access to the must carry programmes. Typically, this credit card sized piece of plastic can have embedded within it, about 2 Kbytes of ROM, 1 Kbyte of EPROM, 48 bytes of RAM and a limited degree of computing ability. The electronics are thus able to keep track of the debit/credit rating of the subscriber.

Line segmentation

In this technique, each video line is divided into segments and it is these that are then interchanged in the manner shown in Fig. 37.27. By varying the *cut-point* in a pseudo-random manner, a very high degree of security can be achieved using only two segments per line. A single 8-bit binary word can identify any one of 254 cut-points (the all 1s and 0s are not used). The

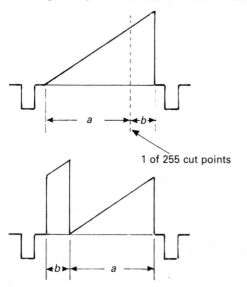

Figure 37.27 TV line scrambling.

concept is particularly economical to implement on video signals that are digital at some stage of processing.

Line shuffle

The lines of a normal video signal are normally transmitted in sequence, from 1 through to 525 or 625. If a frame store were available at each transmitter and receiver, this order could be scrambled in a pseudo-random manner. This might, at some time in the future, provide a very secure system.

Line translation

This technique causes the line blanking period to be varied in a pseudo-random manner, over a period of several frames. It requires the use of a line store, but those based on charge coupled device (CCD) technology are relatively inexpensive. To some extent the degree of security depends upon the number of time shifts permitted within the blanking interval. Systems using as few as three shifts have been shown to be insecure. The key to the PRBS for the time shifting, which is transmitted within the field blanking period, can readily be changed. The technique can be raised to the security level of a two- or three- key system, by combining the PRBS with a programmable personal identity number (PIN) and a smart card.

37.6.3 Practical systems in use

B-CRYPT

The algorithm for this system was devised by GPT Video Systems Ltd, for use with videoconferencing and to overcome the restrictions imposed on the use of the DES algorithm. Like DES, this system operates with two 56-bit codes, one which forms a crypto-variable key calculated by the system and the second an initialisation code. The latter is a random number that is changed and transmitted every 32 ms. The overall encryption key, which is a combination of both codes, is never transmitted over the link.

Digicrypt and videocode

These two similar systems that were developed by Space Communications (Sat-Tel) Ltd, use a line shuffling technique based on a sequence of 32 or 128 lines, respectively. Each encoder and, decoder has line storage for 32 or 128 lines and, in addition to line shuffling, provision is made to randomly displace the line start by ±1.5 μs. An encrypted key is transmitted in the vertical blanking interval (VBI) during lines 6, 7, 8, 9 and 319, 320, 321, 322, to instruct the decoder of the precise sequence in use.

Eurocrypt

This system, developed by CCETT (Centre Commun d'Etudes de Télédiffusion et Télécommunication, France), has much in common with VideoCipher II, but is intended for use with the D2-MAC system (either cable or satellite delivery), where the primary specification provides for scrambling on a cut and rotate basis.

Controlled access is via a smart card which is capable of being programmed over the air and thus ultimately carries all the necessary keys for security, services and management. The final element in the key system is the control word that is transmitted within the signal multiplex, and this is changed every 10 s.

Eurocypher

This system, developed by European Television Encryption Ltd in cooperation with General Instruments Corp., is modelled on the Video-Cipher II management and access system with enhanced security, but modified to be used with the MAC scrambling and encryption system.

An access control module within the decoder has a unique address and its memory holds a series of keys necessary to decode the signals. Other keys, which are changed on a regular basis, are transmitted over the air to provide the final level of access control.

This system is also adaptable to those PAL transmissions that are encrypted using the cut and rotate principle.

MAC scrambling/encryption

Scrambling, using the double cut and rotate principle (luminance and chrominance), was included the MAC specification because as described earlier, this can under certain interference conditions provide an enhancement of 2 dB in the C/N ratio.

The cut point positions in each component of the video signal are controlled by a PRBS generator and an 8-bit control word is used to define 1 of 256 unique cut points. A 60-bit linear feedback shift register is used to generate a PRBS with a very long cycle time. This sequence is reset to a different starting point every 256 frames by using a control word transmitted during line 625.

The scrambling sequence is encrypted using a shared key system to provide controlled access in the manner indicated in Fig. 37.26. The data periods in the MAC packet system provide ample capacity for system management by over-the-air addressing. This allows for programme tiering, parental lock-out, decoder authorisation for specific services and pirate lock-out.

The audio channel digital data stream is also encrypted in similar way using the same PRBS generator.

PALcrypt/Videocrypt

This system, originally designed for PAL DTH services, also uses the cut and rotate principle using a code that can be changed every 10 s. The cut point is calculated by the system control computer and the cut and rotate performed in a one-line memory, with the signal in the digital domain. This avoids the problems of chroma sub-carrier phase errors when each line is rejoined.

This public key system is controlled partly by data transmitted during the VBI and partly by a code held in a smart card which has to be changed periodically.

The audio channel is not generally encrypted.

Sat-Pac

This scheme was designed by Matsushita of Japan for use with PAL systems in Europe and Scandinavia. The method of encryption involves inverting alternate fields of the video signal and shifting the dc level of the blanking signals. This ensures that the sync pulses and colour burst are buried within the video signal level.

The composite blanking signals are then transmitted on an FM 7.56 MHz sub-carrier. A second sub-carrier at 7.02 MHz is used to carry data to enable/disable viewers' decoders according to their subscription levels.

VideoCipher II

This North American system, developed by General Instruments Inc., includes all the necessary elements for the secure delivery of high quality vision, plus stereo audio to both CATV and DTH subscribers. It also carries all the signals necessary for system control and management, including tiering of authorised access, impulse pay per view, on-screen menus to aid subscriber choice, parental lock-out, together with text and message services.

Video security is achieved partly by removing all line and field sync pulses, inverting the video signal and positioning the colour burst at a non-standard level.

The two analogue audio channels are filtered, sampled and digitised at the standard CD rate. Each sample is added modulo-2 to a PRBS generated by the DES algorithm, forward error coded (FEC) and interleaved for transmission over a satellite channel. The FEC coding can detect and correct all single errors, and detect and conceal all double errors using interpolation. Together with addressing and control information, this multiplex is transmitted during the sync pulse period of each line. Since this multiplex becomes part of the 4.2 MHz video signal, no additional audio sub-carriers are needed. This effective bandwidth reduction results in an overall 2 dB improvement in video C/N ratio, relative to the unscrambled signal.

Video scrambling operates on the cut and rotate principle. The active period of each line is sampled and quantised at four times the colour sub-carrier frequency and stored in a multiple line memory. The line data is then split into segments of variable length, under the control of the DES algorithm, and the position of the segments interchanged. This data is then read out of memory and converted back into an analogue format for transmission.

To ensure system security with flexibility, a multi-level key hierarchy is employed. Each decoder has a unique public address and a number of DES keys stored within its microprocessor memory. To receive scrambled/encrypted programmes for each billing period, the decoder first receives a message with the monthly key, together with service attributes (tiering, credit, etc.). This is transmitted over a control channel to the decoder with that unique key. So that only the authorised decoder can process the encrypted signal, this data is added to the decoder memory.

Every programme is encrypted with a different programme key and only those decoders equipped with the monthly key can decipher these programmes. By changing monthly keys, the programme provider can automatically authorise/cancel a complete set of decoders with a single transmission.

At the programme originating centre, the file of decoding and address keys is itself DES encrypted for added security. The system is so engineered that if a decoder is stolen, cloning will not allow access to programmes because pirate decoders are easily de-authorised over the air.

37.7 Videoconferencing

A videoconference is a *meeting* between two or more groups of people in different locations, which is organised via a two-way sound and vision link. The locations can be in very different parts of the world, in which case, satellites can play a vital role in the provision of the link. Currently, the space links for these services are provided by such organisations as Eutelsat and Intelsat, with the large earth stations and back-haul links being provided by the PTTs and other national telecommunication organisations over wide-band land lines. The effectiveness of these services lies in the reduction of travel for personnel and the speed with which international videoconferences can be organised. Typically, up to six persons can be comfortably catered for at each end terminal or studio.

The essential elements of this system are based on closed circuit TV (CCTV). It consists of two *face-to-face* cameras and two colour TV receivers for the incoming signals. A local colour monitor is also provided and the terminal is often housed in a roll-away cabinet. The TV system for each end terminal is either PAL or NTSC encoded. A high resolution *graphics* camera is also

incorporated. Additional provision can be made for an alphanumeric keyboard to insert text data as the need arises, together with a video printer and interface for a VTR. A digital interface is also available to connect a Fax machine or a graphics tablet via a data port.

Local control over the system is via a conventional infra-red remote control, similar to that which is used on many TV receivers. This form of control can also be extended to the *focus* and *zoom* functions on the graphics camera.

The central element in each terminal is an image compression codec. This codes and decodes the vision, sound and data signals that are transmitted over land lines. The system also provides for security of information by using optional encryption/decryption of the data stream.

The main information channel is the face-to-face vision sub-system. Here, it is usual for each camera to provide a full frame view of up to three persons, sitting side by side. The middle strip of each frame from each camera output is selected and then electronically stacked, to provide a three-over-three, full frame view for transmission. This *split screen* approach avoids the corners of the camera images, where resolution and distortion is not of the best. The decoder receives this stacked image and generates two half-sized images. These are then centrally positioned in each video frame, with the top and bottom sections blanked to black level. These are then displayed on two adjacent monitors to give a wide screen image of the distant studio.

The earlier codecs (which are still in use) had two modes of operation: fast for high resolution graphics and slow for moving picture face-to-face operation. The graphics mode used a systematic replenishment coding system, where only every 19th pixel was transmitted. The transmission time for one complete frame was then in the order of 2 s. Initial sampling rates of 12.5 MHz and 4.166 MHz were used for luminance and chrominance respectively. Six-bit PCM coding provided adequate definition for the detailed display of a document.

In the fast mode, the luminance and chrominance signals were sampled at 5 MHz and 2.5 MHz, respectively, and then coded in 8-bit PCM. The moving areas were detected and coded using DPCM which generated a non-uniform bit rate. This data stream was buffered in a memory which was then read out at either 1.544 or 2.048 Mbit/s as required. The transmitted code combined moving area addresses with line and field start data to ensure that the receiver maintained synchronism with the transmission.

The audio channel, which provided for stereo as well as mono signals, used A-law companding and PCM coding, for *sound-in-vision* (SIV). This employed a time multiplex to avoid the necessity for separate sound and control channels. The audio signal was sampled at a rate synchronised to the line frequency of the local camera and the derived data words were inserted, one in each line blanking period. The sound sub-system also included an *echo cancelling* unit. Echoes can be caused by the outputs from the terminal loudspeakers which may be retransmitted by the terminal's microphones. This unit sampled the incoming audio signal, compared it with the studio sound and estimated the echo component, which was then subtracted from the outgoing audio signal.

Current videoconferencing systems now use a codec based on the H.261 algorithm (see Image processing, p. 139).

37.8 Videotelephony (see also Image processing, p. 139)

The image compression needed to squeeze a video signal into the limited bandspace of a telephone line can be achieved using a technique based on the H.261 codec signal processing algorithm. These codecs use the MPEG algorithm and are based on the current 525/625 line TV systems. They are basically designed to run at Integrated Services Digital Network (ISDN) data rates (multiples of 64 Kbit/s, up to 2.048 Mbit/s).

Either TV or standard signals are used to generate a *common image format* (CIF) signal for the transmission channel. This is based on 288 non-interlaced lines per picture and 30 pictures per second. 625 line systems use 576 active lines per frame, so that 288 lines form one field. Therefore 625/25 codecs only have to perform a conversion to meet the 30 Hz picture rate and 525/30 codecs already operate at the correct frame rate so these only have to convert between 240 and 288 active lines.

The CIF is formed by sampling the video signal luminance at 6.75 MHz and chrominance at 3.375 MHz to produce 352, 8-bit samples per line. This is equivalent to an analogue bandwidth of about 3.4 MHz which is too wide for videotelephony.

For this application, a lower image quality is acceptable so that a further degree of compression can be made. This is achieved by using only the luminance component with 176 samples per line and 144 lines per frame. Since this represents half-resolution in each dimension, the standard is referred to as the *quarter common image format* (QCIF).

37.9 Videowalls

A videowall is formed from a group of display devices arranged to provide large area images. In order to maintain the overall aspect ratio, it is usual to arrange the displays as a square matrix of monitors (i.e. 3×3 or 4×4). When CRT monitors are used, these should be constructed in very slim frames to reduce the lattice effect. The displayed images may be from a broadcast, a video recorder, or a laser disc source. In either case, because of the large degree of magnification employed, it is important that these should be high definition sources. All the current broadcast standards are provided for, but in order to ensure best image quality, it is usual to use R, G, B drive signals for each monitor. Each monitor may display images from multiple sources, the same image or just part of a single total image.

The signal sharing is achieved using digital signal processing (DSP) with a semiconductor random access memory (RAM). A video RAM allows almost simultaneous read-out while writing is still in progress. Current systems commonly use 720 luminance samples, with 360 samples for each of the two chrominance components per line, and with 8 bits per samples. The digital signal is distributed between the monitors under microprocessor or computer control, in the following manner. For a nine-monitor wall (3×3), each pixel in the original image has to be displayed three times in each line to stretch the image over three monitors. In a similar way, each line of the video signal has to be repeated three times so that the vertical height is shared between the three rows of the wall. Thus for a total of 576 active video lines, each monitor displays 192 lines repeated in groups of three.

For a projection system that is capable of providing the large screen effect directly, the sharing effect must be reversed to produce nine smaller images. This is achieved by displaying every third pixel and every third row on any particular small section of the screen. Multiple projection systems can be used to provide very large displays and these function in the same way as the multi-monitor videowall.

The use of computer control allows many special effects to be generated. These range from programmable walls, distortion by horizontal or vertical stretching, freeze frame by repeated reading of the same digital data, or by varying the matrix dimension, such as using a 4×4 wall to display four 2×2 images.

37.10 Miscellaneous terms

Achromatic lens. A lens that produces no chromatic aberration and creates an image without colour fringing. Usually consists of two lenses, one made of flint glass and the other from crown glass, with one lens correcting the dispersion produced by the other.

Aperture distortion. The loss of sharpness of fine detail in an electronically scanned image due to the finite dimension of the electron beam.

Astigmatic lens. A lens that has two focal planes mutually at right angles to each other and separated by a small distance, a circular image being formed between these two planes.

CCIR 601. This recommendation defines the digital video standards for both 525 and 625 line systems used for studio purposes. The luminance (*Y*) signal is sampled at 13.5 MHz, and quantised to 8 bits/sample between levels 16 and 235 (zero amplitude is set at level 16).

The two colour difference components, C_b and C_r, are sampled at 6.75 MHz, with 8 bits per sample between levels 16 and 240. Because these components are bipolar, zero amplitude is set at level 128. These samples are cosited with alternate luminance samples. (C_b and C_r are comparable with the *U* and *V* components of PAL but have a different weighting.) This video signal is known as the 4:2:2 format where the figures represent the ratios of the sampling frequencies which are based on 3.375 MHz.

CCIR 656. This recommendation defines how the signals defined in CCIR 601 are interfaced between equipment using two 25-way D type connectors. Alternately a single BNC connector or equivalent optical fibre connector is specified for the serial mode. The serial data rate operates at 243 Mbit/s, while the parallel rate represents a time multiplex running at 27 Mword/s. The data stream format is (C_b, *Y*, C_r) *Y* (C_b, *Y*, C_r) *Y* (......). The words (C_b, *Y*, C_r) refer to the cosited luminance and chrominance samples, and *Y* represents the luminance sample only.

CCIR 601/656 sampling rates

4:1:1 The ratio of sampling frequencies for luminance and chrominance stated in CCIR 601. Since this is based on the frequency of 3.375 MHz, the luminance is sampled at 13.5 MHz and the two chrominance channels at 3.375 MHz.

4:2:2 Luminance sampling frequency of 13.5 MHz and chrominance at 6.75 MHz.

4:4:4 All components sampled at 13.5 MHz. Suitable for either *Y*, *U*, *V* or *R*, *G*, *B* components.

4:4:4:4 As for 4:4:4, but with a key signal used as the fourth component which is also sampled at 13.5 MHz.

Component video. For transmission or processing purposes, the luminance and chrominance components are maintained in separation. Component systems therefore operate with either *Y*, *I*, *Q* or *Y*, *U*, *V* or *Y*, C_r, C_b or *R*, *G*, *B* signals that retain maximum bandwidth.

Compositive video. These systems operate with the video components of luminance and chrominance, frequency multiplexed or encoded into a composite signal. For example, NTSC, PAL or SECAM. The aim of this form of encoding is usually to minimise the bandwidth requirement.

Constant luminance. In an ideal colour TV system, the gamma corrected luminance (*Y*) signal should be identical to that of a monochrome system. But in practice, *Y* is obtained from gamma corrected *R*, *G*, *B* primary signals. This results in some of the luminance energy being lost to the receiver due to bandwidth limiting in the chrominance channel. The effect, which removes some of the high frequency detail from the image, is most noticeable in highly saturated red and blue areas.

Differential gain. A term used to describe the nonlinear amplification or processing in a TV system. If a constant amplitude of colour sub-carrier is superimposed on a stepped video waveform similar to that shown in Fig. 37.28, any nonlinearity will produce changes of sub-carrier amplitude. The differential gain is then expressed as the maximum variation, as a percentage of the video blanking level.

Differential phase. Nonlinear processing, as mentioned above, will create a sub-carrier that is non-sinusoidal. Since the positive and negative half-cycles now have different amplitudes, this is equivalent to a phase error. The maximum phase deviation relative to the reference

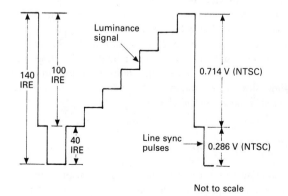

Figure 37.28 The IRE unit.

sub-carrier defines the differential phase error, and is often expressed as a percentage.

Dioptre (diopter) D. This is a term used to quantify the power of a lens or prism and measured in units of the reciprocal of length (m^{-1}). Converging and diverging lenses are considered to have positive and negative values, respectively. For a lens, D is the reciprocal of the focal length and defined as follows: if the deviation of a ray passing through a prism is 1 cm measured at a distance of 1 metre, then the lens is said to have a power of 1 prism dioptre.

Enhanced television systems. Systems which retain the current 525 or 625 line scanning standards while providing improvements in image quality and other features as a result of new processes of analysis, synthesis or signal processing, with or without modifications to the transmission standards.

Extended definition TV. This category refers to enhanced TV systems in which changes may be made to the transmission standards to obtain improved resolution on new receivers, while at same time maintaining compatibility with older receivers.

Ghost cancelling. Multi-path signal reflections from high buildings, aircraft, etc., which arrive at the receiver antenna over varying path lengths, give rise to the superimposed images referred to as ghosts. In ghost cancelling systems, a reference signal is inserted into the VBI at the transmitter. When received at a distance and without ghosting, this reference can be regenerated without distortion. When this is compared with a multi-path distorted signal, an error component can be evaluated. This may then be used to modify the delay and amplitude coefficients of a transversal filter which produces a cancellation to the ghost images. At the same time, this recreates correctly timed line and field sync pulses.

Insertion test signals (ITS). To ensure that the high quality of the original signal is maintained throughout the TV network, standard signals are inserted into the vertical blanking interval (VBI). These are then recovered at a distance and then monitored for distortion. The three most important parameters include the system linearity, frequency and phase response. The waveforms shown in Fig. 37.29(a) and (b) are typical of those used for this purpose. Figure 37.29(a) shows the *pulse and bar* used for exercising the luminance channel. The pulse, which is a sine-squared shape to ensure that it contains very little energy outside of the pass band, represents the most rapid transitions between white and black levels.

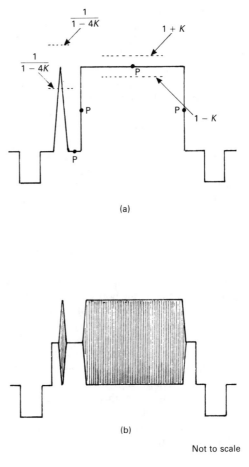

Not to scale

Figure 37.29 Insertion test signals: (a) luminance; (b) chrominance.

Two standard pulses are in use. The T pulse, which is filtered to have a half-amplitude duration of half the period of the nominal cut-off frequency. (For a 625 line TV system with cut-off frequency of 5 MHz, this duration would be 100 ns.) A 2T pulse therefore has a half-amplitude duration of 200 ns. The bar is filtered to have skirts of the same shape so that it covers the same spectrum as the pulse. Both are generated and inserted with the same amplitude. If the pulse is detected as being of lower amplitude, the system rise time and high frequency response is too low. The slope of the detected bar indicates the quality of the system low frequency response.

To test the chrominance channel, the sine-squared pulse and bar are used to modulate the chrominance sub-carrier. This signal is then superimposed on a dc level (pedestal) to produce the waveform shown in Fig. 37.29(b).

IRE units. The Institute of Radio Engineers (IRE), which later became the Institute of Electrical and Electronic Engineers (IEEE), established a unit of measurement for the amplitude of the video signal. The IRE unit was defined as 1% of the video range from peak white to blanking level and without reference to the actual signal voltage. Although defined as a ratio, it became common practice to refer to an IRE unit as being equal to 7.14 mV due to the adoption of the standard 1 V composite video signal. This relationship is indicated in Fig. 37.28 which also shows the 10:4 ratio between vision and sync pulse amplitudes. In other systems such as PAL and SECAM, these values are vision 700 mV and sync 300 mV, but still representing the same 1 V standard video signal.

Kell factor. The resolution of an image is usually measured in terms of lines per picture height (line/ph) in the vertical sense and the maximum frequency needed to display the finest detail in the horizontal plane. For a scanned TV system, the finest detail that can be displayed is governed by the diameter of the spot formed by the electron beam. For a system using 576 active lines, the maximum definition is 576 line/ph. However, statistical and subjective testing shows that this resolution is reduced generally by a factor of about 0.7, the Kell factor. This system therefore has a vertical resolution of $0.7 \times 576 = 403$ line/ph. If the horizontal and vertical resolutions are equal, then with a 4:3 aspect ratio, the horizontal definition becomes

$$\frac{403 \times 4}{3} = 537 \text{ pixels}$$

If these are to be displayed as alternate black and white pixels in 52 μs, this is equivalent to a horizontal frequency of $(537 \times 1)/(2 \times 52) = 5.17$ MHz, or about 79 lines/MHz of bandwidth.

K-Rating. This forms a method of quantifying what is essentially the subjective impairment of a TV signal during processing through a network. A single factor (K) which represents this distortion is derived from the pulse and bar of the ITS signal. The various limits are calculated with reference to the distortion caused by a single long-term echo displaced from the main signal by at least eight times the half-period of the system cut-off frequency. These limits are then used to provide a template that can be placed over a CRT display. By aligning the display with points P in Fig. 37.29(a), the relative amplitudes can be used to establish K. Figure 37.30 shows the mask for a $2T$ pulse, in a 625 line system with 5 MHz cut-off frequency and $K = 5\%$.

Newton's rings. The coloured rings that can be seen around the point of contact between a plane reflecting surface and a convex lens. The effect arises because of the interference between the light rays reflected between the boundaries of the air film that separates the two elements.

PAL M. A variation of the PAL standard adapted for use with a 525 line, 60 Hz field structure.

Quadrature mirror filters (QMF) (see also Quadrature mirror filters, p. 129). These digital filters can be used to partition a wideband signal, such as video, into a number of sub-bands in both the temporal and spatial domains. This allows each component so formed to be optimally processed. The filters are designed in complementary sets or *filter banks* to produce M sub-bands. In the case $M = 2$, this simply generates high and low frequency components. Such filters allow frequency fold-over and aliasing to occur because the complementary receiver processing produces cancellation of these artifacts. This feature allows perfect reconstruction of the original signal, even when the individual filters have characteristics that are less than ideal. Efficient 3D QMF devices can be simply designed by cascading 1D digital filters.

SC-H. This refers to the relationship between the sub-carrier (SC) phase and the line frequency (H). In the PAL system an accurately timed sub-carrier is necessary for correct demodulation. To ensure correct lock, a burst of nominally 10 cycles of sub-carrier is inserted into the composite signal immediately after each line sync pulse. Because of the 1/4 line frequency plus 25 Hz offset in the sub-carrier frequency, the field to sub-carrier phase repeats only every eight fields. Thus this $4 \times 625 = 2500$ lines forms a unique set. If two PAL signals need to be combined, say during an editing process, then it is important that the SC-H phasing should be accurately maintained. The problem is less critical with NTSC signals as there is only a four field or 1050 line sequence. The problem entirely disappears with SECAM because the phase sequence repeats every 625 lines.

Sub-band coding. In any wideband signal, many of the frequencies in the band are only rarely present. A good example of this is the television image, where much of the background remains unchanged from frame to frame and with only relatively small areas of movement. By dividing (filtering) the broad spectrum into narrower sub-bands, it will be found that many of these

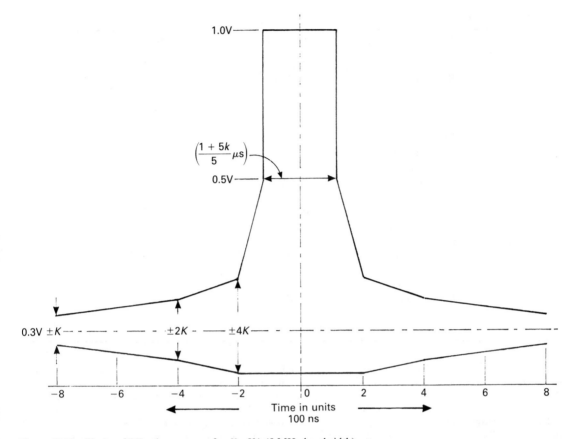

Figure 37.30 Limits of 2*T* pulse response for *K* = 5% (5 MHz bandwidth).

will contain only occasional information. If these subbands are sampled, then by using a variable length coding such as Huffman, which allocates the shortest code words to those values that have the highest occurrence probability (the stationary areas of the image), a useful bandwidth compression can be achieved.

Triple beats. A form of intermodulation interference that affects cable distribution systems. The wideband trunk distribution system has to handle many carriers simultaneously. Any nonlinearity will create intercarrier beats, particularly from a carrier and its two adjacent channels. The triple beat power is normally summed for the whole bandwidth and quoted in dB as an interfering power level relative to the power in each channel.

Video formats. Some of the important variations in the world video standards are explained in Table 37.5. In some cases a *set-up* or blanking variation level is included to ensure complete beam cut-off during the vertical retrace period.

Zone plates. Many of the distortions that arise in a TV system only become apparent when operating with real-time signals. Zone plate generators have been developed in order to provide a degree of dynamic testing to overcome such problems. These provide a series of concentric circles of decreasing pitch that ripple outwards from the centre of the CRT. They thus create a single repeatable sweeping signal function that exercises the system in the horizontal, vertical and temporal dimensions simultaneously. The moiré patterns that arise can give an indication of the extent of defective filtering, distortions and beat notes, modulation transfer function, and general frequency response in the *x*, *y* and *t* domains in real time.

37.11 Useful references

Carnt P.S. and Townsend G.B. (1969) *Colour Television*, Vols 1 and 2, Iliffe Books Ltd, London.

Hutson G.H. (1971) *Colour Television Theory*, McGraw-Hill, London.

Table 37.5   *Variations in video standards*

	Video output levels	IRE Units	Volts
NTSC	Blank to white	100 IRE	1.0
RS-170A	Blank to black	7.5 IRE	0.075
	Blank level		0
	Blank to sync	40 IRE	−0.4
NTSC	Blank to white	100 IRE	0.714
RS-343A	Blank to black	7.5 IRE	0.054
	Blank level		0
	Blank to sync	40 IRE	−0.286
PAL	Blank to white	100 IRE	0.714
	Blank to black	0 IRE	0
	Blank level		0
	Blank to sync	43 IRE	−0.307
SECAM	Blank to white	100 IRE	0.714
	Blank to black	0–7 IRE	0–0.049
	Blank level		0
	Blank to sync	43 IRE	−0.307

IBA (1974) *Technical Reference Book 2*, IBA, London.

Kell R.D. et al. (1934) An experimental television system. *Proc. IRE*, **22**, 1246–1265.

Lewis G.E. (1992) *Communication Services via Satellite*, 2nd edition, Butterworth-Heinemann, Oxford.

Navalpotro J. (1992) The PAL 8-field sequence and SCH-Phase. *Applications note*, Tektronix, Beaverton, OR.

Navalpotro J. (1992) Colorimetry and television camera color measurement. *Applications note*, Tektronix, Beaverton, OR.

Simpson R. (1991) *Videowalls*, Focal Press, Butterworth-Heinemann, London.

Slater J. (1991) *Modern Television Systems – To HDTV and Beyond*, Pitman, London.

Townsend G.B. and Jackson K.G. (1991) *Television and Video Engineers' Reference Book*, Butterworth-Heinemann, Oxford.

Weaver L.E. (1982) *The SECAM Colour Television System*, Tektronix, Beaverton, OR.

38   *Transmission lines and waveguides*

Electromagnetic energy may be considered as being carried by a system of conducting cables, or as being guided by the surrounding electric and electromagnetic fields when it travels through the free space between the cables. This latter concept is based on Huygen's principle which states that each point on a primary wavefront can be considered to be the source of a new secondary spherical wave. Thus by the principle of superposition, the resultant displacement of energy at any point is due to the combined effects of the individual waves at that point. It is this feature that forms the basis of the theory of light wave interference. The concept of free space has a further significance. Free space, for which a vacuum is a good approximation, forms a region which is used to define the absolute standard and physical constants for the characterisation of electromagnetic waves. The *permittivity* (ε), *permeability* (μ) and *velocity* (c) of light for free space are related by the expression;

$$c = 1/\sqrt{\varepsilon_0 \mu_0}$$

38.1   Transmission lines

For analysis purposes, it is convenient to consider the loop length of a transmission line system to be divided into two parts: one considered to be ideal and lossless, while all the inductive, capacitive and resistive effects are referred to the other line in the manner shown in Fig. 38.1(a).

38.1.1 Primary line constants

By redrawing the equivalent circuit in the manner shown in Fig. 38.1(a), the transmission line can be seen as a tee-section low-pass filter which effectively describes its behaviour. From this, it is easy to deduce the important *primary constants*:

- loop resistance R_1 per unit length;

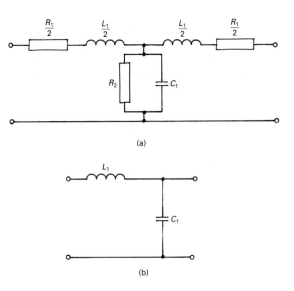

Figure 38.1   Transmission lines equivalent circuits: (a) real line; (b) lossless line.

- loop inductance L_1 per unit length;
- shunt capacitance C_1 per unit length;
- shunt conductance G or leakage resistance R_2 per unit length ($G = 1/R_2$).

While R_1, L_1, C_1 and G all increase with length, R_2 falls. In the loss-free or ideal line where the effects of R and G are negligible, the circuit simplifies to that shown in Fig. 38.1(b).

38.1.2 Propagation constant and characteristic impedance

If the line excitation is sinusoidal, then the signal will propagate losing amplitude due to the resistance, and acquire a phase shift due to the reactive components. It is shown that the propagation constant

$$\gamma = \sqrt{(R_1 + j\omega L_1)(G + j\omega C_1)}$$

has the form $\alpha + j\beta$, where α is termed the attenuation constant (nepers/unit length) and β is the phase constant (radians/unit length).

In general, the solutions for α and β are complicated expressions involving R_1, L_1, C_1 and G, but two cases are important:

(1) when $R_1/L_1 = G/C_1$, $\alpha = \sqrt{R_1 G}$ and $\beta = \omega\sqrt{L_1 C_1}$; and for the low loss case
(2) when $\omega L_1 >> R_1$ and $\omega C_1 >> G$, $\beta = \omega\sqrt{L_1 C_1}$ and $\alpha = (1/2)(R_1/Z_0 + GZ_0)$, where Z_0 is described as the characteristic impedance of the line.

The characteristic impedance (Z_0) of the line is defined as the input impedance of an infinitely long length of line, or the input impedance of a shorter length when terminated in that impedance. Mathematically Z_0 is also given by

$$\sqrt{\frac{R_1 + j\omega L_1}{G + j\omega C_1}}$$

For the low-loss case where resistive effects are negligible, $Z_0 = \sqrt{L_1/C_1}$.

Z_0 is also given by the expression:

$$Z_0 = \sqrt{Z_{sc} \times Z_{oc}}$$

where Z_{sc} and Z_{oc} are the input impedances when the line is terminated with a short circuit and an open circuit, respectively.

Z_0 values for cables commonly in use are 300 and 600 ohms for parallel lines and 50 and 75 ohms for coaxial cables.

38.1.3 Reflection coefficient, standing waves and return loss

When electromagnetic energy flows along a transmission line, the voltage and current distribution along the line depends on the nature of the load at the far end. If the load impedance correctly matches the characteristic impedance of the line, then all the energy travelling on the line will be absorbed by the load. Under mis-match conditions, the load can only absorb an amount of power that is dictated by Ohm's law, any surplus energy being reflected back along the line towards the generator. The forward and reflected waves combine to form *standing waves* all along the line, that are indicative of the degree of mis-match. Figure 38.2(a) shows the voltage and current distribution near to the load for two extreme cases of mis-match, where the load is either an open or short circuit. In either case, there will be no power absorbed in the load, because either the

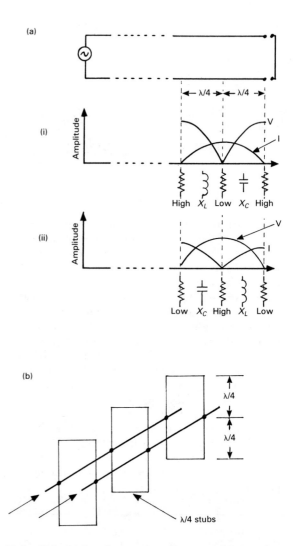

Figure 38.2 (a) Standing waves on (i) open circuit line, (ii) short circuit line; (b) transmission line supported on $\lambda/4$ stubs.

current or the voltage is zero. Hence there will be a total reflection of energy.

The standing wave pattern depends on the wavelength/frequency of the transmitted signal and is repetitive every half wavelength as shown in Fig. 38.2(a).

The impedance $(V/I$ ratio$)$ seen by the signal thus varies all along the line, being purely resistive of very high or very low value at the $\lambda/4$ points, and either capacitive or inductive reactive in between. An open circuit at the end behaves as a very low resistance just $\lambda/4$ away, while the short circuit case behaves in the opposite way.

This impedance transformation property allows transmission lines to be supported on metallic $\lambda/4$ stubs as shown in Fig. 38.2(b) without affecting the signal power flow in any way. If the signal frequency is reduced, the stub becomes less than $\lambda/4$ long at the new frequency and so behaves as an inductive reactance to give rise to low frequency losses. In a similar way, an increase of frequency causes the stub to develop capacitive reactance, to limit the high frequencies.

The reflection coefficient (ρ), which is a function of the degree of mis-match between the load and line impedances, is given by

$$\rho = \frac{Z_L - Z_0}{Z_L + Z_0}$$

and is complex $(re^{j\theta})$.

The return loss represents the ratio of the reflected to incident power at the load and expressed in dB is given by $-10 \log P_r/P_1$ or $-20 \log r$.

38.1.4 Voltage standing wave ratio (VSWR)

The VSWR represents the ratio of the maximum to minimum amplitude values of the standing wave that arises due to a mis-match. Numerically VSWR $= (1 + r)/(1 - r)$.

38.2 Smith's charts

These charts provide a ready solution to many transmission line problems. Although graphical results appear to be less accurate than theoretical calculations, the overall accuracy is well within the range that would be expected from circuits with normal component tolerances and achieved with less difficulty.

Smith's charts use normalised values to ensure maximum range of scaling. They are constructed from three sets of orthogonal circles on a circular chart. Set 1 is a series of concentric circles centred on the chart centre, and represent the VSWR (s) and the reflection coefficient (ρ). Set 2 are centred on the horizontal axis, but are drawn circumferentially tangential to the outer circle in the manner shown in Fig. 38.3, and represent constant resistance values. Set 3 is drawn with centres along a vertical axis tangential to the right-hand edge of the chart and represent constant reactance values. Intersection points between set 2 and set 3 circles therefore represent impedance values. For example, points A and B represent impedances of $0.2 + j0.35$ ohms and $1.5 - j1.5$ ohms, respectively.

Points diametrically opposite to each other across the centre represent impedance to admittance transformations. For example, point B transforms from $1.5 - j1.5$ ohms to $0.33 + j0.33$ S. If this is denormalised for a 50 ohms system, the admittance becomes

$$\frac{0.33 + j0.33}{50} S = 6.6 + j6.6 \, mS$$

The outer circle of the chart is linearly scaled only from 0 to $\lambda/2$ clockwise (backwards) towards the generator and 0 to $\lambda/2$ anticlockwise (forward) towards the load, because the standing wave pattern repeats every half wavelength (not shown in Fig. 38.3).

Example. Find the VSWR and reflection coefficient when a 50 ohms line is terminated in a load $75 - j75$ ohms. Normalise the impedance

$$\frac{75 - j75}{50} = 1.5 - j1.5$$

and mark this point on the chart (point B).

With centre 1.0 draw an arc to cut the right-hand side of the horizontal axis. VSWR $(s) = 4$.

From $|\rho| = (s - 1)/(s + 1)$ calculate $|\rho| = 0.6$.

In a similar way, it is possible to obtain the values of many other parameters. Typical of these, are the distance from the load to the first minimum voltage and the required length of a matching stub.

The Smith's chart may also be used to display a locus that represents the frequency variation of a device input impedance.

38.3 Waveguides

If an infinity of $\lambda/4$ stubs are connected in parallel all along the lines in the manner shown in Fig. 38.2(b), a rectangular box shape develops which will continue to carry energy. Such a structure is described as a *waveguide.*

At frequencies above about 1 GHz or so, the losses in co-axial cable transmission lines become unacceptable and so waveguides are commonly used. Although these are physically larger, mechanically stiffer and more expensive than cables, these disadvantages are outweighed by the very low losses at microwave frequencies.

38.3.1 Cut-off frequency, dominant or fundamental and second-order modes

The elctromagnetic energy propagates through the guide by reflections off the side walls. Figure 38.4(a) shows this behaviour for a range of frequencies, with the lower frequencies being reflected off the walls at the sharpest angles. The guide behaves as a high-pass filter (HPF), because at some low frequency a critical wavelength occurs (*cut-off wavelength,* λ_c) where the energy is simply reflected back and forth across the guide so that propagation ceases. For the *dominant* or fundamental

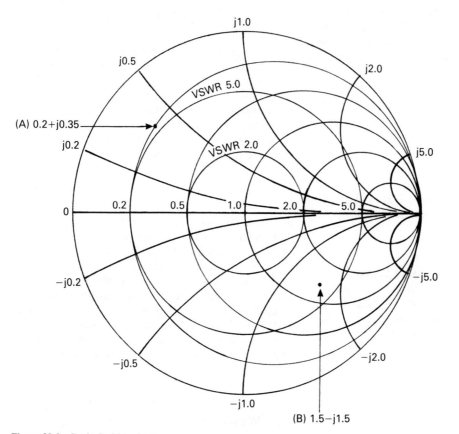

Figure 38.3 Basic Smith's chart.

mode of propagation, the cut-off wavelength is proportional to the transverse dimension (*a*) in Fig. 38.4(b), λ_c being equal to 2*a*. The narrow dimension (*b*) is not so critical and is usually equal to *a*/2.

The velocity of propagation within waveguides is similar to that in co-axial cables. In the latter case, the wave travel is slowed by the charging and discharging of the cable self-capacitance and by a current flowing against its self-inductance. While in the waveguide, the energy has to travel via the rather longer reflective path.

For sinusoidal signals, the dominant mode of propagation gives the longest value of critical wavelength and so determines the lower cut-off frequency. Figure 38.5 shows the distribution of the *E* and *H* field force lines, the *E* field lines terminating on the side walls and the *H* field lines forming complete current loops, both propagating through the guide. The lengths of the *E* field force lines in Fig. 38.5(a) and (b) represent the sinusoidal variation of field intensity.

Higher order modes can propagate as indicated in Fig. 38.6. In fact, the larger the waveguide (in terms of wavelengths), the greater the number of modes that can propagate. However, the presence of very many modes

all travelling at different velocities and following different reflective paths, is clearly undesirable. It is therefore usual to choose the waveguide dimensions large enough to support only the dominant mode. This restricts the use of a given rectangular waveguide to a relatively narrow range of frequencies.

Similar modes of propagation occur in circular waveguides, which tend to have a wider bandwidth. However, these introduce mechanical and electrical problems if they are long and include bends.

38.3.2 Resonant cavities and waveguide couplings

At lower frequencies, a parallel combination of inductance (*L*) and capacitance (*C*) provides a resonant circuit across which maximum voltage will be developed at some frequency. Progressive reduction of both *L* and *C* will cause the resonant frequency to increase. Figure 38.7 shows a limiting condition where *C* has been reduced to a pair of parallel plates, while *L* has been reduced by connecting the plates together with parallel straps that behave as inductors. The ultimate limit is reached where the two plates are connected together

(a)

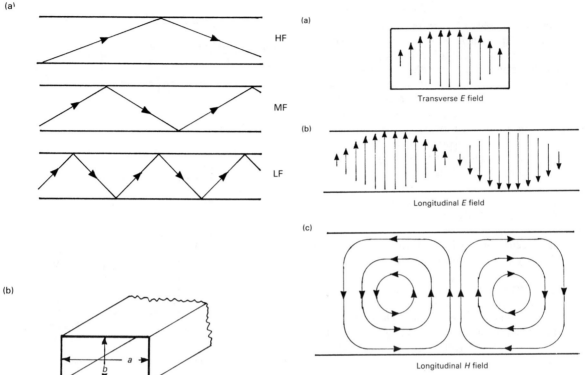

HF

MF

LF

(a)

Transverse *E* field

(b)

Longitudinal *E* field

(c)

Longitudinal *H* field

Figure 38.5 *E* and *H* fields for dominant mode propagation.

(b)

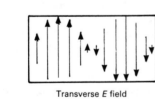

Figure 38.4 (a) Wavelength reflective paths for a range of frequencies; (b) dimensions of rectangular waveguide.

(a)

Transverse *E* field

(b)

Longitudinal *H* field

Figure 38.6 Second-order mode of propagation.

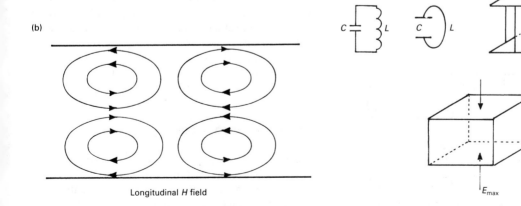

Figure 38.7 Development of resonant cavity.

with the four sides of a box. This structure is a *resonant cavity* and if energised at an appropriate frequency, will develop maximum voltage across the points shown.

Figure 38.8(a) shows that this structure has a resonant wavelength dictated by the length of its diagonal, the electromagnetic field reflecting off the sides as indicated. Figure 38.8(b) shows how the fields oscillate at quarter period intervals when the cavity is energised, the total energy being transferred between the electric and magnetic fields in an oscillatory manner.

The Q factor, which is very high, is typically about 15 000 (unloaded), and is given by:

$Q = 2\pi$ (total energy stored in the cavity/energy lost in 1 cycle due to the resistivity of the walls)

From this it can be deduced that Q is proportional to the ratio of cavity volume to interior surface area.

Waveguide couplings. Of the various ways in which signals may be coupled into cavities and waveguides, those shown in Fig. 38.9 are common. The probe in Fig. 38.9(a) acts as a dipole aerial generating or responding to the E field. The current loop either generates or responds to the H field.

(a)

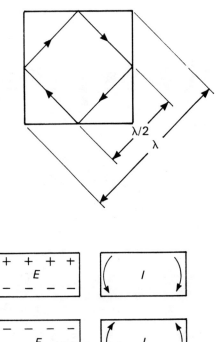

(b)

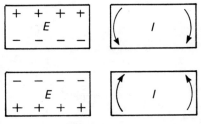

Figure 38.8 Oscillations in resonant cavities: (a) resonant wavelength; (b) oscillating E and H fields.

(a)

(b)

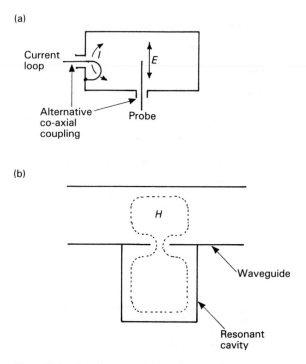

Figure 38.9   Coupling to cavities and waveguides.

Figure 39.9(b) shows how a resonant cavity can be coupled to a waveguide via a hole or *iris*, coupling being achieved through the H field. Cavities of different resonant wavelengths may be coupled to a waveguide in this way, to form a filter circuit. Cavities can be retuned by introducing a metallic screw or dielectric rod into the cavity, from the centre of either of the large area sides. In each case, either the inductance or capacitance of the cavity is increased, to lower the resonant frequency.

38.3.3 Waveguide parameters

Table 38.1 shows the approximate internal dimensions for the popular waveguides in both inches and millimetres, together with the operating frequency ranges. The *breakdown power rating* for these waveguides is approximately proportional to the cross-sectional area, ranging from about 50 mW for WG6 linearly down to about 100 μW for WG22.

38.4 Miscellaneous terms

Balanced lines. Two identical wires are used to carry the signals. The pair is driven and terminated via centre-tapped balanced transformers with only the centre tap earthed. The signal currents in each line are therefore balanced with respect to earth.

Table 38.1 *Dimensions and operating frequency of common waveguides*

Type number	Internal dimensions (a, b in. (mm))	Frequency range (dominant mode GHz)
WG6	6.5, 3.25 (16.5, 8.25)	1.12–1.7
WG8	4.3, 2.15 (10.9, 5.46)	1.7–2.6
WG10	2.84, 1.34 (7.2, 3.4)	2.6–3.95
WG12	1.87, 0.87 (4.75, 2.22)	3.95–5.85
WG14	1.37, 0.62 (3.49, 1.58)	5.85–8.2
WG16	0.9, 0.4 (2.29, 1.02)	8.2–12.4
WG18	0.62, 0.31 (1.58, 0.79)	12.4–18
WG20	0.42, 0.17 (1.07, 0.43)	18–26.5
WG22	0.28, 0.14 (0.71, 0.36)	26.5–40

Balun (BALanced to UNbalanced). A special type of transformer designed to match a balanced line feeder system to an unbalanced system or vice versa.

Delay line. Because electromagnetic energy takes a finite length of time to propagate along a transmission line, this feature can be used to construct an artificial delay line. In a circuit such as shown in Fig.38.1(b), the time delay is given by $T_D = \sqrt{L_1/C_1}$, where T_D is the time delay in seconds and L_1 and C_1 are the inductance and capacitance in henrys and farads, respectively. For n identical sections in series, the total delay becomes nT_D.

Nepers (N). Units based on logarithms to base e (ln) that are used to equate current ratios in transmission lines. For example,

$$N = \ln I_1/I_2 \text{ and decinepers (dN)} = 10 \ln I_1/I_2$$

Quarter wave transformer. A $\lambda/4$ length of transmission with a characteristic impedance of Z_3 can be used to match two other lines of the same type, with characteristic impedances of Z_1 and Z_2 according to the following relationship:

$$Z_3 = \sqrt{Z_1 \times Z_2}$$

Unbalanced lines. Transmission lines of the type characterised by coaxial cables. One line, the inner conductor, is shielded from external influences by the other, which takes the form of an earthed outer screen.

Unshielded twisted pair (UTP). A balanced pair of transmission lines in which the parallel pair are twisted together along their length in order to improve the interference rejection characteristic.

Velocity ratio. The ratio of the velocity of electromagnetic wave propagation in a transmission line system to the velocity which is obtained in free space. Values for typical cables range from about 0.6 to 0.85.

38.5 Useful references

Connors F.R. (1972) *Introductory Topics in Electronics and Telecommunications.* Vol. 3, *Wave Transmission*, Edward Arnold, London.

Edwards T.C. (1981) *Foundations of Microstrip Circuit Design*, Wiley, London.

Hewlett Packard (1972) Waves on transmission lines. *Application Note 16*. Hewlett Packard Inc., Palo Alto, USA.

Mazda F.F. (1989) *Electronic Engineers' Reference Book*. 6th edition, Butterworth-Heinemann, Oxford.

Saed T. (1971) *Micro-wave Engineers' Handbook*, Vols. 1 and 2, Artech House, London.

Videotex(t) is the generic title of the service that provides a user with alphanumeric and graphics type information. This data is distributed either via a telephone or cable network or by a multiplex process within the television broadcast system. In general, videotex(t) data is displayed on a terminal based on the standard TV receiver.

39.1 Closed caption transmissions

A simplified *teletext* type of service has been devised in cooperation with the American National Captioning Institute, to provide a service for viewers with impaired hearing. Although this was specifically designed to operate with NTSC receivers, the service can be applied to any PAL or SECAM receiver that is equipped with a digital control microprocessor.

An area of the screen is allocated for the display of up to eight rows of text, each containing up to 32 characters. Apart from where the characters are displayed, this area is transparent. For NTSC, the data is transmitted during line 21 of field 1 (odd) in NRZ (non-return to zero) format. The data stream at 503 kHz consists of a *clock-run-in* sequence followed by a *start code* consisting of two logic 0s and a logic 1. This is followed by two 8-bit bytes that represent two ASCII characters. A simple parity check is provided by a single bit in each byte. The actual usable data rate is thus 16 bits per 1/30 s or 480 bit/s.

The received codes, including control codes, are decoded and stored in a memory that takes the form of a RAM, plus a ROM character generator. The data is then read out and displayed on the CRT under the control of the system microprocessor. A wide range of *fonts* and many of the normal teletext type of *attributes* are permitted.

39.2 Interactive video

Throughout the 1970s and 1980s, the telephone network was used to provide the necessary return link for a viewer to interact with a TV transmission. In the main, these services were restricted to data and text applications. During the 1990s, it is anticipated that, linked to the multi-media concept, there will be extensive developments in this area of communications.

An interactive link can be formed by using the cellular telephone concept with a low power VHF FM channel of about 200 Hz bandwidth. Each user terminal includes a low power transmitter/receiver which may be linked to a TV transmission where coded information is transmitted either within the programme as pulses of infrared (IR) light or via data signals in the VBI.

By using data compression, low definition TV type images and text can be displayed on small flat panel LCD or LED displays. Such information may be used for in-store shopping guides, advertising and in educational applications.

When using a return channel, the viewer on a cable distribution system can select an image produced by an alternative camera to obtain a different viewpoint of a sporting event. For guidance, these alternative images may be displayed as a small picture-in-picture (PIP) on the main image.

Systems using satellite and microwave broadcast links that include controlled access, plus a low power FM radio return link, are being tested for a number of commercial applications.

39.3 Teletext

Teletext is the generic title for digital data services that can be broadcast by time multiplexing within the VBI without interference to the normal vision service.

At the end of each TV field period, a time must be allowed for the scanning beam to retrace to start the next field. During this period, the electron beam must be *blanked*, or clamped to the black or zero level, so that the retrace action will be invisible. This period is described as the *vertical blanking interval* (VBI). However, the line and field synchronising (sync) pulses and colour burst must be maintained throughout this period. For the 625 line PAL system, there are 25 inactive lines in each field period, while for the 525 line NTSC system, the corresponding number is nominally 20. These unused lines of the VBI are therefore chosen to carry this digital information as a time multiplexed signal component.

The World Standard Teletext (WST) system was originally devised for 625 line TV systems, but it can also be modified to operate with 525 line transmissions. Each *screen*, *frame* or *page* is designed to contain 24 rows of text or graphics symbols, with each row containing up to 40 characters. The pages are organised into *magazines* nominally containing 100 pages and each transmission channel can control up to eight magazines.

The data for each row is synchronously transmitted during the period of one VBI line. Each character or graphics symbol cell in the display is defined by 6 dots or pixels per line and 10 lines deep in each of the two interlaced fields. Each character and row is separated by one dot so that in total, each cell represents an area of 5×18 rectangular pixels. For graphics representation, the cell borders are omitted and the characters are then said to be *contiguous*. The page information is displayed during the middle $40\,\mu s$ period of each TV line. The colour palette of the display is limited to three saturated primary (red, green and blue) and secondary (cyan, magenta and yellow) colours, plus black and white.

Each data line carries the binary digits as a two-level, non-return-to-zero (NRZ) code. The pulses are filtered and shaped to ensure that most of the energy is contained within a bandwidth of about 5 MHz. The timing and signal levels relative to the peak white and black levels are shown in Fig. 39.1.

The data bit rate is chosen to be an even multiple of the line timebase frequency and for the 625 line system runs at $6.375\,\text{Mbit/s}$ $(444 \times 15.625\,\text{kHz})$. The corresponding bit rate for the 525 line system typically lies in the range 4.0 to 5.5 Mbit/s.

The teletext data signal, which consists of 45, 8-bit bytes per line, is multiplexed into the period of about $52\,\mu s$, between a colour burst and the following sync pulse in the manner shown in Fig. 39.1. Odd parity is used for error control, to ensure that there is at least one data transition during each byte. This improves the synchronism of the decoder bit rate clock.

The structure of the two types of data line is shown in Fig. 39.2. Each line carries 360 bits in 45 bytes and each line begins with three identical bytes; two bytes of *clock run-in* consisting of 16 alternate 1s and 0s, followed by a *framing code* byte of 11100100. These are provided to synchronise the decoder bit rate clock and to identify the beginning of each block of data. The fourth and fifth bytes are described as the magazine and row address group and are used to uniquely identify a row in any particular magazine. Three bits are used to identify the particular magazine and five bits to identify a maximum of 31 rows. The following two bytes occur only in the *header row* and identify the particular page number within a magazine.

All these bytes are Hamming code protected so that all single bit errors are correctable and double errors detectable.

The header row also carries four bytes that provide a four-digit time code (not necessarily related to real time) and by invoking this sequence, up to 3200 different versions of the previously identified page can be selected. Therefore a particular page may be identified by its magazine, page and time code. The following two bytes

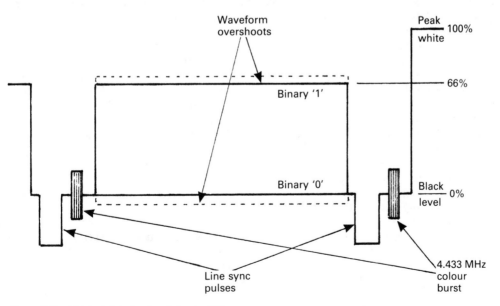

Figure 39.1 Teletext data levels relative to VBI.

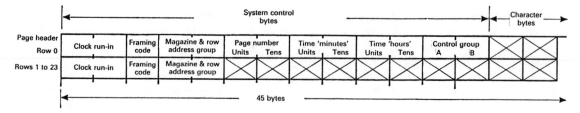

Figure 39.2   Teletext page code structure.

contain 11 usable bits for system control. The header row ends with eight bytes which are used to provide a real-time clock display on each page. Since each displayed line of a page occupies 40 μs and carries 40 characters, each using 6 horizontal pixels or dots (one character per μs), the display or *dot clock* must run at 6 MHz.

The character code set that is commonly used is shown in Table 39.1. This is a variation on the ASCII (American Standard Code for Information Interchange) and the ISO-7 (International Standards Organisation) formats. The bits for each symbol are transmitted in the order b_1 to b_8. This code format, which provides for 96 alphanumeric and 32 control characters, can be extended by dropping the parity check on the character bytes. The use of eight bits then provides for an extended character set of 192 alphanumeric and 64 control codes, which allows characters from different languages to be displayed.

39.4 Viewdata

Viewdata is the generic title of the videotext services provided over the analogue, dial-up telephone network. The systems are compatible with teletext, in that they may use a standard TV receiver and are displayed with the same page format, using the same character set and much of the same decoder control hardware. However, because viewdata employs the telephone system as the communications medium, it provides a user-interactive control feature that is not possible with teletext. The user has a faster access to a much larger database of pages, each of which can be directly called as and when required. To simplify the search procedure, the database is organised on a *tree* and *branch* basis. Since each subscriber has access to a central computer and memory, each is allocated an area of memory location that can additionally be used for E-mail services.

Subscribers' terminals are equipped with modems that provide auto-dial and auto-answer features with full duplex communications.

The data signalling rates are chosen to be compatible with the analogue network, and frequency shift keying (FSK) is employed. The basic character code set is that of the ISO-7 format used for teletext, but with even parity, plus start and stop bits. The code format is thus 10-bit asynchronous. The signalling rates and frequencies employed are shown in Table 39.2.

39.5 Miscellaneous terms

Access time. The time delay between selecting a page and its presentation at the receiver. For viewdata this time is relatively very short. For teletext it depends upon the number of lines used to transmit the signal. For example, if 6 VBI lines per field are used for teletext, then 24 rows will be transmitted in four field periods, or 80 ms. The cycle time for a magazine of 100 pages is therefore 8 s. Thus the average access time will be $8/2 = 4$ s.

Character rounding. A technique used to remove the jagged edges of displayed symbols. This is achieved by adding a half pixel into the corners formed within the symbol shape.

Clock cracker. A test page of 960 characters consisting of the alternating bytes 01111111 and 00000001 which represent a worst-case clock sync pattern.

Closed user group (CUG). Part of the viewdata database may be barred from all but those subscribers equipped with the necessary password needed to gain access to specific areas of the memory.

Data broadcasting. A digital information transmission service broadcast within the VBI but unrelated to teletext. Provided for specific business users under CUG control. If not to be displayed by a standard decoder, this data must be encrypted or transmitted on a row such as packets 29 or 31.

Fastext or full level one features (FLOF). A development that makes use of three extension packets (rows) to provide an enhanced service.

A packet with the address 8/30 (Magazine 8-Row 30) carries designation codes that allow sub-addressing which reduces the access time after the first page.

Table 39.1 *Basic videotext character code set (courtesy BBC, IBA and BREMA)*

b7 b6 b5 / Bits (b4 b3 b2 b1)	Col/Row	0 (000)	1 (001)	2 (010)	2a	3 (011)	3a	4 (100)	5 (101)	6 (110)	6a	7 (111)	7a
0 0 0 0	0	NUL [1]	DLE [1]		▦	0	▦	@	P	─	▦	p	▦
0 0 0 1	1	Alpha Red	Graphics Red	!	▦	1	▦	A	Q	a	▦	q	▦
0 0 1 0	2	Alpha Green	Graphics Green	"	▦	2	▦	B	R	b	▦	r	▦
0 0 1 1	3	Alpha Yellow	Graphics Yellow	£	▦	3	▦	C	S	c	▦	s	▦
0 1 0 0	4	Alpha Blue	Graphics Blue	$	▦	4	▦	D	T	d	▦	t	▦
0 1 0 1	5	Alpha Magenta	Graphics Magenta	%	▦	5	▦	E	U	e	▦	u	▦
0 1 1 0	6	Alpha Cyan	Graphics Cyan	&	▦	6	▦	F	V	f	▦	v	▦
0 1 1 1	7	Alpha White	Graphics White [2]	'	▦	7	▦	G	W	g	▦	w	▦
1 0 0 0	8	Flash	Conceal Display	(	▦	8	▦	H	X	h	▦	x	▦
1 0 0 1	9	Steady [2]	Contiguous Graphics [2]	)	▦	9	▦	I	Y	i	▦	y	▦
1 0 1 0	10	End Box [2]	Separated Graphics	*	▦	:	▦	J	Z	j	▦	z	▦
1 0 1 1	11	Start Box	ESC [1]	+	▦	;	▦	K	←	k	▦	¼	▦
1 1 0 0	12	Normal Height [2]	Black Background [2]	,	▦	<	▦	L	½	l	▦	‖	▦
1 1 0 1	13	Double Height	New Background	-	▦	=	▦	M	→	m	▦	¾	▦
1 1 1 0	14	SO [1]	Hold Graphics	.	▦	>	▦	N	↑	n	▦	÷	▦
1 1 1 1	15	SI [1]	Release Graphics [2]	/	▦	?	▦	O	#	o	▦	█	▦

[1] These control characters are reserved for compatibility with other data codes

[2] These control characters are presumed before each row begins

Codes may be referred to by their column and row e.g. 2/5 refers to %

⸛⸛ Character rectangle

Black represents display colour

White represents background

Table 39.2 *Viewdata signal rates and frequencies*

	Computer to user	User to computer
Logic 1	1.3 kHz	390 Hz
Logic 0	2.1 kHz	450 Hz
Symbol rate	120 ch/s	7.5 ch/s

Packet 26 can be used with a microprocessor control system to provide additional characters to expand the basic language capability.

Packet 27 (link packet) provides a link between pages with a similar context. These pages are then automatically stored in memory and can be accessed by the user with a single key stroke when needed.

Packet 24 (prompt packet) carries keywords that briefly explain the nature of the information on the first four linked pages. This is displayed as a 25th row of text with each keyword colour coded. The indicated page is then selected by the user operating the single key of the appropriate colour.

Level 2 teletext. This introduces the possibility of improved graphics displays by allowing sloping edges rather than the rectangular jagged edges of level 1 systems. A range of pastel shades are also provided through the introduction of half-amplitude colour signals.

Level 3 teletext. These systems allow for the use of dynamically redefinable character sets (DRCS) which extend the range of colours, graphics shapes and languages.

Level 4 teletext. To improve the graphics representation, shapes are provided via a range of elemental geometric figures described as alpha-geometrics. A wide range of in-fill colours are provided and these systems introduce the possibility of limited animation. Each display is formed from a list of instructions or operational codes (Op codes) similar in manner to the operation of a simple computer-aided drawing system.

Level 5 teletext. These systems are intended to provide displays of still pictures to photographic standard and with a full range of colours. The systems will therefore rely on the use of image compression techniques.

Rolling pages. In some cases, the information available for a particular news or similar item will occupy several pages. All these are allocated the same page number and will be delivered to the decoder for viewing in sequence.

Top of pages (TOP). A technique that functions in a similar way to Fastext. The page linking information and page titles are transmitted on dedicated pages rather than in the extension packets. This concept divides each magazine into blocks, groups and direct access pages, such as sport, tennis and results.

Video programming systems (VPS). These are intended to simplify the problem of programming video cassette recorders and are variously known as video programming by Videotext (VPV), or programme delivery control (PDC). The services usually operate via service packet 8/30 or via some other dedicated VBI line. The viewer selects the appropriate TV programme page and then the required programme via a cursor. The data code associated with this selection is then stored in the system memory. Each TV transmission carries a hidden code that identifies its start and finish times. When the decoder recognises the identical codes, it automatically starts and stops the recorder, even if the particular TV programme is transmitted at a different clock time.

39.6 Useful references

Cole H.A. (1983) *Basic Colour Television*, Gower, Aldershot.

Money S.A. (1979) *Teletext and Viewdata*, Butterworth-Heinemann, London.

Mothersole P.L. and White N. (1992) *Broadcast Data Systems*, Butterworth-Heinemann, Oxford.

Woolfe R. (1980) *Videotex – The New Television Information Services*, Heydon, London.

Appendix 1. *CCITT recommendations (numbers and brief descriptions)*

(Note: ter and bis are terms with Latin roots meaning 3 and 2, respectively. Recommendations with these terms in the number have 3 or 2 data rates, the system normally starting up in the higher rate but falling back to the lower speed under noisy conditions).

X 1	International user classes of service in public data networks.
X 2	International user facilities in public data networks.
X 3	Packet assembly/disassembly facility in a public data network.
X 4	General structure of signals of International Alphabet No. 5 code for data transmission over public data networks.
X 20	Interface between data terminal equipment and data circuit terminating equipment for start-stop transmission services on public data networks.
X 20 bis	V21-compatible interface between data terminal equipment and data circuit-terminating equipment for start-stop transmission services on public data networks.
X 21	General purpose interface between data terminal equipment and data circuit-terminating equipment for synchronous operation on public data networks.
X 21 bis	Use on public data networks of data terminal equipments which are designed for integrating to synchronous V-series modems.
X 24	List of definitions of interchange circuits between data terminal equipment and data circuit-terminating equipment on public data networks.
X 25	Interface between data terminal equipment and data circuit-terminating equipment for terminals operating in the packet mode on public data networks.
X 26	Electrical characteristics for unbalanced double-current interchange circuits for general use with integrated circuit equipment in the field of data communications.
X 27	Electrical characteristics for unbalanced double-current interchange circuits for general use with integrated circuit equipment in the field of data communications.
X 28	DTEE/DCE interface for a start-stop-mode data terminal equipment assessing the packet assembly/disassembly facility (PAD) on a public data network situated in the same country.
X 29	Procedures for exchange on control information and user data between a packet mode DTE and a packet assembly/disassembly facility (PAD).
X 30	Standardisation of basic model page-printing machine in accordance with International Alphabet No. 5.
X 31	Characteristics, from the transmission point of view, at the interchange point between data terminal equipment and data circuit-terminating equipment when a 200-baud start-stop data-terminal equipment in accordance with International Alphabet No. 5 is used.
X 32	Answer-back units for 200-bauds start-stop machines in accordance with International Alphabet No. 5.
X 33	Standardisation of an international text for the measurement of the margin of start-stop machines in accordance with International Alphabet No. 5.
X 40	Standardisation of frequency-shift modulated transmission systems for the provision of telegraph and data channels by frequency division of a primary group.
X 50	Fundamental parameters of a multiplying scheme for the international interface between synchronous data networks.
X 51	Fundamental parameters of a multiplexing scheme for the international interface between synchronous data networks using 10-bit envelope structure.
X 60	Common channel signalling for synchronous data applications–data user part.
X 70	Terminal and transit control signalling system for start-stop services on international circuits between antisynchronous data networks.

X 71   Decentralised terminal and transit control signalling system on international circuits between synchronous data networks.

X 75   Terminal and transit call control procedures and data transfer systems on international circuits between packet switched data networks.

X 92   Hypothetical reference connections for public synchronous data networks.

X 95   Network parameters in public data networks.

X 96   Call progress signals in public data networks.

X 121 International numbering plan for public data networks.

V 1   Equivalence between binary notation symbols and the significant conditions of a two-condition code.

V 2   Power levels for data transmission over telephone lines.

V 3   International Alphabet No. 5.

V 4   General structure of signals of International Alphabet No. 5 code.

V 5   Standardisation of data signalling rates for synchronous data transmission in the general switched telephone network.

V 6   Standardisation of data signalling rates for synchronous data transmission on leased telephone-type circuits.

V 10   Electrical characteristics for unbalanced double-current interchange circuits for general use with integrated circuit equipment in the field of data communications.

V 11   Electrical characteristics for balanced double-current interchange circuits for general use with integrated circuit equipment in the field of data communications.

V 15   Use of acoustic coupling for data transmission.

V 16   Medical analogue data transmission modems.

V 19   Modems for parallel data transmission using telephone signalling frequencies.

V 20   Parallel data transmission modems standardised for universal use in the general switched telephone network.

V 21   200-baud modem standardised for use in the general switched telephone network.

V 22   1200 bps full-duplex 2-wire modem standardised for use in the general switched telephone network.

V 22 bis   2400 bps full-duplex 2-wire modem standardised for use in the general switched telephone network with 1200 bps fall-back.

V 23   600/1200-baud modem standardised for use in the general switched telephone network.

V 24   List of definitions for interchange circuits between data-terminal equipment and data circuit-terminating equipment.

V 25   Automatic calling and/or answering on the general switched telephone network, including disabling or echo-suppressors on manually established calls.

V 25 bis   Bit synchronous auto-dialling protocol for use over PSTN.

V 26   2400/bit/s modem standardised for use on four-wire leased circuits.

V 26 bis   2400 1200 bit/s modem standardised for use in the general switched telephone network.

V 27   4800/bit/s modem standardised for use on leased circuits.

V 27 bis   4800 bit/s modem with automatic equaliser standardised for use on leased circuits.

V 27 ter   4800/2400 bit/s modem standardised for use in the general switched telephone network.

V 28   Electrical characteristics for unbalanced double-current interchange circuits.

V 29   9600 bit/s modem for use on leased circuits.

V 31   Electrical characteristics for single-current interchange circuits controlled by contact closure.

V 32   9600 bps full-duplex 2-wire modem standardised for use in the general switched telephone network with 48000 bps fall back.

V 33   14.4 Kbps full-duplex over 4-wire leased circuits.

V 35   Data transmission at 48 kilobit/s using 60 to 108 kHz group band circuits.

V 36   Modems for synchronous data transmission using 60–108 kHz group band circuits.

V 40   Error indication with electromechanical equipment.

V 41   Code-independent error control system.

V 42   Covers error-checking protocols for use on standard dial-up lines. This will enable error-free data transfer at speeds up to 19200 bps without the need for leased lines.

V 50   Standard limits for transmission quality of data transmission.

V 51   Organisation of the maintenance of international telephone-type circuits used for data transmission.

V 52   Characteristics of distortion and error-rate measuring apparatus for data transmission.

V 53   Limits for maintenance of telephone-type circuits used for data transmission.

V 54   Loop test devices for modems.

V 55   Specification for an impulsive noise measuring instrument for telephone type circuits.

V 56   Comparative tests of modems for use over telephone-type circuits.

V 57   Comprehensive data test set for high data signalling rates.

Appendix 2. *Abbreviations and acronyms*

ABM	Adaptive bandwidth management
ABSOC	Advanced Broadcasting Systems of Canada
ACK	Acknowledge
ADC, A/D	Analogue to digital conversion/convertor
ADM	Adaptive delta modulation
ADPCM	Adaptive differential pulse code modulation
AFTN	Aeronautical fixed telecommunications network
AM	Amplitude modulation
AMI	Alternate mark inversion
AM-VSB	Amplitude modulation-vestigial sideband
ANSI	American National Standards Institute
APD	Avalanche photodiode
ARINC	Aeronautical Radio INCorporated
ARPA	Advanced Research Projects Agency
ARQ	Automatic repeat request
ASCII	American Standard Code for Information Interchange
ASIC	Application specific integrated circuit
ASID	Application specific integrated device
ASK	Amplitude shift keying
ASR	Automatic send/receive
ATDM	Asynchronous time division multiplex(er)
ATE	Automatic test equipment
ATSC	Advanced Television Standards Committee
BAA	British Astronomical Association
BABT	British Approvals Board for Telecommunications
BCC	Block check character
BCD	Binary coded decimal
BCH	Bose–Chadhuri–Hocquenghem
BER	Bit error rate
BERT	Bit error rate tester
B-ISDN	Broadband-Integrated Services Digital Network
BPS	b/s, Bits per second
BPSK	Binary phase shift keying
BRISC	Bi-polar/BiCMOS reduced instruction set computer
BSC	Binary synchronous communications
BSI	British Standards Institute
BTAM	Basic telecommunications access method
BTSC	Broadcast Television Systems Committee

CAD	Computer-aided design
CAE	Computer-aided engineering
CAI	Common air interface
CAM	Computer-aided manufacturing
CASE	Computer-aided software engineering
CATG	Computer-aided test generator
CATV	Community antenna television
C_B, C_R	Chrominance, equivalent to U, $(B - Y)$ and V, $(R - Y)$
CCD	Charge coupled device
CCETT	Centre Common d'Etudes de Télédiffusion et Télécommunications
CCIR	Comité Consultatif International des Radiocommunication
CCITT	Comité Consultatif International Télégraphique et Téléphonique
CCSS	Communications channel signalling system
CCTV	Closed circuit television
CCU	Camera control unit
CD	Carrier detect
CDM	Code division multiplex
CDMA	Code division multiple access
CDR	Common data rate
CEI	Commission Electrotechnique Internationale
CELPC	Code excited linear predictive coding
CEN-CENELEC	Comité Européen de Normalisation-Comité Européen de Normalisation Electrotechnique
CEPT	Conference Européene des Administrations des Postes et Télécommunications
CFSK	Coherent frequency shift keying
CHOTS	Corporate Headquarters Office Technology System
CIE	Commission Internationale de L'Eclairage
CIF	Common image format
CIM	Computer-integrated manufacturing
CISC	Complex instruction set computer
CISDR	Common image structure data rate
CISPR	Comité Internationale des Perturbations Radiotechniques
CLUT	Colour look-up table
CMRR	Common mode rejection ratio
CNET	French National Centre for Scientific Research
CNR	C/N carrier to noise ratio

COMSAT	Communications Satellite Corp.
COSHH	Control of substances hazardous to health
COST	Cooperation on science and technology
CPFSK	Continuous phase frequency shift keying
CPM	Continuous phase modulation
CRC	Cyclic redundancy check
CRPE	Centre for Environmental Physics Research
CSDN	Circuit switched data network
CSMA	Carrier sense multiple access
CSMA/CA	CSMA/collision avoidance
CSMA/CD	CSMA/collision detection
CSSB/AM	Companded single sideband/amplitude modulation
CSU	Channel service unit
CTS	Clear to send
CVSD	Continuously variable slope delta-modulation
CW	Continuous wave
DAC, D/A	Digital to analogue convertor
DAMA	Demand assigned multiple access
DARPA	Defense Advanced Research Project Agency
DASS	Digital access signalling standard
DATV	Direct antenna television
dB	decibel
dBc	decibels relative to carrier level
dBi	decibels relative to isotropic source
dBK	decibels relative to 1K
dBm	decibels relative to 1 mW
dBmO	decibels relative to 1 mW at point of zero reference level
dBmOp	as dBmO but psophometrically weighted
dBmOps	as dBmOp but weighted for audio transmissions
dBv	decibels relative to 1 V
dBµ	decibels relative to 1 µV
DBPT	Deutsche BundesPost Telekom
DBS	Direct broadcasting by satellite
DCE	Data circuit equipment
DDS	Digital data service
DES	Data encryption standard
DFB	Distributed feedback
DFT	Discrete Fourier transform
DIN	Deutsches Institut für Normung
DLC	Data link control
DM	Delta modulation
DMA	Direct memory access
DOPSK	Differential offset phase shift keying
DOS	Disk operating system
DPCM	Differential pulse code modulation
DPSK	Differential phase shift keying
DQPSK	Differential quadrature phase shift keying
DRAM	Dynamic random access memory
DSB	Double sideband
DSBSC	Double sideband suppressed carrier
DSI	Digital speech interpolation
DSP	Digital signal processing
DSR	Data set ready
DS-SSMA	Direct sequence-spread spectrum multiple access
DTE	Data terminal equipment

DTH	Direct to home
DTMF	Dual tone multi-frequency
DTR	Data terminal ready
E	Erlang
EAN	European Article Numbering Association
EARN	European Academic Research Network
EAX	Electronic automatic exchange
EBCDC	Extended binary coded decimal code
EBCDIC	Extended binary coded decimal interchange code
EBU	European Broadcasting Union
ECL	Emitter coupled logic
ECS	European Communications Satellite
EDC	Error detection and correction
EDS	Electronic data service
EDTV	Extended definition television
EEPROM	Electrically erasable programmable read only memory
EFP	Electronic film production
EPROM	Erasable programmable read only memory
EIA	Electronics Industry Association
EIRP	Effective isotropic radiated power
EISA	Enhanced industry standard architecture
ELV	Earth launch vehicle
EMC	Electromagnetic compatibility
EMI	Electromagnetic interference
EMP	Electromagnetic pulse
ENG	Electronic news gathering
ENQ	Enquiry
EPLD	Erasable programmable logic device
ERA	Electrically reconfigurable array
ESA	European Space Agency
ESPRIT	European Strategic Programme for Research in Information Technology
ESRO	European Space Research Organisation
ETSI	European Telecommunications Standards Institute
EUMETSAT	EUropean METeorological SATellite Organisation
EUTELSAT	EUropean TELecommunications SATellite organisation
EWP	Electronic white pages
FCC	Federal Communications Commission
FCS	Frame check sequence
FDDI	Fibre distributed data interface
FDM	Frequency division multiplex
FDMA	Frequency division multiple access
FEC	Forward error control
FFT	Fast Fourier transform
FIFO	First in-first out
FIR	Finite impulse response
FLIR	Forward looking infra-red
FM	Frequency modulation
FRAD	Frame relay assembler/disassembler
FSIC	Function specific integrated circuit
FSK	Frequency shift keying
FSS	Fixed satellite service
FTAM	File transfer access and management

GARP	Global Atmospheric Research Programme
G^3	Gadolinium gallium garnet
GLONASS	GLObal NAvigation Satellite System
GMS	Geostationary meteorological satellite
GMSK	Gaussian filtered minimum shift keying
GOES	Geostationary operational environmental satellite
GOMS	Geostationary operational meteorological satellite
GPIB	General purpose interface bus
GPS	Global positioning system
G/T	Gain to noise temperature ratio
HAD	Hole accumulation diode
HBI	Horizontal blanking interval
HDB	High density bi-polar
HDLC	High-level data link control
HDTV	High definition television
HDX	Half duplex transmission
HEMT	High electron mobility transistor
HLR	Home location register
HPB	Half-power bandwidth
HUD	Head-up display
IC	Integrated circuit
ICAO	International Civil Aviation Organisation
IDD	International direct dialling
IDN	Integrated digital network
IEC	International Electrotechnical Commission
IEEE	Institute of Electrical and Electronic Engineers
IFRB	International Frequency Registration Board
IIR	Infinite impulse response
IMD	Inter-modulation distortion
IMSI	International mobile subscribers identity
INTELSAT	INternational TELecommunications SATellite Organisation
INMARSAT	INternational MARitime SATellite Organisation
I/O	Input-output
ISB	Independent side band
ISD	International subscriber dialling
ISDN	Integrated Services Digital Network
ISI	Inter-symbol interference
ISO	International Standards Organisation
IT	Information technology
ITU	International Telecommunications Union
JBIG	Joint Bi-level Image Group
JEDEC	Joint Electronic Device Engineering Council
JESSI	Joint European Submicron SIlicon programme
JISC	Japanese Industrial Standards Committee
JIT	Just in time
JND	Just noticeable difference
JPEG	Joint Photographic Experts Group
LAN	Local area network
LAP, LAP-M	link access protocol – modems
LED	Light emitting diode
LEO	Low earth orbiter
LF	Low frequency
LNA	Low noise amplifier
LNB	Low noise block (convertor)
LNC	Low noise convertor
LORAN	LOng Range Aid to Navigation
LOS	Line of sight
LPC	Linear predictive coding
LRC	Longitudinal redundancy check
LSB	Least significant bit or lower side band
LSI	Large scale integration
LUF	Lowest usable frequency
Luma	Luminance (Y)
LUT	Look-up table, local user terminal
LVA	Large vertical aperture
MAC	Multiplying accumulator
MAC	Multiplexed analogue components
MAC	Multiplexed analogue confusion
MAN	Metropolitan area network
MAP	Manufacturing automation protocol
MARECS	MARitime European Communications Satellite
MASCAM	Masking-pattern adaptive sub-band coding and multiplexing
MATV	Master antenna television
MCTDMA	Multiple carrier time division multiple access
MCW	Modulated continuous wave
MDS	Microwave distribution system
MF	Medium frequency
MFLOPS	Million floating-point operations per second
MFSK	Multiple/minimum frequency shift keying
MHS	Message handling system/service
MIC	Microwave integrated circuit
MIPS	Millions of instruction per second
MMDS	Microwave multi-channel distribution system
MMIC	Monolithic microwave integrated circuit
MNP	Microcom networking protocol
MODEM	Modulator/demodulator
MOVPE	Metallic organic vapour phase epitaxy
MPEG	Motion picture expert group
MQORC	Modified quadrature overlapped raised cosine
MSK	Minimum shift keying
MSRN	Mobile station roaming number
MTBF	Mean time between failures
MTF	Modulation transfer function
MTTR	Mean time to repair
MUF	Maximum usable frequency
MULDEX	Multiplexer/demultiplexer
MUSA	Multiple unit steerable array
MUSE	Multiple sub-Nyquist sampling and encoding
MUSICAM	Masking pattern universal sub-band integrated coding and multiplexing
M^3VDS	Millimetre-wave multi-channel multipoint video distribution system
NAB	National Association of Broadcasters
NACCB	National Accreditation Council for Certification Bodies
NAK	Negative acknowledgement
NAMAS	NAtional Measurement Accreditation Service
NBFM	Narrow band frequency modulation
NBS	National Bureau of Standards

NFS	Network filing system		PSDN	Packet switched data network
NICAM	Near instantaneous companded audio multiplex		PSK	Phase shift keying
			PSS	Packet switched system/service
NICQA	National Inspection Council Quality Assurance		PSTN	Public switched telephone network
NIIR	Russian Scientific Research Institute of Radio Communications		PT	Packet terminal
			PTM	Pulse time modulation
NOAA	National Oceanographic and Atmospheric Administration		PTT	Posts, Telegraphs and Telecommunications
			PWM	Pulse width modulation
NRZ	Non-return to zero			
NRZI	Non-return to zero inverted		QA	Quality assurance/assessment
NTSC	National Television Standards Committee		QAM	Quadrature amplitude modulation
			Q&R	Quality and reliability
NTSC	Never twice the same colour		QASK	Quadrature amplitude shift keying
NTU	Network terminating unit		QCIF	Quarter common image format
NUI	Network user identification		QDU	Quantisation distortion units
			QFSK	Quadrature frequency shift keying
OA	Office automation		QDPSK	Quadrature differential phase shift keying
OAA	Over air addressing			
OEM	Original equipment manufacturer		QORC	Quadrature offset raised cosine
OIRT	Organisation Internationale de Radiodiffusion et Télévision		QOSRC	Quadrature overlapped squared raised cosine
			QPR	Quadrature partial response
OOK	On-off keying		QPSE	Queued packet synchronous exchange
OQPSK	Offset quadrature phase shift keying		QPSK	Quadrature/quaternary phase shift keying
OSI	Open Systems Interconnect			
OTP	One time programmable		RACE	Research in Advanced Communications for Europe
OTS	Orbital test satellite			
OWF	Optimum working frequency		RADAR	Radio detection and ranging
			RAM	Random access memory
PABX	Private automatic branch exchange		RCLP	Residual code linear predictive (coding)
PAD	Packet assembler/disassembler device		RDS	Radio data service
PAL	Phase alternation by line		RDSS	Radio determination satellite service
PAL	Peace at last		RFI	Radio frequency interference
PAL	Programmable array logic		RGB	Red green blue
PAM	Pulse amplitude modulation		RISC	Reduced instruction set computer
PARS	Private advanced radio service		RLC	Run length code
PBX	Private branch exchange		RMS	Root mean square
PCB	Printed circuit board		ROM	Read only memory
PCM	Pulse code modulation		RS	Recommended standard
PCN	Personal communications network		RSA	Rivest Shamir and Adleman
PCS	Personal communications service		RTD	Resistance temperature detector
PDF	Pulse density function		RTS	Request to send
PDM	Pulse duration modulation		RTTY	Radio teletypewriter
PDN	Public data network		RTZ	Return to zero
PEL	Picture element		R–S	Reed–Solomon
PFD	Power flux density			
PFM	Pulse frequency modulation		SAR	Small/synthetic aperture radar
PIXEL	Picture element		SARBE	Search and rescue beacon equipment
PIO	Programmable input output		SAW	Surface acoustic wave
PLA	Programmable logic array		SCADA	Supervisory control and data acquisition
PLC	Programmable logic controller		SCFM	Sub-carrier frequency modulation
PLL	Phase locked loop		SCPC	Single channel per carrier
PLUGE	Picture line up generating equipment		SCPI	Standard commands for programmable instruments
PM	Phase modulation			
PMR	Personal mobile radio		SCRAM	Static C-mos random access memory
POCSAG	Post Office Code Standardisation Advisory Group		SCSI	Small computer system interface
			SDLC	Synchronous data link control
PPM	Pulse position modulation		SDM	Space division multiplex
PRBS	Pseudo-random binary sequence		SDMA	Space division multiple access
PRF	Pulse repetition frequency		SECAM	SEquential Couleur à Mémoire
PROM	Programmable read only memory		SECAM	System essentially contrary to American method
PSD	Power spectral density			

SHF	Super high frequency	USART	Universal synchronous/asynchronous receiver transmitter
SINAD	SIgnal noise and distortion		
SIS	Sound in synchs	USB	Upper side band
SITA	Société Internationale de Télécommunications Aeronautiques	UT	Universal time (GMT)
SIV	Sound in video	VADS	Value added data service
SLIC	Subscribers line interface circuit	VANS	Value added network service
SMATV	Satellite master antenna television	VBI	Vertical blanking interval
SMDS	Switched multi-megabit data service	VCR	Video cassette recorder
SMPTE	Society of Motion Picture and Television Engineers	VDU	Visual display unit
		VF	Voice frequency
SMQORC	Staggered modified quadrature overlapped raised cosine	VFO	Variable frequency oscillator
		VHF	Very high frequency
SNA	Systems network architecture	VLBI	Very long baseline interferometry
SNR, *S/N*	Signal to noise ratio	VLF	Very low frequency
SONET	Synchronous optical network	VLIW	Very long instruction word
SOS	Silicon on sapphire	VLSI	Very large scale integration
SPC	Stored program control	VPN	Virtual private network
SSB	Single side band	VPS	Video programming signal
SSBSC	Single side band suppressed carrier	VRC	Vertical redundancy check
SSI	Small scale integration	VSAT	Very small aperture terminal
SSMA	Spread spectrum multiple access	VSB	Vestigial side band
SSTDMA	Satellite switched time division multiple access	VSBSC	Vestigial side band suppressed carrier
STD	Subscriber trunk dialling	VSWR	Voltage standing wave ratio
SRAM	Static random access memory	VTAM	Virtual telecommunications access method
SVC	Switched virtual circuit	VTR	Video tape recorder
Syn/Sync	Synchronism, Synchronising		
		WAN	Wide area network
TACS	Total access communications system	WARC	World Administrative Radio Conference
TASI	Time assignment speech interpolation	WATTC	World Administrative Telephone and Telegraph Conference
TAT	Trans-Atlantic telephone (telecommunications) cable		
		WDM	Wavelength division multiplex
TCP/IP	Transmission control protocol/Internet protocol	WIMPs	Windows icons mouse and pull-down menus
		WG	Waveguide
TDM	Time division multiplex	WMO	World Meteorological Organisation
TDMA	Time division multiple access	WORM	Write once read many times
TE	Transverse electric, Trans-equatorial	WST	World Standard Teletext
TEM	Transverse electromagnetic	WTDM	Wavelength time division multiplex
TFM	Time frequency multiplex	WWW	World weather watch
THD	Total harmonic distortion		
TQM	Total quality management	X-Off	Transmitter Off
TT&C	Telemetry tracking and control	X-On	Transmitter On
TTL	Transistor transistor logic	XPD	Cross-polar discrimination
TTY	Teletype terminal	XPI	Cross-polar interference
TV	Television		
TVRO	Television receive only	Y	Luminance
TWT	Travelling wave tube	YAG	Yttrium aluminium garnet
TXE	Telephone exchange, electronic	YIG	Yttrium iron garnet
T/H	Track and hold	YIQ	Luminance plus two colour difference components (NTSC)
UART	Universal asynchronous receiver transmitter		
UHF	Ultra high frequency	YUV	Luminance plus two colour difference components (PAL & SECAM)
ULA	Uncommitted logic array		

Index